Selected Titles in This Series

268 Robert Gulliver, Walter Littman, and Roberto Triggiani, Editors, Differential geometric methods in the control of partial differential equations, 2000

267 Nicolás Andruskiewitsch, Walter Ricardo Ferrer Santos, and Hans-Jürgen Schneider, Editors, New trends in Hopf algebra theory, 2000

266 Caroline Grant Melles and Ruth I. Michler, Editors, Singularities in algebraic and analytic geometry, 2000

265 Dominique Arlettaz and Kathryn Hess, Editors, Une dégustation topologique: Homotopy theory in the Swiss Alps, 2000

264 Kai Yuen Chan, Alexander A. Mikhalev, Man-Keung Siu, Jie-Tai Yu, and Efim I. Zelmanov, Editors, Combinatorial and computational algebra, 2000

263 Yan Guo, Editor, Nonlinear wave equations, 2000

262 Paul Igodt, Herbert Abels, Yves Félix, and Fritz Grunewald, Editors, Crystallographic groups and their generalizations, 2000

261 Gregory Budzban, Philip Feinsilver, and Arun Mukherjea, Editors, Probability on algebraic structures, 2000

260 Salvador Pérez-Esteva and Carlos Villegas-Blas, Editors, First summer school in analysis and mathematical physics: Quantization, the Segal-Bargmann transform and semiclassical analysis, 2000

259 D. V. Huynh, S. K. Jain, and S. R. López-Permouth, Editors, Algebra and its applications, 2000

258 Karsten Grove, Ib Henning Madsen, and Erik Kjær Pedersen, Editors, Geometry and topology: Aarhus, 2000

257 Peter A. Cholak, Steffen Lempp, Manuel Lerman, and Richard A. Shore, Editors, Computability theory and its applications: Current trends and open problems, 2000

256 Irwin Kra and Bernard Maskit, Editors, In the tradition of Ahlfors and Bers: Proceedings of the first Ahlfors-Bers colloquium, 2000

255 Jerry Bona, Katarzyna Saxton, and Ralph Saxton, Editors, Nonlinear PDE's, dynamics and continuum physics, 2000

254 Mourad E. H. Ismail and Dennis W. Stanton, Editors, q-series from a contemporary perspective, 2000

253 Charles N. Delzell and James J. Madden, Editors, Real algebraic geometry and ordered structures, 2000

252 Nathaniel Dean, Cassandra M. McZeal, and Pamela J. Williams, Editors, African Americans in Mathematics II, 1999

251 Eric L. Grinberg, Shiferaw Berhanu, Marvin I. Knopp, Gerardo A. Mendoza, and Eric Todd Quinto, Editors, Analysis, geometry, number theory: The Mathematics of Leon Ehrenpreis, 2000

250 Robert H. Gilman, Editor, Groups, languages and geometry, 1999

249 Myung-Hwan Kim, John S. Hsia, Yoshiyuki Kitaoka, and Rainer Schulze-Pillot, Editors, Integral quadratic forms and lattices, 1999

248 Naihuan Jing and Kailash C. Misra, Editors, Recent developments in quantum affine algebras and related topics, 1999

247 Lawrence Wasson Baggett and David Royal Larson, Editors, The functional and harmonic analysis of wavelets and frames, 1999

246 Marcy Barge and Krystyna Kuperberg, Editors, Geometry and topology in dynamics, 1999

245 Michael D. Fried, Editor, Applications of curves over finite fields, 1999

For a complete list of titles in this series, visit the AMS Bookstore at **www.ams.org/bookstore/**.

CONTEMPORARY MATHEMATICS

268

Differential Geometric Methods in the Control of Partial Differential Equations

1999 AMS-IMS-SIAM Joint Summer Research Conference
on Differential Geometric Methods in the Control
of Partial Differential Equations
University of Colorado, Boulder
June 27–July 1, 1999

Robert Gulliver
Walter Littman
Roberto Triggiani
Editors

American Mathematical Society
Providence, Rhode Island

Editorial Board

Dennis DeTurck, managing editor

Andreas Blass   Andy R. Magid   Michael Vogelius

The 1999 AMS-IMS-SIAM Joint Summer Research Conference on "Differential Geometric Methods in the Control of Partial Differential Equations" was held at the University of Colorado, Boulder, June 27–July 1, 1999, with support from the National Science Foundation, grant DMS-9618514.

2000 *Mathematics Subject Classification.* Primary 53–XX, 35–XX, 49–XX, 93–XX.

Any opinions, findings, and conclusions or recommendations expressed in this material are those of the authors and do not necessarily reflect the views of the National Science Foundation.

Library of Congress Cataloging-in-Publication Data
Differential geometric methods in the control of partial differential equations : joint summer research conference, differential geometric methods in the control of partial differential equations, University of Colorado, Boulder, June 27–July 1, 1999 / Robert Gulliver, Walter Littman, Roberto Triggiani, editors.
    p. cm. — (Contemporary mathematics, ISSN 0271-4132 ; 268)
 Includes bibliographical references.
 ISBN 0-8218-1927-5 (alk. paper)
    1. Boundary value problems—Numerical solutions—Congresses. 2. Differential equations, Partial—Numerical solutions—Congresses. 3. Geometry, Differential—Congresses. I. Gulliver, Robert, 1945– II. Littman, Walter. III. Triggiani, R. (Roberto), 1942– IV. Contemporary mathematics (American Mathematical Society) ; v. 268.
QA379.D54   2000
515′.35—dc21   00-046884

Copying and reprinting. Material in this book may be reproduced by any means for educational and scientific purposes without fee or permission with the exception of reproduction by services that collect fees for delivery of documents and provided that the customary acknowledgment of the source is given. This consent does not extend to other kinds of copying for general distribution, for advertising or promotional purposes, or for resale. Requests for permission for commercial use of material should be addressed to the Assistant to the Publisher, American Mathematical Society, P. O. Box 6248, Providence, Rhode Island 02940-6248. Requests can also be made by e-mail to reprint-permission@ams.org.

Excluded from these provisions is material in articles for which the author holds copyright. In such cases, requests for permission to use or reprint should be addressed directly to the author(s). (Copyright ownership is indicated in the notice in the lower right-hand corner of the first page of each article.)

© 2000 by the American Mathematical Society. All rights reserved.
The American Mathematical Society retains all rights
except those granted to the United States Government.
Printed in the United States of America.

∞ The paper used in this book is acid-free and falls within the guidelines
established to ensure permanence and durability.
Visit the AMS home page at URL: http://www.ams.org/

10 9 8 7 6 5 4 3 2 1   05 04 03 02 01 00

Contents

Preface ... vii

Wellposedness of a structural acoustics model with point control
  GEORGE AVALOS ... 1

Intrinsic geometric model for the vibration of a constrained shell
  JOHN CAGNOL AND JEAN-PAUL ZOLÉSIO ... 23

A noise reduction problem arising in structural acoustics: A three-dimensional solution
  MEHMET CAMURDAN AND GUANGCAO JI ... 41

The free boundary problem in the optimization of composite membranes
  S. CHANILLO, D. GRIESER, AND K. KURATA ... 61

Tangential differential calculus and functional analysis on a $C^{1,1}$ submanifold
  MICHEL C. DELFOUR ... 83

Carleman estimates with two large parameters and applications
  MATTHIAS M. ELLER AND VICTOR ISAKOV ... 117

On the prescribed Scalar curvature problem on compact manifolds with boundary
  JOSÉ F. ESCOBAR ... 137

Chord uniqueness and controllability: The view from the boundary, I
  ROBERT GULLIVER AND WALTER LITTMAN ... 145

Nonlinear boundary stabilization of a system of anisotropic elasticity with light internal damping
  MARY ANN HORN ... 177

Carleman estimate with the Neumann boundary condition and its applications to the observability inequality and inverse hyperbolic problems
  VICTOR ISAKOV AND MASAHIRO YAMAMOTO ... 191

Nonconservative wave equations with unobserved Neumann B. C.: Global uniqueness and observability in one shot
  I. LASIECKA, R. TRIGGIANI, AND X. ZHANG ... 227

Uniform stability of a coupled structural acoustic system with thermoelastic effects and weak structural damping
CATHERINE LEBIEDZIK ... 327

Topological derivative for nucleation of non-circular voids. The Neumann problem
TOMASZ LEWIŃSKI AND JAN SOKOŁOWSKI ... 341

Remarks on global uniqueness theorems for partial differential equations
WALTER LITTMAN ... 363

Evolution of a graph by Levi form
ZBIGNIEW SLODKOWSKI AND GIUSEPPE TOMASSINI ... 373

Observability inequalities for the Euler-Bernoulli plate with variable coefficients
PENG-FEI YAO ... 383

Preface

This volume contains selected papers which were presented at the AMS-IMS-SIAM Joint Summer Research Conference on "Differential Geometric Methods in the Control of Partial Differential Equations," which was held at the University of Colorado, Boulder, June 27–July 1, 1999.

The aim of the Conference was to explore the infusion of differential-geometric methods into the analysis of control theory of partial differential equations, particularly in the challenging case of variable coefficients, where the physical characteristics of the medium vary from point to point. While a mutually profitable link had been long established, for at least 30 years, between differential geometry and control of ordinary differential equations, a comparable relationship between differential geometry and control of partial differential equations (PDE's) is a new and promising topic. Very recent research, just prior to the Colorado Conference, supported the expectation that differential geometric methods, when brought to bear on classes of PDE modelling and control problems with variable coefficients, will yield significant mathematical advances.

Although the subject of boundary control of PDE's is about a quarter of a century old, and that of Riemannian geometry is much older still, there has been relatively little interaction between the two. It was just over 10 years ago that the role that bicharacteristics play in boundary control of hyperbolic PDE's was brought to the forefront. This then naturally leads one to think about their geometric equivalent, at least for time-invariant, second-order hyperbolic equations—geodesics, a basic concept in Riemannian geometry. Accordingly, we expect that other local and global geometrical invariants such as various notions of curvature, isoperimetric constants, uniqueness of the geodesics, etc., will be seen to be relevant. Each of these two disciplines—control theory of PDE's, and Riemannian geometry—has been pursued in virtually complete independence, or occasionally even ignorance, of the other. One of the motivations which led us to organize this Conference was the belief that both subjects have much to gain by closer interaction with one another. This, in particular, may turn out to be true in providing intrinsic (coordinate invariant) and mathematically manageable models in shell theory. On the one hand, we anticipate that the reservoir of as yet untapped Riemann-geometric methods and concepts could be applied productively in boundary control. Several of the papers in these Proceedings demonstrate such applications. On the other hand, we expect that certain problems which have arisen recently in the boundary control of PDE will stimulate the disciplines of Riemannian and Lorentzian geometry to undertake new areas of research. We will name just two examples here. First, the study of time-like Lorentzian geodesics has until now been principally motivated by cosmological problems, which, while important, are of a quite different nature from those which arise in the control of hyperbolic PDE's with time-varying coefficients. Second, there are numerous geometric issues which arise in the optimal design of boundary control problems, few, if any, of which appear to have come to the attention of researchers in Riemannian and Lorentzian geometry.

For these and related reasons, the Conference brought together specialists and junior scientists in differential geometry, in PDE's and in boundary control theory, to join in becoming acquainted with the tools of the other disciplines, in overcoming interdisciplinary barriers, and ultimately in thrusting forward this mathematical area of research.

The papers presented at the Conference, in particular those enclosed in this volume, collectively support the claim that the aims of the Conference are being fulfilled. Each paper has a geometric flavor and component; some dig deeply into the interaction between geometric tools and the evolution of control problems. The papers feature the following topics: stabilization for structural acoustic problems and systems of anisotropic elasticity; intrinsic geometric models for shell equations; free boundary problems arising in optimization of composite membranes; evolution of a graph by its Levi form; prescribed scalar curvature on compact manifolds with boundary; sensitivity analysis; exact controllability of hyperbolic equations and related inverse problems; local and global uniqueness of PDE's (or, equivalently, approximate controllability of the adjoint system); Carleman estimates and their implications.

The organizers thus expect that the Colorado Conference will serve as a springboard for further research activities in this area and for further conferences to follow.

The conference web page, including the conference schedule, abstracts of the main talks and a participant list with email addresses, will be maintained for the next few years at:

http://www.ima.umn.edu/ gulliver/confs/control.html

The Conference was funded by the National Science Foundation, whose support is gratefully acknowledged.

The organizers are grateful to all participants for their contributions to the Conference, either by lecturing, by publishing in the present Proceedings, or by actively taking part in the intellectual exchange of ideas debated at the Conference.

Very warm thanks are extended to the AMS staff, and, in particular, to Mr. Wayne Drady and Ms. Donna Salter, whose much appreciated efforts and smooth, professional coordination of a large variety of activities were essential to the success of the conference.

Finally, we wish to thank Ms. Christine Thivierge from the AMS Publication Office, for precious help in connection with the publication of the present volume.

Robert Gulliver, Walter Littman, and Roberto Triggiani

Robert Gulliver, Professor of Mathematics
University of Minnesota
Minneapolis, MN 55455 USA; E-mail: gulliver@math.umn.edu

Walter Littman, Professor of Mathematics
University of Minnesota
Minneapolis, MN 55455 USA; E-mail: littman@math.umn.edu

Roberto Triggiani, Professor of Mathematics
University of Virginia
Charlottesville, VA 22904 USA; E-mail: rt7u@virginia.edu

Wellposedness of a Structural Acoustics Model with Point Control

George Avalos

ABSTRACT. In this work, we analyze a partial differential equation (PDE) model in which control is implemented on the boundary as a point impulse. The PDE mathematically describes a rectangular acoustic chamber with one of its (flexible) walls serving as the control region. The PDE model is comprised of a wave equation coupled to a structurally damped beam equation, where the coupling arises from boundary trace terms. In [3] and [4] it is shown that there is a certain space $\mathcal{X}$ of initial data, strictly contained in the basic energy space $\mathcal{H}$ of the structural acoustics model, such that (unbounded) point observations of the wave velocity can be justified for solutions of the controlled structural acoustics PDE. Our main result here states that for such initial data, the resulting solutions actually stay in the special space $\mathcal{X}$, pointwise in time. Therefore, taking the state space to be $\mathcal{X}$ (instead of the basic space $\mathcal{H}$), the controlled and observed structural acoustics model is *wellposed*, to use the language of infinite dimensional systems theory. In this work, sharp regularity theory for wave equations is combined in an essential way with the techniques used for dealing with unbounded control problems involving analytic dynamics.

1. Introduction

Let $\Omega \in \mathbb{R}^2$ be the rectangle $(0,\pi) \times (0,\pi)$ with boundary Γ. Throughout, we will denote the "active" portion of the boundary to be the edge $\Gamma_0 = (0,\pi)$. On this rectangular region, we will analyze the following system of time dependent partial differential equations (PDE's), in which a wave equation on Ω, satisfied by $z(t,x)$, is coupled through boundary "trace" terms to a structurally damped (parabolic–like) beam equation on Γ_0, satisfied by function $v(t,x)$:

1991 *Mathematics Subject Classification.* Primary 35M20, 93B05.
The research of G. Avalos is partially supported by the NSF Grant DMS–9972349.

(1.1)
$$z_{tt} = \Delta z \text{ on } (0,T) \times \Omega$$
$$\frac{\partial z}{\partial \nu} = \begin{cases} v_t & \text{on } (0,T) \times \Gamma_0 \\ 0 & \text{on } (0,T) \times \Gamma \setminus \Gamma_0 \end{cases}$$

(1.2)
$$v_{tt} = -\Delta^2 v - \Delta^2 v_t - z_t + \delta'(\zeta_0)u \text{ on } \Gamma_0 \times (0,T)$$
$$v(t,0) = v(t,\pi) = \frac{\partial v(t,0)}{\partial x} = \frac{\partial v(t,\pi)}{\partial x} = 0, \quad \forall\, t \in (0,T)$$
$$\begin{cases} [z(t=0), z_t(t=0)] = [z_0, z_1] \\ [v(t=0), v_t(t=0)] = [v_0, v_1] \end{cases}.$$

Above, $[z_0, z_1, v_0, v_1] \equiv X_0$ is given initial data for the problem. Furthermore, the term $\delta'(\zeta_0)u(t)$ is a "control" being applied on the boundary edge Γ_0, with $\delta'(\xi_0)$ being the derivative of the delta function evaluated at a given point $\xi_0 \in \Gamma_0$, and $u(t) \in \mathcal{U} \equiv L^2(0,T)$. Setting

(1.3)
$$\mathcal{H} = H^1(\Omega) \times L^2(\Omega) \times H_0^2(\Gamma_0) \times L^2(\Gamma_0),$$

It is known (see [**2**]) that for $\{X_0, u\} \in \mathcal{H} \times \mathcal{U}$, one has that the corresponding solution $[z, z_t, v, v_t] \in C([0,T]; \mathcal{H})$, By distribution theory then, the system (1.1)–(1.2) is an example of PDE control by means of an *unbounded* input operator. Indeed, $\delta'(\zeta_0) \in \mathcal{L}\left(\mathbb{R}, H^{-\frac{3}{2}-\epsilon}(\Gamma_0)\right)$ for all $\epsilon > 0$; so the term $\delta'(\xi_0)u$ is in a space strictly larger than $L^2(\Gamma_0)$.

In the motivating engineering application, one wishes to control sound waves in a cavity Ω which has a flexible wall Γ_0. To this end, control is implemented by conducting a voltage via a piezoelectric ceramic patch (or patches) located at the point ζ_0 (or points ζ_i) of the wall Γ_0. The controlled PDE (1.1)–(1.2) mathematically captures this structural acoustics interaction (see [**8**], [**17**]).

Moreover, for the purpose of control design a sensor (or sensors) is placed at a point x_0 (or points x_i) within the interior of the cavity Ω, and subsequently measurements are taken of the acoustic pressure at x_0. In mathematical terms, this involves taking the pointwise evaluation $z_t(\cdot, x_0)$ ($x_0 \in \Omega$). Numerical control design experiments with this pointwise observation $z_t(\cdot, x_0)$ have actually been undertaken in [**7**], with respect to the PDE model (1.1)–(1.2), for initial data in the energy space $\mathcal{H}$. However, these numerics constitute only formal manipulations. Indeed, for initial data $X_0 \in \mathcal{H}$, one has from [**2**] that $z_t \in C([0,T]; L^2(\Omega))$ only. Consequently, one cannot use Sobolev embedding theory to have a well–definition of pointwise spatial values of z_t; that is, for $x_0 \in \Omega$, $z_t(\cdot, x_0)$ is a formal expression only.

In [**3**],[**4**] it was shown that for initial data taken from a certain (strict) subspace of $\mathcal{H}$, again Ω a rectangular region, one could indeed take pointwise observations of the acoustic pressure $z_t(\cdot, x_0)$. To precisely characterize this space, we define the self-adjoint, positive, semidefinite operator A_N, $A_N = -\Delta$, with

(1.4)
$$D(A_N) = \left\{ f \in \frac{H^2(\Omega)}{\mathbb{R}} \ni \frac{\partial f}{\partial \nu} = 0 \right\}.$$

Note by **[13]**, we have the following characterization of the domains of fractional powers of A_N:

(1.5) $$D(A_N^\alpha) = \frac{H^{2\alpha}(\Omega)}{\mathbb{R}}, \quad 0 \leq \alpha < \frac{3}{4};$$

(1.6) $$D(A_N^{\frac{3}{4}}) = \left\{ f \in \frac{H^{\frac{3}{2}}(\Omega)}{\mathbb{R}} \ni \frac{\partial f}{\partial \nu} \in L^2_{-\frac{1}{2}}(\Omega) \right\},$$

where, as in **[13]**, $L^2_{-\frac{1}{2}}(\Omega)$ denotes the space of functions $h(x)$ such that $\varrho(x)^{-\frac{1}{2}} h(x) \in L^2(\Omega)$, with $\varrho(x)$ being the distance from x to boundary Γ.

Associated with A_N is the so-called Neumann map $N : L^2(\Gamma_0) \to \frac{L^2(\Omega)}{\mathbb{R}}$, defined by $Ng = h$ iff

(1.7) $$\begin{aligned} \Delta h &= 0 \text{ on } \Omega; \\ \frac{\partial h}{\partial \nu} &= \begin{cases} g & \text{on } \Gamma_0 \\ 0 & \text{on } \Gamma \backslash \Gamma_0 \end{cases} . \end{aligned}$$

By standard elliptic theory, it is known that

(1.8) $$N \in \mathcal{L}\left(H^s(\Gamma_0), \frac{H^{s+\frac{3}{2}}(\Omega)}{\mathbb{R}} \right).$$

With these elliptic operators and $\epsilon > 0$, we now define the space $\mathcal{X} \subset \mathcal{H}$ by

(1.9) $$\mathcal{X} \equiv \left\{ [z_0, z_1, v_0, v_1] \in \frac{H^{\frac{3}{2}}(\Omega)}{\mathbb{R}} \times \frac{H^{\frac{1}{2}}(\Omega)}{\mathbb{R}} \times H_0^2(\Gamma_0) \times H^{\frac{1}{2}-\epsilon}(\Gamma_0) \right. \\ \left. \text{such that } z_0 - Nv_1 \in D(A_N^{\frac{3}{4}}) \right\}.$$

It is shown in **[3]**,**[4]** that for initial data taken from $\mathcal{X}$, and which is smoother in the fourth component v_1 by "2ϵ", one can proceed to take the pointwise evaluation $z_t(\cdot, x_0)$ of the corresponding solution $[\vec{z}, \vec{v}]$. The intent of the present paper is to show that for initial data X_0 taken from $\mathcal{X}$, one has that the corresponding solution $[\vec{z}, \vec{v}]$ evolves in $\mathcal{X}$ continuously. Consequently, using the language of systems theory (see e.g., **[19]**) the controlled PDE (1.1)–(1.2) with the pointwise acoustic pressure observation $z_t(\cdot, x_0)$ in place, will be a *wellposed* system for initial data X_0 taken from $\mathcal{X}$, and $u \in \mathcal{U}$.

We note that the PDE system (1.1)–(1.2) can be modeled by a closed, densely defined operator $\mathcal{A} : D(\mathcal{A}) \subset \mathcal{H} \to \mathcal{H}$, where $\{e^{\mathcal{A}t}\}_{t \geq 0}$ generates a C_0–semigroup on $\mathcal{H}$. In fact, define for all $X_0 \in \mathcal{H}$,

$$\mathcal{A} X_0 = \begin{bmatrix} z_1 \\ \Delta z_0 \\ v_1 \\ -\Delta^2 v_0 - \Delta^2 v_1 - z_1 \end{bmatrix},$$

with $D(\mathcal{A}) = \left\{ [z_0, z_1, v_0, v_1] \in [H^1(\Omega)]^2 \times [H_0^2(\Gamma_0)]^2 : \Delta z_0 \in L^2(\Omega), \right.$

(1.10) $$\left. \frac{\partial z_0}{\partial \nu} = v_1 \text{ on } \Gamma_0, \frac{\partial z_0}{\partial \nu} = 0 \text{ on } \Gamma \backslash \Gamma_0; \Delta^2 v_0 + \Delta^2 v_1 \in L^2(\Gamma_0) \right\}.$$

Then the coupled system (1.1)–(1.2) can be given the abstract representation

$$\frac{d}{dt}\begin{bmatrix} \vec{z} \\ \vec{v} \end{bmatrix} = \mathcal{A}\begin{bmatrix} \vec{z} \\ \vec{v} \end{bmatrix} + \mathcal{B}u;$$

(1.11) $$[\vec{z}(0), \vec{v}(0)] = X_0,$$

where $\mathcal{B}u = [0, 0, 0, \delta'(\zeta_0)u]$. As $\mathcal{A}$ generates a strongly continuous semigroup $\{e^{\mathcal{A}t}\}_{t\geq 0}$ on $\mathcal{H}$, the solution $[\vec{z}, \vec{v}]$ may then be written as

(1.12) $$\begin{bmatrix} \vec{z}(t) \\ \vec{v}(t) \end{bmatrix} = e^{\mathcal{A}t}X_0 + \int_0^t e^{\mathcal{A}(t-s)}\mathcal{B}u(s)ds.$$

In [2] it was shown that the maps

(1.13) $$\{X_0, \mathcal{U}\} \to [\vec{z}, \vec{v}] \in \mathcal{L}(\mathcal{H} \times \mathcal{U}, C([0,T]; \mathcal{H}));$$

(1.14) $$\{X_0, \mathcal{U}\} \to v_t \in \mathcal{L}(\mathcal{H} \times \mathcal{U}, L^2(0, T; H_0^2(\Gamma_0)))$$

(a priori $[\vec{z}, \vec{v}]$ is in the larger space $C([0,T]; [D(\mathcal{A}^*)]')$ only, on account of the unbounded control input). However, this regularity is not enough to justify the physically relevant pointwise evaluations $z_t(\cdot, x_0)$. Our main result here states that for initial data $X_0 \in \mathcal{X}$ and $u \in \mathcal{U}$, the resulting solution $[\vec{z}, \vec{v}]$ of (1.12) (and so of (1.1)–(1.2)) actually evolves continuously in the smaller space $\mathcal{X}$.

THEOREM 1.1. *Let ϵ in (1.9) be small enough (say, $0 < \epsilon \leq \frac{1}{25}$). Then:*
(i) The linear operator $\mathcal{A}$ generates a strongly continuous semigroup on $\mathcal{X}$.
(ii) The mapping

(1.15) $$u \to \int_0^t e^{\mathcal{A}(t-s)}\mathcal{B}u(s)ds \in \mathcal{L}(\mathcal{U}, C([0,T]; \mathcal{X})).$$

REMARK 1.2. Defining formally the observation operator $\mathcal{C}_{x_0}$ on $\mathcal{H}$ by $\mathcal{C}_x[z_0, z_1, v_0, v_1] = z_1$, then the controlled and observed system

$$\frac{d}{dt}\begin{bmatrix} \vec{z} \\ \vec{v} \end{bmatrix} = \mathcal{A}\begin{bmatrix} \vec{z} \\ \vec{v} \end{bmatrix} + \mathcal{B}u;$$

(1.16) $$y(t) = \mathcal{C}_{x_0}\begin{bmatrix} \vec{z}(t) \\ \vec{v}(t) \end{bmatrix},$$

is wellposed for that initial data $[z_0, z_1, v_0, v_1]$ taken from $\mathcal{X}$, which in addition satisfies $v_1 \in H_0^{\frac{1}{2}+\epsilon}(\Gamma_0)$ (again ϵ small enough). That is to say, from Theorem 1.1 we have that $[\vec{z}, \vec{v}] \in C([0,T]; \mathcal{X})$, and moreover, by interpolation between Theorem 1.1 and (1.14), we have $v_t \in L^{\frac{3+2\epsilon}{2\epsilon}}(0, T; H^{\frac{1}{2}+\epsilon}(\Gamma_0))$. In addition, from [3],[4], one has that $y(t) \in L^2(0,T)$. Thus, the solution $[\vec{z}, \vec{v}]$ of (1.1)–(1.2) evolves in the space of chosen initial data; that is, $[\vec{z}, \vec{v}] \in C([0,T]; \mathcal{X})$, and by taking ϵ small enough, $v_t \in L^p(0, T; H^{\frac{1}{2}+\epsilon}(\Gamma_0))$ for all $0 < p < \infty$. Moreover, for this space of initial data, the pointwise observation $z_t(\cdot, x_0)$ is a welldefined quantity. With this established evolution of the trajectory, one can then proceed to employ the machinery of PDE control theory to attain the end of rigorous control design.

REMARK 1.3. Note that the space $\mathcal{X}$ defined in (1.9) bears no relation to $D(\mathcal{A}^k)$, and so the proof of Theorem 1.1 does not depend on soft semigroup arguments. As a matter of fact, the control to state map defined in (1.15) maps u, pointwise in time, into a space strictly larger than $D(\mathcal{A})$. So simply taking the space of initial data to be "smooth" will not ensure that the state $[\vec{z}, \vec{v}]$ evolves in the space, nor that pointwise pressure observations of the acoustic pressure are justified. Actually, our choice of $\mathcal{X}$ is not motivated at all by structural acoustic analysis, but by the work in [20], wherein spatial point evaluations are justified for uncoupled wave equations, in spaces of initial data which are similar to the first two components of the space $\mathcal{X}$.

REMARK 1.4. In [5], we consider the same problem of showing the "wellposedness" of the controlled and observed structural acoustics system (1.16), but for the more general case that Ω is a smooth, arbitrary domain. The methodology in [5], however, is necessarily much different than the present work. In Section 2 of this paper, Fourier expansion and Harmonic analysis techniques are adopted, which exploit the fact that here Ω is a rectangle, so as to obtain various sharp regularity results for the wave equation. For Ω a smooth, arbitrary domain, these Fourier expansion methods do not apply, and so to obtain the analogous results for hyperbolic equations in [5], one must invoke a full apparatus of pseudodifferential and microlocal analysis. Hence, the separate treatment here for the canonical rectangle. Since numerical experimentation and analysis of the 2–D structural acoustics model are usually done for Ω a rectangle (see e.g., [7]), the utility of showing the wellposedness of the structural acoustics PDE (1.1)–(1.2) in the canonical case is self-evident.

The details of proof of Theorem 1.1 can be outlined as follows: In Section 2, we prove sharp regularity results for the wave equation with Neumann forcing data that is "better" than L^2 in regards to smoothness. Subsequently, we interpolate between these new results and the already established regularity results in [16] to obtain the wave regularity result Lemma 2.6, which is suited for the controlled structural acoustics PDE with initial data $X_0 \in \mathcal{X}$. In Section 3, we likewise prove regularity results for the beam component $[v, v_t]$ of the (1.1)–(1.2). In this part of the work, we will use critically the fact that the damped beam component is abstractly modeled by the generator of an *analytic* semigroup A_0, and whose fractional powers can be explicitly characterized in terms of familiar Hilbert spaces. In Section 4, we collect these preliminary results to prove the Theorem 1.1. To use the results of Sections 2 and 3 appropriately, we invoke the cosine operator approach (with respect to the wave dynamics of (1.1)–(1.2)) documented in [12].

2. Analysis of the Wave Equation

Again, Ω is the rectangular domain stated in the introduction. On this geometry we consider the following wave equation with boundary input:

(2.1)
$$\begin{aligned} z_{tt}(x,t) &= \Delta z(x,t) \quad \text{on } \Omega \times (0,T); \\ \frac{\partial z}{\partial \nu} &= \begin{cases} g, & \text{on } \Gamma_0 \times (0,T) \\ 0 & \text{on } \Gamma \setminus \Gamma_0 \times (0,T); \end{cases} \\ z(x,0) &= z_t(x,0) = 0 \quad \text{on } \Omega. \end{aligned}$$

Recalling the operator A_N of (1.4), we let $\{\lambda_{mn}, \Phi_{mn}\}_{m,n=1}^{\infty}$ denote its respective the eigenvalues and orthonormalized eigenfunctions (neglecting the zero eigenvalue and its constant eigenfunction), viz.

$$\text{(2.2)} \quad \begin{aligned} \lambda_{mn} &= m^2 + n^2 & \text{for } m, n = 1, 2, \ldots; \\ \Phi_{mn} &= \tfrac{2}{\pi} \cos nx \cos my & \text{for } m, n = 1, 2, \ldots. \end{aligned}$$

Using then the cosine and sine operators $[\mathcal{C}(t), \mathcal{S}(t)]$ associated with A_N (see e.g. [**12**]), the solution $[z, z_t]$ to (2.1) may be abstractly as

$$\text{(2.3)} \quad z(t) = A_N \int_0^t S(t-\tau) N g(\tau) d\tau;$$

$$\text{(2.4)} \quad z_t(t) = A_N \int_0^t C(t-\tau) N g(\tau) d\tau.$$

In turn, via the eigenvalues and eigenfunctions of A_N, and the elliptic operator N, it is shown in [**14**] (see also [**1**]) that these expressions take the explicit form

$$\text{(2.5)} \quad z(t) = \sum_{m,n=1}^{\infty} \left\{ \frac{1}{\sqrt{n^2+m^2}} \int_0^t \sin\sqrt{n^2+m^2}(t-\tau) g_n(\tau) d\tau \right\} \Phi_{mn};$$

$$\text{(2.6)} \quad z_t(t) = \sum_{m,n=1}^{\infty} \left\{ \int_0^t \cos\sqrt{n^2+m^2}(t-\tau) g_n(\tau) d\tau \right\} \Phi_{mn},$$

where

$$\text{(2.7)} \quad g_n(t) = \frac{2}{\pi}(g(t,\cdot), \cos n(\cdot))_{L^2(\Gamma_0)}.$$

We thus have for all real s

$$\text{(2.8)} \quad \begin{aligned} A_N^s z(t) &= \\ & \sum_{m,n=1}^{\infty} \left\{ (m^2+n^2)^{s-\frac{1}{2}} \int_0^t \sin\sqrt{m^2+n^2}(t-\tau) g_n(\tau) d\tau \right\} \Phi_{mn}; \end{aligned}$$

and

$$\text{(2.9)} \quad \begin{aligned} A_N^s z_t(t) &= \\ & \sum_{m,n=1}^{\infty} \left\{ (m^2+n^2)^{s} \int_0^t \cos\sqrt{m^2+n^2}(t-\tau) g_n(\tau) d\tau \right\} \Phi_{mn}. \end{aligned}$$

Denoting $\widehat{f}$ as the Laplace transform of a time dependent function f with Laplace variable $\lambda = \gamma + i\omega$, and extending g by zero outside the interval $[0, T]$ (as the equation (2.1) has zero initial data), we have upon applying the convolution theorem to (2.8) and (2.9),

$$\text{(2.10)} \quad \widehat{A_N^s z}(\lambda) = \sum_{m,n=1}^{\infty} \left\{ \frac{(m^2+n^2)^s}{\lambda^2 + \lambda_{mn}} \widehat{g_n}(\lambda) \right\} \Phi_{mn};$$

$$\text{(2.11)} \quad \widehat{A_N^s z_t}(\lambda) = \sum_{m,n=1}^{\infty} \left\{ \frac{\lambda (m^2+n^2)^s}{\lambda^2 + \lambda_{mn}} \widehat{g_n}(\lambda) \right\} \Phi_{mn}.$$

As shown in [1] (see (3.30) therein), one has that for $g \in H^\alpha(\Gamma_0)$, $0 \leq \alpha < \frac{3}{2}$,

$$\|g\|^2_{H^\alpha(\Gamma_0)} = \sum_{n=1}^\infty \frac{4}{\pi^2} |(g, \cos n(\cdot))|^2 n^{2\alpha}. \tag{2.12}$$

In obtaining the Sobolev norms for the solution z of (2.1) above, and corresponding to boundary data g of prescribed smoothness, we will have critical need of the following uniform estimates for given singular integrals. Their lengthy and technical proofs are relegated to the Appendix below, as they digress from the main theme of the paper. Throughout, we will denote the symbol $\langle \cdot \rangle$ by $\langle \varpi \rangle = \left(\gamma^2 + |\varpi|^2 \right)^{\frac{1}{2}}$.

LEMMA 2.1. *(i)*

$$\int_1^\infty \frac{n^{\frac{3}{2}} dy}{(y^2 + \gamma^2 + n^2 - \omega^2)^2 + 4\gamma^2 \omega^2} \leq C, \tag{2.13}$$

for all $\omega \in \mathbb{R}$ and $n = 1, 2, \ldots$ (see Proposition 5.1).
(ii) Let $\epsilon > 0$ be arbitrary. Then

$$\int_1^\infty \frac{\langle \omega \rangle^{-1+\epsilon} y^{3-\epsilon} dy}{(y^2 + \gamma^2 + n^2 - \omega^2)^2 + 4\gamma^2 \omega^2} \leq C, \tag{2.14}$$

where constant C is independent of $\omega \in \mathbb{R}$, and $n = 1, 2, 3, \ldots$ (see Proposition 5.2).
(iii)

$$\left(n^2 + \langle \omega \rangle \right)^{-1} \int_1^\infty \frac{\left(y^2 + n^2 \right)^{\frac{1}{2}} \left(\gamma^2 + \omega^2 \right) dy}{(y^2 + \gamma^2 + n^2 - \omega^2)^2 + 4\gamma^2 \omega^2} \leq C, \tag{2.15}$$

for all $\omega \in \mathbb{R}$ and $n = 1, 2, \ldots$ (see Proposition 5.6).

The proofs of these estimates are carried out in Section 5. These estimates are absolutely necessary to attain the following *sharp* regularity estimates for the wave equation (2.1) in the case that Ω is rectangular.

LEMMA 2.2. *(i) Let the boundary data g in (2.1) be in $\mathcal{G}_1$, where*

$$\mathcal{G}_1 \equiv \left\{ g \in H^{\frac{1}{2}-\epsilon}(0, T; L^2(\Gamma_0)) \cap L^2(0, T; H_0^{\frac{3}{4}-\epsilon}(\Gamma_0)) \right\}. \tag{2.16}$$

Then for every $\epsilon > 0$, the corresponding solution $z \in C([0,T]; H^{\frac{3}{2}-\epsilon}(\Omega))$, with the estimates

$$\|z\|_{C([0,T]; H^{\frac{3}{2}-\epsilon}(\Omega))} \leq C_T \|g\|_{\mathcal{G}_1}. \tag{2.17}$$

(ii) Let the boundary data g in (2.1) be in $\mathcal{G}_2$, where

$$\mathcal{G}_2 \equiv \left\{ g \in H^{\frac{1}{2}}(0, T; L^2(\Gamma_0)) \cap L^2(0, T; H_0^1(\Gamma_0)) \cap C([0,T]; L^2(\Gamma_0)), \right.$$
$$\left. \text{with } g(0) = 0 \right\}. \tag{2.18}$$

Then the corresponding velocity $z_t \in C([0,T]; H^{\frac{1}{2}}(\Omega))$, with the estimate

$$\|z_t\|_{C([0,T]; H^{\frac{1}{2}}(\Omega))} \leq C_T \|g\|_{\mathcal{G}_2}. \tag{2.19}$$

Proof of (i): It is enough to show that continuously,

(2.20) $$g \in \mathcal{G}_1 \Rightarrow z \in L^2(0,T; H^{\frac{3}{2}-\epsilon}(\Omega));$$

as we can subsequently appeal to the "lifting theorem" in [**15**] to improve the time regularity to continuity. To show (2.20): We have by the characterization (1.5), the generalized Parseval's relation (see [**11**], p. 212), and the expression (2.10)

$$2\pi \int_0^\infty e^{-2\gamma t} \|z(t)\|^2_{H^{\frac{3}{2}-\epsilon}(\Omega)} dt = \int_{-\infty}^\infty \left\| A_N^{\frac{3}{4}-\frac{\epsilon}{2}} \widehat{z}(\gamma+i\omega) \right\|^2_{L^2(\Omega)} d\omega$$

(2.21) $$= \int_{-\infty}^\infty \sum_{m,n=1}^\infty \frac{(m^2+n^2)^{\frac{3}{2}-\epsilon}}{|(\gamma+i\omega)^2 + \lambda_{mn}|^2} |\widehat{g_n}(\gamma+i\omega)|^2 d\omega.$$

If we can show that

(2.22) $$\left(\langle\omega\rangle^{1-2\epsilon} + n^{\frac{3}{2}-2\epsilon}\right)^{-1} \sum_{m=1}^\infty \frac{m^{3-2\epsilon} + n^{3-2\epsilon}}{|(\gamma+i\omega)^2 + \lambda_{mn}|^2} < C,$$

where the constant C is independent of $\omega \in \mathbb{R}$ and $n = 1, 2, 3, ...$, we would then have from (2.21)

$$2\pi \int_0^T e^{-2\gamma t} \|z(t)\|^2_{H^{\frac{3}{2}-\epsilon}(\Omega)} dt$$

$$\leq C \int_{-\infty}^\infty \left(\langle\omega\rangle^{1-2\epsilon} + n^{\frac{3}{2}-2\epsilon}\right) \sum_{n=1}^\infty |\widehat{g_n}(\gamma+i\omega)|^2$$

$$\cdot \left(\langle\omega\rangle^{1-2\epsilon} + n^{\frac{3}{2}-2\epsilon}\right)^{-1} \sum_{m=1}^\infty \frac{m^{3-2\epsilon} + n^{3-2\epsilon}}{|(\gamma+i\omega)^2 + \lambda_{mn}|^2} d\omega$$

$$\leq C \int_{-\infty}^\infty \left(\langle\omega\rangle^{1-2\epsilon} + n^{\frac{3}{2}-2\epsilon}\right) \sum_{n=1}^\infty |\widehat{g_n}(\gamma+i\omega)|^2 d\omega \quad \text{(after using (2.22))}$$

$$= C \left(\int_{-\infty}^\infty \left\|\lambda^{\frac{1}{2}-\epsilon} \widehat{g}(\lambda)\right\|^2_{L^2(\Gamma_0)} d\omega + \int_{-\infty}^\infty \|\widehat{g}(\lambda)\|^2_{H_0^{\frac{3}{4}-\epsilon}(\Gamma_0)} d\omega \right)$$

(after using the characterization in (2.12); here $\lambda = \gamma + i\omega$)

(2.23) $$= C \left(\int_0^T e^{-2\gamma t} \left\|D_t^{\frac{1}{2}-\epsilon} g\right\|^2_{L^2(\Gamma_0)} dt + \int_0^T e^{-2\gamma t} \|g\|^2_{H_0^{\frac{3}{4}-\epsilon}(\Gamma_0)} dt \right),$$

after another use of Parseval's relation. The desired estimate (2.22), is equivalent, by the definition of Riemann integrability to having that

(2.24) $$\frac{1}{\langle\omega\rangle^{1-2\epsilon} + n^{\frac{3}{2}-2\epsilon}} \int_1^\infty \frac{y^{3-2\epsilon} dy}{(y^2+\gamma^2+n^2-\omega^2)^2 + 4\gamma^2\omega^2} \leq C;$$

(2.25) $$\frac{n^{3-2\epsilon}}{\langle\omega\rangle^{1-2\epsilon} + n^{\frac{3}{2}-2\epsilon}} \int_1^\infty \frac{dy}{(y^2+\gamma^2+n^2-\omega^2)^2 + 4\gamma^2\omega^2} \leq C,$$

where the constant C is independent of $\omega \in \mathbb{R}$ and $n = 1, 2, 3....$ But these estimates are contained in the results of Lemma 2.1(i)–(ii). We thus have (2.24),(2.25) $\Rightarrow$ (2.22) $\Rightarrow$ (2.23) $\Rightarrow$ (2.20). As noted above, L^2–regularity can be lifted to continuity by [**15**]. This concludes the proof of **(i)**.

Proof of (ii): Again, by [15], it is enough to show the estimate

(2.26) $$\|z_t\|_{L^2(0,T;H^{\frac{1}{2}}(\Omega))} \leq C_T \|g\|_{\mathcal{G}_2}.$$

To this end, by the characterization (1.5), we have $H^{\frac{1}{2}}(\Omega) = D(A_N^{\frac{1}{4}})$, and so with the use of Parseval's relation and (2.11) (with $s = \frac{1}{4}$) we have

$$2\pi \int_0^\infty e^{-2\gamma t} \|z_t(t)\|^2_{H^{\frac{1}{2}}(\Omega)} dt$$

$$= \int_{-\infty}^\infty \sum_{n=1}^\infty |\widehat{g_n}(\gamma + i\omega)|^2 \sum_{m=1}^\infty \frac{(\gamma^2 + \omega^2)(m^2 + n^2)^{\frac{1}{2}}}{|(\gamma + i\omega)^2 + \lambda_{mn}|^2} d\omega$$

$$= \int_{-\infty}^\infty (\langle\omega\rangle + n^2) \sum_{n=1}^\infty |\widehat{g_n}(\gamma + i\omega)|^2$$

(2.27) $$\cdot \sum_{m=1}^\infty \frac{(\langle\omega\rangle + n^2)^{-1}(\gamma^2 + \omega^2)(m^2 + n^2)^{\frac{1}{2}}}{|(\gamma + i\omega)^2 + \lambda_{mn}|^2} d\omega.$$

If we can show that

$$(\langle\omega\rangle + n^2)^{-1} \sum_{m=1}^\infty \frac{(\gamma^2 + \omega^2)(m^2 + n^2)^{\frac{1}{2}}}{|(\gamma + i\omega)^2 + \lambda_{mn}|^2} < C,$$

or equivalently

(2.28) $$(\langle\omega\rangle + n^2)^{-1} \int_1^\infty \frac{(y^2 + n^2)^{\frac{1}{2}}(\gamma^2 + \omega^2)}{(y^2 + \gamma^2 + n^2 - \omega^2)^2 + 4\gamma^2\omega^2} dy \leq C,$$

uniformly in $\omega \in \mathbb{R}$ and $n = 1, 2, 3...$, then this estimate and (2.27) would yield

$$2\pi \int_0^T e^{-2\gamma t} \|z_t(t)\|^2_{H^{\frac{1}{2}}(\Omega)} dt$$

$$\leq C \int_{-\infty}^\infty \left(\langle\omega\rangle \sum_{n=1}^\infty |\widehat{g_n}(\gamma + i\omega)|^2 + \sum_{n=1}^\infty n^2 |\widehat{g_n}(\gamma + i\omega)|^2 \right)$$

$$= C \left(\int_{-\infty}^\infty \left\|\lambda^{\frac{1}{2}} \widehat{g}(\lambda)\right\|^2_{L^2(\Gamma_0)} d\omega + \int_{-\infty}^\infty \|\widehat{g}(\lambda)\|^2_{H^1(\Gamma_0)} d\omega \right)$$

(2.29) $$\leq C_T \left(\left\|D_t^{\frac{1}{2}} g\right\|^2_{L^2(0,T;L^2(\Gamma_0))} + \|g\|^2_{L^2(0,T;H^1(\Gamma_0))} \right)$$

(this use above of Parseval's relation is where we require $g(0) = 0$ in (2.18)). But the estimate (2.28) is exactly the result of Lemma 2.1(iii). We thus have (2.28) $\Rightarrow$ (2.29) $\Rightarrow$ (2.26). ∎

REMARK 2.3. One could work further to show that the map

$$g \to [z, z_t] \in \mathcal{L}\left(\mathcal{G}_2, C([0,T]; H^{\frac{3}{2}}(\Omega) \times H^{\frac{1}{2}}(\Omega))\right),$$

where $\mathcal{G}_2$ is the space defined in (2.18). This would require, however, the need to estimate more singular integrals such as those in (2.13), (2.14) and (2.15) (see also the supporting integral estimates in Section 5); and this additional work is not needed in the present context.

For the next result, we recall the sharp regularity estimates derived in [**16**] and [**1**].

THEOREM 2.4. *(see* [**16**], *Theorem 3.1(ii)-(iii) and Theorem 3.2 (i)). Assume that* $g \in \mathcal{G}_3$, *where*

$$\mathcal{G}_3 \equiv \left\{ g \in H^1(0,T; L^2(\Gamma_0)) \cap C([0,T]; H^{\frac{1}{4}-\epsilon}(\Gamma_0)); \; g(0) = 0 \right\}. \tag{2.30}$$

Then the solution z *of (2.1) satisfies the estimate*

$$\|[z, z_t]\|_{C([0,T]; H^{\frac{7}{4}-\epsilon}(\Omega) \times H^{\frac{3}{4}-\epsilon}(\Omega))} \leq C_T \|g\|_{\mathcal{G}_3}. \tag{2.31}$$

THEOREM 2.5. *(see* [**1**], *Theorem 1). Assume that* $g \in L^2(0,T; H^{\frac{1}{4}}(\Gamma_0))$. *Then the solution* $[z, z_t]$ *of (2.1) satisfies the estimate*

$$\|[z, z_t]\|_{C([0,T]; H^1(\Omega) \times L^2(\Omega))} + \|z_t\|_{L^2(0,T; H^{-\frac{1}{4}}(\Gamma))} \leq C_T \|g\|_{L^2(0,T; H^{\frac{1}{4}}(\Gamma_0))}. \tag{2.32}$$

LEMMA 2.6. *Let the boundary data* $g \in \mathcal{G}_*$, *where*

$$\mathcal{G}_* = \left\{ g \in H^{\frac{5}{8}-\frac{\epsilon}{4}}(0,T; L^2(\Gamma_0)) \cap C([0,T]; H^{\frac{1}{2}-\epsilon}(\Gamma_0)); \; g(0) = 0 \right\}. \tag{2.33}$$

Then for $\epsilon > 0$ *small enough (* $\epsilon \leq \frac{1}{25}$, *say), the corresponding solution* $[z, z_t]$ *of (2.1) satisfies the estimate*

$$\|[z, z_t]\|_{C([0,T]; H^{\frac{3}{2}}(\Omega) \times H^{\frac{1}{2}}(\Omega))} \leq C_T \|g\|_{\mathcal{G}_*}. \tag{2.34}$$

Proof: We intend to interpolate between the results in Lemma 2.2, and Theorems 2.4(i) and 2.5.

Proof of z–regularity: Interpolating between Lemma 2.2(i) and Theorem 2.5, we obtain for $0 \leq \theta \leq 1$,

$$\begin{aligned} g &\in H^{\left(\frac{1}{2}-\epsilon\right)\theta}(0,T; L^2(\Gamma_0)) \cap L(0,T; H_0^{\frac{1}{4}(1-\theta)+\theta\left(\frac{3}{4}-\epsilon\right)}(\Gamma_0)) \\ &\Rightarrow z \in C([0,T]; H^{(1-\theta)+\left(\frac{3}{2}-\epsilon\right)\theta}(\Omega)). \end{aligned} \tag{2.35}$$

In turn, interpolating between this and Theorem 2.4(i) yields for $0 \leq \theta, \xi \leq 1$,

$$\begin{cases} g \in H^{\left(\frac{1}{2}-\epsilon\right)\theta(1-\xi)+\xi}(0,T; L^2(\Gamma_0)) \\ \quad \cap C([0,T]; H_0^{\left[\frac{1}{4}(1-\theta)+\theta\left(\frac{3}{4}-\epsilon\right)\right](1-\xi)+\xi\left(\frac{1}{4}-\epsilon\right)}(\Gamma_0)) \\ g(0) = 0; \end{cases}$$

$$\Rightarrow z \in C([0,T]; H^{[(1-\theta)+\left(\frac{3}{2}-\epsilon\right)\theta](1-\xi)+\xi\left(\frac{7}{4}-\epsilon\right)}(\Omega)). \tag{2.36}$$

Solving for ξ in the equation

$$\left((1-\theta) + \left(\frac{3}{2}-\epsilon\right)\theta\right)(1-\xi) + \xi\left(\frac{7}{4}-\epsilon\right) = \frac{3}{2}, \tag{2.37}$$

we obtain

$$\xi = \frac{2 - 2\theta + 4\theta\epsilon}{3 - 2\theta + 4\theta\epsilon - 4\epsilon}. \tag{2.38}$$

Using this value of ξ to solve for θ in the equation

$$\left(\frac{1}{2}-\epsilon\right)\theta(1-\xi) + \xi = \frac{5}{8} - \frac{\epsilon}{4}, \tag{2.39}$$

we obtain

(2.40) $$\theta = \frac{1 + 26\epsilon - 8\epsilon^2}{2 + 16\epsilon - 40\epsilon^2}.$$

Substituting these values of ξ, θ into (2.36) yields now

(2.41) $$\begin{cases} g \in H^{\frac{5}{8} - \frac{\epsilon}{4}}(0, T; L^2(\Gamma_0)) \cap C([0, T]; H_0^{\frac{3+30\epsilon}{8+32\epsilon}}(\Gamma_0)) \\ g(0) = 0; \end{cases}$$
$$\Rightarrow z \in C([0, T]; H^{\frac{3}{2}}(\Omega)).$$

For $\epsilon > 0$ small enough, (say, $\epsilon \leq \frac{1}{25}$), an element g of $\mathcal{G}_*$ will be in $C([0, T]; H^{\frac{3+30\epsilon}{8+32\epsilon}}(\Gamma_0))$, and we thus deduce from (2.41) the estimate

(2.42) $$\|z\|_{C([0,T]; H^{\frac{3}{2}}(\Omega))} \leq C_T \|g\|_{\mathcal{G}_*}.$$

Proof of z_t-regularity. This runs in the same way as for z, and so we omit the full details here. Interpolating between Lemma 2.2(ii) and Theorem 2.5, and subsequently with Theorem 2.4(i) yields

(2.43) $$\begin{cases} g \in H^{\frac{5}{8} - \frac{\epsilon}{4}}(0, T; L^2(\Gamma_0)) \cap C([0, T]; H_0^{\frac{43-66\epsilon+24\epsilon^2}{18-64\epsilon}}(\Gamma_0)) \\ g(0) = 0; \end{cases}$$
$$\Rightarrow z_t \in C([0, T]; H^{\frac{1}{2}}(\Omega)).$$

(In this interpolation, we are using the implicitly the fact from [16], Theorem 3.1(iii), that it is enough that $g \in H^1(0, T; L^2(\Gamma_0))$ and $g(0) = 0$ to yield that $z_t \in C([0, T]; H^{\frac{3}{4} - \epsilon}(\Omega))$.) For $\epsilon > 0$ small enough (say, $\epsilon \leq \frac{1}{15}$), we then have that for all $g \in \mathcal{G}_*$, $g \in C([0, T]; H^{\frac{43-66\epsilon+24\epsilon^2}{18-64\epsilon}}(\Gamma_0))$; and we thus have the estimate from (2.43)

(2.44) $$\|z_t\|_{C([0,T]; H^{\frac{1}{2}}(\Omega))} \leq C_T \|g\|_{\mathcal{G}_*}.$$

Combining now (2.42) and (2.44) yields now the regularity estimate in (2.34). ∎

3. Analysis of the Beam Component

We start by abstractly modeling the elastic component of (1.1)–(1.2). To this end, we define the operator $\mathring{A} : L^2(\Gamma_0) \supset D(\mathring{A}) \to L^2(\Gamma_0)$ as

$$\mathring{A} = \Delta^2, \quad D(\mathring{A}) = H^4(\Gamma_0) \cap H_0^2(\Gamma_0).$$

$\mathring{A}$ is positive definite, self–adjoint with its fractional powers therefore being well–defined. By [13], we have

(3.1) $$D(\mathring{A}^\alpha) = H_0^{4\alpha}(\Gamma_0), \text{ for } 0 \leq \alpha < \frac{5}{8}.$$

Through this elliptic operator, we define

$$A_0 := \begin{bmatrix} 0 & I \\ -\mathring{A} & -\mathring{A} \end{bmatrix} : D(\mathring{A}) \to H_0^2(\Gamma_0) \times L^2(\Gamma_0),$$

(3.2) $$D(A_0) = \left\{ [v_1, v_2] \in \left[H_0^2(\Gamma_0)\right]^2 \text{ such that } v_1 + v_2 \in D(\mathring{A}) \right\}.$$

It is a rather nontrivial result that A_0 generates an *analytic* C_0–semigroup $\{e^{A_0 t}\}_{t \geq 0}$ of contractions on $H_0^2(\Gamma_0) \times L^2(\Gamma_0)$ (see [**9**]). With this operator formalism, the solution $[v, v_t]$ to the damped beam equation (1.2) can be written as

$$(3.3) \quad \begin{bmatrix} v(t) \\ v_t(t) \end{bmatrix} = e^{A_0 t} \begin{bmatrix} v_0 \\ v_1 \end{bmatrix} + \int_0^t e^{A_0(t-\tau)} \begin{bmatrix} 0 \\ -z_t(\tau)|_{\Gamma_0} + \delta'(\zeta_0) u(\tau) \end{bmatrix} d\tau.$$

In analyzing this expression, we will use the underlying analyticity of the semigroup $\{e^{A_0 t}\}_{t \geq 0}$; in particular we will invoke the following regularity results which were established in [**10**]:

THEOREM 3.1.

(i) (see [**10**], Corollary 5.1(i)) The mapping

$$\vec{f} \to \int_0^t e^{A_0(t-\tau)} \overrightarrow{f(\tau)} d\tau$$

is an element of $\mathcal{L}\left(L^2(0, T; L^2(\Gamma_0)), L^2(0, T; D(A_0)) \cap C([0,T]; D((-A_0)^{\frac{1}{2}}))\right)$.

(ii) (see [**10**], Theorem 1.1)(a) For $0 \leq s \leq \frac{1}{2}$, we have the characterization

$$D((-A_0)^s) = H_0^2(\Gamma_0) \times H_0^{4s}(\Gamma_0).$$

(b) For $s > \frac{1}{2}$, we have the characterization

$$D((-A_0)^s) = \{[v_0, v_1] \in H_0^2(\Gamma_0) \times H_0^2(\Gamma_0) \text{ such that } v_0 + v_1 \in D(\mathring{A}^s)\}.$$

(iii) (see [**10**], Corollary 5.1(ii)) For $0 \leq s \leq \frac{1}{2}$, the mapping

$$\begin{bmatrix} v_0 \\ v_1 \end{bmatrix} \to e^{A_0 t} \begin{bmatrix} v_0 \\ v_1 \end{bmatrix}$$

is an element of

$$\mathcal{L}\left(H_0^2(\Gamma_0) \times H_0^{4s}(\Gamma_0), L^2(0, T; D((-A_0)^{s+\frac{1}{2}})) \cap C([0,T]; H_0^2(\Gamma_0) \times H_0^{4s}(\Gamma_0))\right).$$

Using these results, and the definition

$$(3.4) \quad \widetilde{\mathcal{X}} \equiv H^1(\Omega) \times L^2(\Omega) \times H_0^2(\Gamma_0) \times H^{\frac{1}{2}-\epsilon}(\Gamma_0),$$

we now prove the following:

PROPOSITION 3.2. *In* (1.1)–(1.2), *let the initial data* $[z_0, z_1, v_0, v_1] = X_0 \in \widetilde{\mathcal{X}}$. *Then the beam component* $[v, v_t]$ *satisfies*:

(i) $[v, v_t] \in L^2\left(0, T; D((-A_0)^{\frac{5}{8}-\frac{\epsilon}{4}})\right) \cap C\left([0,T]; D((-A_0)^{\frac{1}{8}-\frac{\epsilon}{4}})\right)$, *with the estimate*

$$(3.5) \quad \left\| \begin{bmatrix} v \\ v_t \end{bmatrix} \right\|_{L^2(0,T;D((-A_0)^{\frac{5}{8}-\frac{\epsilon}{4}})) \cap C([0,T];D((-A_0)^{\frac{1}{8}-\frac{\epsilon}{4}}))} \leq C\left(\|X_0\|_{\widetilde{\mathcal{X}}} + \|u\|_{\mathcal{U}}\right).$$

(ii) $v + v_t \in H^{\frac{5}{8}-\frac{\epsilon}{4}}(0, T; L^2(\Gamma_0))$ *with the estimate*

$$(3.6) \quad \|v + v_t\|_{H^{\frac{5}{8}-\frac{\epsilon}{4}}(0,T;L^2(\Gamma_0))} \leq C\left(\|X_0\|_{\widetilde{\mathcal{X}}} + \|u\|_{\mathcal{U}}\right).$$

Proof of (i): First, $-z_t|_{\Gamma_0} + \delta'(\zeta_0)u \in L^2(0, T; H^{-\frac{3}{2}-\epsilon}(\Gamma_0))$, by (2.32) and distribution theory. The Theorem 3.1(ii) will give then that $[0, -z_t|_{\Gamma_0} + \delta'(\zeta_0)u] \in$

$L^2\left(0,T;D((-A_0)^{-\frac{3}{8}-\frac{\epsilon}{4}})\right)$. Next, coupling this regularity with the equality (3.3), we have

$$\begin{bmatrix} v(t) \\ v_t(t) \end{bmatrix}$$
$$= e^{A_0 t}\begin{bmatrix} v_0 \\ v_1 \end{bmatrix} + \int_0^t e^{A_0(t-\tau)}\begin{bmatrix} 0 \\ -z_t(\tau)|_{\Gamma_0} + \delta'(\zeta_0)u(\tau)\end{bmatrix}d\tau$$
$$= e^{A_0 t}\begin{bmatrix} v_0 \\ v_1 \end{bmatrix}$$
$$\quad + (-A_0)^{\frac{3}{8}+\frac{\epsilon}{4}}\int_0^t e^{A_0(t-\tau)}(-A_0)^{-\frac{3}{8}-\frac{\epsilon}{4}}\begin{bmatrix} 0 \\ -z_t(\tau)|_{\Gamma_0} + \delta'(\zeta_0)u(\tau)\end{bmatrix}d\tau$$

(3.7) $\in L^2\left(0,T;D((-A_0)^{\frac{5}{8}-\frac{\epsilon}{4}})\right) \cap C\left([0,T];D((-A_0)^{\frac{1}{8}-\frac{\epsilon}{4}})\right),$

after using Theorem 3.1(i). Moreover, taking the norms of both sides of (3.7) and using again Theorem 3.1, we have

$$\left\|\begin{bmatrix} v \\ v_t \end{bmatrix}\right\|_{L^2(0,T;D((-A_0)^{\frac{5}{8}-\frac{\epsilon}{4}}))\cap C([0,T];D((-A_0)^{\frac{1}{8}-\frac{\epsilon}{4}}))}$$
$$\leq C\left(\left\|\begin{bmatrix} v_0 \\ v_1\end{bmatrix}\right\|_{H_0^2(\Gamma_0)\times H^{\frac{1}{2}-\epsilon}(\Gamma_0)} + \left\|-z_t|_{\Gamma_0} + \delta'(\zeta_0)u\right\|_{L^2(0,T;H^{-\frac{3}{2}-\epsilon}(\Gamma_0))}\right)$$
$$\leq C\left(\|X_0\|_{\widetilde{\mathcal{X}}} + \|u\|_{\mathcal{U}}\right),$$

where for the last inequality we have also used the trace estimate in (2.32). This completes (i).

Proof of (ii): By (i) and Theorem 3.1(ii)(b), we have that

(3.8) $\qquad v + v_t \in L^2(0,T; H_0^{\frac{5}{2}-\epsilon}(\Gamma_0)).$

Moreover, differentiating the expression (3.3), and using (3.8), (2.32) and the characterization (3.1), yields

(3.9) $\quad v_{tt} = -\mathbf{\mathring{A}}^{\frac{3}{8}+\frac{\epsilon}{4}}\mathbf{\mathring{A}}^{\frac{5}{8}-\frac{\epsilon}{4}}(v+v_t) - z_t|_{\Gamma_0} + \delta'(\zeta_0)u \in L^2\left(0,T; H^{-\frac{3}{2}-\epsilon}(\Gamma_0)\right).$

Therefore, we have continuously,

(3.10) $\begin{cases} v + v_t \in H^0(0,T; H_0^{\frac{5}{2}-\epsilon}(\Gamma_0)) \\ v + v_t \in H^1(0,T; H^{-\frac{3}{2}-\epsilon}(\Gamma_0)) \end{cases} \Longrightarrow v + v_t \in H^{\frac{5}{8}-\frac{\epsilon}{4}}(0,T; L^2(\Gamma_0)).$

This concludes (ii), with the accompanying estimate following from (3.10) and (3.5). ∎

4. Proof of Theorem 1.1

To prove Theorem 1.1(i)–(ii), it will suffice by the well-known theorem of J. Ball in [6] to show that the mapping $\{X_0, u\} \to [\overrightarrow{z}, \overrightarrow{v}]$ defined in (1.12) is continuous from $\mathcal{X} \times \mathcal{U}$ into $C([0,T];\mathcal{X})$, where the space $\mathcal{X}$ was defined above in (1.9).

Step 1 (beam regularity). From Proposition 3.2, we have that the beam component $[v, v_t]$ of (1.1)–(1.2) satisfies the estimate

(4.1) $\qquad \|[v, v_t]\|_{C([0,T]; H_0^2(\Gamma_0) \times H^{\frac{1}{2}-\epsilon}(\Gamma_0))} \leq C\left(\|X_0\|_{\widetilde{\mathcal{X}}} + \|u\|_{\mathcal{U}}\right).$

Step 2 (wave regularity). Taking initial data $X_0 = [z_0, z_1, v_0, v_1] \in \mathcal{X}$ and control $u \in \mathcal{U}$, we can write out explicitly the wave component $[z, z_t]$ of (1.1)–(1.2) by means of the Cosine operators associated with the elliptic operator A_N, (see (2.3)–(2.4)). To wit, we have

$$\begin{aligned} z(t) &= \mathcal{C}(t)z_0 + \mathcal{S}(t)z_1 + \int_0^t A_N \mathcal{S}(t-s) N v_t(s) ds \\ &= \mathcal{C}(t)z_0 + \mathcal{S}(t)z_1 + \int_0^t A_N \mathcal{S}(t-s) N \left(v_t(s) - v_1 + v_1 \right) ds \\ &= z^{(0)}(t) + z^{(1)}(t) + z^{(2)}(t); \end{aligned} \quad (4.2)$$

where

$$\begin{aligned} z^{(0)}(t) &= \mathcal{C}(t)(z_0 - Nv_1) + \mathcal{S}(t)z_1; \\ z^{(1)} &= Nv_1; \\ z^{(2)}(t) &= \int_0^t A_N \mathcal{S}(t-s) N \left(v_t(s) - v_1 \right) ds. \end{aligned} \quad (4.3)$$

To estimate $\left[z^{(0)}, z_t^{(0)} \right]$: As $\{z_0 - Nv_1, z_1\} \in D(A_N^{\frac{3}{4}}) \times D(A_N^{\frac{1}{4}})$ and $\mathcal{C}(\cdot) \in \mathcal{L}\left(D(A_N^\alpha), C([0,T]; D(A_N^\alpha))\right)$ and $\mathcal{S}(\cdot) \in \mathcal{L}\left(D(A_N^\alpha), C([0,T]; D(A_N^{\alpha+\frac{1}{2}}))\right)$ (see [**12**]), we then have

$$\left\| \left[z^{(0)}, z_t^{(0)} \right] \right\|_{C([0,T]; D(A_N^{\frac{3}{4}}) \times D(A_N^{\frac{1}{4}}))} \leq \left\| [z_0 - Nv_1, z_1] \right\|_{D(A_N^{\frac{3}{4}}) \times D(A_N^{\frac{1}{4}})} \leq \|X_0\|_{\mathcal{X}}. \quad (4.4)$$

To estimate $z^{(1)}$: By (1.8), we have

$$\left\| z^{(1)} \right\|_{H^{\frac{3}{2}}(\Omega)} \leq C \|v_1\|_{L^2(\Gamma_0)} \leq C \|X_0\|_{\mathcal{X}}. \quad (4.5)$$

To estimate $\left[z^{(2)}, z_t^{(2)} \right]$: We decompose $z^{(2)} = z^{(2.a)} + z^{(2.b)}$, where

$$z^{(2.a)}(t) = \int_0^t A_N \mathcal{S}(t-s) N \left(v(s) + v_t(s) - (v_0 + v_1) \right) ds; \quad (4.6)$$

$$z^{(2.b)}(t) = \int_0^t A_N \mathcal{S}(t-s) N \left(v_0 - v(s) \right) ds. \quad (4.7)$$

By Proposition 3.2 and Theorem 3.1, we have that $v + v_t \in H^{\frac{5}{8}-\frac{\epsilon}{4}}(0,T; L^2(\Gamma_0)) \cap C([0,T]; H^{\frac{1}{2}-\epsilon}(\Gamma_0))$. Subsequently, we can apply Lemma 2.6 (with $g = v + v_t - (v_0 + v_1)$ therein, and again $\epsilon \leq \frac{1}{25}$) to have

$$\begin{aligned} \left\| \left[z^{(2.a)}, z_t^{(2.a)} \right] \right\|_{C([0,T]; H^{\frac{3}{2}}(\Omega) \times H^{\frac{1}{2}}(\Omega))} &\leq C_T \|v + v_t - (v_0 + v_1)\|_{\mathcal{G}_*} \\ &\leq C_T \left(\|X_0\|_{\mathcal{X}} + \|u\|_{\mathcal{U}} \right), \end{aligned} \quad (4.8)$$

where $\mathcal{G}_*$ is as defined in (2.33) (in the last inequality we have also implicitly used Proposition 3.2).

To handle $\left[z^{(2.b)}, z_t^{(2.b)}\right]$, we use Theorem 2.4 (with $g = v_0 - v$ therein) and Proposition 3.2 to have

$$\left\|\left[z^{(2.b)}, z_t^{(2.b)}\right]\right\|_{C([0,T];H^{\frac{3}{2}}(\Omega) \times H^{\frac{1}{2}}(\Omega))} \leq C_T \|(v_0 - v)\|_{\mathcal{G}_3}$$
(4.9)
$$\leq C_T \left(\|X_0\|_{\mathcal{X}} + \|u\|_{\mathcal{U}}\right),$$

where $\mathcal{G}_3$ is as defined in (2.30).

Combining (4.8), (4.9), (4.4) and (4.5) gives now

(4.10) $$\|[z, z_t]\|_{C([0,T];H^{\frac{3}{2}}(\Omega) \times H^{\frac{1}{2}}(\Omega))} \leq C \left(\|X_0\|_{\mathcal{X}} + \|u\|_{\mathcal{U}}\right).$$

Step 3 (Showing that $z - Nv_t \in C([0,T]; D(A_N^{\frac{3}{4}}))$): Using again the decomposition in (4.2), we have

(4.11) $$z(t) - Nv_t(t) = z^{(0)}(t) + z^{(1)}(t) + z^{(2)}(t) - Nv_t(t).$$

Analyzing the right hand side of this expression, the expressions in (4.3) yield

$$\left. \frac{\partial}{\partial \nu} \left(z^{(1)}(t) + z^{(2)}(t) - Nv_t(t) \right) \right|_{\Gamma_0} = 0,$$

and this, combined with (4.5), (4.8), (4.9) and (1.8) gives the deduction that

(4.12) $$\left\| z^{(1)} + z^{(2)} - Nv_t \right\|_{C([0,T];D(A_N^{\frac{3}{4}}))} \leq C \left(\|X_0\|_{\mathcal{X}} + \|u\|_{\mathcal{U}}\right)$$

(here, we have also implicitly interpolated between the fractional powers of $D(A_N)$; see (1.5)–(1.6)). Combining (4.11), (4.4) and (4.12) yields then

(4.13) $$\|z - Nv_t\|_{C([0,T];D(A_N^{\frac{3}{4}}))} \leq C \left(\|X_0\|_{\mathcal{X}} + \|u\|_{\mathcal{U}}\right).$$

The estimates (4.1),(4.10) and (4.13) show that the map defined in (1.12) is an element of $\mathcal{L}\left(\mathcal{X} \times \mathcal{U}, C([0,T]; \mathcal{X})\right)$. Letting then $u = 0$ in (1.12) and subsequently appealing to [6] yields the semigroup generation result Theorem 1.1(i). Letting instead $X_0 = 0$ in (1.12) yields Theorem 1.1(ii). This completes the proof of Theorem 1.1. ∎

5. Appendix: Singular Integral Estimates

Here we prove the critical integral estimates which are used in Section 2 (See Lemma 2.1). The techniques invoked are motivated by the work in [14] and [20]. Thoughout we will have need of the following inequality which was proved in [1] (see (3.33) therein):

PROPOSITION 5.1.

(5.1) $$\int_1^\infty \frac{n^{\frac{3}{2}} dy}{(y^2 + \gamma^2 + n^2 - \omega^2)^2 + 4\gamma^2 \omega^2} \leq C,$$

for all $\omega \in \mathbb{R}$ and $n = 1, 2, \ldots$.

PROPOSITION 5.2. *For arbitrary $0 < \epsilon < 1$, we have the following estimate*

(5.2) $$\mathcal{I} \equiv \langle \omega \rangle^{-1+\epsilon} \int_1^\infty \frac{y^{3-\epsilon} dy}{(y^2 + \gamma^2 + n^2 - \omega^2)^2 + 4\gamma^2 \omega^2} \leq C_\epsilon,$$

where $\langle \varpi \rangle = \left(\gamma^2 + |\varpi|^2 \right)^{\frac{1}{2}}$, and the constant C_ϵ is independent of $\omega \in \mathbb{R}$ and $n = 1, 2, \ldots$.

Proof: We must consider two cases:

Case 1: $\gamma^2 + n^2 - \omega^2 \geq 0$: Easily then,

$$\mathcal{I} \leq \int_1^\infty y^{-1-\epsilon} dy \leq C_\epsilon. \tag{5.3}$$

Case 2: $\gamma^2 + n^2 - \omega^2 < 0$: We set $a_{n\omega}^2 = \omega^2 - n^2 - \gamma^2$ and split the integral I into the intervals

$$\mathcal{I} = \int_1^\infty \leq \int_1^{\frac{a_{n\omega}}{2}} + \int_{\frac{a_{n\omega}}{2}}^{2a_{n\omega}} + \int_{2a_{n\omega}}^\infty = I_1 + I_2 + I_3. \tag{5.4}$$

To estimate I_1:

$$I_1 \leq \langle \omega \rangle^{-1+\epsilon} \int_0^{\frac{a_{n\omega}}{2}} \frac{y^{3-\epsilon} dy}{(y^2 - a_{n\omega}^2)^2 + 4\gamma^2\omega^2} \leq \langle \omega \rangle^{-1+\epsilon} \int_0^{\frac{a_{n\omega}}{2}} \frac{y^{3-\epsilon} dy}{\frac{9}{16} a_{n\omega}^4 + 4\gamma^2\omega^2}$$

$$= \frac{\langle \omega \rangle^{-1+\epsilon}}{4-\epsilon} \frac{\left(\frac{1}{2} a_{n\omega}\right)^{4-\epsilon}}{\frac{9}{16} a_{n\omega}^4 + 4\gamma^2\omega^2} \leq \frac{1}{\gamma^{1-\epsilon}(4-\epsilon) 2^{4-\epsilon}} \frac{a_{n\omega}^{4-\epsilon}}{\frac{9}{16} a_{n\omega}^4 + 4\gamma^2} \leq C. \tag{5.5}$$

To estimate I_3:

$$I_3 = \langle \omega \rangle^{-1+\epsilon} \int_{2a_{n\omega}}^\infty \frac{y^{3-\epsilon} dy}{(y^2 - a_{n\omega}^2)^2 + 4\gamma^2\omega^2} \leq \langle \omega \rangle^{-1+\epsilon} \int_{2a_{n\omega}}^\infty \frac{y^{3-\epsilon} dy}{\frac{9}{16} y^4 + 4\gamma^4}$$

$$\leq \frac{1}{\gamma^{1-\epsilon}} \int_0^1 \frac{y^{3-\epsilon} dy}{\frac{9}{16} y^4 + 4\gamma^4} + \frac{1}{\gamma^{1-\epsilon}} \int_1^\infty \frac{y^{3-\epsilon} dy}{\frac{9}{16} y^4 + 4\gamma^4} \leq C_{\gamma,\epsilon}. \tag{5.6}$$

To estimate I_2: This is where we need the "smoothing term" $\langle \omega \rangle^{-1+\epsilon}$. Initially we have

$$I_2 = \langle \omega \rangle^{-1+\epsilon} \int_{\frac{a_{n\omega}}{2}}^{2a_{n\omega}} \frac{y^{3-\epsilon} dy}{(y^2 - a_{n\omega}^2)^2 + 4\gamma^2\omega^2}$$

$$\leq 2^{1-\epsilon} \langle \omega \rangle^{-1+\epsilon} a_{n\omega}^{1-\epsilon} \int_{\frac{a_{n\omega}}{2}}^{2a_{n\omega}} \frac{y^2 dy}{(y^2 - a_{n\omega}^2)^2 + 4\gamma^2\omega^2}$$

$$\leq C \int_{\frac{a_{n\omega}}{2}}^{2a_{n\omega}} \frac{y^2 dy}{(y^2 - a_{n\omega}^2)^2 + 4\gamma^2\omega^2}. \tag{5.7}$$

Setting $t \equiv y^2 - a_{n\omega}^2$, we then have

$$I_2 \leq C \int_{\frac{a_{n\omega}}{2}}^{2a_{n\omega}} \frac{y^2 dy}{(y^2 - a_{n\omega}^2)^2 + 4\gamma^2\omega^2} = C \int_{-\frac{3}{4} a_{n\omega}^2}^{3a_{n\omega}^2} \frac{\sqrt{t + a_{n\omega}^2}}{t^2 + 4\gamma^2\omega^2} dt$$

$$\leq C a_{n\omega} \int_{-\frac{3}{4} a_{n\omega}^2}^{3a_{n\omega}^2} \frac{dt}{t^2 + 4\gamma^2\omega^2} = C \frac{a_{n\omega}}{2\gamma |\omega|} \arctan\left(\frac{t}{2\gamma |\omega|}\right)\Big|_{-\frac{3}{4} a_{n\omega}^2}^{3a_{n\omega}^2}$$

$$\leq C \frac{\pi}{2\gamma}. \tag{5.8}$$

Collecting (5.4), (5.5), (5.6) and (5.8) now yields the desired inequality (5.2) for Case 2. ∎

PROPOSITION 5.3. *For all $\omega \in \mathbb{R}$ and $n = 1, 2, \ldots$, we have the estimate*

$$(5.9) \qquad \mathcal{I} \equiv \langle \omega \rangle^{-\frac{3}{2}} \int_1^\infty \frac{n^3 dy}{(y^2 + \gamma^2 + n^2 - \omega^2)^2 + 4\gamma^2 \omega^2} \leq C.$$

Proof:
Case 1: $\omega^2 < \gamma^2 + \frac{n^2}{2}$. Then

$$\mathcal{I} = \langle \omega \rangle^{-\frac{3}{2}} n^3 \int_1^\infty \frac{dy}{(y^2 + \gamma^2 + n^2 - \omega^2)^2 + 4\gamma^2 \omega^2}$$

$$(5.10) \qquad \leq n^3 \int_{-\infty}^\infty \frac{dy}{(y^2 + \frac{n^2}{2})^2} = n^3 \frac{C}{n^3} = C.$$

Case 2: $\gamma^2 + \frac{n^2}{2} < \omega^2 < \gamma^2 + n^2$. Then

$$\mathcal{I} = \langle \omega \rangle^{-\frac{3}{2}} n^3 \int_1^\infty \frac{dy}{(y^2 + \gamma^2 + n^2 - \omega^2)^2 + 4\gamma^2 \omega^2}$$

$$\leq \langle \omega \rangle^{-\frac{3}{2}} n^3 \int_0^\infty \frac{dy}{y^4 + 2\gamma^2 n^2}$$

$$(5.11) \qquad \leq C \langle \omega \rangle^{-\frac{3}{2}} n^{\frac{3}{2}} \leq C \left(\frac{n}{\sqrt{2}} \right)^{-\frac{3}{2}} n^{\frac{3}{2}} \leq C.$$

Case 3: $\gamma^2 + n^2 < \omega^2$. Again setting $a_{n\omega}^2 = \omega^2 - n^2 - \gamma^2 > 0$, we split the integral $\mathcal{I}$ into the following appropriate segments,

$$(5.12) \qquad \mathcal{I} = \int_1^\infty = \int_1^{\frac{a_{n\omega}}{2}} + \int_{\frac{a_{n\omega}}{2}}^{2a_{n\omega}} + \int_{2a_{n\omega}}^\infty = I_1 + I_2 + I_3,$$

and subsequently estimate each I_i.

To estimate I_1:

$$(5.13) \quad I_1 \leq \langle \omega \rangle^{-\frac{3}{2}} n^3 \int_0^{\frac{a_{n\omega}}{2}} \frac{dy}{(y^2 - a_{n\omega}^2)^2 + 4\gamma^2 \omega^2} \leq \langle \omega \rangle^{-\frac{3}{2}} n^3 \frac{\frac{a_{n\omega}}{2}}{\frac{9}{16} a_{n\omega}^4 + 4\gamma^2 \omega^2};$$

and if we consider the function

$$(5.14) \qquad f(x) \equiv \frac{x}{\frac{9}{16} x^4 + 4\gamma^2 \omega^2},$$

with global maximizer (on $(0, \infty)$) $\bar{x} = \frac{2\sqrt{2}}{3^{\frac{3}{4}}} \sqrt{\gamma} \sqrt{|\omega|}$, then from (5.13)

$$(5.15) \qquad I_1 \leq C_0 \langle \omega \rangle^{-\frac{3}{2}} \frac{n^3}{\omega^2} \sqrt{|\omega|} \leq C_0 \langle \omega \rangle^{-\frac{3}{2}} \frac{|\omega|^3}{\omega^2} \sqrt{|\omega|} \leq C.$$

To estimate I_3:

$$I_3 = \langle\omega\rangle^{-\frac{3}{2}} n^3 \int_{2a_{n\omega}}^{\infty} \frac{dy}{(y^2 - a_{n\omega}^2)^2 + 4\gamma^2\omega^2}$$

$$\leq \langle\omega\rangle^{-\frac{3}{2}} n^3 \int_{2a_{n\omega}}^{\infty} \frac{dy}{\frac{9}{16}y^4 + 4\gamma^2 n^2}$$

$$\leq \frac{1}{2} \langle\omega\rangle^{-\frac{3}{2}} n^3 \int_{-\infty}^{\infty} \frac{dy}{(\frac{3}{4}y^2 + 2\gamma n)^2}$$

(5.16)
$$= C_0 \langle\omega\rangle^{-\frac{3}{2}} n^{\frac{3}{2}} \leq C.$$

To estimate I_2:
Making the substitution $t = y^2 - a_{n\omega}^2$, we have

$$I_2 = \langle\omega\rangle^{-\frac{3}{2}} n^3 \int_{\frac{a_{n\omega}}{2}}^{2a_{n\omega}} \frac{dy}{(y^2 - a_{n\omega}^2)^2 + 4\gamma^2\omega^2}$$

$$= \frac{1}{2} \langle\omega\rangle^{-\frac{3}{2}} n^3 \int_{-\frac{3}{4}a_{n\omega}^2}^{3a_{n\omega}^2} \frac{dt}{(t^2 + 4\gamma^2\omega^2)\sqrt{t + a_{n\omega}^2}}$$

$$\leq \frac{\langle\omega\rangle^{-\frac{3}{2}} n^3}{a_{n\omega}} \int_{-\frac{3}{4}a_{n\omega}^2}^{3a_{n\omega}^2} \frac{dt}{(t^2 + 4\gamma^2\omega^2)}$$

(5.17)
$$= \frac{\langle\omega\rangle^{-\frac{3}{2}} n^3}{2\gamma|\omega|a_{n\omega}} \left[\arctan\left(\frac{3a_{n\omega}^2}{8|\gamma||\omega|}\right) + \arctan\left(\frac{3a_{n\omega}^2}{2|\gamma||\omega|}\right)\right].$$

(a) If $a_{n\omega} > n^{\frac{1}{2}}$, then from (5.17) and the fact that $|\omega| > n$,

(5.18)
$$I_2 \leq \frac{\pi \langle\omega\rangle^{-\frac{3}{2}} n^3}{2|\gamma|n^{\frac{3}{2}}} \leq C.$$

(b) If $a_{n\omega} \leq n^{\frac{1}{2}}$, then again from (5.17),

(5.19)
$$I_2 \leq \frac{C_\gamma \langle\omega\rangle^{-\frac{3}{2}} n^3 a_{n\omega}}{\omega^2} \leq C_\gamma \omega^{-\frac{7}{2}} n^{\frac{7}{2}} \leq C.$$

Collecting (5.12), (5.15), (5.16), (5.18) and (5.19) completes the consideration of Case 3, and so the proof of the Proposition. ■

PROPOSITION 5.4. *For all $\omega \in \mathbb{R}$ and $n = 1, 2, \ldots$, we have the estimate*

(5.20)
$$\mathcal{I} \equiv \int_0^\infty \frac{y|\omega|\,dy}{(y^2 + \gamma^2 + n^2 - \omega^2)^2 + 4\gamma^2\omega^2} \leq C.$$

Proof: If $\omega = 0$ the result is obvious, and so we assume otherwise. As before, we split the situation into cases.

Case 1: $\gamma^2 + n^2 \geq \omega^2$. Then

(5.21)
$$\mathcal{I} \leq |\omega| \int_1^\infty \frac{y\,dy}{y^4 + 4\gamma^2\omega^2} \leq |\omega| \frac{C}{|\omega|} = C.$$

Case 2: $\gamma^2 + n^2 < \omega^2$. We then set $a_{n\omega}^2 = \omega^2 - n^2 - \gamma^2 > 0$, and again split the integral as

$$\mathcal{I} = \int_1^\infty = \int_1^{\frac{a_{n\omega}}{2}} + \int_{\frac{a_{n\omega}}{2}}^{2a_{n\omega}} + \int_{2a_{n\omega}}^\infty = I_1 + I_2 + I_3.$$

To estimate I_1: We have

$$I_1 \leq |\omega| \int_0^{\frac{a_{n\omega}}{2}} \frac{y\,dy}{(y^2 - a_{n\omega}^2)^2 + 4\gamma^2\omega^2} \leq |\omega| \int_0^{\frac{a_{n\omega}}{2}} \frac{y\,dy}{\frac{9}{16}a_{n\omega}^4 + 4\gamma^2\omega^2}$$

(5.22) $$\leq |\omega| \int_0^{\frac{a_{n\omega}}{2}} \frac{y\,dy}{3a_{n\omega}^2 \gamma |\omega|} = \frac{a_{n\omega}^2}{24 a_{n\omega}^2 \gamma} = C.$$

To estimate I_3:

$$I_3 = |\omega| \int_{2a_{n\omega}}^\infty \frac{y\,dy}{(y^2 - a_{n\omega}^2)^2 + 4\gamma^2\omega^2} \leq |\omega| \int_{2a_{n\omega}}^\infty \frac{y\,dy}{\frac{9}{16}y^4 + 4\gamma^2\omega^2}$$

(5.23) $$\leq |\omega| \frac{C}{|\omega|} = C.$$

To estimate I_2:

$$I_2 = |\omega| \int_{\frac{a_{n\omega}}{2}}^{2a_{n\omega}} \frac{y\,dy}{(y^2 - a_{n\omega}^2)^2 + 4\gamma^2\omega^2} = \frac{|\omega|}{2} \int_{-\frac{3}{4}a_{n\omega}^2}^{3a_{n\omega}^2} \frac{dt}{t^2 + 4\gamma^2\omega^2}$$

(5.24) $$= \frac{1}{2\gamma} \arctan\left(\frac{t}{2\gamma|\omega|}\right)\Big|_{-\frac{3}{4}a_{n\omega}^2}^{3a_{n\omega}^2} \leq C.$$

Collecting (5.22)–(5.24) settles Case 2, and so completes proof of the Proposition. ∎

PROPOSITION 5.5. *For all $\omega \in \mathbb{R}$ and $n = 1, 2, \ldots$, we have the estimate*

(5.25) $$\mathcal{I} \equiv \frac{n\omega^2}{n^2 + \langle\omega\rangle} \int_1^\infty \frac{dy}{(y^2 + \gamma^2 + n^2 - \omega^2)^2 + 4\gamma^2\omega^2} \leq C.$$

Proof:
Case 1: $\gamma^2 + n^2 \geq \omega^2$. We then have

$$\mathcal{I} \leq \frac{n\omega^2}{n^2} \int_1^\infty \frac{dy}{(y^2 + \gamma^2 + n^2 - \omega^2)^2 + 4\gamma^2\omega^2}$$

(5.26) $$\leq \int_1^\infty \frac{n^{-1}(\gamma^2 + n^2)\,dy}{(y^2 + \gamma^2 + n^2 - \omega^2)^2 + 4\gamma^2\omega^2} \leq C,$$

after applying estimate (5.1).

Case 2: $\gamma^2 + n^2 < \omega^2$. We set $a_{n\omega}^2 = \omega^2 - n^2 - \gamma^2 \geq 0$, and split up the integral into the segments

(5.27) $$\mathcal{I} = \int_1^\infty = \int_1^{\frac{a_{n\omega}}{2}} + \int_{\frac{a_{n\omega}}{2}}^{2a_{n\omega}} + \int_{2a_{n\omega}}^\infty = I_1 + I_2 + I_3.$$

To estimate I_1:

(a) If $a_{n\omega} < n$, then

$$
\begin{aligned}
I_1 &\leq \frac{n\omega^2}{n^2 + \langle\omega\rangle} \int_0^{\frac{a_{n\omega}}{2}} \frac{dy}{(y^2 - a_{n\omega}^2)^2 + 4\gamma^2\omega^2} \\
&\leq n^{-1}\omega^2 \int_0^{\frac{a_{n\omega}}{2}} \frac{dy}{4\gamma^2\omega^2} \leq \frac{n^{-1}a_{n\omega}}{8\gamma^2} \leq C.
\end{aligned}
\tag{5.28}
$$

(b) If $a_{n\omega} > n$, then

$$
\begin{aligned}
I_1 &\leq n|\omega| \int_0^{\frac{a_{n\omega}}{2}} \frac{dy}{(y^2 - a_{n\omega}^2)^2 + 4\gamma^2\omega^2} \\
&\leq n|\omega| \int_0^{\frac{a_{n\omega}}{2}} \frac{dy}{\frac{9}{16}a_{n\omega}^4 + 4\gamma^2\omega^2} \\
&\leq n|\omega| \frac{8a_{n\omega}}{9a_{n\omega}^4 + 64\gamma^2\omega^2} \\
&\leq n|\omega| \frac{a_{n\omega}}{6a_{n\omega}^2\gamma|\omega|} \leq \frac{n}{6a_{n\omega}\gamma} \leq C.
\end{aligned}
\tag{5.29}
$$

To estimate I_3:
(c) If $a_{n\omega} < n$, then $\omega^2 < 2n^2 + \gamma^2$, and so

$$
\begin{aligned}
I_3 &= \frac{n\omega^2}{n^2 + \langle\omega\rangle} \int_{2a_{n\omega}}^{\infty} \frac{dy}{(y^2 - a_{n\omega}^2)^2 + 4\gamma^2\omega^2} \\
&\leq \frac{n\omega^2}{n^2 + \langle\omega\rangle} \left(\int_0^1 \frac{dy}{(y^2 - a_{n\omega}^2)^2 + 4\gamma^2\omega^2} + \int_1^{\infty} \frac{dy}{(y^2 - a_{n\omega}^2)^2 + 4\gamma^2\omega^2} \right) \\
&\leq \frac{n\omega^2}{\langle\omega\rangle\, 4\gamma^2\omega^2} \int_0^1 dy + n^{-1}\omega^2 \int_1^{\infty} \frac{dy}{(y^2 - a_{n\omega}^2)^2 + 4\gamma^2\omega^2} \\
&\leq C + (2n + n^{-1}\gamma^2) \int_1^{\infty} \frac{dy}{(y^2 - a_{n\omega}^2)^2 + 4\gamma^2\omega^2} \leq C,
\end{aligned}
\tag{5.30}
$$

after using (5.1).
(d) If $a_{n\omega} > n$, then

$$
\begin{aligned}
I_3 &\leq n|\omega| \int_{2a_{n\omega}}^{\infty} \frac{dy}{(y^2 - a_{n\omega}^2)^2 + 4\gamma^2\omega^2} \\
&\leq n|\omega| \int_{2a_{n\omega}}^{\infty} \frac{dy}{\frac{9}{16}y^4 + 4\gamma^2\omega^2} \\
&\leq n|\omega| \int_{2a_{n\omega}}^{\infty} \frac{dy}{3\gamma y^2 |\omega|} = \frac{n}{6\gamma a_{n\omega}} \leq C.
\end{aligned}
\tag{5.31}
$$

To estimate I_2: Making the substitution $t = y^2 - a_{n\omega}^2$, we have

$$I_2 = \frac{n\omega^2}{n^2 + \langle\omega\rangle} \int_{\frac{a_{n\omega}}{2}}^{2a_{n\omega}} \frac{dy}{(y^2 - a_{n\omega}^2)^2 + 4\gamma^2\omega^2}$$

$$= \frac{1}{2} \frac{n\omega^2}{(n^2 + \langle\omega\rangle)} \int_{-\frac{3}{4}a_{n\omega}^2}^{3a_{n\omega}^2} \frac{dt}{(t^2 + 4\gamma^2\omega^2)\sqrt{t + a_{n\omega}^2}}$$

$$\leq \frac{n\omega^2}{[n^2 + \langle\omega\rangle] a_{n\omega}} \int_{-\frac{3}{4}a_{n\omega}^2}^{3a_{n\omega}^2} \frac{dt}{(t^2 + 4\gamma^2\omega^2)}$$

(5.32) $$= \frac{n\omega^2}{[n^2 + \langle\omega\rangle] 2\gamma a_{n\omega}|\omega|} \left[\arctan\left(\frac{3a_{n\omega}^2}{8|\gamma||\omega|}\right) + \arctan\left(\frac{3a_{n\omega}^2}{|\gamma||\omega|}\right) \right].$$

(e) If $a_{n\omega} < n$, then

(5.32)
$$\leq \frac{n^{-1}\omega^2}{2\gamma a_{n\omega}|\omega|} \left[\arctan\left(\frac{3a_{n\omega}^2}{8\gamma|\omega|}\right) + \arctan\left(\frac{3a_{n\omega}^2}{2\gamma|\omega|}\right) \right]$$

(5.33) $$\leq \frac{n^{-1}|\omega|}{2\gamma a_{n\omega}} \left[\arctan\left(\frac{3a_{n\omega}^2}{8\gamma|\omega|}\right) + \arctan\left(\frac{3a_{n\omega}^2}{\gamma|\omega|}\right) \right] \leq C.$$

(f) If $a_{n\omega} > n$, then

(5.32)
(5.34) $$\leq \frac{n|\omega|}{2\gamma a_{n\omega}|\omega|} \left[\arctan\left(\frac{3a_{n\omega}^2}{8|\gamma||\omega|}\right) + \arctan\left(\frac{3a_{n\omega}^2}{|\gamma||\omega|}\right) \right] \leq \frac{\pi}{2\gamma}.$$

Collecting (5.27)–(5.34) completes the consideration of Case 2, and so the proof of the Proposition. ∎

PROPOSITION 5.6.

$$\left(n^2 + \langle\omega\rangle\right)^{-1} \int_1^\infty \frac{\left(y^2 + n^2\right)^{\frac{1}{2}} \left(\gamma^2 + \omega^2\right) dy}{(y^2 + \gamma^2 + n^2 - \omega^2)^2 + 4\gamma^2\omega^2} \leq C,$$

for all $\omega \in \mathbb{R}$ and $n = 1, 2, \ldots$.

Proof: The result follows readily from combining (5.1), (5.2), (5.20) and (5.25). ∎

References

[1] G. Avalos, *Sharp regularity estimates for solutions of the wave equation and their traces with prescribed Neumann data*, Applied Mathematics and Optimization, Vol. 35 (1997), pp. 203–219.

[2] G. Avalos and I. Lasiecka, *Differential Riccati equation for the active control of a problem in structural acoustics*, Journal of Optimization Theory and Applications, Vol. 91, No. 3 (December, 1996), pp. 695–728.

[3] G. Avalos, I. Lasiecka and R. Rebarber, *Lack of time-delay robustness of stabilization of a structural acoustics model*, SIAM J. Control Optim., Vol. 37, No. 5 (1999), pp. 1394–1418.

[4] G. Avalos, *Pointwise pressure observations of a canonical structural acoustics model*, Journal of Computational and Applied Mathematics, 114 (2000), pp. 1394–1418.

[5] G. Avalos, I. Lasiecka and R. Rebarber, *Wellposedness of a structural acoustics control model with point observation of the pressure*, to appear in Journal of Differential Equations.

[6] J. M. Ball, *Strongly continuous semigroups, weak solutions and the variation of constants formula*, Proc. Amer. Math. Soc. **63** (1977), pp. 370–373.

[7] H. T. Banks, M. A. Demitriou and R. C. Smith, *An H^∞ minimax periodic control in a 2–D structural acoustics model with piezoceramic actuators*. IEEE Trans. on Automatic Control, 41 (1996), pp. 943–959.

[8] J. Beale, *Spectral properties of an acoustic boundary condition*, Indiana Univ. Math. J., 9 (1976), pp. 895–917.

[9] S. Chen and R. Triggiani, *Proof of extensions of two conjectures on structural damping for elastic systems*, Pacific J. Math., **136** (1989), pp. 15–55.

[10] S. Chen and R. Triggiani, *Characterization of domains of fractional powers of certain operators arising in elastic systems and applications*, J. Diff. Eqns., **88** (1990), pp. 279–293.

[11] G. Doetsch, Introduction to the Theory and Application of the Laplace Transform, Springer–Verlag, New York (1974).

[12] H. O. Fattorini, *Ordinary differential equations in linear topological spaces, I and II*, Journal of Differential Equations, **5** (1968); and **6** (1969), pp. 537–565.

[13] P. Grisvard, *Caractérization de quelques espaces d'interpolation*, Arch. Rational Mech. Anal. **25** (1967), pp. 40–63.

[14] I. Lasiecka and R. Triggiani, *A cosine operator approach to modeling $L_2(0,T;L_2(\Gamma))$–boundary input hyperbolic equations*, Appl. Math. Optim. 7 (1981), pp. 35–93.

[15] I. Lasiecka and R. Triggiani, *A lifting theorem for the time regularity of solutions to abstract equations with unbounded operators and applications to hyperbolic equations*, Proc. Amer. Math. Soc. Vol. 104, No. 3 (1988), pp. 745–755.

[16] I. Lasiecka and R. Triggiani, *Regularity theory of hyperbolic equations with non–homogeneous Neumann boundary conditions. II. General boundary data*, Journal of Differential Equations, **94** (1991), pp. 112–164.

[17] P. M. Morse and K. U. Ingard, *Theoretical acoustics*, McGraw–Hill, New York (1968).

[18] A. Pazy, Semigroups of Linear Operators and Applications to Partial Differential Equations, Springer–Verlag, New York (1983).

[19] D. Salamon, *Realization theory in Hilbert spaces*, Math. Systems Theory, Vol. 21 (1989), pp. 147–164.

[20] R. Triggiani, *Interior and boundary regularity of the wave equation with interior point control*, Differential and Integral Equations, Vol. 6, No. 1 (1993), pp. 111–129.

DEPARTMENT OF MATHEMATICS AND STATISTICS, TEXAS TECH UNIVERSITY, LUBBOC K, TEXAS 79409

Current address: Department of Mathematics and Statistics, University of Nebraska-Linco ln, Lincoln, Nebraska 68588-0323

E-mail address: `gavalos@math.unl.edu`

Intrinsic Geometric Model for the Vibration of a Constrained Shell

John Cagnol and Jean-Paul Zolésio

ABSTRACT. This paper deals with the vibration of a constrained thin elastic shell. The constraint comes from a large displacement small deformation applied to the shell. We extend the intrinsic model developed by Michel Delfour and Jean-Paul Zolésio to the case of the constrained shells. The model $p(d, \infty)$ is investigated. The wellposedness of the system and the regularity of the solution at the boundary are established.

1. Introduction

The stabilization of large structures, for the aeronautic industry, has been studied since the early eighties (see [**Bal88**], [**CDKP87**], [**DLP86**], [**Lag83**]). More recently an increase in the development of flexible structures has taken place with the emergence of micro-satellites. The advantages in the use of micro-satellites are noteworthy. They are light weight, and they can be produced in small quantities such as a couple of hundreds. The concept of unfolding, locking and stabilizing systems for appendices such as solar panels, antennae, and cables should be simple, inexpensive and reliable. In response to those constraints, the engineers have proposed a concept known as passive control.

For instance one might want to minimize the vibration of an elastic body by optimizing the shape of that body. We consider a family $\mathcal{O}$ of open domains Ω and a domain functional $J : \mathcal{O} \to \mathbb{R}^+$ which represents a cost function. One would like to analyze the sensitivity of $J(\Omega)$ with respect to the perturbation of the domain Ω. The intrinsic concept related to this analysis turns out to be the shape derivative (see [**DZ92**], [**SZ91**]). It has been solved for many classical linear and non linear boundary value problems on manifolds. We refer to [**Cea81**], [**DP75**], [**MS76**], [**Zol81**] for the concepts of derivative. Existence theorems have been studied in [**Buc95**], [**DZ97**], [**SZ91**], [**Zol92**], [**Zol94**] and the numerical applications have been studied in [**Mas87**], [**Pir82**], [**Rou93**]. Nevertheless specific complications arise for the hyperbolic situation, which is the natural framework for the vibration analysis. In previous work (see [**CZ99**]) the authors have proven the existence of such a derivative for the wave equation. The proof relies on some extra regularity at the boundary called *hidden regularity* (see [**LLT86**]).

That extra regularity is crucial to obtain the shape differentiability under weak regularity of the data. In the case of the vibration of an isotropic body we refer

---

1991 *Mathematics Subject Classification.* Primary 35L, 53B ; Secondary 74B, 93B.

to [**Hor98a**] and [**Hor98b**] where the counterpart of this regularity for a linear elasticity model derives of her results. The case of anisotropic bodies is investigated by Mary Ann Horn in these proceedings. However the case of prestressed body leads to a different setting in which the regularity at the boundary needs to be re-investigated.

In this paper we are interested in the vibration of a constrained shell (for instance the link between solar panels). Michel Delfour and Jean-Paul Zolésio have worked on an intrinsic model relying the oriented distance function (see [**DZ99**]). In this paper we investigate the extension of this model to the constrained shells. That requires the use of a symbolic computation system. Then we apply this model to the passive control problem described earlier. Details on the implementation can be found in [CZ98b]. We shall recall the equation of the vibration and establish the regularity in the interior of the domain as well as the counter part of the hidden regularity. Eventually we investigate the shell model.

2. Physical Setting and Background Material

2.1. The Physical Setting. Let $\Omega^0 \subset \mathbb{R}^3$ be a shell in its unconstrained state, and $\Omega \subset \mathbb{R}^3$ be that shell in a constrained state which is a static equilibrium. Let $T_0 : \mathbb{R}^3 \to \mathbb{R}$ be such that $T_0(\Omega^0) = \Omega$, Hereafter this mapping will be called the *static displacement*. We assume T_0 is such that the coefficients of the matrix $E = DT_0 \circ T_0^{-1}$ belong to $W^{1,\infty}(\mathbb{R}^3)$, which is always the case when $T_0 \in W^{2,\infty}(\Omega^0, \mathbb{R}^3)$.

We suppose Ω is under a vibration. Let τ be the final time and $t < \tau$, we note $\Omega(t)$ the shell at the time t and $T(t)$ the mapping such that $T(t)(\Omega) = \Omega(t)$. We suppose Ω^0 and $\Omega(t)$ are homogeneous; Moreover, for the sake of simplicity, we will suppose they are isotropic. We assume T belongs to $L^2([0,\tau], H^1(\Omega, \mathbb{R}^3)) \cap H^1([0,\tau], L^2(\Omega, \mathbb{R}^3))$ and satisfies the embedding condition of the shell to Γ_D, that is $T_{|\Gamma_D} = I$. Hereafter the part of the boundary with Neumann boundary condition will be denoted Γ_N.

The mapping T is a perturbation of I, we have $T = I + u$ where u is small in H^1. From now on, u will be the parameter to be considered. We note $H^1_{\Gamma_D}(\Omega; \mathbb{R}^3)$ the set of functions of $H^1(\Omega, \mathbb{R}^3)$ that vanishes on Γ_D. The parameter u belongs to

$$(2.1) \qquad \mathbf{H} = L^2([0,\tau], H^1_{\Gamma_D}(\Omega, \mathbb{R}^3)) \cap H^1([0,\tau], L^2(\Omega, \mathbb{R}^3))$$

We note $\mathrm{T} = T \circ T_0$. Let $\tilde{u} = u \circ T_0$, we have $\mathrm{T}(t) = T_0 + \tilde{u}(t)$. We note C the 4th-order elastic tensor, and we note:

$$(2.2) \qquad \bar{\varepsilon}(u) = \frac{1}{2}({}^*DT\,DT - I)$$

$$(2.3) \qquad \varepsilon(u) = \frac{1}{2}({}^*Du + Du)$$

$$(2.4) \qquad \Sigma(u) = \frac{2}{|\det(E)|} E(C..({}^*E\varepsilon(u)E))\,{}^*E$$

The displacement is a large one therefore $\bar{\varepsilon}$, which is not linearized, is to be used for the computation of the energy.

Let ϕ and ψ be the value of u and $\partial_t u$ at $t = 0$. We suppose $\phi \in H^1(\Omega, \mathbb{R}^3)$ and $\psi \in L^2(\Omega, \mathbb{R}^3)$.

2.2. Equation of the Vibration.
The vibration is governed by the subsequent hyperbolic equation

(2.5)
$$\begin{cases} \rho \partial_t^2 u - \operatorname{div}(\Sigma(u)) = 0 & \text{on } [0;\tau[\times\Omega \\ \Sigma(u).n = 0 & \text{on } \Gamma_N \\ u = 0 & \text{on } \Gamma_D \\ u(0) = \phi,\ \partial_t u(0) = \psi & \text{on } \Omega \end{cases}$$

We refer to [**CMZ98**] for the proof of this statement. In the same reference are established some properties about the matrix $E = DT_0 \circ T_0^{-1}$. At this point, it should be reminded that T_0 is a local minimum for the elastic energy and therefore induces properties on E.

$$\operatorname{div}\left(\left(\frac{1}{\det(DT_0)} DT_0(\bar\varepsilon(0)..C)\,{}^*\!DT_0\right)\circ T_0^{-1}\right) = 0$$

$$\operatorname{div}\left(\frac{1}{\det E} E(({}^*\!EE - I)..C)\,{}^*\!E\right) = 0$$

Moreover T_0 is assumed to be a small deformation therefore $\|\bar\varepsilon(0)\|_{L^\infty}$ is small as compared to 1. On the other hand from that last assertion comes the existence of a real $c > -1$ such that

(2.6)
$$\begin{cases} \forall v \in \mathbb{R}^3,\ \|{}^*\!EEv\|_{L^2} \geq (1+c)\|v\|_{L^2} \\ \forall v \in \mathbb{R}^3,\ \|E\,{}^*\!Ev\|_{L^2} \geq (1+c)\|v\|_{L^2} \end{cases}$$

2.3. Example: The Carpentier's Joint.
As an example, we consider the *Carpentier's joint* which is a rectangular strip, with a constant curvature in its width and a null curvature in its length.

FIGURE 1. A Carpentier's joint

The physical motivation for such a shell comes from the aeronautic industry. We consider a sandwich type device consisting of two plates attached to each other with a series of bolts and between them. The metallic strip is in the folded position of 180 degrees. One studies its unfolding in micro-gravity. When the link is destroyed the two plates exert a small torque at the beginning which then opens the device. Upon approaching the open status the plates are in the same plane and the torque increases.

The investigation of the vibration of that preconstrained structure is of interest and falls into the setting that has been described earlier. In the particular case of the Carpentier's joint the static displacement T_0 was computed in [**CM98**]; the shape is obtained through an hybrid program numerical computation-symbolic computation. The result obtained matches the experiment where a "shape in Ω" was found.

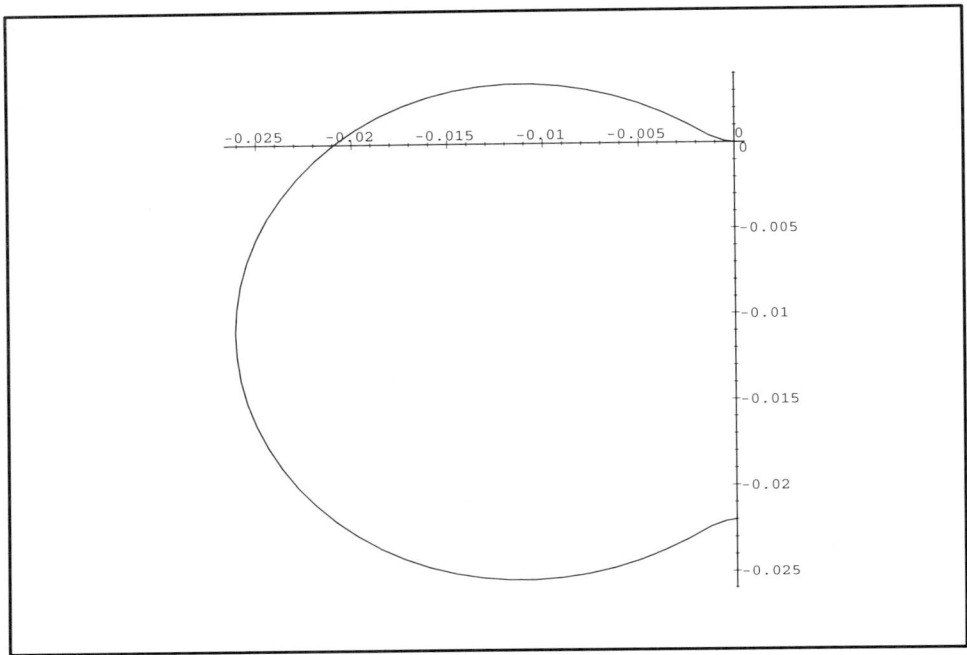

FIGURE 2. Static equilibrium of the constrained shell. View of the shell from the side

2.4. Notations and Conventions. We use the notations of [**Ger86**]. Subsequently, we will make no difference between a matrix and a 2nd-order tensor nor between a vector and a 1st-order tensor.

LEMMA 2.1. *Let A, B, X and Y be four matrices and C be a third-order tensor then $(XAY)..C..B = A..(\,^*X(C..B)\,^*Y)$*

LEMMA 2.2. *Let A and Z be two 3rd-order tensors and U and V be two vectors then $(A.V)..(U.Z) = (Z..A)..(U \otimes V)$*

Let $X = (x_{i,j,k})_{i,j,k}$ be a 3rd-order tensor, following [**AZ98**], let

$$X^\triangleleft = (x_{k,i,j})_{i,j,k}$$

$$X^{\triangleleft\triangleleft} = (x_{j,k,i})_{i,j,k}$$

For a matrix X and a third-order tensor $Y = (Y_{i,j,k})_{i,j,k}$ we note $X \times Y$ (*resp.* $Y \times X$) the third-order tensor $(X \times Y_i)_i$ (*resp.* $(Y_i \times X)_i$) where Y_i is a matrix and $\times$ the multiplication of matrices. For a third-order tensor X let us note $\text{tr}(X)$ the vector whose coefficients are the traces of the matrices X_i.

Hereafter, we will note $P(u) = \rho \partial_t^2 u - \text{div}\,(\Sigma(u))$ and
$$|\varphi|_E = \varphi..C..\varphi$$

3. Existence and Regularity of the Solutions

We consider the hyperbolic system (2.5), we shall prove it is well-posed and we shall establish some regularity of the solution.

THEOREM 3.1 (Well-posedness and Interior Regularity). *The system (2.5) has a unique solution u in $C([0;\tau]; H^1_{\Gamma_D}(\Omega, \mathbb{R}^3)) \cap C^1([0;\tau]; L^2(\Omega, \mathbb{R}^3))$.*

Let us note $Q = (0, \tau) \times \Omega$. Then, for all φ and w in $C^\infty(Q)$
$$\int_0^\tau \int_{\Gamma_D} \langle \Sigma(\varphi).n, w \rangle = \int_0^\tau \int_\Omega \text{div}\,(\Sigma(\varphi))w + \int_0^\tau \int_\Omega \Sigma(\varphi)..\varepsilon(w)$$

When the test w belongs to $L^2((0,\tau), H^1_{\Gamma_D}(\Omega, \mathbb{R}^3))$, its trace $w|_{(0,\tau) \times \Gamma_D}$ belongs to $L^2((0,\tau), H^{\frac{1}{2}}(\Gamma_D, \mathbb{R}^3))$. When u belongs to $L^2((0,\tau), H^1_{\Gamma_D}(\Omega, \mathbb{R}^3))$, by classical density method $\Sigma(u).n$ is defined as an element of $L^2((0,\tau), (H^{\frac{1}{2}}((0,\tau), \mathbb{R}^3))')$ by
$$\int_0^\tau \langle \Sigma(u).n, w|_{\Gamma_D} \rangle_{H^{\frac{1}{2}},(H^{\frac{1}{2}})'} dt = \int_0^\tau \int_\Omega \text{div}\,(\Sigma(\varphi))w\,dx\,dt + \int_0^\tau \int_\Omega \Sigma(\varphi)..\varepsilon(w)\,dx\,dt$$

When u is solution to (2.5) we have $u \in \mathbf{H}$ and $Pu \in L^2(Q)$ we prove the regularity may be improved:

THEOREM 3.2 (Boundary Regularity). *We have the following regularity at the boundary: $\Sigma(u).n \in L^2_{\text{LOC}}((0,\tau) \times \Gamma_D)$*

3.1. Proof of Theorem 3.1. The existence and regularity of the solutions will be derived from the theory of semi-groups. Let Λ be the linear operator in $L^2(\Omega, \mathbb{R}^3)$ defined by
$$D(\Lambda) = H^2(\Omega; \mathbb{R}^3) \cap H^1_{\Gamma_D}(\Omega; \mathbb{R}^3)$$
$$\Lambda = -\frac{1}{\rho}\text{div}\,(\Sigma)$$

LEMMA 3.3. *The operator Λ is self-adjoint.*

PROOF. Let $v \in \mathbf{H}$.
$$\int_0^\tau \int_\Omega \Lambda(u)v = \frac{1}{\rho} \int_0^\tau \int_\Omega \Sigma(u)..\varepsilon(v)$$
from the symmetry of Σ and lemma 2.1 follows $\Sigma(u)..\varepsilon(v) = \varepsilon(u)..\Sigma(v)$. □

LEMMA 3.4. *The operator Λ is coercive in $H^1_{\Gamma_D}(\Omega, \mathbb{R}^3)$.*

PROOF. One has
$$\int_\Omega \Lambda(u)u = \frac{1}{\rho}\int_\Omega \frac{2}{|\det(E)|}({}^*E\varepsilon(u)E)..C..({}^*E\varepsilon(u)E)$$
from [**Ger86**] and the isotropy of Ω, it follows
$$\int_\Omega \Lambda(u)u = \frac{1}{\rho}\int_\Omega \frac{1}{|\det(E)|}(\lambda({}^*E\varepsilon(u)E)..I + 2\mu({}^*E\varepsilon(u)E)..({}^*E\varepsilon(u)E))$$
since $\lambda \geq 0$ the following inequality holds
$$\int_\Omega \Lambda(u)u \geq \frac{2\mu}{\rho}\int_\Omega \frac{1}{|\det(E)|}({}^*E\varepsilon(u)E)..({}^*E\varepsilon(u)E)$$

We have $({}^*E\varepsilon(u)E)..({}^*E\varepsilon(u)E) = \varepsilon(u)..(E\,{}^*E\varepsilon(u)E\,{}^*E)$ therefore, using (2.6), there exists a non negative constant k such that
$$\int_\Omega \Lambda(u)u \geq \frac{2k\mu}{\rho} \int_\Omega \frac{1}{|\det(E)|} \varepsilon(u)..\varepsilon(u)$$
Korn's inequality and $u_{|\Gamma_D} = 0$ yields the existence of a non negative real C such that
$$\int_\Omega \Lambda(u)u \geq \left(\frac{2kC\mu}{\rho} \frac{1}{\max_\Omega |\det(E)|}\right) \|u\|_{H^1(\Omega)}$$
hence Λ is coercive. $\square$

We note
$$U = \begin{pmatrix} u \\ \partial_t u \end{pmatrix}, \quad U_0 = \begin{pmatrix} \phi \\ \psi \end{pmatrix} \quad \text{and} \quad A = \begin{pmatrix} 0 & 1 \\ -\Lambda & 0 \end{pmatrix}$$
then (2.5) is equivalent to

(3.1) $\qquad \begin{cases} \partial_t U = AU \\ U(0) = U_0 \end{cases}$

Following the theory of semi-groups, we have

PROPOSITION 3.5. *The operator A is the infinitesimal generator of a strongly continuous semi-group of contraction S on H. Moreover ${}^*A = -A$.*

COROLLARY 3.6. *System (2.5) has a unique solution u and*
$$u \in C([0;\tau]; H^1_{\Gamma_D}(\Omega, \mathbb{R}^3)) \cap C^1([0;\tau]; L^2(\Omega, \mathbb{R}^3))$$

3.2. Proof of Theorem 3.2. We consider a flow mapping T_s and the associated vector field V (*cf.* [**CZ99**]). We note Ω_s the image of Ω by T_s, the same notation is used for Q, Σ, etc...

Let V be such that $\langle V(0), n\rangle$ vanish on $\overline{\Gamma_N}$ and $\langle V(0), n\rangle \geq \alpha > 0$ on $\overline{\Gamma_D}$. Let u be the solution to (2.5). We shall prove that $\Sigma(u).n \in L^2((0,\tau) \times \Gamma_D)$

Let us compute the derivative with respect to s, at $s = 0$, of
$$\frac{1}{2} \int_{Q_s} \frac{1}{|\det(E)|} ({}^*E\varepsilon(\varphi \circ T_s^{-1})E)..C..({}^*E\varepsilon(\varphi \circ T_s^{-1})E)$$
via two different ways, given by the two subsequent lemmas. The point here is to have the distributed integral to be defined when φ belongs to $\mathbf{H}$ with $P\varphi \in L^2(Q)$.

LEMMA 3.7. *One has*
$$\frac{\partial}{\partial s}\left(\frac{1}{2}\int_{Q_s} \frac{1}{|\det(E)|}({}^*E\varepsilon(\varphi \circ T_s^{-1})E)..C..({}^*E\varepsilon(\varphi \circ T_s^{-1})E)\,dx\,dt\right)_{s=0} =$$
$$\rho\int_Q \frac{1}{|\det(E)|}\left(\operatorname{div}(V(0))|\partial_t\varphi|^2 + \langle D\varphi.\partial_t V(0), \partial_t\varphi\rangle - \langle D\varphi.V(0), P\varphi\rangle\right)\,dx\,dt$$
$$-\int_{(0,\tau)\times\Gamma} \frac{1}{|\det(E)|}\langle D\varphi.V(0), \Sigma(\varphi)n\rangle\,d\Gamma\,dt$$
$$-\int_{(0,\tau)\times\Gamma} \frac{1}{|\det(E)|}\left(\frac{1}{2}(({}^*E\varepsilon(\varphi)E)..C..({}^*E\varepsilon(\varphi)E) - \rho|\partial_t\varphi|^2)\langle V(0), n\rangle\right)\,d\Gamma\,dt$$

PROOF. From [**CZ98a**], the left hand side of the equality is equal to

$$\int_Q \frac{1}{|\det(E)|} \left({}^*E \frac{\partial}{\partial s}(\varepsilon(\varphi \circ T_s^{-1}))_{s=0} E \right) .. C ..({}^*E\varepsilon(\varphi)E) \, dx \, dt$$

$$+ \frac{1}{2} \int_{(0,\tau) \times \Gamma} \frac{1}{|\det(E)|} ({}^*E\varepsilon(\varphi)E) .. C ..({}^*E\varepsilon(\varphi)E) \langle V(0), n \rangle \, d\Gamma \, dt$$

we have

$$\frac{\partial}{\partial s} D(\varphi \circ T_s^{-1}) = -D(D\varphi . V(0))$$

hence the previous expression may be rewritten

$$- \int_Q \frac{1}{|\det(E)|} ({}^*E\varepsilon(D\varphi . V(0))E) .. C ..({}^*E\varepsilon(\varphi)E) \, dx \, dt$$

$$+ \frac{1}{2} \int_{(0,\tau) \times \Gamma} \frac{1}{|\det(E)|} ({}^*E\varepsilon(\varphi)E) .. C ..({}^*E\varepsilon(\varphi)E) \langle V(0), n \rangle \, d\Gamma \, dt$$

we obtain

$$\int_Q \frac{1}{|\det(E)|} \langle D\varphi . V(0), \text{div}\,(\Sigma(\varphi)) \rangle \, dx \, dt$$

$$- \int_0^\tau \int_\Gamma \frac{1}{|\det(E)|} \left(\langle D\varphi . V(0), \Sigma(\varphi) n \rangle - \frac{1}{2}({}^*E\varepsilon(\varphi)E) .. C ..({}^*E\varepsilon(\varphi)E) \langle V(0), n \rangle \right) d\Gamma \, dt$$

that is

$$\rho \int_Q \frac{1}{|\det(E)|} \langle D\varphi . V(0), \partial_{tt}\varphi \rangle \, dx \, dt - \rho \int_Q \frac{1}{|\det(E)|} \langle D\varphi . V(0), P\varphi \rangle \, dx \, dt$$

$$- \int_0^\tau \int_\Gamma \frac{1}{|\det(E)|} \left(\langle D\varphi . V(0), \Sigma(\varphi) n \rangle + \frac{1}{2}({}^*E\varepsilon(\varphi)E) .. C ..({}^*E\varepsilon(\varphi)E) \langle V(0), n \rangle \right) d\Gamma \, dt$$

the integral over Q may be undefined for $\varphi \in \mathbf{H}$ such that $P\varphi \in L^2(Q)$ however

$$\int_Q \frac{1}{|\det(E)|} \langle D\varphi . V(0), \partial_{tt}\varphi \rangle \, dx \, dt$$

is equal to

$$- \int_Q \frac{1}{|\det(E)|} \langle \partial_t D\varphi . V(0), \partial_t \varphi \rangle \, dx \, dt + \left[\int_\Gamma \frac{1}{|\det(E)|} \langle D\varphi . V(0), \partial_t \varphi \rangle \, d\Gamma \right]_0^T$$

we get

$$\int_Q \frac{1}{|\det(E)|} \langle D\varphi . V(0), \partial_{tt}\varphi \rangle \, dx \, dt$$

$$= - \int_Q \frac{1}{|\det(E)|} \left(\langle V(0), \partial_t {}^*D\varphi \partial_t \varphi \rangle + \langle D\varphi . \partial_t V(0), \partial_t \varphi \rangle \right) dx \, dt$$

$$= - \int_Q \frac{1}{|\det(E)|} \left(\langle V(0), \nabla |\partial_t \varphi| \rangle + \langle D\varphi . \partial_t V(0), \partial_t \varphi \rangle \right) dx \, dt$$

$$= \int_Q \frac{1}{|\det(E)|} \left(\text{div}\,(V(0)) |\partial_t \varphi|^2 + \langle D\varphi . \partial_t V(0), \partial_t \varphi \rangle \right) dx \, dt$$

$$+ \int_0^\tau \int_\Gamma |\partial_t \varphi|^2 \langle V(0), n \rangle \, d\Gamma \, dt$$

the identity derives. $\square$

LEMMA 3.8. *One has*
$$\frac{\partial}{\partial s}\left(\frac{1}{2}\int_{Q_s}\frac{1}{|\det(E)|}(^*E\varepsilon(\varphi\circ T_s^{-1})E)..C..(^*E\varepsilon(\varphi\circ T_s^{-1})E)\,dx\,dt\right)_{s=0}=$$
$$\int_Q \frac{1}{|\det(E)|}(\frac{1}{2}tr(E^{-1}.DE.V(0))(^*E\varepsilon(\varphi)E)..C..(^*E\varepsilon(\varphi)E)$$
$$+(^*DE.V(0)\varepsilon(\varphi)E)..C..(^*E\varepsilon(\varphi)E)+\frac{1}{2}{}^*E\varepsilon(D\varphi.DV(0))E)..C..(^*E\varepsilon(\varphi)E)$$
$$+(^*E\varepsilon(\varphi)DE.V(0))..C..(^*E\varepsilon(\varphi)E)+\frac{1}{2}\mathrm{div}\,(V(0))(^*E\varepsilon(\varphi)E)..C..(^*E\varepsilon(\varphi)E))\,dx\,dt$$

PROOF. We make the change of variable T_s in the left hand side, let $\gamma_s = \det(DT_s)$, we get
$$\frac{\partial}{\partial s}(\frac{1}{2}\int_Q \frac{1}{|\det(E\circ T_s)|}((^*E\circ T_s)(\varepsilon(\varphi\circ T_s^{-1})\circ T_s)(E\circ T_s)..C$$
$$..((^*E\circ T_s)(\varepsilon(\varphi\circ T_s^{-1})\circ T_s)(E\circ T_s)\gamma_s)_{s=0}$$
the lemma derives from
$$\frac{\partial}{\partial s}(^*E\circ T_s)_{s=0}={}^*DE.V(0)$$
$$\frac{\partial}{\partial s}\left(\varepsilon(\varphi\circ T_s^{-1})\circ T_s\right)_{s=0}=\varepsilon(D\varphi.DV(0))$$
$$\frac{\partial}{\partial s}(E\circ T_s)_{s=0}=DE.V(0)$$
$$\frac{\partial}{\partial s}\left(\frac{1}{|\det(E\circ T_s)|}\right)_{s=0}=\frac{1}{|\det(E)|}\mathrm{tr}(E^{-1}.DE.V(0))$$
$$\frac{\partial}{\partial s}(\gamma_s)_{s=0}=\mathrm{div}\,(V(0))$$
□

LEMMA 3.9. *On* Γ_D
$$\langle\Sigma(\varphi).n,D\varphi.V(0)\rangle=(^*E\varepsilon(\varphi)E)..C..(^*E\varepsilon(\varphi)E)\,\langle V(0),n\rangle$$
$$=|(D\varphi.n)\,{}^*n|_E^2\,\langle V(0),n\rangle$$

PROOF. Since $\varphi|_{(0,\tau)\times\Gamma}=0$ we have $D\varphi=(D\varphi.n).{}^*n$ therefore
$$\langle\Sigma(\varphi).n,D\varphi.V(0)\rangle=\langle E(C..(^*E\varepsilon(\varphi))E)\,{}^*En,(D\varphi.n)\,{}^*nV(0)\rangle$$
hence
$$\langle\Sigma(\varphi).n,D\varphi.V(0)\rangle=\langle E(C..(^*E\varepsilon(\varphi))E)\,{}^*En,D\varphi.n\rangle\,\langle V(0),n\rangle$$
it follows
$$\langle\Sigma(\varphi).n,D\varphi.V(0)\rangle=(^*D\varphi E(C..(^*E\varepsilon(\varphi)E))\,{}^*E)..(n\otimes n)\,\langle V(0),n\rangle$$
thus
$$\langle\Sigma(\varphi).n,D\varphi.V(0)\rangle=(^*ED\varphi E)..C..(^*E\varepsilon(\varphi)E)\,\langle V(0),n\rangle$$
from the property of C we derive
$$\langle\Sigma(\varphi).n,D\varphi.V(0)\rangle=(^*E\varepsilon(\varphi)E)..C..(^*E\varepsilon(\varphi)E)\,\langle V(0),n\rangle$$
$$=|{}^*E(D\varphi.n)\,{}^*nE|^2\,\langle V(0),n\rangle$$
Using the very same property and $D\varphi(D\varphi.n)\,{}^*n$ we obtain the other part of the equality. □

PROPOSITION 3.10. *Let*
$$\begin{aligned}
B(\varphi) = & \ 2\int_Q \frac{1}{|\det(E)|}(\rho\mathrm{div}\,(V(0))|\partial_t\varphi|^2 \\
& +\rho\,\langle D\varphi.\partial_t V(0),\partial_t\varphi\rangle - \langle D\varphi.V(0),P\varphi\rangle \\
& -\frac{1}{2}tr(E^{-1}.DE.V(0))(\,^*E\varepsilon(\varphi)E)..C..(\,^*E\varepsilon(\varphi)E) \\
& -(\,^*DE.V(0)\varepsilon(\varphi)E)..C..(\,^*E\varepsilon(\varphi)E) \\
& -\frac{1}{2}\,^*E\varepsilon(D\varphi.DV(0))E)..C..(\,^*E\varepsilon(\varphi)E) \\
& -(\,^*E\varepsilon(\varphi)DE.V(0))..C..(\,^*E\varepsilon(\varphi)E) \\
& -\frac{1}{2}\mathrm{div}\,(V(0))(\,^*E\varepsilon(\varphi)E)..C..(\,^*E\varepsilon(\varphi)E))
\end{aligned}$$
then
$$\int_{(0,\tau)\times\Gamma_D}\frac{1}{|\det(E)|}|(D\varphi.n)\,^*n|^2_E\,\langle V(0),n\rangle = B(\varphi)$$

PROOF. Lemmas 3.7 and 3.8 yield
$$-\int_0^\tau\int_\Gamma \frac{1}{|\det(E)|}\,\langle D\varphi.V(0),\Sigma(\varphi)n\rangle$$
$$+\int_0^\tau\int_\Gamma \frac{1}{|\det(E)|}\left(\frac{1}{2}((\,^*E\varepsilon(\varphi)E)..C..(\,^*E\varepsilon(\varphi)E) - \rho|\partial_t\varphi|^2)\,\langle V(0),n\rangle\right)$$
$$+\int_Q \frac{1}{|\det(E)|}\left(\rho\mathrm{div}\,(V(0))|\partial_t\varphi|^2 + \rho\,\langle D\varphi.\partial_t V(0),\partial_t\varphi\rangle - \langle D\varphi.V(0),P\varphi\rangle\right)\,dx\,dt$$
$$=\int_Q \frac{1}{|\det(E)|}(\frac{1}{2}tr(E^{-1}.DE.V(0))(\,^*E\varepsilon(\varphi)E)..C..(\,^*E\varepsilon(\varphi)E)$$
$$+(\,^*DE.V(0)\varepsilon(\varphi)E)..C..(\,^*E\varepsilon(\varphi)E) + \frac{1}{2}\,^*E\varepsilon(D\varphi.DV(0))E)..C..(\,^*E\varepsilon(\varphi)E)$$
$$+(\,^*E\varepsilon(\varphi)DE.V(0))..C..(\,^*E\varepsilon(\varphi)E) + \frac{1}{2}\mathrm{div}\,(V(0))(\,^*E\varepsilon(\varphi)E)..C..(\,^*E\varepsilon(\varphi)E))$$

We have $\langle V(0),n\rangle = 0$ on $\overline{\Gamma_N}$ moreover $\partial_t\varphi = 0$ on Γ_D therefore the integral over $(0,\tau)\times\Gamma$ is equal to
$$-\int_{(0,\tau)\times\Gamma_D}\frac{1}{|\det(E)|}\left(\langle D\varphi.V(0),\Sigma(\varphi)n\rangle - \frac{1}{2}(\,^*E\varepsilon(\varphi)E)..C..(\,^*E\varepsilon(\varphi)E)\,\langle V(0),n\rangle\right)$$
lemma 3.9 gives
$$-\frac{1}{2}\int_{(0,\tau)\times\Gamma_D}\frac{1}{|\det(E)|}|(D\varphi.n)\,^*n|^2_E\,\langle V(0),n\rangle$$
therefore
$$\int_{(0,\tau)\times\Gamma_D}\frac{1}{|\det(E)|}|(D\varphi.n)\,^*n|^2_E\,\langle V(0),n\rangle = B(\varphi)$$
$\square$

REMARK 3.11. *The real $B(\varphi)$ is defined when $\varphi\in\mathbf{H}$ and $P(\varphi)\in L^2(Q)$.*

Let $(\varphi^m)_m$ be a sequence of functions of $C^\infty(Q)$ that vanish on Γ_D such that $\varphi^m \to u$ in $H^1(Q)$ and $P\varphi^m \to Pu$ in $L^2(Q)$. We refer to [**DZ98a**] for the proof of the density where a similar one is given. Since $B(u)$ exists,

$$\|(\varepsilon(\varphi^m)_{|(0,\tau)\times\Gamma_D})|_E \sqrt{\langle V(0),n\rangle}\|_{L^2((0,\tau)\times\Gamma_D)}$$

is bounded. The assumption $\langle V(0),n\rangle \geq \alpha > 0$ on Γ_D yields $|\varepsilon(\varphi^m)_{|(0,\tau)\times\Gamma_D}|_E$ is bounded; we derive the existence of ξ in in $L^2((0,\tau)\times\Gamma_D)$ and of a subsequence such that

$$\varepsilon(\varphi^m)|_{|(0,\tau)\times\Gamma_D} \rightharpoonup \xi \text{ weakly in } L^2((0,\tau)\times\Gamma_D) \text{ as } m_k \to +\infty$$

we derive

$$\Sigma(\varphi^m).n \rightharpoonup \frac{2}{|\det(E)|} E(C..({}^*E\xi E)) {}^*E.n \text{ weakly in } L^2((0,\tau)\times\Gamma_D) \text{ as } m_k \to +\infty$$

but by Green's theorem we have $\Sigma(\varphi^m).n \rightharpoonup \Sigma(u).n$ in $L^2((0,\tau), H^{\frac{1}{2}}(\Gamma_D))'$ then

$$\Sigma(u).n = \frac{2}{|\det(E)|} E(C..({}^*E\xi E)) {}^*E.n \in L^2((0,\tau)\times\Gamma_D)$$

which proves theorem 3.2.

When $\overline{\Gamma_D} \cap \overline{\Gamma_N} \neq \emptyset$ and
- $\langle V(0),n\rangle$ vanish on $\overline{\Gamma_N}$
- $\langle V(0),n\rangle \geq 0$ on $\overline{\Gamma_D}$
- $\langle V(0),n\rangle \geq \alpha > 0$ on a subset of Γ_D with a non zero measure

we derive by the same means that $\Sigma(u).n \in L^2_{\text{LOC}}((0,\tau)\times\Gamma_D)$

4. Extension of the Shell Model to the Prestressed Case

4.1. Background Material. Let b be the oriented distance functions defined by

$$\forall x \in \mathbb{R}^3, \ b_\Omega(x) = d_\Omega(x) - d_{\mathbb{R}^3\setminus\Omega}(x)$$

and p be the orthogonal projection from $\mathbb{R}^3$ onto Ω. That projection exists in a tubular neighborhood of Ω (cf. [**DZ98b**]).

We will note b^0 and p^0 the corresponding function for Ω^0 and b_X and p_X the corresponding functions the set X. A similar notation will be used for all subsequent functions depending and constants on Ω and will not be pointed out again.

We suppose Ω satisfies the

HYPOTHESIS 4.1 (Shell Form Assumption). *There exists $\mathcal{O} \subset \mathbb{R}^3$ such that $\partial\mathcal{O}$ is a manifold and $\omega \subset \partial\mathcal{O}$ such that $\Omega = \{x \in \mathbb{R}^3 \text{ s.t. } |b_\mathcal{O}(x)| < h, \ p(x) \in \omega\}$.*

This hypothesis is backed up by the thinness of the shell and the smallness of the deformation.

Let us introduce the following notations

$$\Gamma_z = \{x \in \Omega \mid b_\mathcal{O}(x) = z\}$$

$$\Gamma = \Gamma_0$$

$$S_h = \cup_{-h<z<h}\Gamma_h$$

$$T_z : x \mapsto x + z\nabla b$$

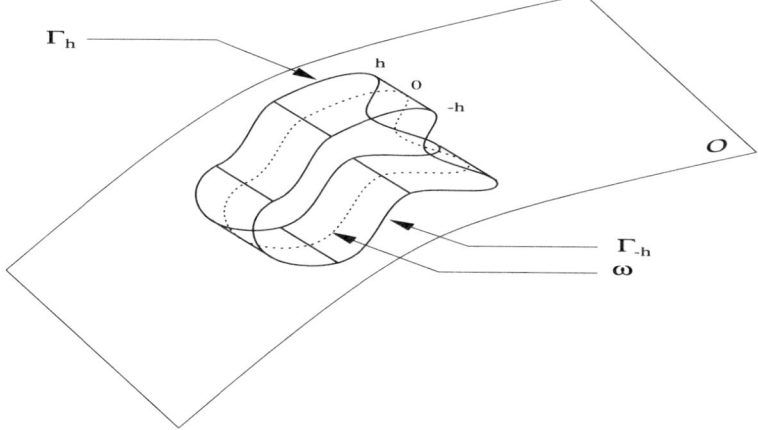

FIGURE 3. The Shell Form Assumption

Let λ_1 and λ_2 be the main curvature of Γ, they are the eigenvalues of $D^2 b$ not associated to ∇b. We note

$$H = \triangle b = \operatorname{tr}(D^2 b) = \lambda_1 + \lambda_2$$

$$K = \frac{1}{2}(H^2 - D^2 b .. D^2 b) = \lambda_1 \lambda_2$$

We consider

$$\mathcal{U}_h^d = \left\{ U \in H^1(S_h, \mathbb{R}^3) \mid \exists (u_i), U = \sum_{i=0}^{d} b^i \, u_i \circ p \right\}$$

$$\mathcal{U}_h = \cup_{d \in \mathbb{N}} \, \mathcal{U}_h^d$$

Moreover we suppose

HYPOTHESIS 4.2. *This diagram commutes*

$$\begin{array}{ccc} \Gamma_z^0 & \xrightarrow{T_0} & \Gamma_z \\ T_z^0 \uparrow & & \uparrow T_z^{-1} \\ \Gamma^0 & \xrightarrow{T_0} & \Gamma \end{array}$$

4.2. The Model. The Federer's measure decomposition yields

(4.1) $$\int_0^\tau \int_{-h}^{h} \int_{\Gamma_z} (\Sigma(u) .. \varepsilon(w) - \rho \partial_t u \partial_t w) \, d\Gamma_z \, dz \, dt = 0$$

In order to compute that integral, let us consider $\int_{\Gamma_z} \Sigma(u) .. \varepsilon(w) \, d\Gamma_z$, we perform a change of variable and obtain

$$\int_{\Gamma_z} \Sigma(u) .. \varepsilon(w) \, d\Gamma_z = \int_{\Gamma} (\Sigma(u) \circ T_z) .. (\varepsilon(w) \circ T_z) \det(DT_z) \, d\Gamma$$

LEMMA 4.3. *We have* $\det(DT_z) = 1 + Hz + Kz^2$. *We will note* $j(z)$ *that expression.*

PROOF. One has $T_z = I + z\nabla b$ thus $DT_z = I + zD^2b$. Let V_1, V_2, ∇b be the eigenvectors or b associated to λ_1, λ_2 and 0. Those vectors are the eigenvectors of $(I + zD^2b)$ and the associated eigenvalues are $1 + \lambda_1$, $1 + \lambda_2$ and 1. Therefore $\det(DT_z) = (1 + \lambda_1)(1 + \lambda_2) = 1 + Hz + Kz^2$. □

LEMMA 4.4. *We have* $E \circ T_z = (I + zD^2b).E.((I + zD^2b^0)^{-1} \circ T_0^{-1} \circ T_z)$.

PROOF. From hypothesis 4.2 we have $T_0 = T_z \circ T_0 \circ T_z^0$ therefore
$$DT_0 = ((DT_z) \circ T_0 \circ T_z^0).((DT_0) \circ T_z^0).(DT_z^0)$$
therefore
$$E = ((DT_z) \circ T_0 \circ T_z^0 \circ T_0^{-1}).((DT_0) \circ T_z^0 \circ T_0^{-1}).((DT_z^0) \circ T_0^{-1})$$
since $T_0 \circ T_z^0 \circ T_0^{-1} = T_z^{-1}$ and $T_z^0 \circ T_0^{-1} = T_0^{-1} \circ T_z^{-1}$ we get
$$E \circ T_z = (DT_z).E.((DT_z^0) \circ T_0^{-1} \circ T_z))$$
then $DT_z = I + zD^2b$ and $DT_z^0 = I + zD^2b^0$ which yields the result. □

Let us define f and g by
$$f(z) = z\frac{1 + Hz}{j(z)} \quad \text{and} \quad g(z) = z^2\frac{1}{j(z)}$$

LEMMA 4.5. *With this notation we have*
$$(I + zD^2b)^{-1} = I - f(z)D^2b + g(z)(D^2b)^2$$

PROOF. Let $n \geq 3$, one has
$$(4.2) \qquad (D^2b)^n = -\lambda_1\lambda_2\frac{\lambda_1^{n-2} - \lambda_2^{n-2}}{\lambda_1 - \lambda_2}D^2b + \frac{\lambda_1^{n-1} - \lambda_2^{n-1}}{\lambda_1 - \lambda_2}(D^2b)^2$$
where the functions are replaced by their analytic expansion when $\lambda_1\lambda_2 = 0$ or $\lambda_1 = \lambda_2$. The identity
$$(I + zD^2b)^{-1} = \sum_{n=0}^{+\infty}(-1)^n(D^2b)^nz^n$$
and (4.2) give lemma 4.5. □

PROPOSITION 4.6. *We note* $g_0 = g^0 \circ T_0^{-1}$ *and* $f_0 = f^0 \circ T_0^{-1}$ *then*
$$E \circ T_z = (I + zD^2b).E.$$
$$\left(I - f_0(z)\left(D^2b^0 \circ T_0^{-1} \circ T_z\right) + g_0(z)\left((D^2b^0)^2 \circ T_0^{-1} \circ T_z\right)\right)$$

PROOF. This proposition is a consequence of lemmas 4.4 and 4.5. □

For the sake of simplicity we suppose that Ω^0 is a plate In the case of the Carpentier's joint described earlier this means the transversal curvature is zero.

COROLLARY 4.7. *We have*
  i) $E \circ T_z = (I + zD^2b).E$
  ii) $^*E \circ T_z = {^*E}.(I + zD^2b)$

PROOF. If Ω^0 is a plate then $D^2b^0 = 0$ thus (i) follows. Hence $E = (I + zD^2b) \circ T_z^{-1}.E \circ T_z^{-1}$, therefore $^*E = {^*E} \circ T_z^{-1}.(I + zD^2b) \circ T_z^{-1}$, thus (ii) follows. □

The shell Ω is supposed isotropic. Following [**Ger86**], one has

$$C..X = \frac{\lambda}{2}\text{tr}(X)I + \mu X$$

where X is a matrix, therefore

$$\begin{aligned}\Sigma \circ T_z &= \frac{\lambda}{\det(E \circ T_z)}\text{tr}((^*E \circ T_z)(\varepsilon \circ T_z)(E \circ T_z))\,(E\,^*E) \circ T_z \\ &\quad + 2\mu \frac{1}{\det(E \circ T_z)}((E\,^*E) \circ T_z)(\varepsilon \circ T_z)((E\,^*E) \circ T_z)\end{aligned}$$

Following [**AZ98**] we consider the differential operators D_N and $\mathcal{D}_{\text{SHELL}}$ defined by

$$\text{D}_\text{N} = \begin{pmatrix} \partial_z \\ 1 \\ f \\ g \end{pmatrix} \quad \text{and} \quad \mathcal{D}_{\text{SHELL}}v = \begin{pmatrix} (v \otimes n) \\ (D_\Gamma v) \\ (D_\Gamma v.D^2 b) \\ (D_\Gamma v.(D^2 b)^2) \end{pmatrix}$$

If $\theta : z \mapsto \theta(z)$ is a scalar function and v is a vector then $\text{D}_\text{N}\theta(z)$ is a vector and $\mathcal{D}_{\text{SHELL}}v$ is a 3rd-order tensor. The subsequent identity holds

(4.3) $$(Du) \circ T_z = \sum_{i=0}^{d} \text{D}_\text{N} z^i . \mathcal{D}_{\text{SHELL}} u_i$$

Then $(^*E \circ T_z)(\varepsilon \circ T_z)(E \circ T_z)$ is equal to the symmetrized of

$$^*E(I + zD^2 b)\left(\sum_{i=0}^{d} \text{D}_\text{N} z^i .(\mathcal{D}_{\text{SHELL}} u_i)\right)(I + zD^2 b)E$$

which can be rewritten

$$\sum_{i=0}^{d} \text{D}_\text{N} z^i .(^*E(\mathcal{D}_{\text{SHELL}} u_i)E) + \sum_{i=0}^{d} z^2 \text{D}_\text{N} z^i .(^*ED^2 b(\mathcal{D}_{\text{SHELL}} u_i)D^2 bE)$$

$$+ \sum_{i=0}^{d} z\text{D}_\text{N} z^i .(^*ED^2 b(\mathcal{D}_{\text{SHELL}} u_i)E + {}^*E(\mathcal{D}_{\text{SHELL}} u_i)D^2 bE)$$

we get

$$\text{tr}((^*E \circ T_z)(\varepsilon \circ T_z)(E \circ T_z)) = \sum_{i=0}^{d} \text{D}_\text{N} z^i .(^*E\text{tr}(\mathcal{D}_{\text{SHELL}} u_i)E)$$

$$+ \sum_{i=0}^{d} z\text{D}_\text{N} z^i .\text{tr}(^*ED^2 b(\mathcal{D}_{\text{SHELL}} u_i)E) + \sum_{i=0}^{d} z^2 \text{D}_\text{N} z^i .\text{tr}(^*ED^2 b(\mathcal{D}_{\text{SHELL}} u_i)D^2 bE)$$

Since

$$\text{D}_\text{N} z^i = \begin{pmatrix} iz^{i-1} \\ z^i \\ z^i f(z) \\ z^i g(z) \end{pmatrix}$$

the real $\text{tr}((^*E \circ T_z)(\varepsilon \circ T_z)(E \circ T_z))$ may be written as a function of 10 terms depending on z with coefficients independent of z. Moreover

$$(E\,^*E) \circ T_z = (I + zD^2 b)E\,^*E(I + zD^2 b)$$

hence
$$(E^*E) \circ T_z = E^*E + z(D^2bE^*E + E^*ED^2b) + z^2 D^2bE^*ED^2b$$

therefore the first term of $\Sigma \circ T_z$ may be written as function of 16 terms depending on z with coefficients being matrices independent of z. The same method applies for the second term of $\Sigma \circ T_z$, it follows

$$\Sigma \circ T_z = \sum_{i=0}^{d} \sum_{\alpha=-1}^{4} a_{\alpha,i} z^{\alpha+i} \frac{1}{j(z)} + b_{\alpha,i} z^{\alpha+i} \frac{f(z)}{j(z)} + c_{\alpha,i} z^{\alpha+i} \frac{g(z)}{j(z)}$$

the coefficients $a_{\alpha,i}$, $b_{\alpha,i}$ and $c_{\alpha,i}$ can be computed with a symbolic computation system, they are presented in [**CZ98b**]. From those coefficients we define the matrices $A_{i,1}$, $A_{i,2}$ and $A_{i,3}$ such that

(4.4) $$\Sigma \circ T_z = \sum_{i=0}^{d+4} A_{i,1} z^{i-1} \frac{1}{j(z)} + A_{i,2} z^{i-1} \frac{f(z)}{j(z)} + A_{i,3} z^{i-1} \frac{g(z)}{j(z)}$$

We will refer to the third-order tensor A_i which is $3 \times 3 \times 3$. Let $V_i(z)$ be the vector

$$z^i \begin{pmatrix} 1 \\ f(z) \\ g(z) \end{pmatrix}$$

then (4.4) may be rewritten

$$\Sigma \circ T_z = \sum_{i=0}^{d+4} \frac{1}{j(z)} A_i^{\triangleleft\triangleleft} . V_i(z)$$

We suppose $w = b^k.(w_k \circ p)$ then

$$\varepsilon(w) \circ T_z = D_N z^k . \mathcal{D}_{\text{SHELL}} w$$

therefore

$$(\Sigma \circ T_z)..(\varepsilon(w) \circ T_z) \det(DT_z) = \sum_{i=0}^{d+4} \left(\frac{1}{j(z)} A_i^{\triangleleft\triangleleft}.V_i(z) \right) .. \left(D_N z^k . \mathcal{D}_{\text{SHELL}} w_k \right) j(z)$$

then, lemma 2.2 yields

$$(\Sigma \circ T_z)..(\varepsilon(w) \circ T_z) \det(DT_z) = \sum_{i=0}^{d+4} ((\mathcal{D}_{\text{SHELL}} w_k)..A_i^{\triangleleft\triangleleft})..(D_N z^k \otimes V_i(z))$$

Let us define the matrix tensor $L_i^k(u,w)$ by $L_i^k(u,w) = (\mathcal{D}_{\text{SHELL}} w_k)..A_i^{\triangleleft\triangleleft}$ and the matrix $q_i^k(z)$ by $D_N z^k \otimes V_i(z)$, that is

$$q_i^k(z) = \begin{pmatrix} kz^{i+k-1} & kz^{i+k-1}f(z) & kz^{i+k-1}g(z) \\ z^{i+k} & z^{i+k}f(z) & z^{i+k}g(z) \\ z^{i+k}f(z) & z^{i+k}f(z)^2 & z^{i+k}f(z)g(z) \\ z^{i+k}g(z) & z^{i+k}f(z)g(z) & z^{i+k}g(z)^2 \end{pmatrix}$$

then

$$(\Sigma \circ T_z)..(\varepsilon(w) \circ T_z) \det(DT_z) = \sum_{i=0}^{d+4} L_i^k(u,w)..q_i^k(z)$$

hence
$$\int_0^\tau \int_{-h}^h \int_{\Gamma_z} \Sigma(u)..\varepsilon(w)\, d\Gamma_z\, dz\, dt = \sum_{i=0}^{d+4} \int_0^\tau \int_\Gamma L_i^k(u,w)..(Q_i^k(h) - Q_i^k(-h))\, dz\, dt$$

where Q_i^k is the matrix whose terms are the anti-derivative of the terms of q_i^k. The terms of that matrix can be computed explicitly as the functions to be integrated are rational fractions.

On the other hand
$$\partial_t u \circ T_z = \sum_{i=0}^d b^i \circ T_z.(u_i \circ p \circ T_z) = \sum_{i=0}^d z^i u_i$$

similarly $\partial_t w = z^k w_k$ therefore equation (4.1) yields for all $k \in \mathbb{N}$

(4.5)
$$\sum_{i=0}^{d+4} \int_0^\tau \int_\Gamma \left(-\frac{1+(-1)^{i+k}}{i+k+1}\chi_{i\leq d} h^{i+k+1}\rho \partial_t u_i \partial_t w_k \right.$$
$$\left. + L_i^k(u,w)..(Q_i^k(h) - Q_i^k(-h)) \right) d\Gamma dt = 0$$

where $\chi_{i \leq d} = 1$ if $i \leq d$ and 0 otherwise.

The point of this model is to carry on the exact computation till the end. It is $p(d,\infty)$ as no truncation of the series has been made in the expansion. As a consequence, the coerciveness in the shell model will be straightforward.

4.3. Existence and Regularity of the Solutions for the Shell Model.

LEMMA 4.8. *The function $\|\cdot\|_d$ defined on $\mathcal{U}_h^d$ by*
$$\|U\|_d = \left\|\sum_{i=0}^d b^i u_i \circ p\right\|_{H^1(\Omega)}$$
is a norm on $\mathcal{U}_h^d$, which is a Banach space for that norm.

DEFINITION 4.9. *Let $\bar{\Lambda}$ be the operator defined by*
$$D(\bar{\Lambda}) = H^2(\Omega) \cap \mathcal{U}_h^d$$
$$\int_\Gamma \bar{\Lambda}(u)w = -\frac{1}{\rho}\int_\Gamma \sum_{i=0}^{d+4} L_i^k(u,w)..(Q_i^k(h) - Q_i^k(-h)))d\Gamma dt = 0$$

We have $\int_\Gamma \bar{\Lambda}(u)w = -\frac{1}{\rho}\int_\Omega \Sigma(u)..\varepsilon(w)$. The subsequent lemmas follow

LEMMA 4.10. $\bar{\Lambda}$ *is self-adjoint.*

LEMMA 4.11. $\int_\Gamma \bar{\Lambda}(u)u \geq C\|u\|_{H^1(\Omega)} \geq C\|u\|_d$

We note
$$\bar{U} = \begin{pmatrix} u \\ \partial_t u \end{pmatrix}, \quad \bar{U}_0 = \begin{pmatrix} \phi \\ \psi \end{pmatrix} \quad \text{and} \quad \bar{A} = \begin{pmatrix} 0 & 1 \\ -\Lambda & 0 \end{pmatrix}$$

then (2.5) is equivalent to $\partial_t \bar{U} = \bar{A}\bar{U}$, $\bar{U}(0) = \bar{U}_0$. Following [**BDPDM92**, prop. 2.12], we have

PROPOSITION 4.12. *The operator $\bar{A}$ is the infinitesimal generator of a strongly continuous semi-group of contraction $\bar{S}$. Moreover $^*\bar{A} = -\bar{A}$.*

The subsequent proposition derives from [**BDPDM92**, prop 3.3] and proposition 4.12.

PROPOSITION 4.13. *Equation (4.5) has a unique solution U in $H^2(\Omega) \cap \mathcal{U}_h^d$.*

References

[AZ98] Jean-Christophe Aguilar and Jean-Paul Zolésio, *Coque fluide intrinsèque sans approximation géométrique (intrinsic fluid shell without geometrical approximation)*, Comptes Rendus de l'Académie des Sciences, Paris, series I, Partial Differential Equations **326** (1998), no. 11, 1341–1346.

[Bal88] A. V. Balakrishnan, *Stability enhancement of flexible structures by nonlinear boundary-feedback control*, Boundary Control and Boundary Variations (Jean-Paul Zolésio, ed.), Lectures Notes Control and Information Science, no. 100, IFIP, Springer-Verlag, 1988, Nice, France, pp. 18–37.

[BDPDM92] Alain Bensoussan, Giuseppe Da Prato, Michel C. Delfour, and Sanjoy K. Mitter, *Representation and control of infinite dimensional systems*, Systems & Control: Foundations & Applications, vol. I, Birkhäuser, 1992.

[Buc95] Dorin Bucur, *Contrôle par rapport au domaine dans les edp*, Ph.D. thesis, Ecole des Mines de Paris, 1995.

[CDKP87] Gong Chen, Michel C. Delfour, Allan M. Krall, and Guy Payre, *Modeling, stabilization and control of serially connected beams*, SIAM Journal on Control and Optimization **2** (1987), no. 3, 526–546.

[Cea81] Jean Cea, *Problems of shape optimal design*, Optimization of distributed parameter structures II, N.A.T.O. Asi Series. Series E: Applied Sciences, no. 50, N.A.T.O., 1981, pp. 1005–1048.

[CM98] John Cagnol and Jean-Paul Marmorat, *Static equilibrium of hyperelastic thin shell: Symbolic and numerical computation*, Mathematics and Computers in Simulation **46** (1998), no. 2, 103–115.

[CMZ98] John Cagnol, Jean-Paul Marmorat, and Jean-Paul Zolésio, *Elastic vibration of a thin shell around a large static equilibrium*, Proceedings of the 5th MMAR (S. Domek, R. Kaszyński, and L. Tarasiejski, eds.), vol. 1, Szczecin Technical University Press, 1998, Międzyzdroje, Poland, pp. 161–166.

[CZ98a] John Cagnol and Jean-Paul Zolésio, *Hidden shape derivative in the wave equation*, System Modelling and Optimization (Peter Kall, Irena Lasiecka, and Mike Polis, eds.), Chapman & Hall/CRC Research Notes in Mathematics, vol. 396, IFIP, CRC Press LLC, Boca Raton, 1998, Detroit, Michigan, USA., pp. 42–52.

[CZ98b] John Cagnol and Jean-Paul Zolésio, *The tensor A_{SHELL} for a constrained shell: Explicit computation with a symbolic computation system*, Tech. report, CMA, Ecole des Mines de Paris, France, 1998.

[CZ99] John Cagnol and Jean-Paul Zolésio, *Shape derivative in the wave equation with dirichlet boundary condition*, Journal of Differential Equations **158** (1999), no. 2, 175–210.

[DLP86] Michel C. Delfour, John Lagnese, and Michael P. Polis, *Stabilization of hyperbolic systems using concentrated sensors and actuators*, IEEE Trans. Autom. Control **AC-31** (1986), 1091–1096.

[DP75] Alain Dervieux and Bernadette Palmerio, *Une formule de Hadamard dans des problemes d'identification de domaines*, Comptes Rendus de l'Académie des Sciences, Paris, series A (1975), 1697–1700.

[DZ92] Michel C. Delfour and Jean-Paul Zolésio, *Structure of shape derivatives for non smooth domains*, Journal of Functional Analysis **104** (1992), no. 1, 1–33.

[DZ97] Raja Dziri and Jean-Paul Zolésio, *Shape existence in navier-stokes flow with heat convection*, Annali della Scuola Normale Superiore di Pisa, Scienze Fisiche e Matematiche, serie IV **XXIV** (1997), no. 1, 165–192.

[DZ98a] Michel C. Delfour and Jean-Paul Zolésio, *Hidden boundary smoothness in hyperbolic tangential problems of nonsmooth domains*, System Modelling and Optimization (Peter Kall, Irena Lasiecka, and Mike Polis, eds.), Chapman & Hall/CRC Research Notes in Mathematics, vol. 396, IFIP, CRC Press LLC, Boca Raton, 1998, Detroit, Michigan, USA. To appear, pp. 53–61.

[DZ98b] Michel C. Delfour and Jean-Paul Zolésio, *Shape analysis via distance functions: Local theory*, Boundaries, Interfaces, and Transitions (Michel C. Delfour, ed.), CRM Proceedings & Lecture Notes, vol. 13, American Mathematical Society, 1998, Banff, Alberta, Canada, pp. 91–123.

[DZ99] Michel C. Delfour and Jean-Paul Zolésio, *Shapes and geometries: analysis, differential calculus and optimization*, Springer-Verlag, 1999, Forthcoming.

[Ger86] Paul Germain, *Mécanique*, marketing ed., vol. I, Ellipses, 1986, Ecole Polytechnique.

[Hor98a] Mary Ann Horn, *Implications of sharp trace regularity on boundary stabilization of the system of linear elasticity*, Jounral of Mathematical Analysis and Applications **223** (1998), 126–150.

[Hor98b] Mary Ann Horn, *Sharp trace regularity for the solutions of the dynamic elasticity*, Jornal of Mathematical Systems, Estimation Control **8** (1998), no. 2, 217–219.

[Lag83] John Lagnese, *Boundary stabilization of linear elastodynamic systems*, SIAM Journal on Control and Optimization **21** (1983), no. 6, 968–984.

[LLT86] Irena Lasiecka, Jacques-Louis Lions, and Roberto Triggiani, *Non homogeneous boundary value problems for second order hyperbolic operators*, Journal de Mathématiques pures et Appliquées **65** (1986), no. 2, 149–192.

[Mas87] Mohamed Masmoudi, *Outils pour la conception optimale de formes*, thèse d'etat, Université de Nice, 1987.

[MS76] François Murat and Jacques Simon, *Sur le contrôle par un domaine géométrique*, Tech. Report 76015, Université Pierre et Marie Curie, Paris, France, 1976.

[Pir82] Olivier Pironneau, *Optimal shape design for elliptic systems*, System Modeling and Optimization, Lectures Notes Control and Information Science, no. 38, IFIP, Springer-Verlag, 1982, New York, New York, USA, pp. 42–66.

[Rou93] Bernard Rousselet, *Introduction to shape sensitivity, three dimensional and surface systems*, Optimization of Large Structural Systems (George I. N. Rozvany, ed.), N.A.T.O. Asi Series. Series E: Applied Sciences, vol. 231, N.A.T.O., Kluwer Academic Publishers, March 1993, pp. 397–432.

[SZ91] Jan Sokolowski and Jean-Paul Zolésio, *Introduction to shape optimization*, Springer series in Computational Mathematics, no. 16, Springer-Verlag, 1991.

[Zol81] Jean-Paul Zolésio, *The material derivative (or speed) method for shape optimization*, Optimization of distributed parameter structures II, N.A.T.O. Asi Series. Series E: Applied Sciences, no. 50, N.A.T.O., 1981, pp. 1089–1151.

[Zol92] Jean-Paul Zolésio, *Shape formulation of free boundary problems with non linearized Bernoulli condition*, Boundary Control and Boundary Variation (Jean-Paul Zolésio, ed.), Lecture Notes in Control and Information Sciences, vol. 178, IFIP WG 7.2, Springer-Verlag, 1992, 1990, Sophia Antipolis, France, pp. 362–392.

[Zol94] Jean-Paul Zolésio, *Weak shape formulation of free boundary problems*, Annali della Scuola Normale Superiore di Pisa **21** (1994), 11–44.

UNIVERSIÉ LÉONARD DE VINCI, FST, DER-CS, 92916 PARIS LA DÉFENSE CEDEX, FRANCE
E-mail address: John.Cagnol@devinci.fr

CNRS, UMR 6618, 1361 ROUTE DES LUCIOLES, 06560 SOPHIA ANTIPOLIS, FRANCE
Current address: Centre de Mathématiques Appliquées, Ecole des Mines de Paris, 2004 route des Lucioles, BP 93, 06902 Sophia Antipolis Cedex, France
E-mail address: Jean-Paul.Zolesio@sophia.inria.fr

A Noise Reduction Problem Arising In Structural Acoustics: A Three-Dimensional Solution

Mehmet Camurdan and Guangcao Ji

ABSTRACT. We consider a coupled PDE system arising in noise reduction problems. In a three dimensional chamber, the acoustic pressure (unwanted noise) is represented by a hyperbolic wave equation. The floor of the chamber is subject to the action of a piezo-ceramic patch (smart material). The goal is to reduce the acoustic pressure by means of the vibrations of the floor which is modeled by a hyperbolic Kirchhoff equation. These two hyperbolic equations are coupled by appropriate trace operators. This overall model differs from those previously studied in the literature in two ways. The first difference is that the elastic chamber floor is here more realistically modeled by a hyperbolic Kirchhoff equation, rather than by a parabolic Euler-Bernoulli equation with Kelvin-Voight structural damping as in past literature, see [C.1]. Thus, the hyperbolic/parabolic coupled system of past literature is replaced here by a hyperbolic/hyperbolic coupled model, whose optimal regularity was studied in [C-T.1]. The second difference is the generalization of the two dimensional acoustic model in [C.1] into a three dimensional one. The main claim of this paper is a uniform stabilization of the coupled PDE system by a (physically appealing) boundary dissipation.

1. Introduction

In this paper we study the uniform stabilization of two coupled hyperbolic equations arising in the noise reduction problem for structural acoustic 3-D models. The acoustic pressure (unwanted noise) within a three dimensional chamber is mathematically represented by a hyperbolic wave equation, whereas a hyperbolic Kirchhoff equation models the elastic displacements of the two dimensional moving floor of the chamber. Such a floor is subject to the action of a piezo-ceramic patch (smart material), which is mathematically modeled as the distributional derivative of a Dirac mass concentrated on a smooth curve on the floor. The interaction between the chamber and the moving floor is represented by appropriate trace operators acting on the interface between the floor and the chamber.

More precisely, let Ω be a three dimensional, open and bounded domain (the chamber) in $\mathbb{R}^3$ with smooth, connected boundary Γ. The boundary consists of the closure of two open and disjoint portions Γ_0 and Γ_1: $\Gamma = \overline{\Gamma_0 \cap \Gamma_1}$. We assume that

---

1991 *Mathematics Subject Classification.* 35Q72, 73D35.

© 2000 American Mathematical Society

Γ_0 is a flat, two dimensional, planar surface, which is smooth, open, and bounded domain in $\mathbb{R}^2$, and that it is smoothly attached to Γ_1.

The acoustic medium within Ω is described by the wave equation in the variable z. The structural vibrations of the elastic floor Γ_0 are modeled by the variable v. We assume that v satisfies a Kirchhoff equation on Γ_0, coupled with the wave equation which is satisfied by z in the interior of the domain Ω. Then the PDE model in the variables z and v is as follows:

(1.1)
$$\begin{cases} \text{Wave Equation} \\ \begin{cases} z_{tt} = \Delta z & \text{on } Q \\ \frac{\partial z}{\partial \nu}|_{\Gamma_1} = -k_1 z_t & \text{on } \Sigma_1 \\ \frac{\partial z}{\partial \nu}|_{\Gamma_0} = -k_1 z_t - v_t & \text{on } \Sigma_0 \end{cases} \\ \text{Kirchhoff Equation} \\ \begin{cases} v_{tt} - \gamma \Delta v_{tt} + \Delta^2 v - z_t = \frac{\partial \delta_\beta}{\partial \nu} u(t) & \text{on } \Sigma_0 \\ v|_{\partial \Sigma_0} = 0; \Delta v|_{\partial \Gamma_0} = -k_2 \frac{\partial v_t}{\partial \nu} & \text{on } \partial \Sigma_0 \end{cases} \\ \text{with initial conditions} \\ z(0, \cdot) = z_0; \; z_t(0, \cdot) = z_1; \; v(0, \cdot) = v_0; \; v_t(0, \cdot) = v_1 \text{ in } \Omega \times \Omega \times \Gamma_0 \times \Gamma_0 \end{cases}$$

where

$\gamma > 0$, $k_1 \geq 0$, $k_2 \geq 0$;
$Q \equiv (0, T] \times \Omega$, $\Sigma_1(0, T] \equiv \times \Gamma_1$, $\Sigma_0 \equiv (0, T] \times \Gamma_0$, $\partial \Sigma_0 \equiv (0, T] \times \partial \Gamma_0$.

The symbol β, which appears in the term $\frac{\partial \delta_\beta}{\partial \nu}$ in the Kirchhoff equation, represents a smooth curve contained in Γ_0. The control $u(t)$ influences the coupled system through $\frac{\partial \delta_\beta}{\partial \nu}$, which is, as in [**J-T.1**], the distributional derivative (with respect to the normal to β) of the Dirac mass concentrated on the curve β. Problem (1.1) is a hyperbolic/hyperbolic 3-D model. By contrast, the usual 2-D model of past literature ([**A.1**], [**A-L.1**], and [**A-L.2**]) is a hyperbolic/parabolic model, where the v-equation on Σ_0 is a structurally damped Euler-Bernoulli equation, rather than a Kirchhoff equation. Moreover, in [**A.1**] and [**A-L.1**], $u(t)$ acts through a finite number of the distributional derivatives of a Delta mass concentrated at points of the 1-D floor: $\delta'(x_i)$. Problem (1.1) is well-posed in the space Y defined below in (2.7), see Theorem 1.1. The inclusion of $\frac{\partial \delta_\beta}{\partial \nu}$ in the *uncoupled* Kirchhoff equation without the coupling term was considered in [**J-T.1**]. The curve β satisfies some geometric assumptions, namely, β in Γ_0 is the union of a finite number of curves $\beta_i, 1, 2, \ldots, n$, each of which, in turn, obeys one of the following conditions: 1) There exist $n_i \geq 1$ such that β_i is of class C^{n_i+3} and the curvature of β vanishes at most at one end of β_i where the tangent line has contact of order n; 2) β_i is a segment. [**J-T.1**] estimates the Fourier transform of the distribution $\frac{\partial \delta_\beta}{\partial \nu}$ as to fall into the Laplace-Fourier setting of [**T.1**], in particular, in order to invoke the basic estimate in of Lemma 1.1 [**T.1**]. This way, [**J-T.1**] proves that the Kirchhoff problem on flat Γ_0 of dim(Γ_0)=2 with forcing term $\frac{\partial \delta_\beta}{\partial \nu}$ has the same regularity space as the Kirchhoff equation on one dimensional Γ_0 with forcing term $\delta(x_0)u$, $x_0 \in \Gamma_0$; i.e., the space Y defined (2.7) below.

The basic structure of acoustic flow models has been known for a long time (see [**M-I.1**]). Some related mathematical questions regarding the spectral properties or the strong stabilization of the model in [**M-I.1**] are studied in [**B.1**] and [**L-L.1**]. Smart material technology has suggested the introduction of a dissipation acting at the edge of the floor via moments or shears. This motivates one to consider the damped coupled model (1.1) where the dissipation is through the bending moment of the hyperbolic Kirchhoff equation. As noted above, this model differs in a critical way from other models recently studied in noise reduction problems (see [**A.1**], [**A-L.1**], and [**A-L.2**]) in that the elastic dynamics of the moving floor here is more realistically represented by a hyperbolic Kirchhoff equation, rather than by a structurally damped Euler-Bernoulli equation with the so-called Kelvin-Voight damping as in aforementioned references.

The results existing in the literature on the stabilization of structural acoustic models refer precisely to these hyperbolic/parabolic models where the floor is strongly damped by means of structural damping (see [**A.1**] and [**A-L.1**]). In the case of structural damping present in the model, the component of the uncoupled system corresponding to the plate equation represents an analytic semigroup, see [**C-T.2**]. This provides, in addition to exponential stability properties for the plate equation, a lot of regularity properties which facilitate the analysis of stability for the entire structure. The situation is drastically different when the analytic plate equation is actually replaced by a more realistic hyperbolic Kirchhoff equation, as in our present paper, equation (1.1).

The main result of this paper is the following theorem which states that the energy $E_\gamma(t)$ (see (3.1)) associated with the coupled system (1.1) decays to zero exponentially.

THEOREM 1.1. *Let Ω be a bounded open domain in $\mathbb{R}^3$ as assumed above. Define $y(t) = [z(t), z_t(t), v(t), v_t(t)]$. Then the damped coupled PDE system (1.1) with $u(t) \equiv 0$ can be written as*

$$(1.2) \qquad \dot{y} = A_F y \text{ on } [D(A_F^*)]'; \quad y(0) = y_0 \in Y_0,$$

where Y_0 is the closed subspace of Y defined in (2.18) below.

(i) In (1.2), A_F, which is explicitly defined in (2.20), is the infinitesimal generator of a s.c. contraction semigroup $e^{A_F t}$ in Y_0 (see Theorem 2.2).

(ii) The semigroup $e^{A_F t}$ is uniformly exponentially stable on Y_0; that is, there exist constants $\delta > 0$ and $M \geq 1$, which are independent of $\gamma > 0$, in (1.1) such that

$$(1.3) \qquad \left\| e^{A_F t} \right\|_{L(Y_0)} \leq M e^{-\delta t}, \; t \geq 0; \; equivalently, E_\gamma(t) \leq M e^{-\delta t} E_\gamma(0).$$

2. Abstract Models: $\gamma > 0$

2.1. Undamped Problem: $k_1 = k_2 = 0; \; \gamma > 0$. Throughout the paper, $\gamma > 0$ is fixed. The case of undamped coupled equations (that is, $k_1 = k_2 = 0$) is analyzed in a companion paper where a sharp regularity result is obtained (see [**C-T.1**]). The following operators and the abstract setting for problem (1.1) are also quoted from [**C-T.1**]:

i) $\bar{A} : L_2(\Gamma_0) \supset D(\bar{A}) \to L_2(\Gamma_0)$ is the positive self-adjoint operator

$$\bar{A}f = \Delta^2 f, \; D(\bar{A}) = \{f \in H^4(\Gamma_0) : f|_{\partial\Gamma_0} = \Delta f|_{\partial\Gamma_0} = 0\};$$

$$(2.1) \quad \bar{A}^{1/2} = -\Delta, \; D(\bar{A}^{1/2}) = H^2(\Gamma_0) \cap H_0^1(\Gamma_0); \; \|f\|_{D(\bar{A}^{1/2})}^2 = \int_{\Gamma_0} |\Delta f|^2 \, d\Gamma_0.$$

For future reference, we note the following equivalent spaces:

(2.2) $\quad D(\bar{A}^{\frac{1}{2}-\epsilon}) \equiv H^{2-4\epsilon}(\Gamma_0)$ (equivalence in norms for small $\epsilon > 0$).

ii)

(2.3) $\mathbb{A} : L_2(\Gamma_0) \supset D(\mathbb{A}) \to L_2(\Gamma_0)$, $\mathbb{A} = (I + \gamma \bar{A}^{\frac{1}{2}})^{-1}\bar{A}$, $D(\mathbb{A}) = D(\bar{A}^{\frac{1}{2}})$.

The operator $\mathbb{A}$ is positive self-adjoint on the space $D(\bar{A}_\gamma^{\frac{1}{4}})$ topologized by the inner product

$$(x, y)_{D(\bar{A}_\gamma^{\frac{1}{4}})} = \left((I + \gamma \bar{A}^{\frac{1}{2}})x, y\right)_{L_2(\Gamma_0)}, \quad \forall x, y \in D(\bar{A}_\gamma^{\frac{1}{4}});$$

(2.4) $\quad D(\bar{A}_\gamma^{\frac{1}{4}}) = H_0^1(\Gamma_0); \quad \|y\|_{D(\bar{A}_\gamma^{1/4})}^2 = \int_{\Gamma_0} \left(|y|^2 + \gamma |\nabla y|^2\right) d\Gamma_0.$

iii) $A_N : L_2^0(\Omega) \equiv L_2(\Omega) / N(A_N) \to L_2^0(\Omega)$ is the positive self-adjoint operator

(2.5)
$$A_N f = -\Delta f, \quad D(A_N) = \{f \in H^2(\Omega) : \frac{\partial f}{\partial \nu}\big|_\Gamma = 0\}; \quad \|f\|_{D(A_N^{1/2})}^2 = \int_\Omega |\nabla f|^2 \, d\Omega.$$

where $N(A_N)$ is the one-dimensional null space (space of constants) of A_N in $L_2(\Omega)$.

iv) Define the Neumann map N for $g \in L_2(\Gamma)$ as (see [**L-T.3**]):

(2.6) $\quad h = Ng \in H^{3/2}(\Omega) \iff \begin{cases} \Delta h = 0 & \text{on } \Omega \\ \frac{\partial h}{\partial \nu}\big|_\Gamma = g & \text{on } \Gamma \end{cases};$

then $N^* A_N h = -h\big|_\Gamma$, $\forall h \in D(A_N)$.

v) Finally, consider the following spaces with equivalent norms:

$$Y \equiv D(A_N^{\frac{1}{2}}) \times L_2(\Omega) \times D(\bar{A}^{\frac{1}{2}}) \times D(\bar{A}_\gamma^{\frac{1}{4}})$$

(2.7) $\quad \equiv H^1(\Omega) \times L_2(\Omega) \times [H^2(\Gamma_0) \cap H_0^1(\Gamma_0)] \times H_0^1(\Gamma_0).$

Hence, problem (1.1) with $k_1 = k_2 = 0$ can be written abstractly as

(2.8) $\quad \dot{y} = Ay + Bu$ on $[D(A^*)]'$; $y(0) = y_0$; $y(t) = [z(t), z_t(t), v(t), v_t(t)]$

(2.9) $\quad A = \begin{bmatrix} 0 & I & 0 & 0 \\ -A_N & 0 & 0 & A_N N(\cdot|_{\Gamma_0}) \\ 0 & 0 & 0 & I \\ 0 & -(I + \gamma \bar{A}^{\frac{1}{2}})^{-1} N^* A_N & -\mathbb{A} & 0 \end{bmatrix} = -A^*$

where, with $y = [y_1, y_2, y_3, y_4]$,

$D(A) = \{y \in Y : y_2 \in D(A_N^{\frac{1}{2}}), y_3 \in D(\bar{A}^{\frac{3}{4}}), y_4 \in D(\bar{A}^{\frac{1}{2}}), [y_1 - N(y_4|_{\Gamma_0})] \in D(A_N)\}.$

A^* is the Y-adjoint of A and $D(A) = D(A^*)$; while the operator $B : U \to [D(A^*)]'$, $U = \mathbb{R}$ and its adjoint $B^* : D(A^*) \to U$ are

(2.10) $\quad Bu = \begin{bmatrix} 0 \\ 0 \\ 0 \\ (I + \gamma \bar{A}^{\frac{1}{2}})^{-1} \frac{\partial \delta_\beta}{\partial \nu} u \end{bmatrix}; \quad B^* \begin{bmatrix} y_1 \\ y_2 \\ y_3 \\ y_4 \end{bmatrix} = -\frac{\partial_\beta y_4}{\partial \nu}\bigg|_\beta, \quad y \in D(A^*).$

By the skew-adjointness of A on Y (see (2.9)), we see that A generates a s.c. unitary group e^{At} on Y:

$$e^{A^*t} = e^{-At}; \quad \|e^{A^*t}\|_{L(Y)} \equiv \|e^{-At}\|_{L(Y)} \equiv 1 \text{ and}$$
(2.11) $\quad (Ax, x)_Y = (A^*x, x)_Y \equiv 0, \ \forall x \in D(A) = D(A^*).$

Thus the solution to the *undamped* problem (1.1) with $k_1 = k_2 = 0$ can be written as:

(2.12) $\quad y(t) = e^{At} y_0 + (Lu)(t)$ where $(Lu)(t) = \int_0^t e^{A(t-\tau)} Bu(\tau) d\tau.$

Now we quote the main theorem (a sharp regularity result) of [**C-T.1**] (Theorem 1.2), which is extended to include the PDE system (1.1) in section 4 of [**C-T.1**].

THEOREM 2.1. *With reference to the coupled P.D.E. system (1.1), we have that for each $0 < T < \infty$, the operator L defined in (2.12) satisfies the following regularity property:*

(2.13) $\quad L : L_2(0,T) \to C([0,T]; Y), \ continuously.$

Sharp (optimal) regularity results (abstract trace regularity) for the mixed PDE problems have significant implications in the study of associated control theory problems, enabling us to invoke a large body of abstract results on quadratic control theory, min-max game theory, etc. For instance, the abstract results in [**L-T.3**], [**L-T.7**], [**M-T.1**], [**T.2**] can now be readily applied to the coupled PDE problem (1.1) over a finite time interval. In the case of infinite time interval, however, in order to invoke the abstract theory of these references, additional control theoretic hypotheses such as the Finite Cost Condition and the Detectability Condition are needed. The Finite Cost Condition amounts to the property of uniform stability with $L_2(0, \infty; U)$ feedback control. However, the uniform stabilization of the undamped ($k_1 = k_2 = 0$) problem (1.1) on the space of regularity Y (as given in (2.7) for our coupled PDE system) fails. This is actually a general pathology of hyperbolic or Petrowski type dynamics with point controls acting through δ or δ' (see [**L-T.3**]). As the issue of uniform stabilization is of paramount importance, to remedy this situation, we modify the original conservative dynamics $k_1 = k_2 = 0$ by adding damping terms $k_1 > 0$ and $k_2 > 0$ as to make it uniformly stable on the regularity space Y while preserving the same regularity in $C([0,T]; Y)$.

2.2. Damped Problem: $k_1 > 0$, $k_2 > 0$; $\gamma > 0$.

We are mainly interested in the physically appealing boundary stabilization in which energy decay rates are achieved by introducing some form of dissipation on the boundary. Our main goal is to show that the boundary damping added to the wave equation and the boundary dissipation applied through the bending moments at the edge of the plate are enough to provide the desired uniform decay rates of the natural energy function associated with the model. For this purpose, we consider the coupled system (1.1) with strictly positive constants k_1 and k_2, which we actually take equal to 1 for convenience. Since our main interest is a uniform stability result, we also let $u(t) \equiv 0$ in this section.

Now we introduce two more operators, namely, the Dirichlet operator D and the Green's operator G_2 (see [**L-T.1**] and [**L-T.3**]):

$$w = Dh \Leftrightarrow \left\{\Delta w = 0 \text{ in } \Gamma_0;\ w = h \text{ on } \partial\Gamma_0\right\};$$

(2.14) $\quad f = G_2 g \Leftrightarrow \left\{\Delta^2 f = 0 \text{ in } \Gamma_0;\ f = 0,\ \Delta f = g \text{ on } \partial\Gamma_0\right\};\ G_2 = -\bar{A}^{-\frac{1}{2}}D.$

Next let ν be the unit normal on the boundary $\partial\Gamma_0$ and notice from (2.14) and [**L-T.3**] that

(2.15) $\quad G_2^* \bar{A} g = -D^* \bar{A}^{\frac{1}{2}} g = \dfrac{\partial g}{\partial \nu};\ g \in D(\bar{A}^{1/2}).$

Proceeding as in [**L-T.2**] and [**L-T.3**], the damped problem (1.1) with $k_1 = k_2 = 1$ and $u(t) \equiv 0$ can be written abstractly as (see (2.3), (2.6), (2.14), and (2.15))

(2.16) $\quad z_{tt} = -A_N z - A_N N N^* A_N^* z_t + A_N N(v_t|_{\Gamma_0});$

$$v_{tt} = -\mathbb{A}v - \mathbb{A}G_2 G_2^* \bar{A} v_t - (I + \gamma\bar{A}^{\frac{1}{2}})^{-1} N^* A_N z_t$$

(2.17) $\quad = -(I + \gamma\bar{A}^{\frac{1}{2}})^{-1}\bar{A}v - (I + \gamma\bar{A}^{\frac{1}{2}})^{-1}\bar{A}^{\frac{1}{2}}DD^*\bar{A}^{\frac{1}{2}}v_t + (I + \gamma\bar{A}^{\frac{1}{2}})^{-1}(z_t|_{\Gamma_0}).$

Motivated by the stabilization problem of the wave equation with purely Neumann feedback on the entire boundary, we introduce the following subspace Y_0 of Y (see (2.7):

(2.18) $\quad Y_0 = \left\{y = [y_1, y_2, y_3, y_4] \in Y : \int_\Gamma y_1\, d\Gamma + \int_\Omega y_2\, d\Omega + \int_{\Gamma_0} y_3\, d\Gamma_0 = 0\right\}.$

We equip the subspace Y_0 with the norm of Y, see (2.7)). Since we have focused on $L_2(\Omega)/N(A_N)$, we may use the *gradient norm* in $H^1(\Omega)$. The subspace Y_0 is then closed and complete in the Y-topology (by direct trace theory in the first and third coordinates). Then the first order equation corresponding to (2.16) and hence to the damped problem (1.1) with $u(t) \equiv 0$ is

(2.19)
$\quad \dot{y} = A_F y$ on $[D(A_F^*)]';\ y(0) = y_0 \in Y_0;\ y(t) = [z(t), z_t(t), v(t), v_t(t)]$ and

(2.20) $\quad A_F = \begin{bmatrix} 0 & I & 0 & 0 \\ \Delta & 0 & 0 & 0 \\ 0 & 0 & 0 & I \\ 0 & -(I+\gamma\bar{A}^{\frac{1}{2}})^{-1}N^*A_N & -\mathbb{A} & -\mathbb{A}G_2 G_2^*\bar{A} \end{bmatrix}$

with dense domain

$D(A_F) = \Big\{y = [y_1, y_2, y_3, y_4] \in Y_0 : y_1 \in H^2(\Omega),\ y_2 \in H^1(\Omega),\ y_4 \in D(\bar{A}^{1/2});$

$\quad \dfrac{\partial y_1}{\partial \nu}\Big|_{\Gamma_1} = -y_2\Big|_{\Gamma_1}$ and $\dfrac{\partial y_1}{\partial \nu}\Big|_{\Gamma_0} = -y_2\Big|_{\Gamma_0} - y_4\Big|_{\Gamma_0};$

(2.21) $\quad N^*A_N(y_2|_{\Gamma_0}) + \bar{A}y_3 + \bar{A}G_2 G_2^*\bar{A}y_4 \in [D(\bar{A}_\gamma^{\frac{1}{4}})]'\Big\}.$

LEMMA 2.2. *The closed, complete subspace Y_0 is invariant under the operator A_F in (2.20).*

Proof: Let $y = [y_1, y_2, y_3, y_4] \in D(A_F) \subset Y_0$. Hence, (2.18) and (2.21) hold true. Now let us define g such that:

(2.22) $\quad g = [g_1, g_2, g_3, g_4] = A_F y;\ g_1 = y_2;\ g_2 = \Delta y_1;\ g_3 = y_4.$

We must show that $g \in Y_0$. In fact, first notice from (2.22) and (2.21) that

$$(2.23) \quad \int_\Omega g_2 \, d\Omega = \int_\Omega \Delta y_1 d\Omega = \int_\Gamma \frac{\partial y_1}{\partial \nu}\Big|_\Gamma d\Gamma = -\int_\Gamma y_2 d\Gamma - \int_{\Gamma_0} y_4 d\Gamma_0.$$

We also get from (2.22) that

$$(2.24) \quad \int_\Gamma g_1 \, d\Gamma = \int_\Gamma y_2 \, d\Gamma \text{ and } \int_{\Gamma_0} g_3 \, d\Gamma_0 = \int_{\Gamma_0} y_4 \, d\Gamma_0.$$

Therefore, it follows from (2.23) and (2.24) that

$$\int_\Gamma g_1 \, d\Gamma + \int_\Omega g_2 \, d\Omega + \int_{\Gamma_0} g_3 \, d\Gamma_0 = 0.$$

Hence $g = A_F y \in Y_0$ and Y_0 is invariant under the operator A_F. $\square$
Next notice that the Y_0-adjoint A_F^* of A_F is

$$(2.25) \quad A_F^* = \begin{bmatrix} 0 & -I & 0 & 0 \\ -\Delta & 0 & 0 & 0 \\ 0 & 0 & 0 & -I \\ 0 & (I+\gamma \bar{A}^{\frac{1}{2}})^{-1} N^* A_N & \mathbb{A} & -\mathbb{A} G_2 G_2^* \bar{A} \end{bmatrix}$$

with dense domain

$$D(A_F^*) = \{y = [y_1, y_2, y_3, y_4] \in Y_0 : y_1 \in H^2(\Omega),\ y_2 \in H^1(\Omega),\ y_4 \in D(\bar{A}^{1/2});$$

$$\frac{\partial y_1}{\partial \nu}\Big|_{\Gamma_1} = y_2|_{\Gamma_1} \text{ and } \frac{\partial y_1}{\partial \nu}\Big|_{\Gamma_0} = y_2|_{\Gamma_0} + y_4|_{\Gamma_0};$$

$$(2.26) \quad N^* A_N(y_2|_{\Gamma_0}) + \bar{A} y_3 - \bar{A} G_2 G_2^* \bar{A} y_4 \in [D(\bar{A}_\gamma^{\frac{1}{4}})]'\}.$$

LEMMA 2.3. *The space Y_0 is invariant under the operator A_F^* in (2.25).*

Proof: The proof is the same as that of Lemma 2.2. Choose $y = [y_1, y_2, y_3, y_4] \in D(A_F^*) \subset Y_0$. Hence, (2.18) and (2.26) hold true. Now defining $g = A_F^* y$, we get that

$$(2.27) \quad g_1 = -y_2;\ g_2 = -\Delta y_1;\ g_3 = -y_4.$$

Notice from (2.27) and (2.26) that

$$(2.28) \quad \int_\Omega g_2 \, d\Omega = -\int_\Omega \Delta y_1 d\Omega = -\int_\Gamma \frac{\partial y_1}{\partial \nu}\Big|_\Gamma d\Gamma = \int_\Gamma y_2 d\Gamma + \int_{\Gamma_0} y_4 d\Gamma_0.$$

Therefore, it follows from (2.27) and (2.28) that

$$\int_\Gamma g_1 \, d\Gamma + \int_\Omega g_2 \, d\Omega + \int_{\Gamma_0} g_3 \, d\Gamma_0 = 0.$$

Hence $g = A_F^* y \in Y_0$ and Y_0 is invariant under the operator A_F^*. $\square$

THEOREM 2.4. *A_F is a maximal dissipative operator on Y_0 and hence the infinitesimal generator of a s.c. semigroup $e^{A_F t}$ of contractions on Y_0.*

Proof: Since A_F is densely defined, it is enough to show dissipativity of both of A_F and A_F^* on Y_0 (see Corollary 1.4.4, p15 in [**P**]). For this purpose, let $y \in D(A_F)$ so that y satisfies the boundary conditions in (2.21). Note from (2.20) that (see also (2.18) and (2.4))

$$
\begin{aligned}
(A_F y, y)_{Y_0} &= (y_2, y_1)_{H^1(\Omega)} + (\Delta y_1, y_2)_{L_2(\Omega)} + (y_4, y_3)_{D(\bar{A}^{\frac{1}{2}})} \\
&\quad - \left((I + \gamma \bar{A}^{\frac{1}{2}})^{-1} N^* A_N y_2 + \mathbb{A} y_3 + \mathbb{A} G_2 G_2^* \bar{A} y_4, y_4\right)_{D(\bar{A}_\gamma^{\frac{1}{4}})} \\
&= (\nabla y_2, \nabla y_1)_{L_2(\Omega)} - (\nabla y_1, \nabla y_2)_{L_2(\Omega)} + \left(\frac{\partial y_1}{\partial \nu}, y_2\right)_{L_2(\Gamma)} \\
&\quad + (y_4, y_3)_{D(\bar{A}^{\frac{1}{2}})} - \left(N^* A_N y_2\big|_{\Gamma_0}, y_4\right)_{L_2(\Gamma_0)} \\
&\quad - (y_3, y_4)_{D(\bar{A}^{\frac{1}{2}})} - \left(G_2^* \bar{A} y_4, G_2^* \bar{A} y_4\right)_{L_2(\partial\Gamma_0)}.
\end{aligned}
\tag{2.29}
$$

It then follows from (2.21), (2.6), (2.15), and (2.29) that

$$
\begin{aligned}
(A_F y, y)_{Y_0} &= -(y_2, y_2)_{L_2(\Gamma)} - (y_4, y_2|_{\Gamma_0})_{L_2(\Gamma_0)} + (y_2|_{\Gamma_0}, y_4)_{L_2(\Gamma_0)} \\
&\quad - \left(\frac{\partial y_4}{\partial \nu}, \frac{\partial y_4}{\partial \nu}\right)_{L_2(\partial\Gamma_0)} \\
&= -(y_2, y_2)_{L_2(\Gamma)} - \left(\frac{\partial y_4}{\partial \nu}, \frac{\partial y_4}{\partial \nu}\right)_{L_2(\partial\Gamma_0)} \leq 0.
\end{aligned}
\tag{2.30}
$$

By (2.30), A_F is dissipative on Y_0. The dissipativity of A_F^* on Y_0 follows similarly. Choose $y \in D(A_F^*)$ so that y satisfies the boundary conditions in (2.26). Then using (2.25), (2.26), (2.6), and (2.15), we get that

$$
(A_F^* y, y)_{Y_0} = -(y_2, y_2)_{L_2(\Gamma)} - \left(\frac{\partial y_4}{\partial \nu}, \frac{\partial y_4}{\partial \nu}\right)_{L_2(\partial\Gamma_0)} \leq 0.
\tag{2.31}
$$

By (2.31), A_F^* is dissipative on Y_0. □

3. Uniform Stability of $e^{A_F t}$

Our main goal is to show the uniform exponential stability (1.3) of the s.c. contraction semigroup $e^{A_F t}$ in the space Y_0 described in (2.18), corresponding to the coupled damped PDE system (1.1) with $k_1 = k_2 = 1$ and $u(t) \equiv 0$. Accordingly, the 'energy' of the damped system is identified with the norm of Y_0 (see (2.18)) where $y_0 = [z_0, z_1, v_0, v_1] \in Y_0$:

$$
E_\gamma(t) = \|e^{A_F t} y_0\|_Y^2 = E_z(t) + E_{v\gamma}(t);
\tag{3.1}
$$

$$
E_z(t) = \int_\Omega \left(|\nabla z(t)|^2 + z_t^2(t)\right) d\Omega = \|z(t)\|_{D(A_N^{\frac{1}{2}})}^2 + \|z_t(t)\|_{L_2(\Omega)}^2;
\tag{3.2}
$$

$$
E_{v\gamma}(t) = \int_{\Gamma_0} \left((\Delta v(t))^2 + v_t^2(t) + \gamma|\nabla v_t(t)|^2\right) d\Gamma_0 = \|v(t)\|_{D(\bar{A}^{\frac{1}{2}})}^2 + \|v_t(t)\|_{D(\bar{A}_\gamma^{\frac{1}{4}})}^2.
\tag{3.3}
$$

The norms of $D(A_N^{\frac{1}{2}})$, $D(\bar{A}^{\frac{1}{2}})$, and $D(\bar{A}_\gamma^{\frac{1}{4}})$ are described in (2.5), (2.1), and (2.4), respectively.

REMARK 3.1. By Poincare's inequality, which holds true by the boundary conditions in (1.1), $\int_{\Gamma_0} v_t^2(t) d\Gamma_0$ is a lower order term when compared with $E_{v\gamma}(t)$ in case of $\gamma > 0$. However, it is not a lower order term with respect to the energy function when $\gamma = 0$.

LEMMA 3.2. *With respect to the coupled system (1.1) with $k_1 = k_2 = 1$ and $u(t) \equiv 0$ and fixed $\gamma > 0$, we have the following results:*

$$(3.4) \quad E_\gamma(\tilde{t}) + 2\int_0^{\tilde{t}}\int_\Gamma z_t^2\, d\Gamma\, dt + 2\int_0^{\tilde{t}}\int_{\partial\Gamma_0}\left|\frac{\partial v_t}{\partial \nu}\right|^2 d(\partial\Gamma_0) dt = E_\gamma(0),\ \forall \tilde{t} > 0,$$

$$(3.5) \quad \|(z_t|_\Gamma)\|^2_{L_2(0,\infty;L_2(\Gamma))} + \left\|\frac{\partial v_t}{\partial \nu}\right\|^2_{L_2(0,\infty;L_2(\partial\Gamma_0))} \leq \frac{1}{2}E_\gamma(0).$$

Proof: Notice that $e^{(A_F t)}y_0 = [z, z_t, v, v_t]$ is the solution to the problem (1.1) with $k_1 = k_2 = 1$ and $u(t) = 0$. Initially for $y_0 \in D(A_F)$, we get by following the step (2.30) in the proof of Theorem 2.2 that $\forall \tilde{t} > 0$,

$$(3.6) \quad \begin{aligned} \frac{d}{dt}E_\gamma(t) &= \frac{d}{dt}\|e^{A_F t}y_0\|^2_{Y_0} = 2\left(A_F e^{A_F t}y_0, e^{A_F t}y_0\right)_Y \\ &= -2\int_\Gamma z_t^2\, d\Gamma - 2\int_{\partial\Gamma_0}\left|\frac{\partial v_t}{\partial \nu}\right|^2 d(\partial\Gamma_0) \leq 0. \end{aligned}$$

Then (3.4) follows by integrating (3.6), which we then extend to $y_0 \in Y_0$ by density. (3.5) follows immediately from (3.4). □

From (3.4) in Lemma 3.2, it is clear that $E_\gamma(t)$ is non-increasing.

Orientation: Our strategy is to study the Kirchhoff equation on Γ_0 and the wave equation on Ω separately and then combine the results. In both cases, we run multipliers on the corresponding equation. Both PDE's classical multiplier techniques and further methods (those of microlocal analysis) are employed in the analysis of both equations. These techniques will give an estimate of the energy of the system by the dissipation terms plus several lower order terms. Thereon, a standard application of compactness/uniqueness argument gives the final result.

3.1. PDE Estimates for Wave Equation in (1.1).

We will focus on the following wave equation coming from the coupled P.D.E. system (1.1):

$$(3.7) \quad \begin{cases} z_{tt} = \Delta z & \text{on } Q, \\ \frac{\partial z}{\partial \nu}\big|_{\Gamma_1} = -z_t & \text{on } \Sigma_1, \\ \frac{\partial z}{\partial \nu}\big|_{\Gamma_0} = -g - z_t & \text{on } \Sigma_0, \end{cases} \quad \text{where } g\ (= v_t \in L_2(0,T;H^2(\Gamma_0))) \in L_2(\Sigma_0).$$

Let us introduce the following domains where $T > \alpha > 0$ and α is arbitrarily small:

$$Q_\alpha \equiv (\alpha, T-\alpha) \times \Omega;\ \Gamma_\alpha \equiv (\alpha, T-\alpha) \times \Gamma;\ \Gamma_{0\alpha} \equiv (\alpha, T-\alpha) \times \Gamma_0.$$

The main result of this section is the following estimate concerning the wave equation (3.7):

THEOREM 3.3. *Let $h(x)$ be a smooth (C^2) vector field on Ω, satisfying the following coercivity property:*

$$(3.8) \quad \int_\Omega H(x)w(x) \cdot w(x) d\Omega \geq \rho \int_\Omega w^2(x) d\Omega,\ \forall\, w(x) \in L_2(\Omega)$$

where $H(x)$ is the transpose of the Jacobian matrix of $h(x)$ and $\rho > 0$ is a strictly positive constant. Notice that such a vector field is always available. Then, with respect to the uncoupled wave equation (3.7), we have the following inequality:

$$C_T\left\{\|z_t\|^2_{L_2(\Sigma_\alpha)} + \|g\|^2_{L_2(\Sigma_{0\alpha})} + \|z\|^2_{L_2(0,T;H^{\frac{1}{2}+\epsilon_1}(\Omega))} + \|z\|^2_{H^{\frac{1}{2}+\epsilon_1}(0,T;L_2(\Omega))}\right\}$$

$$(3.9) \quad \geq C_{\alpha\epsilon_1}\left(\int_\alpha^{T-\alpha} E_z(t)\,dt - \big[E_z(\alpha) + E_z(T-\alpha)\big]\right).$$

We have the following corollary of Theorem 3.3:

COROLLARY 3.1. *Let $h(x)$ be a smooth (C^2) vector field on Ω chosen as in Theorem 3.3. Then, with respect to the wave equation part of (1.1) with $u(t) \equiv 0$, we have the following inequality (which is actually (3.9) with $g = v_t$):*

$$C_T\left\{\|z_t\|^2_{L_2(\Sigma_\alpha)} + \|v_t\|^2_{L_2(\Sigma_{0\alpha})} + \|z\|^2_{L_2(0,T;H^{\frac{1}{2}+\epsilon_1}(\Omega))} + \|z\|^2_{H^{\frac{1}{2}+\epsilon_1}(0,T;L_2(\Omega))}\right\}$$

$$(3.10) \quad \geq C_{\alpha\epsilon_1}\left(\int_\alpha^{T-\alpha} E_z(t)\,dt - \big[E_z(\alpha) + E_z(T-\alpha)\big]\right).$$

Proof: By Theorem 2.4, A_F is the generator of a s.c. semigroup $e^{A_F t}$ of contractions on $Y_0 \subset Y$. Consequently, the solution of the abstract system (2.19) and hence that of the PDE system (1.1) with $k_1 = k_2 = 1$ and $u(t) \equiv 0$, is given as $[z, z_t, v, v_t] = e^{A_F t} y_0 \in C([0,T]; Y_0)$, $y_0 \in Y_0 \subset Y$. Therefore, $v_t \in C([0,T]; H^1(\Gamma_0))$ (compare with (3.7)) so that Theorem 3.3 holds true for the Kirchhoff equation part of the coupled system (1.1) where g is replaced by v_t. □

We now recall the well-known steps to obtain Theorem 3.3.

Step 1: Let h be a smooth vector field on Ω. We denote the transpose of its Jacobian as H. Also, ν is the unit vector normal to the boundary Γ. Then the following identity holds true for the wave equation (3.7):

$$(3.11) \quad \int_{\Sigma_\alpha} \frac{\partial z}{\partial \nu} h \cdot \nabla z\, d\Sigma_\alpha + \frac{1}{2}\int_{\Sigma_\alpha} z_t^2 h \cdot \nu\, d\Sigma_\alpha - \frac{1}{2}\int_{\Sigma_\alpha} |\nabla z|^2 h \cdot \nu\, d\Sigma_\alpha$$

$$= \int_{Q_\alpha} H\nabla z \cdot \nabla z\, dQ_\alpha + \frac{1}{2}\int_{Q_\alpha}\left[z_t^2 - |\nabla z|^2\right]\text{div}h\, dQ_\alpha + \left[\big(z_t, h\cdot\nabla z\big)_{L_2(\Omega)}\right]_\alpha^{T-\alpha}.$$

Step 2: With respect to the wave equation (3.7), the following equalities hold true:

$$\int_{Q_\alpha}(z_t^2 - |\nabla z|^2)\text{div}h\, dQ_\alpha = \int_{\Sigma_\alpha} z_t z\,\text{div}h\, d\Sigma_\alpha + \int_{\Sigma_{0\alpha}} gz\,\text{div}h\, d\Sigma_{0\alpha}$$

$$(3.12) \quad + \int_{Q_\alpha} z\nabla(\text{div}h)\cdot\nabla z\, dQ_\alpha + \left[\big(z_t, z\,\text{div}h\big)_{L_2(\Omega)}\right]_\alpha^{T-\alpha};$$

$$(3.13)$$

$$\int_{Q_\alpha}(z_t^2 - |\nabla z|^2)\, dQ_\alpha = \int_{\Sigma_\alpha} z_t z\, d\Sigma_\alpha + \int_{\Sigma_{0\alpha}} gz\, d\Sigma_{0\alpha} + \left[\big(z_t, z\big)_{L_2(\Omega)}\right]_\alpha^{T-\alpha}.$$

Step 3: We will need the following result which gives a bound for the tangential gradient $\frac{\partial z}{\partial \tau} = \nabla_\tau z$ (see [**L-T.9**]) :

Consider the wave equation (3.7). For arbitrarily small $\alpha > 0$, there exist a constant $C_{T\alpha\epsilon_1} > 0$ such that

$$(3.14)$$

$$\int_\alpha^{T-\alpha}\int_\Gamma \left|\frac{\partial z}{\partial \tau}\right|^2 d\Sigma_\alpha \leq C_{T\alpha\epsilon_1}\left\{\int_0^T\int_\Gamma\left(\left|\frac{\partial z}{\partial \nu}\right|^2 + z_t^2\right)d\Sigma_\alpha + \|z\|^2_{H^{1/2+\epsilon_1}(Q_\alpha)}\right\}.$$

Step 4: Let us now analyze the basic trace identity (3.11).
With reference to the wave equation (3.7), the following estimate is true:

$$(3.15) \quad \text{LHS of (3.11)} \leq C_{Th\alpha\varepsilon_1}\left\{\int_{\Sigma_\alpha} z_t^2 \, d\Sigma_\alpha + \int_{\Sigma_{0\alpha}} g^2 \, d\Sigma_{0\alpha} + \|z\|^2_{H^{1/2+\epsilon_1}(Q_\alpha)}\right\}.$$

Let us now consider the right hand side of (3.11):
With reference to the wave equation (3.7), the following estimate is true:

$$\text{RHS of (3.11)} \geq \frac{\rho-\epsilon}{2}\int_\alpha^{T-\alpha} E_z(t)\,dt - C_{h,\epsilon}\left[E_z(\alpha) + E_z(T-\alpha)\right]$$

$$(3.16) \quad -C_{h,\epsilon}\left(\int_{Q_\alpha} z^2\,dQ_\alpha + \int_{\Sigma_\alpha} z^2\,d\Sigma_\alpha + \int_{\Sigma_\alpha} z_t^2\,d\Sigma_\alpha + \int_{\Sigma_{0\alpha}} g^2\,d\Sigma_{0\alpha}\right).$$

Final Step of Proof of Theorem 3.3: Combining the inequalities of Step 4, we get that

$$C_T\left\{\|z_t\|^2_{L_2(\Sigma_\alpha)} + \|g\|^2_{L_2(\Sigma_{0\alpha})} + \|z\|^2_{L_2(\Sigma_\alpha)} + \|z\|^2_{L_2(Q)} + \|z\|^2_{H^{\frac{1}{2}+\epsilon_1}(\Omega)}\right\}$$

$$(3.17) \quad \geq C_{h\alpha\epsilon_1}\left(\int_\alpha^{T-\alpha} E_z(t)\,dt - \left[E_z(\alpha) + E_z(T-\alpha)\right]\right).$$

Next by trace theory we get that

$$(3.18) \quad \|z\|^2_{L_2(\Gamma)} \leq C\|z\|^2_{H^{\frac{1}{2}+\varepsilon}(\Omega)}.$$

Finally, (3.17) and (3.18) finish the proof of Theorem 3.3 since $\|z\|^2_{L_2(0,T;H^{\frac{1}{2}+\varepsilon}(\Omega))}$ and $\|z\|^2_{L_2(Q)}$ can be absorbed by $\|z\|^2_{H^{\frac{1}{2}+\epsilon_1}(Q)}$. $\square$

3.2. PDE Estimates for Kirchhoff Equation in (1.1).
In this section, we will consider the following uncoupled Kirchhoff equation:

$$(3.19) \quad \begin{cases} v_{tt} - \gamma\Delta v_{tt} + \Delta^2 v = f & \text{on } \Sigma_0, \\ v|_{\partial\Gamma_0} = 0 & \text{on } \partial\Sigma_0, \\ \Delta v|_{\partial\Gamma_0} = -\frac{\partial v_t}{\partial \nu} & \text{on } \partial\Sigma_0, \\ v(0,\cdot) = v_0; v_t(0,\cdot) = v_1 & \text{in } \Gamma_0; \end{cases} \quad \text{where } f\,(=z_t) \in L_2(\Sigma_0).$$

The main result of this section is the following theorem:

THEOREM 3.4. *Given arbitrary $\epsilon > 0$ and $\beta > 0$ such that $T/2 > \beta > 0$, we have, with respect to the Kirchhoff equation (3.19), the following inequality:*

$$C_{\varepsilon T}(1+\gamma)^2\left\{\|f\|^2_{L_2(\Sigma_0)} + \left\|\frac{\partial v_t}{\partial \nu}\right\|^2_{L_2(\partial\Sigma_0)} + \|v\|^2_{C([0,T];H^{2-\epsilon}(\Gamma_0))} + \|v_t\|^2_{L_2(0,T;H^{-\epsilon}(\Gamma_0))}\right\}$$

$$\geq \int_\beta^{T-\beta} E_{v\gamma}(t)\,dt - \epsilon\left[E_{v\gamma}(\beta) + E_{v\gamma}(T-\beta)\right]$$

$$(3.20) \quad -\gamma^{\frac{1}{2}}\left[E_{v\gamma}(\beta) + E_{v\gamma}(T-\beta)\right].$$

The proof of Theorem 3.4 will be given in the subsequent steps. Before we proceed, let us introduce the following domains with β as introduced above:

$$\Gamma_{0\beta} = (\beta, T-\beta) \times \Gamma_0; \quad \partial\Gamma_{0\beta} = (\beta, T-\beta) \times \partial\Gamma_0.$$

COROLLARY 3.2. *With respect to the Kirchhoff equation part of the coupled PDE's (1.1) with $k_1 = k_2 = 1$ and $u(t) \equiv 0$ in (1.1), we have the following inequality:*

$$C_{\varepsilon T}(1+\gamma)^2 \left\{ \|z_t\|^2_{L_2(\Sigma_0)} + \left\|\frac{\partial v_t}{\partial \nu}\right\|^2_{L_2(\partial \Sigma_0)} + \|v\|^2_{C([0,T];H^{2-\epsilon}(\Gamma_0))} + \|v_t\|^2_{L_2(0,T;H^{-\epsilon}(\Gamma_0))} \right\}$$

$$\geq \int_\beta^{T-\beta} E_{v\gamma}(t)dt - \epsilon [E_{v\gamma}(\beta) + E_{v\gamma}(T-\beta)]$$

(3.21)
$$-\gamma^{\frac{1}{2}}[E_{v\gamma}(\beta) + E_{v\gamma}(T-\beta)].$$

Proof: Notice that the Kirchhoff equation part of (1.1) with $u(t) \equiv 0$ is the same as (3.19) with $f = z_t|_{\Gamma_0} \in L_2(0, \infty; L_2(\Gamma_0))$, (see Lemma 3.2). Hence, the result follows from Theorem 3.4. □

The following is the usual inequality obtained by multipliers method.

LEMMA 3.5. *Let v be the solution of (3.19). The following inequality holds.*

$$\int_{\Sigma_{0\beta}} |v_t|^2 d\Sigma_{0\beta} + \gamma \int_{\Sigma_{0\beta}} |\nabla v_t|^2 d\Sigma_{0\beta} + \int_{\Sigma_{0\beta}} |\Delta v|^2 d\Sigma_{0\beta}$$

$$\leq C \Bigg\{ \Big| \int_{\partial \Sigma_{0\beta}} \Delta v \frac{\partial(h \cdot \nabla v)}{\partial \nu} d(\partial\Sigma_{0\beta}) \Big| + \Big| \int_{\partial \Sigma_{0\beta}} \frac{\partial(\Delta v)}{\partial \nu} h \cdot \nabla v \, d(\partial\Sigma_{0\beta}) \Big|$$

$$+ \int_{\partial \Sigma_{0\beta}} |\Delta v|^2 d(\partial\Sigma_{0\beta}) + (1+\gamma) \int_{\partial \Sigma_{0\beta}} \Big|\frac{\partial v}{\partial \nu}\Big|^2 d(\partial\Sigma_{0\beta})$$

$$+ \gamma \int_{\partial \Sigma_{0\beta}} \Big|\frac{\partial v_t}{\partial \nu}\Big|^2 d(\partial \Sigma_{0\beta}) + \int_{\Sigma_{0\beta}} |f|^2 d\Sigma_{0\beta}$$

$$+ \Big| \big[(v_t, h\cdot \nabla v)_{L_2(\Gamma_0)}\big]_\beta^{T-\beta} \Big| + \gamma \Big| \big[(\Delta v_t, (h\cdot \nabla v))_{L_2(\Gamma_0)}\big]_\beta^{T-\beta} \Big|$$

(3.22)
$$+ \Big|\big[(v_t, v)_{L_2(\Gamma_0)}\big]_\beta^{T-\beta}\Big| + \gamma\Big|\big[(\nabla v_t, \nabla v)_{L_2(\Gamma_0)}\big]_\beta^{T-\beta}\Big| \Bigg\},$$

where the constant C is independent of γ and T, and the two dimensional vector field $h(x) \equiv x - x_0$ for some $x_0 \in R^2$.

The main difficulty in the analysis of the inequality (3.22) arises from the second term on the RHS of (3.22). Lemma 3.6 stated below gives a technical result which estimates $\frac{\partial(\Delta v)}{\partial \nu}$ in the second term of RHS of (3.22). In addition, the proposition below (see [**L-T.8**]) is also essential in our analysis.

PROPOSITION 3.3. [**L-T.8**] *Let v be the solution of (3.19) and let $0 < \beta < T/2$. Then we have that*

$$\left\|\frac{\partial^2 v}{\partial \tau^2}\right\|^2_{L_2(\partial \Sigma_{0\beta})} + \left\|\frac{\partial^2 v}{\partial \nu^2}\right\|^2_{L_2(\partial \Sigma_{0\beta})} + \left\|\frac{\partial^2 v}{\partial \nu \partial \tau}\right\|^2_{L_2(\partial \Sigma_{0\beta})}$$

(3.23)
$$\leq C_{T\beta\epsilon}\left\{ \|f\|^2_{H^{-(\frac{1}{2}-\epsilon)}(\Sigma_0)} + \left\|\frac{\partial v_t}{\partial \nu}\right\|^2_{L_2(\partial\Sigma_0)} + \|v\|^2_{L_2(0,T;H^{3/2+\epsilon}(\Gamma_0))} \right\},$$

where $0 < \epsilon < 1/2$ and $\partial\Sigma_{0\beta} \equiv (\beta, T-\beta) \times \partial\Gamma_0$ and $\frac{\partial}{\partial \tau}$ denotes a tangential derivative for the solution of (3.19).

COROLLARY 3.4. *With respect to the Kirchhoff equation (3.19), we have the following estimate:*

$$\left|\int_{\partial\Sigma_{0\beta}} \Delta v \frac{\partial(h\cdot\nabla v)}{\partial\nu} d(\partial\Sigma_{0\beta})\right| + \int_{\partial\Sigma_{0\beta}} |\Delta v|^2 d(\partial\Sigma_{0\beta}) + (1+\gamma)\int_{\partial\Sigma_{0\beta}} \left|\frac{\partial v}{\partial\nu}\right|^2 d(\partial\Sigma_{0\beta})$$

$$(3.24) \qquad \leq C_{T\beta\epsilon}\left\{\|f\|^2_{H^{-(\frac{1}{2}-\epsilon)}(\Sigma_0)} + \left\|\frac{\partial v_t}{\partial\nu}\right\|^2_{L_2(\partial\Sigma_0)} + \|v\|^2_{L_2(0,T;\,H^{3/2+\epsilon}(\Gamma_0))}\right\}.$$

Proof: First notice by trace theory that

$$(3.25) \qquad \int_{\partial\Sigma_{0\beta}} \left|\frac{\partial v}{\partial\nu}\right|^2 d(\partial\Sigma_{0\beta}) \leq C\|v\|^2_{L_2(0,T;\,H^{3/2+\epsilon}(\Gamma_0))}.$$

Next, since by the boundary conditions in (3.19) $v\big|_{\partial\Gamma_0} = 0$ and $\frac{\partial v}{\partial\tau}\big|_{\partial\Gamma_0} = 0$ we have by direct computation

$$(3.26) \qquad \frac{\partial(h\cdot\nabla v)}{\partial\nu} = h\cdot\nu\frac{\partial^2 v}{\partial\nu^2} + h\cdot\tau\frac{\partial^2 v}{\partial\nu\partial\tau} + \frac{\partial(h\cdot\nu)}{\partial\nu}\frac{\partial v}{\partial\nu}$$

and hence

$$\int_{\partial\Sigma_{0\beta}} \left(\frac{\partial(h\cdot\nabla v)}{\partial\nu}\right)^2 d(\partial\Sigma_{0\beta}) \leq C\Big\{\int_{\partial\Sigma_{0\beta}} \left|\frac{\partial^2 v}{\partial\nu^2}\right|^2 d(\partial\Sigma_{0\beta})$$

$$(3.27) \qquad\qquad + \int_{\partial\Sigma_{0\beta}} \left|\frac{\partial^2 v}{\partial\nu\partial\tau}\right|^2 d(\partial\Sigma_{0\beta}) + \int_{\partial\Sigma_{0\beta}} \left|\frac{\partial v}{\partial\nu}\right|^2 d(\partial\Sigma_{0\beta})\Big\}.$$

Finally, by the boundary conditions $\Delta v = -\frac{\partial v_t}{\partial\nu}$ in (3.19)

$$\left|\int_{\partial\Sigma_{0\beta}} \Delta v \frac{\partial(h\cdot\nabla v)}{\partial\nu} d(\partial\Sigma_{0\beta})\right|$$

$$(3.28) \qquad \leq \frac{1}{2}\Big\{\int_{\partial\Sigma_{0\beta}} \left|\frac{\partial v_t}{\partial\nu}\right|^2 d(\partial\Sigma_{0\beta}) + \int_{\partial\Sigma_{0\beta}} \left(\frac{\partial(h\cdot\nabla v)}{\partial\nu}\right)^2 d(\partial\Sigma_{0\beta})\Big\}.$$

Now using (3.27) in (3.28) and then (3.23) and (3.25) gives

$$\left|\int_{\partial\Sigma_{0\beta}} \Delta v \frac{\partial(h\cdot\nabla v)}{\partial\nu} d(\partial\Sigma_{0\beta})\right|$$

$$(3.29) \leq C_{T\beta\epsilon}\left\{\|f\|^2_{H^{-(\frac{1}{2}-\epsilon)}(\Sigma_0)} + \left\|\frac{\partial v_t}{\partial\nu}\right\|^2_{L_2(\partial\Sigma_0)} + \|v\|^2_{L_2(0,T;\,H^{3/2+\epsilon}(\Gamma_0))}\right\}.$$

Using (3.29), (3.25), and the boundary conditions in (3.19) yields (3.24) as desired. □

LEMMA 3.6. *With the same assumptions as in Proposition 3.3,*

$$\left\|\frac{\partial}{\partial\nu}(\Delta v)\right\|^2_{H^{-1}(\partial\Sigma_{0\beta})} \leq C_T\Big\{\|f\|^2_{H^{-1}(0,T;H^{-\frac{1}{2}}(\Gamma_0))} + (1+\gamma)^2\left\|\frac{\partial v_t}{\partial\nu}\right\|^2_{L_2(\partial\Sigma_0)}$$

$$(3.30) \qquad + (1+T)^2(1+\gamma)^2\|v\|^2_{C([0,T];\,H^{2-\epsilon}(\Gamma_0))} + \|v_t\|^2_{L_2(0,T;H^{-\epsilon}(\Gamma_0))}\Big\}$$

where $H^{-1}(\partial\Sigma_{0\beta})$ is dual to $H^1(\partial\Sigma_{0\beta}) \equiv H^{1,1}((\beta, T-\beta)\times\partial\Gamma_0)$ with respect to the pivot space $L_2(\partial\Sigma_{0\beta})$, $0 < \epsilon < 1/2$, and C_T does not depend on γ.

Proof: Let θ be a smooth cutoff function defined in $[0, T]$ with support $[\beta/2, T - \beta/2]$ and $\theta \equiv 1$ in $[\beta, T - \beta]$. Note that θ is an operator of order zero and it commutes with Δ. Let

$$(3.31) \qquad\qquad \hat{v} = \theta v.$$

To prove (3.30), it suffices to show that

$$\left\|\frac{\partial}{\partial \nu}(\Delta \hat{v})\right\|^2_{H^{-1}(\partial \Sigma_{0\beta})} \leq C\Big\{ \|f\|^2_{H^{-1}(0,T;H^{-\frac{1}{2}}(\Gamma_0))} + \|\Delta v\|^2_{L_2(\partial \Sigma_0)} + \gamma^2 \left\|\frac{\partial v_t}{\partial \nu}\right\|^2_{L_2(\partial \Sigma_0)}$$

(3.32)
$$+ (1+T)^2(1+\gamma)^2 \|v\|^2_{C([0,T];H^{2-\epsilon}(\Gamma_0))} + \|v_t\|^2_{L_2(0,T;H^{-\epsilon}(\Gamma_0))} \Big\}.$$

Let

(3.33) $$P \equiv \frac{\partial^2}{\partial t^2} - \gamma \Delta \frac{\partial^2}{\partial t^2} + \Delta^2 \text{ and } F \equiv \theta f + [P, \theta] v$$

where $[\cdot, \cdot]$ denotes the commutator. By (3.31) and (3.33), $\hat{v}$ satisfies the equation

(3.34) $$\hat{v}_{tt} - \gamma \Delta \hat{v}_{tt} + \Delta^2 \hat{v} = F.$$

Next recall the operator $\bar{A}$ from (2.1) and the Green's operator G_2 from (2.14). Therefore, also taking the boundary conditions of the Kirchhoff equation (3.19), we find that $\hat{v}$ (see (3.31)) satisfies the following abstract equation (see [**L-T.2**]):

(3.35) $$(1 + \gamma \bar{A}^{1/2})\hat{v}_{tt} + \bar{A}\hat{v} - \bar{A} G_2 \Delta \hat{v}|_{\partial \Gamma_0} = F,$$

so that

(3.36) $$\bar{A}^{1/2}\hat{v} = -\bar{A}^{-1/2}(1 + \gamma \bar{A}^{1/2})\hat{v}_{tt} + \bar{A}^{1/2} G_2 \Delta \hat{v}|_{\partial \Gamma_0} + \bar{A}^{-1/2} F.$$

Note that equation (3.35) is satisfied in a weak sense and it is our task to prove the regularity result (3.32) from this equation. First since Dh satisfies the second elliptic equation we can use the sharp regularity results for elliptic equation (see [**K.1**]) to get

(3.37) $$\left\|\frac{\partial}{\partial \nu} \bar{A}^{1/2} G_2 h\right\|_{H^{-1}(\partial \Gamma_0)} = \left\|\frac{\partial}{\partial \nu} Dh\right\|_{H^{-1}(\partial \Gamma_0)} \leq C \|h\|_{L_2(\partial \Gamma_0)}$$

where in the equality we used (2.14). Hence

$$\left\|\frac{\partial}{\partial \nu}(\Delta \hat{v})\right\|_{H^{-1}(\partial \Sigma_{0\beta})} = \left\|\frac{\partial}{\partial \nu} \bar{A}^{1/2} \hat{v}\right\|_{H^{-1}(\partial \Sigma_{0\beta})}$$

$$\leq C\Big\{ \|\Delta \hat{v}\|_{L_2(\partial \Sigma_{0\beta})} + \left\|\frac{\partial}{\partial \nu} \bar{A}^{-1/2}(1+\gamma \bar{A}^{1/2})\hat{v}_{tt}\right\|_{H^{-1}(\partial \Sigma_{0\beta})}$$

(3.38)
$$+ \left\|\frac{\partial}{\partial \nu} \bar{A}^{-1/2} F\right\|_{H^{-1}(\partial \Sigma_{0\beta})} \Big\}.$$

Now we need to estimate the second and the third terms on RHS of (3.38). For the second term on RHS of (3.38), since $\bar{A}^{-1/2}(I + \gamma \bar{A}^{1/2}) = \gamma I + \bar{A}^{-1/2}$, we have that

$$\left\|\frac{\partial}{\partial \nu} \bar{A}^{-1/2}(1+\gamma \bar{A}^{1/2})\hat{v}_{tt}\right\|_{H^{-1}(\partial \Sigma_{0\beta})}$$

$$\leq C\Big\{ \gamma \left\|\frac{\partial}{\partial \nu} \hat{v}_{tt}\right\|_{H^{-1}(\partial \Sigma_{0\beta})} + \left\|\frac{\partial}{\partial \nu} \bar{A}^{-1/2} \hat{v}_{tt}\right\|_{H^{-1}(\partial \Sigma_{0\beta})} \Big\}$$

$$\leq C\Big\{ \gamma \left\|\frac{\partial}{\partial \nu} \hat{v}_t\right\|_{L_2(\partial \Sigma_0)} + \left\|\frac{\partial}{\partial \nu} \bar{A}^{-1/2} \hat{v}_t\right\|_{L_2(\partial \Sigma_0)} \Big\}$$

(3.39) $$\leq C\Big\{ \gamma \left\|\frac{\partial}{\partial \nu} \hat{v}_t\right\|_{L_2(\partial \Sigma_0)} + \|\hat{v}_t\|_{L_2(0,T;H^{-1/2}(\Gamma_0))} \Big\}$$

where for the second inequality we used the fact that $supp\, \hat{v} \subset (0, T)$ and the third inequality is due to the regularity result for elliptic equations in [**K.1**]. For the third term on RHS of (3.38), by (3.33), we obtain that

$$\left\|\frac{\partial}{\partial\nu}\bar{A}^{-1/2}F\right\|_{H^{-1}(\partial\Sigma_{0\beta})} \leq C\left\{\|\theta f\|_{H^{-1}(0,T;H^{-1/2}(\Gamma_0))} + \left\|\frac{\partial}{\partial\nu}\bar{A}^{-1/2}[P,\theta]v\right\|_{H^{-1}(\partial\Sigma_0)}\right\}$$

$$(3.40) \qquad \leq C\left\{\|f\|_{H^{-1}(0,T;H^{-\frac{1}{2}}(\Gamma_0))} + \left\|\frac{\partial}{\partial\nu}\bar{A}^{-1/2}[P,\theta]v\right\|_{H^{-1}(\partial\Sigma_0)}\right\}$$

where in the first inequality we once again used the regularity result for elliptic equations in [**K.1**]. The first term on RHS of (3.40) is good already. To estimate the second term on RHS of (3.40) we need the following fact about $\|\cdot\|_{H^{-1}(\partial\Sigma_0)}$ norm.

$$\begin{aligned}
\|h_t\|_{H^{-1}(\partial\Sigma_0)} &\leq \|h_t\|_{H^{-1}(0,T;L_2(\partial\Gamma_0))} = \sup_{\phi\in H^1(0,T;L_2(\partial\Gamma_0))}\frac{|(h_t,\phi)_{L_2(\partial\Sigma_0)}|}{\|\phi\|_{H^1(0,T;L_2(\partial\Gamma_0))}}\\
&= \sup_{\phi\in H^1(0,T;L_2(\partial\Gamma_0))}\left\{\frac{|-(h,\phi_t)_{L_2(\partial\Sigma_0)} + [(h,\phi)_{L_2(\partial\Gamma_0)}]_0^T|}{\|\phi\|_{H^1(0,T;L_2(\partial\Gamma_0))}}\right\}\\
&\leq C\{\|h\|_{L_2(\partial\Sigma_0)} + \|h(T)\|_{L_2(\partial\Gamma_0)} + \|h(0)\|_{L_2(\partial\Gamma_0)}\}.
\end{aligned}$$

The commutator $[P,\theta]v$ produces two terms v_t and $\gamma\Delta v_t$ with smooth coefficients which may depend on t. Thus we get the estimate on the second term in RHS of (3.40):

$$\begin{aligned}
\left\|\frac{\partial}{\partial\nu}\bar{A}^{-1/2}[P,\theta]v\right\|_{H^{-1}(\partial\Sigma_0)} &\leq C_T\left\{\left\|\frac{\partial}{\partial\nu}\bar{A}^{-1/2}v_t\right\|_{H^{-1}(\partial\Sigma_0)} + \gamma\left\|\frac{\partial}{\partial\nu}\bar{A}^{-1/2}\Delta v_t\right\|_{H^{-1}(\partial\Sigma_0)}\right\}\\
&\leq C_T\left\{\left\|\frac{\partial}{\partial\nu}\bar{A}^{-1/2}v\right\|_{L_2(\partial\Sigma_0)} + \left\|\frac{\partial}{\partial\nu}\bar{A}^{-1/2}v(T)\right\|_{L_2(\partial\Gamma_0)}\right.\\
&\quad + \left\|\frac{\partial}{\partial\nu}\bar{A}^{-1/2}v(0)\right\|_{L_2(\partial\Gamma_0)} + \gamma\left\|\frac{\partial}{\partial\nu}\bar{A}^{-1/2}\bar{A}^{1/2}v\right\|_{L_2(\partial\Sigma_0)}\\
&\quad + \gamma\left\|\frac{\partial}{\partial\nu}\bar{A}^{-1/2}\bar{A}^{1/2}v(T)\right\|_{L_2(\partial\Gamma_0)} + \left.\gamma\left\|\frac{\partial}{\partial\nu}\bar{A}^{-1/2}\bar{A}^{1/2}v(0)\right\|_{L_2(\partial\Gamma_0)}\right\}\\
&\leq C_T\left\{\|v\|_{L_2(0,T;H^{-1/2+\epsilon}(\Gamma_0))} + \|v\|_{C([0,T];H^{-1/2+\epsilon}(\Gamma_0))}\right.\\
&\quad + \left.\gamma\|v\|_{L_2(0,T;H^{3/2+\epsilon}(\Gamma_0))} + \gamma\|v\|_{C([0,T];H^{3/2+\epsilon}(\Gamma_0))}\right\}\\
(3.41)\qquad &\leq C_T(1+T)(1+\gamma)\|v\|_{C([0,T];H^{2-\epsilon}(\Gamma_0))}
\end{aligned}$$

where in the third inequality we used the trace theory.

Squaring inequalities (3.38), (3.39), (3.40) and (3.41), and then using back substitution yields (3.32). This concludes the proof of Lemma 3.6. $\square$

COROLLARY 3.5. *With respect to the Kirchhoff equation (3.19), we have the following estimate:*

$$\left|\int_{\partial\Sigma_0}\frac{\partial(\Delta v)}{\partial\nu}h\cdot\nabla v\,d(\partial\Sigma_0)\right| \leq C_T\left\{\|f\|^2_{H^{-1}(0,T;H^{-\frac{1}{2}}(\Gamma_0))} + (1+\gamma)^2\left\|\frac{\partial v_t}{\partial\nu}\right\|^2_{L_2(\partial\Sigma_0)}\right.$$
$$(3.42) \qquad \left. + (1+T)^2(1+\gamma)^2\|v\|^2_{C([0,T];H^{2-\epsilon}(\Gamma_0))} + \|v_t\|^2_{L_2(0,T;H^{-\epsilon}(\Gamma_0))}\right\}$$

where $0 < \epsilon < 1/2$ and C_T does not depend on γ.

Proof: First notice that

$$\left|\int_{\partial \Sigma_0} \frac{\partial(\Delta v)}{\partial \nu} h \cdot \nabla v \, d(\partial \Sigma_0)\right|$$

$$\leq \left\|\frac{\partial}{\partial \nu}(\Delta v)\right\|_{H^{-1}(\partial \Sigma_0)} \left\|h \cdot \nabla v\right\|_{H^1(\partial \Sigma_0)}$$

(3.43) $\quad \leq C\left\{\left\|\frac{\partial}{\partial \nu}(\Delta v)\right\|_{H^{-1}(\partial \Sigma_0)}^2 + \left\|h \cdot \nabla v_t\right\|_{L_2(\partial \Sigma_0)}^2 + \left\|\frac{\partial(h \cdot \nabla v)}{\partial \tau}\right\|_{L_2(\partial \Sigma_0)}^2\right\}.$

Notice from the boundary condition $v|_{\partial \Gamma_0} = 0$ in (3.19) that $h \cdot \nabla v_t|_{\partial \Gamma_0} = h \cdot \nu \frac{\partial v_t}{\partial \nu}$. The term $\frac{\partial(h \cdot \nabla v)}{\partial \tau}$ is estimated in the same fashion as we did in (3.26). Then the result follows immediately after applying Proposition 3.3 and Lemma 3.6 to (3.43). □

PROPOSITION 3.6. *Let v be the solution of (3.19). The following inequality holds.*

$$\left|\left[(v_t, h \cdot \nabla v)_{L_2(\Gamma_0)}\right]_\beta^{T-\beta}\right| + \gamma\left|\left[(\Delta v_t, (h \cdot \nabla v))_{L_2(\Gamma_0)}\right]_\beta^{T-\beta}\right| + \left|\left[(v_t, v)_{L_2(\Gamma_0)}\right]_\beta^{T-\beta}\right|$$

$$+ \gamma\left|\left[(\nabla v_t, \nabla v)_{L_2(\Gamma_0)}\right]_\beta^{T-\beta}\right|$$

$$\leq \epsilon\left(E_{v\gamma}(\beta) + E_{v\gamma}(T-\beta)\right) + C_\epsilon \|v\|_{C([0,T]; H^1(\Gamma_0))}^2$$

(3.44) $\quad + C\gamma^{\frac{1}{2}}\left(E_{v\gamma}(\beta) + E_{v\gamma}(T-\beta)\right),$

where the constants are independent of γ and T.

Proof: By using Schwarz inequality the inequality (3.44) is clear for the first and third terms on the LHS of (3.44) since they contain $\|v_t\|_{L_2(\Gamma_0)}$ and lower order terms (with respect to the energy norm, see (3.3) and Remark (3.1)). We deal with the fourth term on RHS of (3.44) as follows:

$$\gamma\left|\left[(\nabla v_t \cdot \nabla v)_{L_2(\Gamma_0)}\right]_\beta^{T-\beta}\right| \leq \gamma^{\frac{3}{2}}\left[\|\nabla v_t(\beta)\|_{L_2(\Gamma_0)}^2 + \|\nabla v_t(T-\beta)\|_{L_2(\Gamma_0)}^2\right]$$

$$+ \gamma^{\frac{1}{2}}\left[\|\nabla v(\beta)\|_{H^2(\Gamma_0)}^2 + \|\nabla v(T-\beta)\|_{H^2(\Gamma_0)}^2\right]$$

(3.45) $\quad \leq \gamma^{\frac{1}{2}}\left(E_{v\gamma}(\beta) + E_{v\gamma}(T-\beta)\right).$

As for the second term in (3.44), we have by Green's first theorem, then by Poincare's inequality, and finally by trace theory that

$$\gamma\left|\left[(\Delta v_t, (h \cdot \nabla v))_{L_2(\Gamma_0)}\right]_\beta^{T-\beta}\right| = \gamma\left|\left[\left(\frac{\partial v_t}{\partial \nu}, (h \cdot \nabla v)\right)_{L_2(\partial \Gamma_0)}\right.\right.$$

$$\left.\left. - (\nabla v_t, \nabla(h \cdot \nabla v))_{L_2(\Gamma_0)}\right]_\beta^{T-\beta}\right|$$

$$\leq C\left\{\left[\gamma\left\|\frac{\partial v_t}{\partial \nu}\right\|_{H^{-\frac{1}{2}}(\partial \Gamma_0)}\|(h \cdot \nabla v)\|_{H^{\frac{1}{2}}(\partial \Gamma_0)}\right]_\beta^{T-\beta} + \left[\gamma\|v_t\|_{H^1(\Gamma_0)}^2\|v\|_{H^2(\Gamma_0)}^2\right]_\beta^{T-\beta}\right\}$$

$$\leq C\gamma\left[\|v_t(\beta)\|_{H^1(\Gamma_0)}^2 + \|v(\beta)\|_{H^2(\Gamma_0)}^2 + \|v_t(T-\beta)\|_{H^1(\Gamma_0)}^2 \|v(T-\beta)\|_{H^2(\Gamma_0)}^2\right]$$

(3.46) $\quad \leq C\gamma^{\frac{1}{2}}\left[E_{v\gamma}(\beta) + E_{v\gamma}(T-\beta)\right].$

The inequalities (3.45) and (3.46) conclude the inequality (3.44), finishing the proof of the Proposition 3.6. □

Final Step of Proof of Theorem 3.4: First notice from (3.3) that

$$\text{LHS of (3.22)} = \int_{\beta}^{T-\beta} E_{v\gamma}(t) dt. \tag{3.47}$$

Next we apply Corollary (3.4), Corollary (3.5), and Proposition (3.6) to the RHS of (3.22) and let $\|f\|_{L_2(\Sigma_0)}$ absorb $\|f\|_{H^{-(1/2-\epsilon)}(\Sigma_0)}$ and $\|f\|_{H^{-1}(0,T;H^{-1/2}(\Gamma_0))}$, hence completing the proof of Theorem (3.4). □

3.3. Final PDE Estimate and Absorption of Lower Order Terms.

Combining Corollary 3.1 and Corollary 3.2 over the interval $(\alpha, T - \alpha)$ and using the definition of energy (3.1) in associated with the PDE system (1.1), we get the following PDE estimate:

LEMMA 3.7. *For T big enough, there exists a constant, $C_{T\gamma} > 0$ such that $\forall \gamma \geq 0$*

$$E_\gamma(T) \leq C_{T\gamma} \Big\{ \Big\|\frac{\partial v_t}{\partial \nu}\Big\|^2_{L_2(\partial\Sigma_0)} + \|z_t\|^2_{L_2(\Sigma_0)} + \|v\|^2_{C([0,T];H^{2-\epsilon}(\Gamma_0))}$$
$$+ \|v_t\|^2_{L_2(0,T;H^{-\epsilon}(\Gamma_0))} + \|z\|^2_{H^{\frac{1}{2}+\epsilon}(Q)} \Big\}. \tag{3.48}$$

Note that we keep track of dependence of the constant $C_{T\gamma}$ only on T and γ, the others being insignificant. Also for convenience, we take $\epsilon = \epsilon_1$.

Proof: First we choose constants α and β in Corollaries 3.1 and 3.2, respectively, equal to each other: $\alpha = \beta$. Now we multiply both sides of the inequality (3.10) in Corollary 3.1 by a small constant $\epsilon_0 > 0$ (to be determined soon) to get that

$$\varepsilon_0 C_T \Big\{ \|z_t\|^2_{L_2(\Sigma)} + \|z\|^2_{L_2(0,T;H^{\frac{1}{2}+\epsilon}(\Omega))} + \|z\|^2_{H^{\frac{1}{2}+\epsilon}(0,T;L_2(\Omega))} \Big\} + \varepsilon_0 C_T \|v_t\|^2_{L_2(\Sigma_{0\alpha})}$$
$$\geq \varepsilon_0 \int_\alpha^{T-\alpha} E_z(t)\, dt - \varepsilon_0 \big[E_z(\alpha) + E_z(T-\alpha)\big]. \tag{3.49}$$

After choosing $\epsilon = \epsilon_0$, we add the inequality (3.49) to (3.21) in Corollary 3.2 to get that

$$\big(\epsilon C_T + C_T(1+\gamma)^2\big) \Big\{ \Big\|\frac{\partial v_t}{\partial \nu}\Big\|^2_{L_2(\partial\Sigma_0)} + \|z_t\|^2_{L_2(\Sigma_0)} + \|v\|^2_{C([0,T];H^{2-\epsilon}(\Gamma_0))}$$
$$+ \|v_t\|^2_{L_2(0,T;H^{-\epsilon}(\Gamma_0))} + \|z\|^2_{H^{\frac{1}{2}+\epsilon}(Q)} \Big\} + \epsilon C_T \|v_t\|^2_{L_2(\Sigma_{0\alpha})}$$
$$\geq \epsilon \int_\alpha^{T-\alpha} E_z(t) + E_{v\gamma}(t)\, dt$$
$$-\epsilon\big[E_\gamma(\alpha) + E_\gamma(T-\alpha)\big] - \gamma^{\frac{1}{2}}\big[E_\gamma(\alpha) + E_\gamma(T-\alpha)\big]. \tag{3.50}$$

By letting $\gamma^{\frac{1}{2}} < \epsilon_0 = \epsilon < \frac{1}{1+C_T}$, we see from (3.50) and (3.3) that

$$\left(\epsilon C_T + C_T(1+\gamma)^2\right)\left\{\left\|\frac{\partial v_t}{\partial \nu}\right\|^2_{L_2(\partial \Sigma_0)} + \|z_t\|^2_{L_2(\Sigma_0)} + \|v\|^2_{C([0,T]; H^{2-\epsilon}(\Gamma_0))}\right.$$

$$\left.\|v_t\|^2_{L_2(0,T;H^{-\epsilon}(\Gamma_0))} + \|z\|^2_{H^{\frac{1}{2}+\epsilon}(Q)}\right\}$$

$$\geq \epsilon \int_\alpha^{T-\alpha} E_z(t)\, dt + (1-\epsilon C_T)\int_\alpha^{T-\alpha} E_{v\gamma}(t)dt - (\epsilon + \gamma^{\frac{1}{2}})[E_\gamma(\alpha) + E_\gamma(T-\alpha)]$$

(3.51)
$$\geq \epsilon \int_\alpha^{T-\alpha} E_\gamma(t)\, dt - 2\epsilon[E_\gamma(\alpha) + E_\gamma(T-\alpha)].$$

Next, it follows from (3.4) that

(3.52) $\quad \epsilon \int_\alpha^{T-\alpha} E_\gamma(t)\, dt \geq \epsilon(T-2\alpha)\left(E_\gamma(0) - 2\left\{\left\|\frac{\partial v_t}{\partial \nu}\right\|^2_{L_2(\partial \Sigma_0)} + \|z_t\|^2_{L_2(\Sigma)}\right\}\right).$

It then follows from (3.51), (3.52), and the fact that the energy is non-increasing, that there exists a positive constant, denoted as $C_{T\gamma} > 0$, such that

$$C_{T\gamma}\left\{\left\|\frac{\partial v_t}{\partial \nu}\right\|^2_{L_2(\partial \Sigma_0)} + \|z_t\|^2_{L_2(\Sigma_0)} + \|v\|^2_{C([0,T]; H^{2-\epsilon}(\Gamma_0))}\right.$$

$$\left.+\|v_t\|^2_{L_2(0,T;H^{-\epsilon}(\Gamma_0))} + \|z\|^2_{H^{\frac{1}{2}+\epsilon}(Q)}\right\}$$

$$\geq \epsilon(T-2\alpha-4)E_\gamma(0) \geq \epsilon(T-2\alpha-4)E_\gamma(T).$$

Hence Lemma 3.7 is true when $T > 2\alpha + 4$ and $\gamma < \left(\frac{1}{1+C_T}\right)^2$. $\square$

REMARK 3.8. The proof of Lemma 3.7 shows that the energy, $E_\gamma(T)$, is bounded by lower order terms (with respect to the energy norm (see (3.1) when γ is small enough. However, Lemma 3.7 is true for all $\gamma \geq 0$: when $\gamma > 0$, the term $\|v_t\|_{H^1(\Gamma_0)}$ is a lower order term and it is, therefore, easier to show that the energy inequality (3.48) of Lemma 3.7 holds true. The difficulty arises when $\gamma = 0$ since, in that case, $\|v_t\|_{H^1(\Gamma_0)}$ is no longer a lower order term. Our proof shows that the energy inequality (3.48) holds uniformly for $\gamma \leq \left(\frac{1}{1+C_T}\right)^2$.

In the last lemma, we have an estimate for the energy of the system (1.1) by the dissipative terms plus lower order terms (with respect to the norm of the energy). Our last step is to absorb these lower order terms by means of a standard compactness/uniqueness argument.

PROPOSITION 3.7. *With respect to the coupled P.D.E. system (1.1), for T big enough, there exists a constant $C_{T\gamma} > 0$ $\forall \gamma \geq 0$ such that:*

$$C_{T\gamma}\left(\int_\Sigma z_t^2\, d\Sigma + \int_{\partial \Sigma_0}\left(\frac{\partial v_t}{\partial \nu}\right)^2 d(\partial \Sigma_0)\right) \geq \|v\|^2_{C([0,T]; H^{2-\epsilon}(\Gamma_0))} + \|z\|^2_{H^{\frac{1}{2}+\epsilon}(Q)}$$

$$+\|v_t\|^2_{L_2(0,T;H^{-\epsilon}(\Gamma_0))}.$$

Proof: It follows from the well-known compactness/uniqueness argument. $\square$
Final Step of the Proof of Theorem 1.1: By Lemma 3.7 and Proposition 3.7, we see that when T is large enough, there exists a positive constant $C_{T\gamma} > 0$ $\forall \gamma \geq 0$ such that

$C_{T\gamma}E_\gamma(T) \leq \|z_t\|^2_{L_2(\Sigma)} + \left\|\frac{\partial v_t}{\partial \nu}\right\|^2_{L_2(\partial \Sigma_0)}$. Now recalling the equality (3.4), we get that

$(2C_{T\gamma}+1)E_\gamma(T) \le E_\gamma(0)$ and hence $E_\gamma(T) < E_\gamma(0)$, implying that $\|e^{A_F t}\|_{L(Y_0)} < 1$. Therefore, A_F is the infinitesimal generator of a uniformly stable semigroup on Y_0 and the proof of Theorem 1.1 is finished. □

4. Well-posedness of Coupled P.D.E's with $u \ne 0$

In the previous section, we have taken $u \equiv 0$. Now we consider the coupled PDE system (1.1) on the time interval $[0,\infty]$ instead of on $[0,T]$ with $u(t) \in L_2(0,\infty)$. Now recall from (2.12) that the solution to (1.1) with zero initial conditions can be written as:

$$(4.1) \quad [z(t), z_t(t), v(t), v_t(t)] = \int_0^t e^{A_F(t-\tau)} Bu(\tau)\, d\tau = (Lu)(t).$$

The main result of this section is the following well-posedness theorem involving the regularity of the operator $L : L_2(0,\infty) \to Y_0$.

THEOREM 4.1. *With reference to the system (1.1), the continuous relations below hold true (see (4.1)):*

$$(4.2) \quad L : L_2(0,\infty) \longrightarrow L_2(0,\infty;Y_0).$$

Equivalently, (by duality) $B^* e^{A_F^* t}$ can be extended continuously from Y to $L_2(0,\infty)$:

$$(4.3) \quad \int_0^\infty |B^* e^{A_F^* t} y_0|_\mathbb{R}^2\, dt \le C\|y_0\|_{Y_0}^2, \quad \forall y_0 \in Y_0.$$

Proof: Since the damped PDE system (1.1) is exponentially uniformly stable (see Theorem 1.1), the result follows immediately, hence completing the proof of Theorem 4.1. □

References

[A.1] G. Avalos, The exponential stability of a coupled hyperbolic/parabolic system arising in structural acoustics, *Abstract and Applied Analysis*, 1:203-219, 1996.

[A-L.1] G. Avalos and I. Lasiecka, Differential Riccati equation for the active control of a problem in structural acoustics, *J. Optimiz. Theory and Applications*, 96, 1996.

[A-L.2] G. Avalos and I. Lasiecka, Uniform decay rates of solutions to a structural acoustic model with nonlinear dissipation, *Applied Mathematics and Computations*, 1998.

[B.1] J. T. Beale, Spectral properties of an acoustic boundary condition, *Indiana Univ. Math. J.* 25, n. 9 (1976), 895-917.

[C-T.1] M. Camurdan and R. Triggiani, Sharp regularity of a coupled system of a wave and a Kirchhoff equation with point control, arising in noise reduction *Differential Integral Equations* 12 (1999), no. 1, 101–118.

[C.1] M. Camurdan, Uniform stabilization of a coupled hyperbolic system arising in structural acoustics, *Abstract and Applied Analysis* 3 (1998), no. 3-4, 377–400.

[C-T.2] S. Chen and R. Triggiani, Proof of extensions of two conjectures on structural damping for elastic systems, *Pacific J. of Mathematics*, 136(1):15-55, 1989.

[J-T.1] S. Jafford and M. Tucsnak, Regularity of plate equations with control concentrated in interior curves, R.I. No 325, Centre de Mathematiques Appliquees, Ecole Polytechnique, June 1995.

[K.1] B. Kellog, Properties of elliptic B.V.P. in "Mathematical Foundations of the Finite Element Method", *Academic Press*, 1972.

[L-L.1] W. Littman and B. Liu, On the spectral properties and stabilization of acoustic flow, *IMA, Univ. of Minnesota, IMA Preprint Series #1436*, November 1996.

[L-T.1] I. Lasiecka and R. Triggiani, Regularity theory for a class of Euler-Bernoulli equations: A cosine operator approach, *Boll. Un. Mat. Ital.*, 7:199-228, 1989.

[L-T.2] I. Lasiecka and R. Triggiani, Exact controllability and uniform stabilization of Kirchhoff plates with boundary control only on $\Delta w|_\Sigma$ and homogeneous boundary displacement, *J. Diff. Equations*, 93:62-101, 1991.

[L-T.3] I. Lasiecka and R. Triggiani, Differential and algebraic Riccati equations with application to boundary/point control problems: continuous theory and approximation theory, *Lecture Notes in Control and Information Sciences*, 164: Springer Verlag, 1991.

[L-T.4] I. Lasiecka and R. Triggiani, Regularity theory of hyperbolic equations under boundary terms, *Appl. Math. and Optimiz.*, 10:275-285, 1983.

[L-T.5] I. Lasiecka and R. Triggiani, Regularity theory of hyperbolic equations with non-homogeneous Neumann boundary conditions, Part 1:The L_2-boundary case, *Annali di Matematica Pura e Applicata (4)*, 285-367, 1990, Part 2: General boundary data, *J. Diff. Equations*, 94:112-164, 1991.

[L-T.6] I. Lasiecka and R. Triggiani, Recent advances in regularity of second-order hyperbolic mixed problems, and applications, invited paper for *Book Series, Dynamics Reported*, Springer.

[L-T.7] I. Lasiecka and R. Triggiani, Control theory for partial differential equations: continuous and approximation theories, *Encyclopedia of Mathematics and its Applications* 74, Cambridge University Press, Cambridge, 2000.

[L-T.8] I. Lasiecka and R. Triggiani, Sharp Trace Estimates of Solutions to Kirchhoff and Euler-Bernoulli Equations, *Appl. Math. and Optim.*. 28:277-306, 1993.

[L-T.9] I. Lasiecka and R. Triggiani, Uniform stabilization of the wave equation with Dirichlet or Neumann feedback control without geometrical conditions *Appl. Math. and Optim.*. 25:189-224, 1992.

[M-T.1] C. McMillan and R. Triggiani, Min-max game theory and algebraic Riccati equations for boundary control problems with continuous input-solution map, Part 2: the general case, *Appl. Math. and Optim.*. 29:1-65, 1994.

[M-I.1] P. M. Mores and K. U. Ingard, *Theoretical Acoustics*, McGraw-Hill, New York, 1968.

[O-T.1] N. Ourada and R. Triggiani, Uniform Stabilization of the Euler-Bernoulli Equation With Feedback Operator Only in the Neumann Boundary Condition, *Diff. and Int. Eqns, Vol. 4*, 277-292, 1991.

[P] A. Pazy, *Semigroups of Linear Operators and Applications to PDE's*, Springer-Verlag, New York, 1983.

[R.1] D. L. Russell, Controllability and stabilizability theory for linear partial differential equations, recent progress and open questions, *Siam Review*, 20:639-739, 1978.

[R.2] D. L. Russell, Mathematical models for elastic beam and their control-theoretic properties, *Semigroup Theory and Applications, Pitman Research Notes*, 152:177-217, 1986.

[T.1] R. Triggiani, Regularity with interior control, Part 2:Kirchhoff equations, *J. Diff. Eqns.*, 103:394-420, 1993.

[T.2] R. Triggiani, Min-max game theory for partial differential equations with boundary/point control and disturbance, *Springer-Verlag LNICS*, 197:70-89, 1994.

University of Virginia, Applied Mathematics, Charlottesville, VA 22903

Texas Tech University, Department of Mathematics and Statistics, Lubbock, TX 79409

E-mail address: gji@math.ttu.edu

The Free Boundary Problem in the Optimization of Composite Membranes

S. Chanillo, D. Grieser, K. Kurata

June 6, 2000

Abstract

In this paper, continuing our earlier article [CGIKO], we study qualitative properties of solutions of a certain eigenvalue optimization problem. Especially we focus on the study of the free boundary of our optimal solutions on general domains.

1 Introduction and Summary of results

In this note, we study some qualitative properties of solutions of a certain eigenvalue optimization problem. Our note is a continuation and also a summarization of the key results of our earlier article [CGIKO]. Our problem can be stated in physical terms as :

Problem(P) Build a body of a prescribed shape out of given materials of varying densities, in such a way that the body has a prescribed mass and with the property that the fundamental frequency of the resulting membrane (with fixed boundary) is lowest possible.

The physical problem can be re-formulated as a more general mathematical problem. More precisely, we are given $\Omega \subset \mathbf{R}^n$, a bounded domain with Lipschitz boundaries and numbers $\alpha > 0, A \in [0, |\Omega|]$ (with $|\cdot|$ denoting volume). For any measurable set $D \subset \Omega$, let χ_D be the characteristic function and $\lambda_\Omega(\alpha, D)$ the lowest eigenvalue λ of the problem,

$$\begin{aligned} -\Delta u + \alpha \chi_D u &= \lambda u \quad \text{on} \quad \Omega \\ u &= 0 \quad \text{on} \quad \partial\Omega. \end{aligned} \quad (1)$$

2000 *Mathematics Subject Classification*. Primary 35Pxx, 51–XX.

Define,
$$\Lambda_\Omega(\alpha, A) = \inf_{D \subset \Omega, |D|=A} \lambda_\Omega(\alpha, D). \tag{2}$$

Any minimizer for (2) will be called an optimal configuration for the data (Ω, α, A). If D is an optimal configuration and $u = u_{\alpha,D}$ satisfies (1), then $(u_{\alpha,D}, D)$ will be called an optimal pair (or solution). The mathematical problem then reads,

Problem (M) Study existence, uniqueness and qualitative properties of optimal pairs.

We will hereon always work under the nomalization
$$\int_\Omega u^2 = 1, \ u \geq 0. \tag{3}$$

Furthermore, changing D by sets of measure zero does not affect $\lambda_\Omega(\alpha, D)$ or u, thus sets D that differ by sets of measure zero will be said to be equal.

A basic tool that we use to analyse our problem is a variational characterization of eigenvalues, precisely,
$$\lambda_\Omega(\alpha, D) = \inf_{u \in H_0^1(\Omega)} R_\Omega(u, D), \quad R_\Omega(u, D) = \frac{\int_\Omega |\nabla u|^2 + \alpha \int_\Omega \chi_D u^2}{\int_\Omega u^2}.$$

The minimizer u is well known to exist and is an eigenfunction. Thus, for $\Lambda_\Omega(\alpha, A)$ we have
$$\Lambda_\Omega(\alpha, A) = \inf_{u \in H_0^1(\Omega), |D|=A} R_\Omega(u, D).$$

The theorem that follows is basic to the questions we hope to treat in this paper. The proof of this theorem is to be found in [CGIKO]. To state our theorem we will need to introduce some notation. First, we will consistently use the notation $\{u = t\}$ for $\{x \in \Omega; u(x) = t\}$, and $\{u \leq t\}$ for $\{x \in \Omega; u(x) \leq t\}$.

Theorem 1 ([CGIKO]) *For any $\alpha > 0$ and $A \in [0, |\Omega|]$, there exists an optimal pair. Moreover any optimal pair $(u_{\alpha,D}, D)$ has the following properties:*
(a) $u_{\alpha,D} \in C^{1,\delta}(\Omega) \cap W^{2,2}(\Omega) \cap C^\gamma(\overline{\Omega})$ for every $\delta < 1$ and some $\gamma > 0$.
(b) D is a sub-level set of $u_{\alpha,D}$, i.e. there exists $t > 0$ such that
$$D = \{x \in \Omega; u_{\alpha,D}(x) \leq t\}.$$

(c) *The value of α for which $\Lambda_\Omega(\alpha, A) = \alpha$, is unique.*
(d) *Every level set $\{u_{\alpha,D} = s\}, s \neq t$ has measure zero. If in addtion $\Lambda_\Omega(\alpha, A) \neq \alpha$, the free boundary, $\mathcal{F} = \{u_{\alpha,D} = t\}$ has measure zero.*

We use the notation $\overline{\alpha}_\Omega(A)$ for the unique value of α in part (c) above, that is,
$$\overline{\alpha}_\Omega(A) = \Lambda_\Omega(\overline{\alpha}_\Omega(A), A) \tag{4}$$

We also see right away that our problem to determine an optimal pair $(u_{\alpha,D}, D)$ which seemed linear is now a non-linear problem,
$$\begin{aligned} -\Delta u_{\alpha,D} + \alpha \chi_{\{u_{\alpha,D} \leq t\}} u_{\alpha,D} &= \Lambda_\Omega(\alpha, A) u_{\alpha,D} \quad \text{on} \quad \Omega \\ u_{\alpha,D} &= 0 \quad \text{on} \quad \partial\Omega. \end{aligned} \tag{5}$$

Another important remark is that because any optimal configuration D is a sub-level set and $u_{\alpha,D} = 0$ on $\partial\Omega$, the set D will always contain a tubular neighborhood of $\partial\Omega$, i.e. D always contains a boundary layer.

For notational convenience, from now on we will drop the subscript Ω and write $\Lambda(\alpha, A)$ for $\Lambda_\Omega(\alpha, A)$, and use the simpler notation u for $u_{\alpha,D}$, in situations where no confusion arises. $\|\cdot\|_\infty$ will denote the supremum norm in $L^\infty(\Omega)$ and $\|\cdot\|_2$ the norm in $L^2(\Omega)$.

The main focus in this paper will be on the free boundary $\{u = t\}$ on a general domain Ω.

Before we state the theorems that we prove in this paper, we continue summarizing some of the salient results of [CGIKO]. It is proved there that problem (M) generalizes problem (P). In fact for $\alpha \leq \overline{\alpha}_\Omega(A)$, the solutions of problem (M) and (P) are in one to one correspondence. Another natural questions that arises is if D inherits natural symmetries that Ω possesses, and if given Ω, does there exist a unique optimal configuration D. The answer is negative on general domains unless Ω has very strong topological restrictions. Symmetrization and rearrangement invariant integral methods allow one to prove the next theorem.

Theorem 2 ([CGIKO]) *Assume Ω is symmetric and convex with respect to the hyperplane $\{x_1 = 0\}$. That is for each fixed $x' = (x_2, \cdots, x_n)$ the set*
$$\{x_1 : (x_1, x') \in \Omega\}$$
is either empty or an interval of the form $(-c, c)$. Then any optimal solution (u, D) is symmetric with respect to the hyperplane $\{x_1 = 0\}$ and u is decreasing in x_1 for $x_1 \geq 0$.

Theorem 2 implies the next corollary, which is the only uniqueness result proved in [CGIKO].

Corollary 1 *Let $\Omega = \{|x| < 1\}$ be the ball. Then the optimal configuration is unique for any α, A and furthermore D is an annular region,*

$$D = \{x; r(A) < |x| < 1\}.$$

On other domains we encounter the phenomenon of symmetry breaking. Specifically in [CGIKO] we show symmetry breaking phenomena on annular domains in $\mathbf{R}^2$ and on dumbbell shaped domains. We have

Theorem 3 ([CGIKO]) *Fix any $\alpha > 0$ and $\delta \in (0,1)$. Let $\Omega_a = \{x \in \mathbf{R}^2; a < |x| < a+1\}$. Then there exists $a_0 = a_0(\alpha, \delta)$, such that whenever $a > a_0$ and D is an optimal configuration for Ω_a with parameters α and $A = \delta|\Omega_a|$, then D is not rotationally symmetric.*

Because Ω_a is rotationally invariant, Theorem 3 implies that there are infinitely many choices for the optimal configuration D on annuli. On dumbbell shaped domains we have, in addition to symmetry breaking, some extra information on D. We define the dumbbell shaped domain Ω_h by

$$\Omega_h = B_1((-2,0)) \cup B_1((2,0)) \cup ((-2,2) \times (-h,h)), \tag{6}$$

where $B_r(p) = \{x \in \mathbf{R}^2; |x-p| < r\}$. We call the disks $B_r(p)$, the lobes of the dumbbell and the strip $(-2,2) \times (-h,h)$ the handle. We have

Theorem 4 ([CGIKO]) *For any given $\alpha > 0$ and $A \in (0, 2\pi)$, there exists $h_0 > 0$ such that for domains Ω_h of (6) with $h < h_0$,*
(a) Any optimal pair (u, D) is not symmetric with respect to the x_2−axis.
(b) If $A > \pi$, then for any optimal pair (u, D), D^c is totally contained in one of the lobes $B_1((\pm 2, 0))$.

We end our summary of results from [CGIKO] with a theorem on convex domains.

Theorem 5 ([CGIKO]) *Suppose Ω is convex and has a smooth boundary. Then there exists $\alpha_0(\Omega, A) > 0$, such that for any $\alpha < \alpha_0$ and any optimal configuration D, one has*
(a) $\partial D \cap \Omega$ is real-analytic.
(b) D^c is convex.

Theorem 5 should be compared with the basic theorem of Brascamp-Lieb [BL] that establishes the convexity of level lines of the first eigenfunction on convex domains, when $\alpha = 0$. Theorem 5 extends the result of [BL] to some values of $\alpha > 0$, but it is completely open if Theorem 5 extends to the case of all $\alpha > 0$. The proof of Theorem 5 uses Theorems 6 and 8 below, which are proved in this paper. But Theorem 5 is not used in this paper, so there is no circular reasoning.

We now turn to the results that we prove in this paper. Our focus will primarily be on general domains and in particular on the free boundary for the optimal pair (u, D). In a general domain Ω the first eigenfunction ψ (with standard L^2 normalization) that is,

$$-\Delta \psi = \mu_1 \psi \text{ on } \Omega \tag{7}$$
$$\psi = 0 \text{ on } \partial\Omega, \quad \int_\Omega \psi^2 = 1,$$

is real-analytic in the interior and this places very strong restrictions on the exceptional sets, i.e. places on the level sets of ψ, where $\nabla \psi = 0$. If we were in $\mathbf{R}^2$, the exceptional set would consist of points. In this analysis unique continuation plays a role, since we easily see that if $w = \psi_{x_i}$ then from (7), $-\Delta w = \mu_1 w$, and thus unique continuation yields some information on the zero set of w. Thus in our problem it is clear the free boundary $\{u = t\}$ on a general domain will possess an exceptional set, since ψ in general has one, but any attempt to understand the fine structure of the exceptional set, Hausdorff measure, rectifiability etc., through a unique continuation approach is difficult. The reason being, unique continuation will not apply, since u is only weakly regular. This prevents us from obtaining an equation for $w_1 = u_{x_i}$. In addition a further difficulty is that (5) is an equation of the type $-\Delta u + V(x)u = 0$, and this prevents us from obtaining a homogeneous equation that is satisfied by $w_2 = u - t$ to which we may apply unique continuation to study the level surface $\{u = t\}, t \neq 0$.

Another approach is to view our problem (5) as a perturbation in α from the problem (7). This approach suffers from the fact that we do not get additional information for large α. One may view this again as a difficulty arising from lack of continuation properties in α. Two results in this direction are proved here.

Theorem 6 *For $s \geq 0$, let $[\Omega]^s = \{\psi \leq s\}$, where ψ is the normalized first eigenfunction of problem (7). Fix $A \in [0, |\Omega|]$ and choose t_Ω such that*

$|[\Omega]^{t_\Omega}| = A$. Then for any $\delta > 0$, there is $\alpha_0 = \alpha_0(\delta, \Omega)$ such that if $\alpha < \alpha_0$ and D is an optimal configuration for (α, A), then $|t - t_\Omega| < \delta$ and

$$[\Omega]^{t_\Omega - \delta} \subset D \subset [\Omega]^{t_\Omega + \delta}.$$

The basic lemma that is used to prove Theorem 6 can be used to analyse the limiting behavior as $A \to |\Omega|$. We have

Theorem 7 *Let Ω be a smooth bounded domain. Let $\alpha > 0$ be fixed. Let ψ be the function of (7) and let*

$$M = \max_\Omega \psi.$$

Then for any $\delta > 0$, there is $A_0 = A_0(\delta, \alpha, \Omega) < |\Omega|$, such that whenever $A > A_0$ and D is an optimal configuration for (α, A), then

$$D^c \subset \{\psi > M - \delta\}.$$

The meaning of Theorem 6 is that the free boundary for our optimization problem, that is the set $\{u = t\}$, is "trapped" between the levels $t_\Omega - \delta$ and $t_\Omega + \delta$ of the first eigenfunction ψ for the domain Ω. However, this information is too weak to conclude anything fine about the free boundary even for small $\alpha > 0$. Theorem 7 on the other hand indicates that as $A \to |\Omega|$, D^c coalesces onto the set where ψ achieves its maximum, ψ being the first eigenfunction on Ω, see (7). Now, keeping in mind part (b) of Theorem 4, it is likely that D^c may coalesce onto a strict subset of the set where ψ achieves its maximum.

Even though we have been unable to apply unique continuation to study the free boundary for large α, it is still possible to apply the Hopf lemma [GT, Lemma 3.4] and get some information on the free boundary. The most basic result on the regularity of the free boundary is the following.

Theorem 8 *If D is an optimal configuration, $x \in \partial D$ and $\nabla u(x) \neq 0$ then ∂D is a real analytic hypersurface near x.*

Using this and a refined analysis of the set of points where the gradient vanishes, we obtain:

Theorem 9 *Let $\alpha \geq \Lambda(\alpha, A)$. Let $\mathcal{F} = \{u = t\}$ denote the free boundary set. Then there is a subset $\mathcal{E}$ of $\mathcal{F}$ such that*

(a) $\mathcal{E}$ is a G_δ set.
(b) $\mathcal{F} \setminus \mathcal{E}$ is a real-analytic, $n-1$ dimensional sub-manifold of $\mathbf{R}^n$.
(c) If moreover $\alpha > \Lambda(\alpha, A)$, then for every $x_0 \in \mathcal{F}$ and every $\epsilon > 0$, the ball $B_\epsilon(x_0)$ contains points of both $\{u > t\}$ and $\{u < t\}$.

We refer to the set $\mathcal{E}$ as the exceptional set. From the construction of the exceptional set, we will deduce further geometric information regarding the free boundary. This is the content of Proposition 2, which we do not state here (see section 2).

We now give a sufficient condition that ensures that the hypothesis of Theorem 9, $\alpha \geq \Lambda(\alpha, A)$ is fulfilled.

Proposition 1 Let $\alpha > \mu_1(\Omega)$, where $\mu_1(\Omega)$ is the first Dirichlet eigenvalue for $-\Delta$ on Ω. Then there exists $A_0 = A_0(\alpha)$, such that $\alpha \geq \Lambda(\alpha, A)$ for all $A < A_0$. Furthermore, for $C_1 = ||\psi||_\infty^{-2}$ and for fixed $A \in (0, C_1)$, there exists α_0 such that $\alpha \geq \Lambda(\alpha, A)$ for all $\alpha \geq \alpha_0$.

We lastly investigate the effect of the curvature of $\partial \Omega$ on the free boundary and the "thickness" of the optimal configuration. As observed earlier in the remarks after Theorem 1, D always contains a tubular neighborhood of the boundary. The theorem that follows demonstrates in the model case of an annulus in $\mathbf{R}^2$, that at places where $\partial \Omega$ has large "negative" curvature, one finds that D is "thin". To state our result we will set up some notation. Let

$$\Omega_\epsilon = \{x \in \mathbf{R}^2; \epsilon < |x| < 1\}, \quad B = \{x \in \mathbf{R}^2; |x| < 1\}.$$

For fixed ϵ_0, let $A < \pi(1 - \epsilon_0^2)$. For $\epsilon < \epsilon_0$ and any fixed $\alpha > 0$, let (u_ϵ, D_ϵ) denote the optimal pair for Ω_ϵ, with lowest eigenvalue $\Lambda_{\Omega_\epsilon}(\alpha, A)$, with constraint $|D_\epsilon| = A$. We now claim that for $\mu_1(B)$, the first Dirichlet eigenvalue of the unit disk, we have

$$\Lambda_{\Omega_\epsilon}(\alpha, A) > \mu_1(B). \tag{8}$$

Since $\Omega_\epsilon \subset B$, using (u_ϵ, D_ϵ) as a trial pair in the variational characterization for $\Lambda_{\Omega_\epsilon}(\alpha, A)$, we have

$$\Lambda_{\Omega_\epsilon}(\alpha, A) = \int_{\Omega_\epsilon} |\nabla u_\epsilon|^2 + \alpha \int_{\Omega_\epsilon} \chi_{D_\epsilon} u_\epsilon^2 > \int_{\Omega_\epsilon} |\nabla u_\epsilon|^2 \geq \mu_1(B).$$

This establishes, (8). As a consequence of (8), imposing the hypothesis $\alpha \leq \mu_1(B)$, ensures that for every $\epsilon \geq 0$, $\alpha < \overline{\alpha}_{\Omega_\epsilon}(A)$ (see the definition (4)).

Next, if $\alpha < \overline{\alpha}_{\Omega_\epsilon}(A)$, and if D_ϵ is radially distributed, Theorem 2 of [CGIKO], yields that D_ϵ has the form,

$$D_\epsilon = \{x \in \mathbf{R}^2; \epsilon < |x| < r_\epsilon \text{ or } R_\epsilon < |x| < 1\} \tag{9}$$

for some r_ϵ, R_ϵ, $\epsilon < r_\epsilon < R_\epsilon < 1$. Thus if $\alpha \leq \mu_1(B)$ we may assume that if D_ϵ is radially distributed, then the set D_ϵ has the form described by (9) for every $\epsilon > 0$. We have

Theorem 10 *Assume $\alpha \leq \mu_1(B)$, and $A > 0$ is prescribed. Given this choice of α and A, let (u_ϵ, D_ϵ) be an optimal pair with $|D_\epsilon| = A$. Assume D_ϵ is radially distributed and hence of the form (9). Then,*

$$\limsup_{\epsilon \to 0} r_\epsilon = 0.$$

Thus the implication is that D_ϵ thins out on the boundary layer in contact with $\{x; |x| = \epsilon\}$, the inner boundary of $\partial \Omega_\epsilon$. As $\epsilon \to 0$, the curvature of $\{x; |x| = \epsilon\}$ is increasing and negative as seen from Ω_ϵ. Thus in this model case one may conclude that diam (D) is small on parts of D which are in contact with pieces of $\partial \Omega$, where the curvature of $\partial \Omega$ is large and where $\partial \Omega$ is concave as seen from Ω.

The paper [CGIKO] discusses the historical antecedents of this problem and the interested reader is referred to this paper for a discussion. Furthermore the optimization problem discussed here, is amenable to being modelled on a computer. Details of the numerical simulation are available in [CGIKO] and the interested reader may find the shape of the optimal configuration in many types of domains obtained by these numerical studies.

Finally, we thank the referee for bringing to our attention the results in the work of Alt and Caffarelli [AC] and also suggesting Problem 7 in the section on open problems.

2 Proofs of the Theorems

In this section we prove Theorems 6-10 and Proposition 1. We begin with the proof of Theorem 6. We need a preparatory Lemma, that is well-known in perturbation theory and in the Physics literature [B, Appendix 39, p. 469]. We want a slightly more precise form, though the technique of proof is standard and the basic idea follows from [B].

Lemma 1 *Fix $D \subset \Omega$. Let $u_{\alpha,D}$ be the (positive, L^2 normalized) first eigenfunction of $-\Delta + \alpha\chi_D$ with eigenvalue $\lambda(\alpha, D)$. Then there is a constant $C = C_\Omega$ such that for $0 \leq \alpha \leq 1$ (ψ, μ_1 refers to (7)),*
(a) $0 \leq \lambda(\alpha, D) - \mu_1 \leq \alpha$,
(b) $\|u_{\alpha,D} - \psi\|_{H^2(\Omega)} \leq C\alpha$,
(c) $\|u_{\alpha,D} - \psi\|_{L^\infty(\Omega)} \leq C\alpha$.

Proof of Lemma 1: Recall we have set $u_{\alpha,D} = u$. Note

$$\lambda(\alpha, D) \leq \int_\Omega (|\nabla\psi|^2 + \alpha\chi_D\psi^2) \leq \mu_1 + \alpha\int_\Omega \psi^2$$
$$\leq \mu_1 + \alpha.$$

Thus, $\lambda(\alpha, D) - \mu_1 \leq \alpha$. Next,

$$\lambda(\alpha, D) = \int_\Omega (|\nabla u|^2 + \alpha\chi_D u^2) \geq \mu_1 + \alpha\int_\Omega \chi_D u^2.$$

Thus, $\lambda(\alpha, D) - \mu_1 \geq 0$, and we have (a). To prove (b), let $\{\psi_k\}_{k=1}^\infty$ be an orthogonal basis of eigenfunction of $-\Delta$ with Dirichlet boundary conditions (Note $\psi_1 = \psi$ of problem (6)). The corresponding eigenvalues will be denoted by $\{\mu_k\}_{k=1}^\infty$, where it is well-known that μ_1 is simple. Expanding u, we have $u = \sum_{j=1}^\infty \beta_j\psi_j$, and thus $(-\Delta - \mu_1)u = \sum_{j=2}^\infty \beta_j(\mu_j - \mu_1)\psi_j$ and

$$\|(\Delta + \mu_1)u\|_2^2 = \sum_{j=2}^\infty \beta_j^2(\mu_j - \mu_1)^2, \tag{10}$$

where $\|\cdot\|_2$ denotes the $L^2(\Omega)$ norm. From $-\Delta u + \alpha\chi_D u = \lambda(\alpha, D)u$, we get

$$(-\Delta - \mu_1)u = (\lambda(\alpha, D) - \mu_1)u - \alpha\chi_D u.$$

Therefore applying (a), $\|(\Delta + \mu_1)u\|_2 \leq C\alpha$. Since μ_1 is simple, there exists $\delta > 0, \delta = \delta(\Omega)$, such that $\mu_j - \mu_1 \geq \delta > 0$ for $j \geq 2$. Then from (10) we get

$$\sum_{j=2}^\infty \beta_j^2 \leq \frac{C\alpha^2}{\delta^2}. \tag{11}$$

We re-write u as $u = \beta_1\psi_1 + \Psi$, and (11) gives $\|\Psi\|_2 \leq C\alpha\delta^{-1}$. Now $1 = \|u\|_2 = \beta_1^2 + \|\Psi\|_2^2$, thus

$$|\beta_1 - 1| \leq \frac{\|\Psi\|_2^2}{1 + \beta_1} \leq \frac{C\alpha^2}{\delta^2}.$$

Here we used the fact that $\beta_1 = (u, \psi_1) > 0$, because both u and ψ_1 are positive. Therefore,

$$\|u - \psi_1\|_2^2 = (\beta_1 - 1)^2 + \|\Psi\|_2^2 \leq \frac{C\alpha^2}{\delta^2} \leq C\alpha^2. \tag{12}$$

All the remaining consequences follow from (12). From (6),

$$-\Delta(u - \psi_1) = (\lambda(\alpha, D) - \mu_1)u + \mu_1(u - \psi_1) - \alpha\chi_D u. \tag{13}$$

We re-write (13) as

$$-\Delta(u - \psi_1) - \mu_1(u - \psi_1) = (\lambda(\alpha, D) - \mu_1)u - \alpha\chi_D u = g.$$

Now from [GT, Theorem 8.15] again, it follows that

$$\|u - \psi_1\|_\infty \leq C\|u - \psi_1\|_2 + C\alpha.$$

Using (12) on the right side,

$$\|u - \psi_1\|_\infty \leq C\alpha,$$

which is part (c). Using $\|g\|_\infty \leq C\alpha$ and part (c), we conclude $\|\Delta(u - \psi_1)\|_\infty \leq C\alpha$, which is (b). □

Theorem 6 and 7 are now consequences of Lemma 1.

Proof of Theorem 6: Apply Lemma 1 (c) to the optimal pair (u, D). Choose $\alpha_0 = \delta/(2C)$, so that $\|u - \psi_1\|_\infty \leq \delta/2$ for $\alpha \leq \alpha_0$. From Theorem 1, if $x \in D, u(x) \leq t$, and so $\psi(x) \leq t + \delta/2$ and hence $D \subset [\Omega]^{t+\delta/2}$. In a similar way we establish $[\Omega]^{t-\delta/2} \subset D$. Thus we have

$$[\Omega]^{t-\delta/2} \subset D \subset [\Omega]^{t+\delta/2}.$$

From the statement above, we get $|[\Omega]^{t-\delta/2}| \leq A \leq |[\Omega]^{t+\delta/2}|$, and thus by continuity, there exists t_Ω such that $A = |[\Omega]^{t_\Omega}|$, and $|t_\Omega - t| < \delta/2$. From this assertion the assertions of Theorem 6 follow. □

Proof of Theorem 7: We begin by showing that a slight modification of the proof of Lemma 1 yields,

$$\|u - \psi_1\|_\infty \leq C_{\alpha,\Omega}(|\Omega| - A). \tag{14}$$

We show first,
$$|\mu_1 - (\Lambda(\alpha, A) - \alpha)| \leq C_{\alpha,\Omega}(|\Omega| - A). \tag{15}$$
We re-write our equation for u as
$$-\Delta u - \alpha \chi_{D^c} u = (\Lambda - \alpha)u, \quad \Lambda = \Lambda(\alpha, A). \tag{16}$$
From (16) we have
$$\mu_1 - \alpha \int_\Omega \chi_{D^c} u^2 \leq \int_\Omega |\nabla u|^2 - \alpha \int_\Omega \chi_{D^c} u^2 = \Lambda - \alpha.$$
Thus,
$$\mu_1 - (\Lambda - \alpha) \leq \alpha \int_\Omega \chi_{D^c} u^2 \leq C|D^c| = C(|\Omega| - A).$$
Next, we have
$$\Lambda - \alpha \leq \int_\Omega (|\nabla \psi|^2 - \alpha \chi_{D^c} \psi^2) \leq \mu_1 - \alpha \int_\Omega \chi_{D^c} \psi^2$$
which yields $0 \leq \mu_1 - (\Lambda - \alpha)$. The assertion (15) follows. Using (15) and the equation (16) we can proceed as in Lemma 1 to obtain (14). If $|\Omega| - A < \delta/(2C_{\alpha,\Omega})$, from (14) we see $[\Omega]^{t-\delta/2} \subset D = \{u \leq t\} \subset [\Omega]^{t+\delta/2}$. Since $|D| = A$, we have $A \leq |[\Omega]^{t+\delta/2}| = |\{\psi \leq t + \delta/2\}|$. Thus if in addition $A > A_0$, we can arrange the situation so as to have $M - t \leq \delta/2$. So $[\Omega]^{M-\delta} \subset [\Omega]^{t-\delta/2} \subset D$. The conclusion $[\Omega]^{M-\delta} \subset D$ is readily seen to be equivalent to the assertion made in Theorem 7. □

We need some preparatory lemmas before we prove Theorem 9. As usual $u = u_{\alpha,D}$ will denote the solution to our optimization problem and consequently u will satisfy (5).

Lemma 2 *(a) Fix any $\alpha > 0$. Let the free boundary set be $\mathcal{F}$, $\mathcal{F} = \{u = t\}$. We let $D^+ = \{x; u(x) > t\}$. Assume D^+ satisfies an interior sphere condition with respect to $x_0 \in \mathcal{F}$, that is there exists a ball B, $B \subset D^+$ and $\partial B \cap \mathcal{F} = \{x_0\}$. Then $|\nabla u(x_0)| \neq 0$.*
(b) Let $D^- = \{x; u(x) < t\}$ (in the situation $\alpha \neq \overline{\alpha}_\Omega(A)$, $D^- = D$). Assume that $\alpha \geq \Lambda_\Omega(\alpha, A)$. Let D^- satisfy an interior sphere condition with respect to $x_0 \in \mathcal{F}$, that is there exists a ball B, $B \subset D^-$ and $\partial B \cap \mathcal{F} = \{x_0\}$. Then $|\nabla u(x_0)| \neq 0$.

Proof of Lemma 2: The proof of both parts of our lemma rely on Hopf's lemma [GT, Lemma 3.4]. We prove (a). Set $\phi = t - u$. We observe that in the ball $B \subset D^+$, u satisfies from (5)

$$-\Delta u = \Lambda(\alpha, A)u.$$

Thus $\Delta \phi = \Lambda(\alpha, A)u \geq 0$ in B, and $\phi < 0$ on B with $\phi(x_0) = 0$. Hopf's lemma then yields $|\nabla \phi(x_0)| = |\nabla u(x_0)| \neq 0$.

(b). The proof of this part is similar to part (a). Since $B \subset D^-$, ¿from (5) we see on B we have

$$-\Delta u + \alpha u = \Lambda u.$$

Since $\alpha \geq \Lambda$, we easily see $\Delta u \geq 0$ on B. Thus on $B \subset D^-$ we have $\Delta \phi \leq 0, \phi > 0$ on B and $\phi(x_0) = 0$. Thus Hopf's lemma again yields $|\nabla \phi(x_0)| = |\nabla u(x_0)| \neq 0$. □

Lemma 3 *Let $h(\eta, p), \eta \in \mathbf{R}, p = (p_1, \cdots, p_n) \in \mathbf{R}^n$ be a locally bounded function. Let $w \in C^1(\Omega)$ satisfy*

$$\Delta w = h(w, \nabla w). \tag{17}$$

Assume furthermore h is smooth in the variable p. Assume at the point $x_0 \in \Omega$, $\nabla w(x_0) \neq 0$. Then there exists a ball B, $x_0 \in B$, such that the set,

$$\{x \in B; w(x) = w(x_0)\} = \mathcal{S}$$

is a smooth hypersurface of $\mathbf{R}^n$. If in addition h is real-analytic in the variable p, the set $\mathcal{S}$ is also real-analytic.

Proof of Lemma 3: Since $h(w, \nabla w)$ is locally bounded, it follows by elliptic estimates that $w \in C^{1,\gamma} \cap W^{2,s}$, $s < \infty$. Thus by the implicit function theorem, since $\nabla w(x_0) \neq 0$, we conclude that $\mathcal{S}$ is a $C^{1,\gamma}$ hypersurface for all $\gamma < 1$. Now we shall improve the regularity of the hypersurface $\mathcal{S}$. By a rotation of coordinates we may assume $w_{x_i}(x_0) = 0, i = 1, \cdots, n-1$ and $w_{x_n}(x_0) \neq 0$. Let $x' = (x_1, \cdots, x_{n-1})$ and consider the map,

$$\Psi : B \to \mathbf{R}^n, \quad \Psi(x', x_n) = (x', w(x', x_n)).$$

We denote points in the image of Ψ, by $y = (y', y_n)$ where $y' = (y_1, \cdots, y_{n-1})$ and $y_n = w(x', x_n)$. Let Ψ^{-1} denote the inverse map to Ψ, which will exist if B is picked to be small. We have,

$$\Psi^{-1}(y', y_n) = (y', F(y', y_n)).$$

THE FREE BOUNDARY PROBLEM

Now,
$$F(x', w(x', x_n)) = x_n.$$

Differentiating the equation above we get the equations,
$$F_{y_i} + w_{x_i} F_{y_n} = 0, i = 1, \cdots, n-1, \quad \text{and} \quad F_{y_n} w_{x_n} = 1. \tag{18}$$

By the chain rule,
$$\frac{\partial}{\partial x_i} = \frac{\partial}{\partial y_i} + w_{x_i} \frac{\partial}{\partial y_n}, \quad \frac{\partial}{\partial x_n} = w_{x_n} \frac{\partial}{\partial y_n}.$$

From (18), $w_{x_i} = -F_{y_i}/F_{y_n}, i = 1, \cdots, n-1$ and $w_{x_n} = 1/F_{y_n}$. Thus,

$$\begin{aligned}
\Delta w &= \sum_{i=1}^{n-1} \left(\frac{\partial}{\partial y_i} + w_{x_i} \frac{\partial}{\partial y_n} \right) \left(\frac{-F_{y_i}}{F_{y_n}} \right) + w_{x_n} \frac{\partial}{\partial y_n} \left(\frac{1}{F_{y_n}} \right) \\
&= \sum_{i=1}^{n-1} \left(\frac{\partial}{\partial y_i} - \frac{F_{y_i}}{F_{y_n}} \frac{\partial}{\partial y_n} \right) \left(\frac{-F_{y_i}}{F_{y_n}} \right) + \frac{1}{F_{y_n}} \frac{\partial}{\partial y_n} \left(\frac{1}{F_{y_n}} \right) \\
&= LF.
\end{aligned}$$

Next we freeze the coefficients of L at x_0. Since $w_{x_i}(x_0) = 0, i = 1, \cdots, n-1$ and $w_{x_n}(x_0) = a \neq 0$, we see from (18), $F_{y_i} = 0$ and $F_{y_n} = a^{-1}$ at $\Psi(x_0)$. At $y_0 = \Psi(x_0)$, we have

$$\begin{aligned}
LF &= -\frac{1}{F_{y_n}} \sum_{i=1}^{n-1} F_{y_i y_i} - \frac{1}{F_{y_n}^3} F_{y_n y_n} \\
&= -\frac{1}{F_{y_n}} \left[\sum_{i=1}^{n-1} F_{y_i y_i} + \frac{1}{F_{y_n}^2} F_{y_n y_n} \right].
\end{aligned}$$

Thus if B is picked suitably small, LF is an elliptic, quasi-linear operator. Since w satisfies (17), F satisfies

$$LF = h\left(y_n, \mathcal{A}(\frac{-F_{y_i}}{F_{y_n}}, \frac{1}{F_{y_n}}) \right) \tag{19}$$

where $\mathcal{A}$ is a fixed matrix in $O(n)$, the rotation group, associated with our rotation of coordinates, and $\mathcal{A}(z_i, z_n)$, denotes the product of $\mathcal{A}$ and the vector $(z_1, \cdots, z_n), (z_i) = (z_1, \cdots, z_{n-1})$. Since $w \in C^{1,\gamma} \cap W^{2,s}$ for all $\gamma < 1, s < \infty$, we see the coefficients of L are in $C^{0,\gamma} \cap W^{1,s}$, $s < \infty$. Differentiating equation (19) in any of the variables $y_i, i \neq n$, we may apply a standard bootstrap argument using well-known elliptic estimates for example [GT, Theorem 9.11] to conclude from the fact that h is smooth in the variable p, that F is smooth in any of the variables $y_i, i = 1, \cdots, n-1$. If one has in addition that h is real-analytic in the variables p, then applying the results of Morrey [M, Theorem C] or Friedman [Fr, Theorems 1, 4], we can conclude that F is real-analytic in the variables $y_i, i = 1, \cdots, n-1$. Now the defining equation for $\mathcal{S}$ is given by $x_n = F(x_1, \cdots, x_{n-1}, w(x_0))$. Since F is smooth (real-analytic) depending on the regularity of h in the p variables, it follows that $\mathcal{S}$ is a smooth (real-analytic) manifold, depending on the fact that h is smooth (real-analytic) in the p variables. □

We are now ready to prove Theorems 8 and 9.

Proof of Theorem 8: The PDE satisfied by u, that is (5), can be written as
$$\Delta u = h(u),$$
where $h(\eta) = -\Lambda(\alpha, A)\eta + \alpha \chi_G(\eta)\eta$, $G = \{\eta; \eta \leq t\}$. Thus $h \in L^\infty_{loc}(\mathbf{R})$ and the hypotheses of Lemma 3 apply to u. The theorem follows. □

Proof of Theorem 9: We construct the exceptional set $\mathcal{E}$. Let
$$K_n = \{x \in \Omega; \text{ distance } (x, \mathcal{F}) = \frac{1}{n}\}.$$

Now define
$$F_n = \{x \in \mathcal{F}; \text{ distance } (x, K_n) = \frac{1}{n}\}.$$

The sets K_n and F_n are closed sets for all $n \in \mathbf{N}$. We define the exceptional set by
$$\mathcal{E} = \mathcal{F} \setminus (\cup_{n=1}^\infty F_n).$$

Since each set $\mathcal{F} \cap F_n^c$ is open in $\mathcal{F}$, $\mathcal{E}$ is a G_δ set. This proves Theorem 9 (a).

We shall now prove that for each point $x_0 \in \mathcal{F} \setminus \mathcal{E}$, we can construct either an interior ball $B \subset D^+$, such that $\partial B \cap \mathcal{F} = \{x_0\}$ or an interior ball $B \subset D^-$, such that $\partial B \cap \mathcal{F} = \{x_0\}$. Since we are assuming $\alpha \geq \Lambda(\alpha, A)$, Lemma 2 ensures that for each $x_0 \in \mathcal{F} \setminus \mathcal{E}$, $\nabla u(x_0) \neq 0$. By Theorem 8, there exists a ball B_0, centered at x_0, such that, $\mathcal{F} \cap B_0$ is a real analytic manifold. So we now verify our claims regarding the interior spheres. Fix a point $x_0 \in \mathcal{F} \setminus \mathcal{E}$. Then $x_0 \in F_n$ for some n. Let $z_0 \in K_n$, such that $|x_0 - z_0| = 1/n$. We claim the ball $B_{1/n}(z_0)$ is totally contained in D^+ or D^-. Suppose there are points $z_1 \in D^+$ and $z_2 \in D^-$ in $\overline{B_{1/n}(z_0)}$. Then by the continuity of u, the line segment joining z_1 to z_2 which also lies inside $B_{1/n}(z_0)$ will contain a point of $\mathcal{F}$. Thus distance $(z_0, \mathcal{F}) < 1/n$ and hence $z_0 \notin K_n$. Now pick a ball B', $B' \subset B_{1/n}(z_0)$, and B' centered along the radius joining x_0 to z_0 and with $\partial B' \cap \partial B_{1/n}(z_0) = \{x_0\}$. Clearly $\overline{B' \setminus \{x_0\}}$ is contained either in D^+ or D^- and $\overline{\partial B'} \cap \mathcal{F} = \{x_0\}$. The hypotheses of Lemma 2 are fulfilled. Theorem 9 (b) now follows.

To prove Theorem 9 (c), we can argue via the strong maximum principle. Since $\alpha > \Lambda(\alpha, A)$, applying Theorem 1 (d), we see that $D = \{u < t\}$. If for some $\epsilon > 0$, the ball $B_\epsilon(x_0)$, $x_0 \in \mathcal{F}$ contains no points of D^- and only points $\{u \geq t\}$, then on $B_\epsilon(x_0)$, u satisfies

$$-\Delta u = \Lambda(\alpha, A) u \geq 0.$$

So $\Delta u \leq 0$ on $B_\epsilon(x_0)$ and $u \geq t$ on $B_\epsilon(x_0)$ with $u(x_0) = t$ which is a contradiction. We may argue as in Lemma 2, part (b) and show on $B_\epsilon(x_0)$ there are also points of D^+ for every $\epsilon > 0$. □

To discuss further geometric properties of $\mathcal{E}$, we introduce,

$$K_\epsilon = \{x \in \Omega;\ \text{distance}\ (x, \mathcal{F}) = \epsilon\},$$

$$F_\epsilon = \{x \in \mathcal{F};\ \text{distance}\ (x, K_\epsilon) = \epsilon\}.$$

We have

Proposition 2 *For every $\alpha > 0$,*
(a) $F_{\epsilon_1} \subset F_{\epsilon_2}$ for $\epsilon_1 > \epsilon_2$.
   As a consequence of (a), we have
(b) $\cup_{n \geq 1} F_n = \cup_{\epsilon > 0} F_\epsilon$, $F_n \subset F_m$ for $m > n$, and $\mathcal{E} = \mathcal{F} \setminus \cup_{\epsilon > 0} F_\epsilon$.
(c) If $\alpha > \Lambda(\alpha, A)$, then for every point $x_0 \in \mathcal{E}$, any ball $B_\epsilon(x_0)$ contains points $y_+ \in D^+$ and $y_- \in D^-$ such that distance $(y_\pm, \mathcal{F}) < |y_\pm - x_0|$.
(d) Furthermore, if $z_0 \in \mathcal{F}$, such that $|y_+ - z_0| = $ distance $(y_+, \mathcal{F})$, then D^+ satisfies an interior sphere condition in the sense of Lemma 2 with respect to z_0. Thus by the proof of Theorem 9, there is a neighborhood B of z_0, such that $\{x \in B : u(x) = u(z_0)\}$ is a real-analytic hypersurface. A similar statement holds for y_-.

The meaning of (b) is that since $\mathcal{E} = \mathcal{F} \cap (\cap_{n=1}^{\infty} F_n^c)$ and $F_n^c \supset F_m^c$ for $m > n$, the exceptional set is really a G_δ set formed by the intersection of the nested open sets F_n^c. The meaning of (c), (d) is that the behaviour of the free boundary in the neighborhood of the exceptional set at least in $\mathbf{R}^2$ is that of isolated singularities with a conical structure, with the cone having its vertex at $x_0 \in \mathcal{E}$. The cone locally divides $\mathbf{R}^2$ into at least two components, one component is contained in D^+ and the other is contained in D^-. This geometric picture is only heuristic since it still needs to be rigorously established that the component of $\mathcal{F}$ that contains z_0 also contains $x_0 \in \mathcal{E}$. Only then can we conclude there is a true conical singularity at x_0.

Proof of Proposition 2: Fix $x \in F_{\epsilon_1}$. Then by definition, one can find $z \in K_{\epsilon_1}$, such that $|z - x| = \epsilon_1$. Since $z \in K_{\epsilon_1}$, the ball $B_{\epsilon_1}(z)$ will contain no points of $\mathcal{F}$ in the interior. Now consider the radial line connecting z and x, and locate on this line a point y such that $|y - x| = \epsilon_2$. The ball $B_{\epsilon_2}(y) \subset B_{\epsilon_1}(z)$, and x is the sole point in $\mathcal{F}$ on $\partial B_{\epsilon_2}(y)$. Now by definition $y \in K_{\epsilon_2}$, and since $x \in \mathcal{F}$ and $|x - y| = \epsilon_2$, $x \in F_{\epsilon_2}$. This proves (a), and (b) is then an elementary consequence of set theory.

Next, by Theorem 9 (c) we know that the ball $B_\epsilon(x_0)$ with $x_0 \in \mathcal{E}$ (in general for any point in $\mathcal{F}$ actually), contains points $y_\pm$ in $D^\pm$ respectively. Let $|y_+ - x_0| = \delta$. We claim distance $(y_+, \mathcal{F}) < \delta$. If distance $(y_+, \mathcal{F}) \geq \delta$, then $y_+ \in K_\tau$, $\tau = $ distance $(y_+, \mathcal{F})$, $\tau \geq \delta$. Since $|y_+ - x_0| = \delta \leq \tau$, it follows that $x_o \in F_\tau$. Thus $x_0 \notin \mathcal{E}$. This proves part (c), since an argument similar to the one above takes care of the point $y_- \in D^-$.

Lastly we prove part (d). Now, $|y_+ - z_0| = \tau$, $z_0 \in \mathcal{F}$. Thus by the argument employed in Theorem 9(b) it is easily seen, that the open ball $B_\tau(y_+)$ contains only points of D^+. Again employing the argument of Theorem 9 (b), we can find a ball $B' \subset B_\tau(y_+)$ such that $\partial B' \cap \mathcal{F} = \{z_0\}$ and hence B' is the desired interior ball. □

We now prove Proposition 1.

Proof of Proposition 1: Arguing as in Lemma 1(a), but now using the fact that $\|\psi\|_\infty \leq C$, [GT, Theorem 8.15], we have

$$\Lambda(\alpha, A) \leq \int_\Omega |\nabla \psi|^2 + \alpha \int_\Omega \chi_D \psi^2 \leq \mu_1 + \alpha \|\psi\|_\infty^2 A.$$

Thus,

$$\Lambda(\alpha, A) \leq \mu_1 + \alpha C_0 A, \quad C_0 = \|\psi\|_\infty^2. \tag{20}$$

Since $\alpha > \mu_1$, we can find $A_0 > 0$ such that $\alpha C_0 A_0 \leq \alpha - \mu_1$. It follows from (20) that for $A < A_0$, $\alpha \geq \Lambda(\alpha, A)$. The second part of Proposition 1 also follows from (20). Select $C_1 = C_0^{-1}$. If $A < C_1$, $C_0 A = 1 - \epsilon, \epsilon > 0$. Thus for $\alpha > \alpha_0$, $\mu_1 \leq \epsilon \alpha$, and hence by (20), $\Lambda(\alpha, A) < \alpha$. $\square$

Proof of Theorem 10: We set $\Lambda_\epsilon = \Lambda_{\Omega_\epsilon}(\alpha, A)$, $\Lambda = \Lambda_B(\alpha, A)$. D will denote the optimal configuration for B, that is $D = \{x; r_0 < |x| < 1\}, \pi(1 - r_0^2) = A$. We also need to consider the first Dirichlet eigenvalue λ_ϵ, of the problem,

$$-\Delta g + \alpha \chi_D g = \lambda_\epsilon g \quad \text{in} \quad \Omega_\epsilon$$
$$g|_{\partial \Omega_\epsilon} = 0.$$

We claim

$$0 \leq \Lambda_\epsilon - \Lambda \leq C |\log \epsilon|^{-1}. \tag{21}$$

Extending u_ϵ to $\{|x| < \epsilon\}$ by setting $u_\epsilon = 0$ on $\{|x| < \epsilon\}$ and using this extended function as a trial function with D_ϵ as a trial configuration on $B = \{|x| < 1\}$, we see $\Lambda \leq \Lambda_\epsilon$ and so we have the left side in (21). Next by Theorem 2 in Swanson [S],

$$0 \leq \lambda_\epsilon - \Lambda \leq c |\log \epsilon|^{-1}. \tag{22}$$

In fact $0 \leq \lambda_\epsilon - \Lambda$ is simply a consequence of domain monotonicity. By the variational characterization $\Lambda_\epsilon \leq \lambda_\epsilon$, and thus from (22) we easily have (21).

We now establish $\limsup_{\epsilon \to 0} r_\epsilon = 0$, by contradiction. Assume there is a sequence $\epsilon_j \searrow 0$, $\lim_{j \to \infty} r_{\epsilon_j} = \delta > 0$. Then $R_{\epsilon_j} \to b_\delta$ and since $|D_{\epsilon_j}| = A$, the limit set $\mathcal{D}_\delta$ is:

$$\mathcal{D}_\delta = \{x; |x| < \delta \text{ or } b_\delta < |x| < 1\}, \quad |\mathcal{D}_\delta| = A.$$

We set,
$$D_{\delta,\epsilon} = \{x; \epsilon < |x| < \delta \text{ or } b_\delta < |x| < 1\}.$$

We use the notation $\lambda_\epsilon(D_{\delta,\epsilon})$ for the first Dirichlet eigenvalue on Ω_ϵ for the problem,

$$-\Delta w + \alpha \chi_{D_{\delta,\epsilon}} w = \lambda_\epsilon(D_{\delta,\epsilon}) w \quad \text{on} \quad \Omega_\epsilon \tag{23}$$
$$w = 0 \quad \text{on} \quad \partial \Omega_\epsilon, \quad \int_{\Omega_\epsilon} w^2 = 1.$$

$\lambda(\mathcal{D}_\delta)$ will denote the first Dirichlet eigenvalue for $-\Delta + \alpha\chi_{\mathcal{D}_\delta}$ on B. We claim,
$$|\Lambda_{\epsilon_j} - \lambda(\mathcal{D}_\delta)| \to 0 \quad \text{as} \quad \epsilon_j \to 0. \tag{24}$$

We have
$$\begin{aligned} |\Lambda_{\epsilon_j} - \lambda(\mathcal{D}_\delta)| &\leq |\Lambda_{\epsilon_j} - \lambda_{\epsilon_j}(D_{\delta,\epsilon_j})| + |\lambda_{\epsilon_j}(D_{\delta,\epsilon_j}) - \lambda(\mathcal{D}_\delta)| \\ &= J_1 + J_2. \end{aligned}$$

By using Theorem 2 in [S], one can conclude $J_2 \leq C|\log \epsilon_j|^{-1} \to 0$ as $\epsilon_j \to 0$. We now show $J_1 \to 0$. Let (u_ϵ, D_ϵ) be the optimal pair for Ω_ϵ. Then, using the eigenfunction w from (23), and the uniform bounds $\|w\|_{L^\infty(\Omega_\epsilon)} \leq C$, C independent of $\epsilon > 0$, which follows from [GT, Theorem 8.15], we have

$$\begin{aligned} \Lambda_\epsilon &\leq \int_{\Omega_\epsilon}(|\nabla w|^2 + \alpha\chi_{D_\epsilon}w^2) \\ &= \int_{\Omega_\epsilon}(|\nabla w|^2 + \alpha\chi_{D_{\delta,\epsilon}}w^2) + \alpha\int_{\Omega_\epsilon}(\chi_{D_\epsilon} - \chi_{D_{\delta,\epsilon}})w^2 \\ &\leq \lambda_\epsilon(D_{\delta,\epsilon}) + C\alpha|D_\epsilon \triangle D_{\delta,\epsilon}|, \end{aligned}$$

where $D_\epsilon \triangle D_{\delta,\epsilon}$ is the symmetric difference of the sets $D_\epsilon, D_{\delta,\epsilon}$. Thus for the sequence ϵ_j, we easily have $|D_\epsilon \triangle D_{\delta,\epsilon}| \to 0$ as $j \to \infty$. We conclude

$$\Lambda_{\epsilon_j} \leq \lambda_{\epsilon_j}(D_{\delta,\epsilon_j}) + o(1), \quad j \to \infty.$$

Likewise using the function u_ϵ in the argument above, we have

$$\begin{aligned} \lambda_{\epsilon_j}(D_{\delta,\epsilon_j}) &\leq \Lambda_{\epsilon_j} + C\alpha|D_{\epsilon_j} \triangle D_{\delta,\epsilon_j}| \\ &\leq \Lambda_{\epsilon_j} + o(1), \quad j \to \infty. \end{aligned}$$

Thus,
$$|\lambda_{\epsilon_j}(D_{\delta,\epsilon_j}) - \Lambda_{\epsilon_j}| \to 0 \quad \text{as} \quad j \to \infty.$$

Thus $J_1 \to 0$ as $\epsilon_j \to 0$. This establishes (24). We infer from (21) and (24) that as $j \to \infty$,

$$|\Lambda - \lambda(\mathcal{D}_\delta)| \leq |\Lambda - \Lambda_{\epsilon_j}| + |\Lambda_{\epsilon_j} - \lambda(\mathcal{D}_\delta)| \to 0.$$

Hence $\Lambda = \lambda(\mathcal{D}_\delta)$. However, this contradicts Corollary 1 of our introduction, a proof of which is supplied in [CGIKO]. $\square$

3 Open Problems and Conjectures

A number of open problems and conjectures can be stated based on the numerical data in [CGIKO] and the rigorous results there and also on results proved here. We will outline some.

Problem 1 : *(Uniqueness of the optimal configuration) The only domain for which we have established the uniqueness of the optimal configuration is the ball, see Corollary 1. Is D unique if Ω is convex?*

Problem 2 :*(Continuation Problem) Theorem 5 states that on a convex domain Ω, D^c is convex for small $\alpha > 0$. Is it possible to continue along the values $\alpha > 0$, to obtain convexity of D^c for all $\alpha > 0$?*

Problem 3 :*(The free boundary on general domains) The free boundary $\mathcal{F}$ on general domains will contain an exceptional set $\mathcal{E}$ as constructed in the proof of Theorem 8. What is the Hausdorff dimension of $\mathcal{E}$? Is at least $\mathcal{F}$ a rectifiable set? One suspects $\mathcal{E}$ consists of points, where real-analytic arcs intersect if $\Omega \subset \mathbf{R}^2$.*

Problem 4 :*(Monotonicity of D) Suppose $A < A'$, then does this imply $D_{\alpha,A} \subset D_{\alpha,A'}$? If symmetry breaking occurs this statement needs to be modified. Nevertheless on domains where the optimal configuration is unique, does the above monotonicity hold?*

Problem 5 :*(Symmetry breaking on annuli) When Ω is an annulus, what is the shape of D precisely? The proof of Theorem 3 in our introduction as presented in [CGIKO] and the numerical computations presented in [CGIKO] suggest that D^c lies between two rays $\theta = 0$ and $\theta = \beta$. In fact the results in [CGIKO] suggest that, $\beta = \pi/N, N = N(\alpha, |D|/|\Omega|)$, and $N \to \infty$ as $|D| \to |\Omega|$.*

Problem 6 :*(Influence of the boundary curvature) In Theorem 10 we saw in a model case that the diameter of D is affected by the curvature of $\partial\Omega$. Investigate this phenomena on general domains Ω.*

Problem 7 :*(The reduced Free Boundary) Federer has introduced the notion of the reduced boundary of a set, [F], 4.5.5. Alt and Caffarelli in [AC] show* *that for their minimization problem, the free boundary of the minimizer is regular on the reduced boundary of the set, where the minimizer is positive. In the same token is it true in our case, that the set $\mathcal{F} \setminus \mathcal{E}$ coincides with the reduced boundary of the sub-level set:*

$$D = \{x \in \Omega; u_{\alpha,D} \leq t\}.$$

The difficulty in establishing this equivalence in our situation, seems to be that unlike [AC], here $t > 0$, and this is again related in another way with failure of unique continuation.

We refer the interested reader to [CGIKO] for further problems and conjectures.

REFERENCES

[AC] H. Alt, L. Caffarelli, Existence and regularity for a minimum problem with free boundary, J. Reine Angew. Math., 325(1981), 105-144.

[B] M. Born, Atomic Physics, Dover Publications, 1989.

[BL] H. J. Brascamp, E. Lieb, On extensions of the Brunn-Minkowski and Prekopa-Leindler theorems, including inequalities for log concave functions, and with an application to the diffusion equation, J. Funct. Analysis, 22(1976), 366-389.

[CGIKO] S. Chanillo, D. Grieser, M. Imai, K. Kurata and I. Ohnishi, Symmetry breaking and other phenomena in the Optimization of Eigenvalues of Composite Membranes, to appear in Comm. Math. Phys.

[F] H. Federer, Geometric Measure Theory, Springer-Verlag, Berlin-Heidelberg-New York, 1969.

[Fr] A. Friedman, On the regularity of the solutions of non-linear elliptic and parabolic systems of partial differential equations, Journ. Math. Mech., 7(1958), 43–60.

[GT] D. Gilbarg and N. Trudinger, Elliptic Partial Differential Equations of Second Order, Springer-Verlag, 1983.

[M] C.B. Morrey, On the analyticity of solutions of analytic non-linear elliptic systems of PDE I, Amer. J. of Math. 80(1958), 198–218.

[S] C.A. Swanson, Asymptotic variational formulae for eigenvalues, Canad. Math. Bull., Vol.6, No.1(1963), 15–25.

ADDRESS:
Sagun Chanillo
Department of Mathematics
Rutgers University
New Brunswick, NJ 08903, USA
chanillo@math.rutgers.edu

Daniel Grieser
Department of Mathematics
Humboldt-Universität Berlin
Unter den Linden 6
10099 Berlin, Germany
grieser@mathematik.hu-berlin.de

Kazuhiro Kurata
Department of Mathematics
Tokyo Metropolitan University
Minami-Ohsawa 1-1, Hachioji-shi, Tokyo, Japan
e-mail: kurata@comp.metro-u.ac.jp

Tangential differential calculus and functional analysis on a $C^{1,1}$ submanifold

Michel C. Delfour

ABSTRACT. A completely intrinsic differential calculus on smooth submanifolds of codimension one has been developed in [8] by a marriage of the notion of tangential derivative and the oriented distance function. The function defined on the submanifold is extended to an Euclidean neighborhood by composition with the projection onto the submanifold which can be expressed in terms of the oriented distance function. In this paper this extension is used to introduce an intrinsic characterization of Sobolev spaces on bounded open domains in a $C^{1,1}$ submanifold which coincides with the definition through local $C^{1,1}$ diffeomorphisms. This approach leads to significant simplifications in the associated calculus and functional analysis. As one of the illustrations, we give what is, to our best knowledge, the first completely intrinsic proof of Korn inequality in the $C^{1,1}$ case which is a central part of the theory of thin and asymptotic shells.

1. Introduction

This paper presents a completely intrinsic approach to Sobolev spaces on bounded open domains in $C^{1,1}$ submanifolds of codimension one of the Euclidean space. This results in significant simplifications in the associated functional analysis and brings insight and simplifications in the theory of vectorial partial differential equations on such manifolds (thin and asymptotic shells, thin fluid shells, modeling of oceans, etc) where local coordinates and Christoffel symbols can be completely avoided. As one illustration of the techniques developed in that context, we give what is, to our best knowledge, the first completely intrinsic proof of Korn inequality on a bounded open Lipschitzian subset of a $C^{1,1}$ submanifold which is central to the theory of thin and asymptotic shells. This general approach has been successfully tested in the theory of thin and asymptotic shells ([6, 7, 8, 1]) with an important potential for developing a completely intrinsic framework for their optimal design and control ([10, 2, 3, 4]). The framework is also basic to new 3-D approximations

---

1991 *Mathematics Subject Classification.* Primary: 73K15, 73O62; Secondary: 73K10, 49Q.

This research has been supported by National Sciences and Engineering Research Council of Canada research grant A–8730 and by a FCAR grant from the Ministère de l'Éducation du Québec. The author is pleased to acknowledge discussions with Jean-Paul Zolésio (CNRS) of the Centre de Mathématiques Appliquées, École des Mines, Sophia-Antipolis, France.

of differential equations for thin shells which preserve the asymptotic 2-D behavior of the solution.

The starting point was the work in [5] and the subsequent development in [8] of an intrinsic differential calculus on smooth submanifolds of codimension one by a marriage of the notion of *tangential derivative* and the expression of the projection onto the submanifold in term of the *oriented distance function*. The tangential gradient of a function is obtained via the gradient of a *special extension* to an ambient Euclidean neighborhood of the submanifold: the composition of the function with the *projection* onto the submanifold.

In § 2 and 3 we recall the main notation and results from [8] for sufficiently smooth functions and submanifolds. The main section § 4 contains new results for functions defined on a bounded (relatively) open domain ω of a $C^{1,1}$ submanifold of $\mathbf{R}^N$ of codimension one: the density Theorem 4.2 in $L^p(\omega)$, $1 \leq p < \infty$, the technical Theorems 4.4 and 4.5 which fully characterize the basic isomorphism between the L^p functions on ω and on its Euclidean neighborhood, the new Definition 4.1 of the Sobolev space $W^{1,p}(\omega)$, Theorem 4.6 which makes sense of a tangential gradient for $C^{0,1}(\omega)$ functions, and Theorem 4.7 for $W^{1,p}(\omega)$ functions together with a density result which is the analogue of the one for $L^p(\omega)$. This is completed with Theorem 4.8 which gives the equivalence of our intrinsic definition of $W^{1,p}(\omega)$ with the one via local $C^{1,1}$ diffeomorphisms. Further results on intermediary Sobolev spaces, smoothness of the relative boundary, and continuous extensions are also included. The last § 5 is devoted to the new intrinsic proof of Korn's inequality for a bounded open Lipschitzian domain in a $C^{1,1}$ submanifold of codimension one.

2. Oriented distance function and geometry

2.1. The oriented distance function and smoothness of the boundary.
In this paper the underlying submanifold of $\mathbf{R}^N$ is assumed to be the boundary $\Gamma \stackrel{\text{def}}{=} \partial \Omega$ of an open subset Ω of class $C^{1,1}$ in $\mathbf{R}^N$, $\Gamma \neq \varnothing$. The boundary and its properties can be completely described via the *oriented distance function* of Ω,

(2.1) $$b_\Omega(x) \stackrel{\text{def}}{=} d_\Omega(x) - d_{\complement\Omega}(x), \quad \forall x \in \mathbf{R}^N.$$

(2.2) $$\complement\Omega = \{y \in \mathbf{R}^N : y \notin \Omega\}, \quad d_A(x) \stackrel{\text{def}}{=} \inf_{y \in A} |y - x|, \quad \Rightarrow \Gamma = b_\Omega^{-1}\{0\}.$$

THEOREM 2.1. *Let Ω be a subset of $\mathbf{R}^N$ with boundary $\Gamma \stackrel{\text{def}}{=} \partial\Omega \neq \varnothing$.*

(i) *Ω is of class $C^{1,1}$ if and only if for each $x \in \Gamma$, there exists a neighborhood $V(x)$ such that $b_\Omega \in C^{1,1}(V(x))$.*

(ii) *Let $k \geq 2$ be an integer, ℓ, $0 \leq \ell \leq 1$, be a real number. Ω is of class $C^{k,\ell}$ if and only if for each $x \in \Gamma$, there exists a neighborhood $V(x)$ such that $b_\Omega \in C^{k,\ell}(V(x))$.*

On Γ the gradient ∇b_Ω coincides with the *unit exterior normal* n to Ω and the Hessian matrix $D^2 b_\Omega$ with the *second fundamental form* of Γ ([8]). The *projection* $p_\Gamma(x)$ of a point x onto Γ and the *orthogonal projector* $P(x)$ onto the tangent hyperplane $T_{p_\Gamma(x)}\Gamma$ to Γ are given by

(2.3) $$p_\Gamma(x) = x - b_\Omega(x)\nabla b_\Omega(x), \quad P(x) = I - \nabla b_\Omega(x){}^*\nabla b_\Omega(x).$$

2.2. Flow of the gradient of b_Ω.
Given a real number $k > 0$, define the tubular neighborhood of $\partial\Omega$

$$U_k(\partial\Omega) \stackrel{\text{def}}{=} \{x \in \mathbf{R}^N : |b_\Omega(x)| < k\}. \tag{2.4}$$

THEOREM 2.2. *Let Ω be a bounded $C^{1,1}$ domain in $\mathbf{R}^N$ and let $h > 0$ be such that $b_\Omega \in C^{1,1}(U_{2h+2\varepsilon}(\partial\Omega))$ for some arbitrarily small $\varepsilon > 0$.*

(i) *Let ψ be a C^∞-function such that*

$$\psi(x) = \begin{cases} 1 & x \in U_{2h}(\partial\Omega) \\ \in [0,1] & x \in U_{2h+\varepsilon}(\partial\Omega) \setminus U_{2h}(\partial\Omega) \\ 0 & x \in \mathbf{R}^N \setminus U_{2h+\varepsilon}(\partial\Omega). \end{cases} \tag{2.5}$$

The equation

$$\frac{dx}{dz}(z) = \psi(x(z))\,\nabla b_\Omega(x(z)), \quad x(0) = X \in \mathbf{R}^N, \tag{2.6}$$

has a unique absolutely continuous solution $x(z; X)$. Define

$$T_z(X) \stackrel{\text{def}}{=} x(z; X). \tag{2.7}$$

For all z

$$\nabla(b_\Omega \circ T_z) = \nabla b_\Omega = \nabla b_\Omega \circ p_{\partial\Omega} \text{ in } U_{2h+\varepsilon}(\partial\Omega) \tag{2.8}$$

$$p_{\partial\Omega} \circ T_z = p_{\partial\Omega} \text{ and } \nabla b_\Omega \circ T_z = \nabla b_\Omega \text{ in } U_{2h+\varepsilon}(\partial\Omega). \tag{2.9}$$

For all η, $0 < \eta < 2h$, and z, $|z| \leq 2h - \eta$,

$$b_\Omega \circ T_z = b_\Omega + z \tag{2.10}$$

$$T_z(X) = X + z\,\nabla b_\Omega(X), \quad \forall X \in \overline{U_\eta(\partial\Omega)}. \tag{2.11}$$

(ii) *For all η, $0 < \eta < 2h$, and z, $|z| \leq 2h - \eta$,*

$$DT_z = I + z\,D^2 b_\Omega \tag{2.12}$$

$$DT_z^{-1} \nabla b_\Omega = \nabla b_\Omega = DT_z \nabla b_\Omega. \tag{2.13}$$

$$D^2 b_\Omega = D^2 b_\Omega \circ T_z \,[I + z\,D^2 b_\Omega] \tag{2.14}$$

$$D^2 b_\Omega \circ T_z = [I - z\,D^2 b_\Omega \circ T_z]\,D^2 b_\Omega \tag{2.15}$$

$$[I + z\,D^2 b_\Omega]^{-1} = I - z\,D^2 b_\Omega \circ T_z. \tag{2.16}$$

$$\frac{dDT_z}{dz} = D^2 b_\Omega \circ T_z\, DT_z = D^2 b_\Omega, \quad DT_0 = I \tag{2.17}$$

and

$$j_z(x) \stackrel{\text{def}}{=} \det(DT_z(x)) = \sum_{i=0}^{N-1} \kappa_i(x)\, z^i \tag{2.18}$$

$$\frac{dj_z}{dz} = \Delta b_\Omega \circ T_z\, j_z, \quad j_0 = 1. \tag{2.19}$$

Moreover

$$DT_z \in L^\infty(\overline{U_\eta(\partial\Omega)})^{N \times N},\ j_z \in L^\infty(\overline{U_\eta(\partial\Omega)}). \tag{2.20}$$

(iii) *For all z, $|z| < 2h$, the set*

$$\Omega_z \stackrel{\text{def}}{=} T_z(\Omega) \tag{2.21}$$

is of class $C^{1,1}$, $b_{\Omega_z} \in C^{1,1}(U_{2h_z}(\partial\Omega_z))$ for some h_z, $0 < h_z \leq 2h+2\varepsilon-|z|$,

(2.22) $$\partial\Omega_z = \{x \in \mathbf{R}^N : b_\Omega(x) = z\}$$

(2.23) $$\text{int } \Omega_z = \{x \in \mathbf{R}^N : b_\Omega(x) < z\} \quad \text{int } \complement\Omega_z = \{x \in \mathbf{R}^N : b_\Omega(x) > z\}$$

(2.24) $$b_{\Omega_z} = b_\Omega - z \text{ in } \mathbf{R}^N \text{ and } b_{\Omega_z} \in C^{1,1}(U_{2h+2\varepsilon}(\partial\Omega))$$

(2.25) $$\nabla b_{\Omega_z} = \nabla b_\Omega \text{ in } U_{2h+2\varepsilon}(\partial\Omega)$$

(2.26) $$\forall p_z \in \Pi_{\partial\Omega_z}(x), \exists p \in \Pi_{\partial\Omega}(x), \quad p_z = T_z(p)$$

(2.27) $$p_{\partial\Omega_z} = T_z \circ p_{\partial\Omega} \text{ in } U_{2h+2\varepsilon}(\partial\Omega).$$

For all η, $0 < \eta < 2h$, and z, $|z| \leq 2h - \eta$,

(2.28) $$b_\Omega \circ T_{-z} = b_\Omega - z = b_{\Omega_z} \text{ in } \overline{U_\eta(\partial\Omega)}.$$

3. Differential calculus via tangential derivatives

In this section we recall the main definitions from [8]. For simplicity assume that Γ is bounded, that there exist $h > 0$ and $\varepsilon > 0$ such that $b \in C^2(U_{2h+2\varepsilon}(\Gamma))$, and that the scalar and vector functions are sufficiently smooth. We shall also drop the indices and use the simpler notation $\Gamma = \partial\Omega$, $b = b_\Omega$, $p = p_\Gamma$.

3.1. Definitions of the tangential gradient via the projection. The tangential gradient of a scalar function $f : \Gamma \to \mathbf{R}$ is defined through an appropriately smooth extension F of f in a neighborhood of Γ using the fact that the resulting expression on Γ is independent of the choice of the extension F. The choice of the *canonical extension* $f \circ p$ of the function f will be the basis of a simple differential calculus on Γ which uses the Euclidean differential calculus in an ambient neighborhood of Γ. First recall the following definition and properties. Given $f \in C^1(\Gamma)$, let $F \in C^1(U_h(\Gamma))$ be a C^1-extension of f. Define

$$g(F) \stackrel{\text{def}}{=} \nabla F|_\Gamma - \frac{\partial F}{\partial n} n = P(x)\nabla F(x) \quad \text{on } \Gamma$$

It is easy to show that $g(F) = 0$ for $f = 0$ and that $g(F)$ is independent of the choice of the extension. This justifies the following natural notation and terminology.

DEFINITION 3.1. Assume that there exist $h > 0$ and $\varepsilon > 0$ such that $b_\Omega \in C^2(U_{2h+2\varepsilon}(\Gamma))$. Given an extension $F \in C^1(U_h(\Gamma))$ of $f \in C^1(\Gamma)$, the *tangential gradient* of f on Γ is defined as

$$\nabla_\Gamma f \stackrel{\text{def}}{=} \nabla F|_\Gamma - \frac{\partial F}{\partial n} n.$$

THEOREM 3.1. *Assume that b satisfies the conditions of Definition 3.1, then for any extension $F \in C^1(U_h(\Gamma))$ of f*

(i) $\nabla_\Gamma f = (P\nabla F)|_\Gamma$ and $n \cdot \nabla_\Gamma f = \nabla b \cdot \nabla_\Gamma f = 0$.
(ii) $\nabla(f \circ p) = [I - b\, D^2 b]\, \nabla_\Gamma f \circ p$ and $\nabla(f \circ p)|_\Gamma = \nabla_\Gamma f$.
(iii) $\nabla b \cdot \nabla(f \circ p) = 0$ in $U_h(\Gamma)$ and $n \cdot \nabla_\Gamma f = 0$ on Γ.

3.1.1. *First order derivatives.* The *tangential Jacobian matrix* of a vector function $v \in C^1(\Gamma)^M$, $M \geq 1$, is defined in the same way as the gradient

(3.1) $$\boxed{D_\Gamma v \stackrel{\text{def}}{=} D(v \circ p)|_\Gamma \text{ or } (D_\Gamma v)_{ij} = (\nabla_\Gamma v_i)_j}$$

The equivalent definition from an extension $V \in C^1(U_h(\Gamma))^M$ of v can easily be recovered from Theorem 3.1

(3.2) $\quad \boxed{D_\Gamma v = DV|_\Gamma - DVn\,{}^*n = (DV\,P)|_\Gamma} \quad \boxed{D(v \circ p) = D_\Gamma v \circ p\,[I - b\,D^2 b]}$

Note that

(3.3) $\qquad\qquad\qquad \boxed{D(v \circ p)\,\nabla b = 0 \text{ and } D_\Gamma v\,n = 0.}$

For a vector function $v \in C^1(\Gamma)^N$ define the *tangential divergence* as

(3.4) $\qquad\qquad\qquad \boxed{\mathrm{div}_\Gamma\, v \stackrel{\mathrm{def}}{=} \mathrm{div}(v \circ p)|_\Gamma}$

and it is easy to show that

$$\boxed{\mathrm{div}_\Gamma\, v = \mathrm{tr}[DV|_\Gamma - DV\,n\,{}^*n] = \mathrm{div}\, V|_\Gamma - DV\,n \cdot n.}$$

The *tangential linear strain tensor* of linear elasticity is given by

(3.5) $\qquad \boxed{\varepsilon_\Gamma(v) \stackrel{\mathrm{def}}{=} \frac{1}{2}(D_\Gamma v + {}^*D_\Gamma v)} \Rightarrow \boxed{\varepsilon_\Gamma(v) = \varepsilon(v \circ p)_{|\Gamma}.}$

For the normal n, $n \circ p = n = \nabla b = \nabla b \circ p$, $b \circ p = 0$, and

(3.6) $\qquad\qquad \boxed{D_\Gamma(n) = D_\Gamma(\nabla b) = D^2 b|_\Gamma = {}^*D_\Gamma(\nabla b) = {}^*D_\Gamma(n)}.$

3.1.2. *Second order derivatives.* Assume that Γ is of class C^3. The simplest second order derivative is the *Laplace-Beltrami operator* of a function $f \in C^2(\Gamma)$

(3.7) $\qquad\qquad \boxed{\Delta_\Gamma f \stackrel{\mathrm{def}}{=} \mathrm{div}_\Gamma(\nabla_\Gamma f)} \quad \Rightarrow \quad \boxed{\Delta_\Gamma f = \Delta(f \circ p)|_\Gamma}$

The *tangential Hessian matrix* of second order derivatives is defined as

(3.8) $\qquad\qquad\qquad\qquad \boxed{D_\Gamma^2 f \stackrel{\mathrm{def}}{=} D_\Gamma(\nabla_\Gamma f)}$

The tangential Hessian matrix is not symmetrical and does not coincide with the restriction of the Hessian matrix of the canonical extension:

(3.9) $\qquad \boxed{D_\Gamma^2 f - (D^2 b\,\nabla_\Gamma f)\,{}^*n = D^2(f \circ p)|_\Gamma = {}^*D_\Gamma^2 f - n\,{}^*(D^2 b\,\nabla_\Gamma f)}$

3.2. Projected Jacobian matrices. A vector function V in a neighborhood $S_h(\Gamma)$ of Γ can always be decomposed into its *normal component* $V_n = V \cdot \nabla b$ along the normal field ∇b and its *tangential part* $V_\Gamma = PV$ along the level sets of the function b, where $P = I - \nabla b\,{}^*\nabla b$ is the orthogonal projector. Similarly a matrix or tensor valued functions A defined in the neighborhood $S_h(\Gamma)$ of Γ can also be decomposed as follows

(3.10) $\qquad A = PAP + (PA\nabla b)\,{}^*\nabla b + \nabla b\,{}^*(P\,{}^*A\nabla b) + (A\nabla b \cdot \nabla b)\,\nabla b\,{}^*\nabla b.$

Introduce the notation

(3.11) $\qquad\qquad\qquad\qquad \boxed{A^P \stackrel{\mathrm{def}}{=} PAP}$

In the theory of shells most equations are expressed in terms of tensors in the tangent plane to Γ. For instance the *projected Jacobian matrix* and the *projected linear strain tensor*

(3.12) $\qquad\qquad \boxed{D_\Gamma^P(v) = PD_\Gamma(v)P \text{ and } \varepsilon_\Gamma^P(v) = P\varepsilon_\Gamma(v)P}$

will occur rather than $D_\Gamma(v)$ and $\varepsilon_\Gamma(v)$. $D_\Gamma(v)$ maps $T_x\Gamma$ into $\mathbf{R}^N$ and $D_\Gamma^P(v)$ maps $T_x\Gamma$ into $T_x\Gamma$. By definition it is easy to check that

(3.13) $$\boxed{D_\Gamma(v) = D_\Gamma^P(v) + n\,{}^*({}^*D_\Gamma(v)\,n)}$$

(3.14) $$\boxed{D_\Gamma^P(v_\Gamma) = D_\Gamma(v_\Gamma) + n\,{}^*(D^2b\,v_\Gamma) \quad \text{and} \quad D_\Gamma^P(v) = D_\Gamma^P(v_\Gamma) + v_n\,D^2b}$$

(3.15) $$\boxed{\begin{aligned}\varepsilon_\Gamma^P(v_\Gamma) &= \varepsilon_\Gamma(v_\Gamma) + \frac{1}{2}[(D^2b\,v_\Gamma)\,{}^*n + n\,{}^*(D^2b\,v_\Gamma)] \\ \varepsilon_\Gamma^P(v) &= \varepsilon_\Gamma^P(v_\Gamma) + v_n\,D^2b\end{aligned}}$$

Finally

(3.16) $$\boxed{\operatorname{div}_\Gamma v = \operatorname{tr} D_\Gamma(v) = \operatorname{tr} D_\Gamma^P(v)}$$

For the tangential Hessian matrix

$$D_\Gamma^2 f = P\,D_\Gamma^2 f - n\,{}^*(D^2b\,\nabla_\Gamma f)$$

and from identity (3.9) we get the symmetry of the projected Hessian

(3.17) $$\boxed{\begin{aligned}{}^*D_\Gamma^P(\nabla_\Gamma f) &= {}^*(P\,D_\Gamma^2 f) = P\,D_\Gamma^2 f = D_\Gamma^P(\nabla_\Gamma f) \\ D_\Gamma^P(\nabla_\Gamma f) &= \varepsilon_\Gamma^P(\nabla_\Gamma f) = {}^*D_\Gamma^P(\nabla_\Gamma f)\end{aligned}}$$

4. Sobolev spaces

In this section ω is a bounded (relatively) open domain in the submanifold $\Gamma = \partial\Omega$ associated with a domain Ω of class $C^{1,1}$ in $\mathbf{R}^N$. § 4.1 first gives a density result for $L^p(\omega)$, $1 \leq p < \infty$, in Theorem 4.2. The main result is Theorem 4.5 which establishes the uniformity as h goes to zero of the continuity constants for the family of h-dependent isomorphisms between the space $L^p(\omega)$ and the corresponding closed linear subspace $L_\Gamma^p(S_h(\omega))$ of $L^p(S_h(\omega))$ for the Euclidean neighborhood $S_h(\omega)$ of ω, where

(4.1) $\quad \forall k,\ 0 < k < 2h + 2\varepsilon,\quad \boxed{S_k(\omega) \stackrel{\text{def}}{=} \{x \in \mathbf{R}^N : |b(x)| < k \text{ and } p(x) \in \omega\}.}$

This leads to a new characterization of $L^p(\omega)$ as the set of functions f such that $f \circ p \in L^p(S_k(\omega))$ for some $k > 0$. In § 4.2 a similar characterization is used to make sense of and define the Sobolev space $W^{1,p}(\omega)$ as the set of functions f such that $f \circ p \in W^{1,p}(S_k(\omega))$ for some $k > 0$. Theorem 4.6 makes sense of a tangential gradient for $C^{0,1}(\omega)$ functions and Theorem 4.7 for $W^{1,p}(\omega)$ functions together with a density result which is the analogue for $W^{1,p}(\omega)$ of the one for $L^p(\omega)$. Theorem 4.7 establishes the uniformity as h goes to zero of the continuity constants for the family of h-dependent isomorphisms between the space $W^{1,p}(\omega)$ and the corresponding closed linear subspace of $W^{1,p}(S_h(\omega))$. This is completed with Theorem 4.8 which gives the equivalence of our intrinsic definition with the one obtained via local $C^{1,1}$ diffeomorphisms. § 4.4 extends the approach to intermediary Sobolev spaces. § 4.5 deals with the smoothness of the (relative) boundary γ, and continuous extensions for Lipschitzian domains.

4.1. Local coordinates and Federer's formula.

From Theorem 2.2 a local coordinates system can be introduced using the level sets of the function b and the lines normal to Γ. The isomorphism T allows to go back and forth between ambient Euclidean coordinates and local coordinates.

THEOREM 4.1. *Assume that Γ satisfies the conditions of Theorem 2.2.*

(i) *The map*
$$(X, z) \mapsto T(X, z) \stackrel{\text{def}}{=} T_z(X) = X + z \nabla b(X) : \Gamma \times\,]-h, h[\to U_h(\Gamma)$$
and its inverse $T^{-1}(x) = (p(x), b(x))$ are both Lipschitz continuous.

(ii) T *(resp. T^{-1}) maps $H_{N-1} \times m$ (resp. m_N) measurable sets onto m_N (resp. $H_{N-1} \times m$) measurable sets. Hence the map*

(4.2)
$$F \mapsto f \stackrel{\text{def}}{=} F \circ T \quad (\text{resp. } f \mapsto F \stackrel{\text{def}}{=} f \circ T^{-1})$$

maps m_N measurable functions F onto $H_{N-1} \times m$ measurable functions $f = F \circ T$ (resp. $H_{N-1} \times m$ measurable functions f onto m_N measurable functions $F = f \circ T^{-1}$).

(iii) *The maps*[1]

(4.3)
$$F \mapsto f \stackrel{\text{def}}{=} F \circ T : C^{0,1}(U_h(\Gamma)) \to C^{0,1}(\Gamma \times\,]-h, h[)$$

(4.4)
$$F \mapsto f \stackrel{\text{def}}{=} F \circ T : C^{0,1}(\overline{U_h(\Gamma)}) \to C^{0,1}(\overline{\Gamma \times\,]-h, h[})$$

are linear bijections.

(iv) *The results of parts (i) to (iii) also apply to any bounded (relatively open) subset ω of Γ with $S_h(\omega)$ in place of $U_h(\Gamma)$.*

The next step is the extension of the bijection (4.4) for $C^{0,1}$ functions to a continuous isomorphism for L^p functions. We proceed by a density argument and first prove the density of $C^{0,1}(\overline{\omega})$ in $L^p(\omega)$. For a bounded (relatively) open subset ω of Γ, $(\omega; H_{N-1})$ is a positive measure space and the spaces

$$L^p(\omega) \stackrel{\text{def}}{=} L^p(\omega; H_{N-1}), \quad 1 \le p \le \infty$$

are well-defined Banach spaces with the usual norms. Similarly $(\omega \times\,]-h, h[; H_{N-1} \times m)$ is a positive measure product space and

$$L^p(\omega \times\,]-h, h[) \stackrel{\text{def}}{=} L^p(\omega \times\,]-h, h[; H_{N-1} \times m), \quad 1 \le p \le \infty$$

are also well-defined Banach spaces with the usual norms.

THEOREM 4.2. *Let Γ satisfy the assumptions of Theorem 2.2 and let ω be a bounded (relatively) open subset of Γ. Given $1 \le p < \infty$ and $f \in L^p(\omega)$, there exists a sequence $\{F_n\}$ in $C_c^\infty(S_{2h}(\omega))$ such that*
$$f_n \stackrel{\text{def}}{=} F_n|_\omega \to f \text{ in } L^p(\omega) \;\Rightarrow\; \boxed{\overline{C^{0,1}(\overline{\omega})}^{L^p} = L^p(\omega)}$$

PROOF. Since $L^p(\omega)$ is a measure space, there exists $\{g_k\}$ in $C_c(\omega)$ such that $g_k \to f$ in $L^p(\omega)$ as $k \to \infty$. For any $\varepsilon > 0$, there exists k such that
$$\|g_k - f\|_{L^p(\omega)} \le \varepsilon/2.$$
By construction $g_k \circ p \in C(\overline{S_h(\omega)})$ and there exists $\{F_n\}$ in $C_c^\infty(S_{2h}(\omega))$ such that
$$F_n \to g_k \circ p \text{ in } C(\overline{S_h(\omega)}) \text{ as } k \to \infty.$$

---

[1] $C^{0,1}(\Omega)$ denotes the space of Lipschitz functions on Ω and $C^{0,1}(\overline{\Omega})$ denotes the subspace of Lipschitz functions on Ω which are bounded and uniformly continuous on Ω.

Choose n such that
$$\|F_n - g_k \circ p\|_{C(\overline{S_h(\omega)})} \leq \varepsilon/(2H_{N-1}(\omega)^{1/p}).$$
and define $f_n = F_n|_\omega$. Then
$$\|f_n - g_k\|_{C(\overline{\omega})} \leq \varepsilon/(2H_{N-1}(\omega)^{1/p})$$
$$\|f - f_n\|_{L^p(\omega)} \leq \|f - g_k\|_{L^p(\omega)} + \|g_k - f_n\|_{L^p(\omega)}$$
$$\leq \varepsilon/2 + \varepsilon/(2H_{N-1}(\omega)^{1/p}) H_{N-1}(\omega)^{1/p} = \varepsilon.$$
Finally $f_n \circ p = F_n|_\omega \circ p = F_n \circ p \in C^{0,1}(\overline{S_{2h}(\omega)})$ and $f_n \in C^{0,1}(\overline{\omega})$. □

We now specialize Federer's formula ([**12**]) for the decomposition of the integral in $S_h(\omega)$ along the level sets of a Lipschitzian function to the function b which is not only Lipschitz continuous but also $C^{1,1}$ in $S_h(\omega)$.

THEOREM 4.3 ([**11**] Prop 3, p. 118). *Let $g : \mathbf{R}^N \to \mathbf{R}$ be a Lipschitz continuous function such that $|\nabla g| > 0$ a.e. For any Lebesgue summable function $F : \mathbf{R}^N \to \mathbf{R}$, we have the decomposition of the integral along the level sets of g*
$$\int_{\{g>t\}} F \, dx = \int_t^\infty \Big(\int_{\{g=z\}} \frac{F}{|\nabla g|} dH_{N-1} \Big) dz.$$

We first show that $D^2 b$ and hence j_z exists H_{N-1} almost everywhere on Γ. We then use the change of variable T_z to bring the second integral on $b^{-1}\{z\}$ to Γ.

THEOREM 4.4. *Let Γ satisfy the assumptions of Theorem 2.2.*

(i) *If $D^2 b(x)$ exists at the point $x \in U_h(\Gamma)$, then $D^2 b(p(x))$ exists and*
$$D^2 b(p(x)) = D^2 b(x) [I - b(x) D^2 b(x)]^{-1}.$$

For all $|z| < h$

(4.5) $\quad [I + z D^2 b][I - b D^2 b] \circ T_z = I - b D^2 b = [I - b D^2 b] \circ T_z [I + z D^2 b]$

(4.6) $\quad [I - b D^2 b]^{-1} \circ (T_z \circ p) = [I + z D^2 b] \circ p.$

(ii) *$D^2 b(x)$ exists in H_{N-1} almost all point $x \in \Gamma$, $D^2 b \in L^\infty(\Gamma)^{N \times N}$, and for all z, $|z| < h$, $j_z \in L^\infty(\Gamma)$.*

(iii) *For any Lebesgue summable function $F \in L^1(U_h(\Gamma))$*
$$\int_{U_h(\Gamma)} F \, dx = \int_{-h}^h \int_\Gamma F \circ T_z \, j_z \, dH_{N-1} \, dz = \int_\Gamma \int_{-h}^h F \circ T_z \, j_z \, dz \, dH_{N-1}.$$

PROOF. (i) From Theorem 2.2 $\nabla b = \nabla b \circ p$ in $U_h(\Gamma)$. If $D^2 b(x)$ exists at $x \in U_h(\Gamma)$, then it is readily seen that $Dp(x)$ exists and that
$$Dp(x) = I - \nabla b(x)^* \nabla b(x) - b(x) D^2 b(x).$$
Indeed for any direction v
$$\frac{1}{t}[p(x+tv) - p(x)] = \frac{1}{t}[x + tv - b(x+tv)\nabla b(x+tv) - x + b(x)\nabla b(x)]$$
$$= v - b(x+tv) \frac{\nabla b(x+tv) - \nabla b(x)}{t} - \frac{b(x+tv) - b(x)}{t} \nabla b(x)$$
and the directional derivative of p at x in the direction v is given by
$$dp(x; v) = v - b(x) D^2 b(x) v - \nabla b(x) \cdot v \, \nabla b(x)$$
$$= [I - b(x) D^2 b(x) - \nabla b(x)^* \nabla b(x)] v = Dp(x) v.$$

Consider the vector function $f(x) = \nabla b(x)$ and compute its directional derivative in the direction v

$$df(x;v) = \lim_{t\searrow 0} \frac{f(x+tv) - f(x)}{t} = D^2 b(x) v.$$

Define $g(x) = \nabla b(p(x))$. Since $f = g$, $dg(x;v) = df(x;v)$ exists. The differential quotient can be written as

$$q_t \stackrel{\text{def}}{=} \frac{1}{t}[g(t+tv) - g(x)]$$
$$= \frac{1}{t}\left[\nabla b(p(x) + t\frac{p(x+tv) - p(x)}{t}) - \nabla b(p(x))\right]$$
$$= \frac{1}{t}[\nabla b(p(x) + t dp(x;v)) - \nabla b(p(x))]$$
$$+ \frac{1}{t}\left[\nabla b(p(x) + t\frac{p(x+tv) - p(x)}{t}) - \nabla b(p(x) + t dp(x;v))\right].$$

The second term on the right-hand side of the last identity goes to zero as t goes to zero by Lipschitz continuity of ∇b since it is bounded by

$$\frac{c}{t}\left|p(x) + t\frac{p(x+tv) - p(x)}{t} - p(x) - t dp(x;v)\right|$$
$$\leq c\left|\frac{p(x+tv) - p(x)}{t} - dp(x;v)\right| \to 0 \text{ as } t \searrow 0$$

and $Dp(x)$ exists. Therefore the limit of the first term exists and

$$dg(x;v) = \lim_{t\searrow 0} \frac{1}{t}[\nabla b(p(x) + t dp(x;v)) - \nabla b(x)]$$
$$= df(p(x); dp(x;v)) = df(p(x); Dp(x)v).$$

But since $dg(x;v) = df(x;v) = D^2 b(x) v$, we have $df(p(x); Dp(x)v) = D^2 b(x) v$. Moreover since

$$Dp(x)v = [I - b(x) D^2 b(x)] v - \nabla b(x)^* \nabla b(x) v$$

we have for tangential directions v_Γ, $Dp(x) v_\Gamma = [I - b(x) D^2 b(x)] v_\Gamma$. Now $I - b(x) D^2 b(x)]$ is invertible. It maps $\nabla b(x)$ onto $\nabla b(x)$ and tangential vectors onto tangential vectors. Indeed

$$\frac{1}{t}[\nabla b(x + t\nabla b(x)) - \nabla b(x)] = \frac{\nabla b(x) - \nabla b(x)}{t} = 0 \Rightarrow D^2 b(x) \nabla b(x) = 0.$$

For any tangential w_Γ ($w_\Gamma \cdot n(x) = 0$), $[I - b(x) D^2 b(x)]^{-1} w_\Gamma$ is also tangential and

$$D^2 b(x)[I - b(x) D^2 b(x)]^{-1} w_\Gamma = df(p(x); w_\Gamma)$$
$$\Rightarrow \boxed{df(p(x); w_\Gamma) = D^2 b(x)[I - b(x) D^2 b(x)]^{-1} w_\Gamma}$$

This characterizes $df(p(x); w)$ for tangential vectors w. For general vectors w

$$\frac{1}{t}[\nabla b(p(x+tw)) - \nabla b(p(x))] = \frac{1}{t}[\nabla b(p(x+tw)) - \nabla b(p(x+tw_\Gamma))]$$
$$+ \frac{1}{t}[\nabla b(p(x+tw_\Gamma)) - \nabla b(p(x))].$$

The limit of the second term is $df(p(x); w_\Gamma)$ where w_Γ is the tangential component of w. As for the first term T_1 it goes to zero

$$T_1 \stackrel{\text{def}}{=} \frac{1}{t}[\nabla b(p(x+tw)) - \nabla b(p(x+tw_\Gamma))] = \frac{1}{t}[\nabla b(x+tw) - \nabla b(x+tw_\Gamma)]$$

since $\nabla b \circ p = \nabla b$. Furthermore

$$T_1 = \frac{1}{t}[\nabla b(x+tw) - \nabla b(x)] - \frac{1}{t}[\nabla b(x+tw_\Gamma) - \nabla b(x)]$$

and since $D^2 b(x)$ exists as t goes to zero

$$T_1 \to D^2 b(x)w - D^2 b(x)w_\Gamma = D^2 b(x)(w - w_\Gamma) = D^2 b(x)(w_n \nabla b(x)) = 0,$$

where w_n is the normal component of w. Finally

$$df(p(x); w) = df(p(x); w_\Gamma) = D^2 b(x)[I - b(x)D^2 b(x)]^{-1} w_\Gamma$$
$$= D^2 b(x)[I - b(x)D^2 b(x)]^{-1} w$$

since $[I - b(x)D^2 b(x)]^{-1} \nabla b(x) = \nabla b(x)$ and $D^2 b(x) \nabla b(x) = 0$. In conclusion

$$\forall w, \quad df(p(x); w) = D^2 b(x)[I - b(x)D^2 b(x)]^{-1} w$$
$$\Rightarrow \forall w, \quad D^2 b(p(x))w = D^2 b(x)[I - b(x)D^2 b(x)]^{-1} w$$
$$\Rightarrow \boxed{D^2 b(p(x)) = D^2 b(x)[I - b(x)D^2 b(x)]^{-1}}$$

From Theorem 2.2 for all $|z| < h$, $\nabla b = \nabla b \circ T_z$ and $p = p \circ T_z$. Moreover ∇b and a fortiori p belong to $C^{0,1}(U_{2h+\varepsilon}(\Gamma))^N$. From [13] (Lem. 3.1, pp. 65–66) $T_z : \mathbf{R}^N \to \mathbf{R}^N$ is bijective and T_z and T_z^{-1} are both Lipschitz continuous. Hence

$$Dp = D(p \circ T_z) = Dp \circ T_z DT_z$$
$$I - bD^2 b - \nabla b^* \nabla b = [I - bD^2 b - \nabla b^* \nabla b] \circ T_z [I + zD^2 b]$$
$$= \{[I - bD^2 b] \circ T_z - \nabla b^* \nabla b\}[I + zD^2 b]$$
$$= [I - bD^2 b] \circ T_z [I + zD^2 b] - \nabla b^* \nabla b$$
$$\Rightarrow I - bD^2 b = [I - bD^2 b] \circ T_z [I + zD^2 b].$$

So the first set of identities follows by transposition and symmetry. For the second one we use the fact that $D^2 b \circ p$ is well-defined and that $b \circ p = 0$

$$I = [I - bD^2 b] \circ p = [I - bD^2 b] \circ T_z \circ p [I + zD^2 b] \circ p.$$

(ii) By contradiction. If there exists $A \subset \Gamma$ such that $H_{N-1}(A) > 0$, where $D^2 b(x)$ does not exists, then for all $y \in U_h(\Gamma)$ such that $p(x) = y$, $D^2 b(y)$ does not exist. Otherwise this would contradict part (i). Therefore $D^2 b$ does not exist in

$$S_h(A) \stackrel{\text{def}}{=} \{x \in U_h(\Gamma) : p(x) \in A\} = \{x \in \mathbf{R}^N : |b(x)| < h, p(x) \in A\}.$$

To complete the proof it is sufficient to show that

$$H_{N-1}(A) > 0 \quad \Rightarrow \quad m_N(S_h(A)) > 0.$$

This contradicts the fact that, for the Lipschitz continuous function ∇b, $D^2 b$ exists m_N almost everywhere in $U_h(\Gamma)$ by Rademacher's Theorem. By Federer's formula

$$\int_{S_h(A)} dx = \int_{-h}^{h} H_{N-1}(b^{-1}\{z\} \cap S_h(A)) \, dz = \int_{-h}^{h} H_{N-1}(T_z(A)) \, dz$$

since $b^{-1}\{z\} \cap S_h(A) = T_z(A)$. But we have seen in Theorem 2.2 of that $T_z : \Gamma = b^{-1}\{0\} \to \Gamma_z = b^{-1}\{z\}$ is a bijection and that T_z and T_z^{-1} are both Lipschitz continuous. As a consequence they transport sets of zero measure onto sets of zero measure. Therefore if $H_{N-1}(A) > 0$, then for all $|z| < h$ $H_{N-1}(T_z(A)) > 0$ and a fortiori $m_N(S_h(A)) > 0$.

(iii) Let $g = b$. Then $|\nabla b| = 1$ in $U_h(\Gamma)$ and the formula becomes

$$\int_{\{b>t\}} F\, dx = \int_t^\infty \Big(\int_{\{b=z\}} F\, dH_{N-1}\Big) dz.$$

Since $b \in C^{1,1}(U_h(\Gamma))$ for some $h > 0$, then for $|z| < h$ $T_z(\Gamma) = \{x \in \mathbf{R}^N : b(x) = z\}$ and we can use the change of variable formula

$$\int_{T_z(\Gamma)} F\, dH_{N-1} = \int_\Gamma F \circ T_z\, \omega_z\, dH_{N-1}$$

where ω_z is the canonical density given by

$$\omega_z(x) = |{}^*(DT_z(x))^{-1} \nabla b(x)|\, \det(DT_z(x)) = j_z(x)$$

and j_z belongs to $L^\infty(\Gamma)$. As a result for all $|z| < h$

$$\int_{T_z(\Gamma)} F\, dH_{N-1} = \int_\Gamma F \circ T_z\, j_z\, dH_{N-1}$$

and for $-h < s < t < h$

$$\int_{s<b<t} F\, dx = \int_s^t \Big(\int_\Gamma F \circ T_z\, j_z\, dH_{N-1}\Big) dz = \int_\Gamma \Big(\int_s^t F \circ T_z\, j_z\, dz\Big) dH_{N-1}. \qquad \square$$

We can now give the isomorphism between L^p spaces.

THEOREM 4.5. *Let Γ satisfy the assumptions of Theorem 2.2 and let ω be a bounded (relatively) open subset of Γ.*

(i) *Given $p \geq 1$, the map*

$$F \mapsto f \stackrel{\text{def}}{=} F \circ T : L^p(S_h(\omega)) \to L^p(\omega \times\,]-h, h[) \simeq L^p(-h, h; L^p(\omega))$$

is a continuous linear bijection. Moreover the averaging *map*

(4.7) $$F \mapsto \overline{f} \stackrel{\text{def}}{=} \frac{1}{2h} \int_{-h}^h F \circ T_z\, dz : L^p(S_h(\omega)) \to L^p(\omega)$$

is linear and continuous.

(ii) *If a real function f on ω is H_{N-1}-measurable, then $f \circ p$ is Lebesgue measurable in $S_h(\omega)$. Given $1 \leq p < \infty$ and $f \in L^p(\omega)$, then $f \circ p \in L^p(S_h(\omega))$ and there exists $c > 0$ such that, as h goes to zero, for all $f \in L^p(\omega)$*

$$(1 - ch^2)\|f\|_{L^p(\omega)}^p \leq \frac{1}{2h} \int_{S_h(\omega)} |f \circ p|^p\, dx \leq (1 + ch^2)\|f\|_{L^p(\omega)}^p$$

and

$$\lim_{h \searrow 0} \frac{1}{2h} \|f \circ p\|_{L^p(S_h(\omega))}^p = \lim_{h \searrow 0} \frac{1}{2h} \int_{S_h(\omega)} |f \circ p|^p\, d\Gamma = \|f\|_{L^p(\omega)}^p$$

$$\lim_{h \searrow 0} \frac{1}{2h} \int_{S_h(\omega)} f \circ p\, dx = \int_\omega f\, d\Gamma.$$

(iii) *For all $f \in L^\infty(\omega)$*

$$\|f \circ p\|_{L^\infty(S_h(\omega))} = \|f\|_{L^\infty(\omega)}.$$

(iv) *The set*

$$L^p_\Gamma(S_h(\omega)) \stackrel{\text{def}}{=} \{f \circ p : \forall f \in L^p(\omega)\}$$

is closed in $L^p(S_h(\omega))$, $1 \leq p \leq \infty$. Moreover given $1 \leq p < \infty$, for each $f \in L^p(\omega)$, there exists a sequence $\{F_n\}$ in $C_c^\infty(S_{2h}(\omega))$ such that

$$f_n \circ p \to f \circ p \text{ in } L^p(S_h(\omega)), \quad f_n \stackrel{\text{def}}{=} F_n|_\omega$$

(v) *Given $p \geq 1$, for all k, $0 < k \leq 2h + \varepsilon$,*

$$L^p(\omega) = \{f : \omega \to \mathbf{R} : f \circ p \in L^p(S_k(\omega))\}.$$

PROOF. (i) This can be proved by a density argument combined with Theorem 2.2 (iv), Theorem 4.2 and Federer's formula of Theorem 4.4 (iii). It can also be proved by adapting classical arguments (cf. for instance [**13**], Lem. 3.1, pp. 65–66) and using Federer's formula. The average $\overline{f}$ of $f = F \circ T$ with respect to z is a well-defined element of $L^p(\omega)$. To prove the continuity we use Federer's formula

$$\int_\omega |\overline{f}|^p d\Gamma = \int_\omega \left| \frac{1}{2h} \int_{-h}^h (F \circ T_z) \, dz \right|^p d\Gamma$$

$$\leq \frac{1}{2h} \int_\omega \int_{-h}^h |F \circ T_z|^p dz \, d\Gamma$$

$$\leq \frac{c}{2h} \int_\omega \int_{-h}^h |F \circ T_z|^p j_z \, dz \, d\Gamma \leq \frac{c}{2h} \int_{S_h(\omega)} |F|^p dx.$$

(ii) From Theorem 4.1 (ii) for an $(H_{N-1} \times m)$-measurable function $u : \omega \times]{-h}, h[\to \mathbf{R}$, the function $u \circ T^{-1}$ is Lebesgue measurable in $S_h(\omega)$. In particular for an H_{N-1}-measurable function f on ω, the map $u(X, z) = f(X)$ is $(H_{N-1} \times m)$-measurable and hence $(u \circ T^{-1})(x) = f(p(x)) = (f \circ p)(x)$ is measurable. So if $f \in L^p(\omega)$, then $f \circ p$ is Lebesgue measurable, by Federer's formula

$$\int_{S_h(\omega)} |f \circ p|^p \, dx = \int_\omega \int_{-h}^h j_z |f|^p \, dz \, d\Gamma$$

since $p \circ T_z = p$. But

$$\int_{-h}^h j_z \, dz = \sum_{i=0}^{N-1} \kappa_i \int_{-h}^h z^i \, dz = \sum_{i=0}^{N-1} \kappa_i \frac{h^{i+1}}{i+1}(1 - (-1)^{i+1}).$$

So for i odd, $(1 - (-1)^{i+1}) = 0$ and we only get odd powers of h. Also $\kappa_0 = 1$ and the κ_i's are bounded. Hence there exists $c > 0$ such that as h goes to zero

$$(1 - ch^2)\|f\|^p_{L^p(\omega)} \leq \frac{1}{2h} \int_{S_h(\omega)} |f \circ p|^p \, dx \leq (1 + ch^2)\|f\|^p_{L^p(\omega)}$$

and the limit exists as h goes to zero. For the last formula use the transformation T of part (i) with $F = f \circ p$. Then the average $\overline{f}$ of F reduces to f and

$$\overline{f} = \frac{1}{2h}\int_{-h}^{h} \overline{f}\, dz = \frac{1}{2h}\int_{-h}^{h} (f\circ p)\circ T_z\, dz \in L^p(\omega)$$

$$\Rightarrow \lim_{h\searrow 0}\frac{1}{2h}\int_{S_h(\omega)} f\circ p\, dx = \lim_{h\searrow 0}\int_\omega f\, \frac{1}{2h}\int_{-h}^{h} j_z\, dz\, d\Gamma = \int_\omega f\, d\Gamma.$$

(iii) From (ii) if f belongs to $L^\infty(\omega)$, then $f\circ p$ is Lebesgue measurable. Recall that the $L^\infty(\omega)$ norm is defined as the infimum over all α such that

$$H_{N-1}(\{X\in\omega : |f(X)| > \alpha\}) = 0.$$

Therefore for all $\alpha \geq \|f\|_{L^\infty(\omega)}$,

$$H_{N-1}(\{X\in\omega : |f(X)| > \alpha\}) = 0.$$

Consider the set

$$\{x\in S_h(\omega) : |f(p(x))| > \alpha\} = p^{-1}\{X\in\omega : |f(X)| > \alpha\}$$
$$= S_h(\{X\in\omega : |f(X)| > \alpha\})$$

In view of Theorem 4.1 (i) and (ii) for the isomorphism T, to say that

$$m_N(\{x\in S_h(\omega)) : |f(p(x))| > \alpha\}) = 0$$

is equivalent to say that

$$m_1 \times H_{N-1}(\{(z,X)\in\,]-h,h[\,\times\omega : |f(p(X + z\nabla b(X)))| > \alpha\}) = 0.$$

But $p(X + z\nabla b(X)) = X$ and

$$m_1\times H_{N-1}(\{(z,X)\in\,]-h,h[\,\times\omega : |f(p(X + z\nabla b(X)))| > \alpha\})$$
$$= m_1\times H_{N-1}(\{(z,X)\in\,]-h,h[\,\times\omega : |f(X)| > \alpha\})$$
$$= 2h\, H_{N-1}(\{X\in\omega : |f(X)| > \alpha\})$$

and $m_N(\{x\in S_h(\omega)) : |f(p(x))| > \alpha\}) = 0$ if and only if $H_{N-1}(\{X\in\omega : |f(X)| > \alpha\}) = 0$. Therefore

$$m_N(\{x\in S_h(\omega)) : |f(p(x))| > \alpha\}) = 0$$
$$\Rightarrow \alpha \geq \|f\circ p\|_{L^\infty(S_h(\omega))}$$

and by taking the infimum over all such α's

$$\|f\|_{L^\infty(\omega)} \geq \|f\circ p\|_{L^\infty(S_h(\omega))}$$

Conversely for all $\alpha \geq \|f\circ p\|_{L^\infty(S_h(\omega))}$

$$m_N(\{x\in S_h(\omega)) : |f(p(x))| > \alpha\}) = 0$$
$$\Rightarrow H_{N-1}(\{X\in\omega : |f(X)| > \alpha\}) = 0$$
$$\Rightarrow \alpha \geq \|f\|_{L^\infty(\omega)} \quad \Rightarrow \|f\circ p\|_{L^\infty(S_h(\omega))} \geq \|f\|_{L^\infty(\omega)}$$
$$\Rightarrow \|f\circ p\|_{L^\infty(S_h(\omega))} = \|f\|_{L^\infty(\omega)}.$$

(iv) From part (ii) the equivalence of norms gives the closure. In the proof of Theorem 4.2 we have already constructed a sequence such that $f_n \to f$ in $L^p(\omega)$. The result now follows from the equivalence of norms of the L^p-spaces.

(v) From part (ii) if $f \in L^p(\omega)$, then $f \circ p \in L^p(S_h(\omega))$. To show that $f \circ p \in L^p(S_{2h+\varepsilon}(\omega))$, consider the map

$$y \mapsto R(y) \stackrel{\text{def}}{=} p(y) + \frac{h}{2h+\varepsilon} b(y)\nabla b(y) : S_{2h+\varepsilon}(\omega) \to S_h(\omega).$$

It is clearly well-defined since $p(R(y)) = p(y)$ and $|b(R(y))| = |hb(y)/(2h+\varepsilon)| < h$. Similarly the map

$$x \mapsto S(x) \stackrel{\text{def}}{=} p(x) + \frac{2h+\varepsilon}{h} b(x)\nabla b(x) : S_h(\omega) \to S_{2h+\varepsilon}(\omega)$$

is well-defined and $R(S(x)) = x$ and $S(R(y)) = y$. Moreover both R and $R^{-1} = S$ are Lipschitz continuous. Again from [13] (Lem. 3.1, pp. 65–66), the map

$$u \mapsto u \circ R : L^p(S_h(\omega)) \to L^p(S_{2h+\varepsilon}(\omega))$$

is an isomorphism. In particular choosing $u = v \circ p \in L^p(S_h(\omega))$, then $u \circ R = v \circ p \circ R \in L^p(S_{2h+\varepsilon}(\omega))$. But

$$v(p(R(y))) = v(p(p(y) + \frac{h}{2h+\varepsilon} b(y)\nabla b(y))) = v(p(y))$$

since ∇b, b and p are defined in $S_{2h+\varepsilon}(\omega)$. This implies that

$$v \circ p = (v \circ p) \circ R \in L^p(S_{2h+\varepsilon}(\omega)).$$

Now by restriction for all k, $0 < k \leq 2h+\varepsilon$, $v \circ p \in L^p(S_k(\omega))$. Conversely if $f : \omega \to \mathbf{R}$ is a function such that $f \circ p \in L^p(S_k(\omega))$ for some k, $0 < k \leq 2h+\varepsilon$, then from part (i)

$$(f \circ p) \circ T \in L^p(-k, k; L^p(\omega))$$

$$\Rightarrow f = \frac{1}{2k}\int_{-k}^{k} f\, dz = \frac{1}{2k}\int_{-k}^{k} (f \circ p) \circ T_z\, dz \in L^p(\omega). \qquad \square$$

4.2. Intrinsic definition of tangential Sobolev spaces. For Ω of class C^2 and $\Gamma = \partial\Omega$, we have seen in § 3 that we can associate with a function $f \in C^1(\Gamma)$ a tangential gradient $\nabla_\Gamma f$ such that

$$\nabla(f \circ p) = [I - bD^2 b]\nabla_\Gamma f \circ p \quad \text{and} \quad \nabla_\Gamma f = \nabla(f \circ p)|_\Gamma.$$

This suggests to use a definition of Sobolev spaces on Γ induced by the Sobolev spaces defined on the Euclidean domain $U_h(\Gamma)$. For instance

$$W^{1,p}(\Gamma) \stackrel{\text{def}}{=} \{f \in L^p(\Gamma) : f \circ p \in W^{1,p}(U_h(\Gamma))\}$$

endowed with the $W^{1,p}$-norm of $f \circ p$. Clearly $C^1(\Gamma) \subset W^{1,p}(\Gamma)$ and in view of Theorem 4.5 (ii)

$$\int_\Gamma |f|^p\, d\Gamma = \lim_{h \searrow 0} \frac{1}{2h} \int_{U_h(\Gamma)} |f \circ p|^p\, dx$$

$$\int_\Gamma |\nabla_\Gamma f|^p\, d\Gamma = \lim_{h \searrow 0} \frac{1}{2h} \int_{U_h(\Gamma)} |\nabla_\Gamma f \circ p|^p\, dx$$

$$= \lim_{h \searrow 0} \frac{1}{2h} \int_{U_h(\Gamma)} |[I - bD^2 b]^{-1}\nabla(f \circ p)|^p\, dx$$

$$= \lim_{h \searrow 0} \frac{1}{2h} \int_{U_h(\Gamma)} |\nabla(f \circ p)|^p\, dx.$$

This clearly suggests the intrinsic norm
$$\int_\Gamma |f|^p + |\nabla_\Gamma f|^p \, d\Gamma = \lim_{h \searrow 0} \frac{1}{2h} \int_{U_h(\Gamma)} |f \circ p|^p + |\nabla(f \circ p)|^p \, dx$$
which is independent of h.

All this makes sense for Γ's which are of class C^2 and functions f which are $C^1(\Gamma)$. We now extend the definition of the tangential gradient to arbitrary functions in $W^{1,p}(\omega)$, where ω is a bounded (relatively) open subset of a $C^{1,1}$ submanifold Γ which satisfies the assumptions of Theorem 2.2. We shall also prove a density theorem for Sobolev spaces analogous to the one of Theorem 4.2 for $L^p(\omega)$.

DEFINITION 4.1 (Tangential Sobolev spaces). Let Γ satisfy the assumptions of Theorem 2.2, ω be a bounded (relatively) open subset in Γ, and $1 \le p \le \infty$. Define

(4.8) $$\boxed{W^{1,p}(\omega) \stackrel{\text{def}}{=} \left\{ f : \omega \to \mathbf{R} : \begin{array}{c} \exists k, \, 0 < k \le 2h + \varepsilon \\ \text{such that } f \circ p \in W^{1,p}(S_k(\omega)) \end{array} \right\}.}$$

The definition of $W^{1,p}(\omega)$ is independent of h as can be readily seen from the next lemma. It parallels the characterization of $L^p(\omega)$ given in Theorem 4.5 (v).

LEMMA 4.1. *Let ω be a bounded open subset of Γ for which the assumptions of Definition 4.1 are satisfied and consider a function $f : \omega \to \mathbf{R}$. Given $1 \le p \le \infty$, if there exists $0 < k \le 2h + \varepsilon$ such that $f \circ p \in W^{1,p}(S_k(\omega))$, then*

(4.9) $$f \circ p \in W^{1,p}(S_{2h+\varepsilon}(\omega))$$

and for all $0 < k \le 2h + \varepsilon$

(4.10) $$W^{1,p}(\omega) = \{ f \in L^p(\omega) : f \circ p \in W^{1,p}(S_k(\omega)) \}.$$

PROOF. Consider the map
$$y \mapsto R(y) \stackrel{\text{def}}{=} p(y) + \frac{k}{2h+\varepsilon} b(y) \nabla b(y) : S_{2h+\varepsilon}(\omega) \to S_k(\omega).$$
It is clearly well-defined since $p(R(y)) = p(y)$ and $|b(R(y))| = |kb(y)/(2h+\varepsilon)| < k$. Similarly the map
$$x \mapsto S(x) \stackrel{\text{def}}{=} p(x) + \frac{2h+\varepsilon}{k} b(x) \nabla b(x) : S_k(\omega) \to S_{2h+\varepsilon}(\omega)$$
is well-defined and $R(S(x)) = x$ and $S(R(y)) = y$. Moreover both R and $R^{-1} = S$ are Lipschitz continuous. Again from [13] (Lem. 3.1, pp. 65–66), for $p \ge 1$
$$u \mapsto u \circ R : W^{1,p}(S_k(\omega)) \to W^{1,p}(S_{2h+\varepsilon}(\omega))$$
is an isomorphism. In particular choosing $u = v \circ p \in W^{1,p}(S_k(\omega))$, then $u \circ R = v \circ p \circ R \in W^{1,p}(S_{2h+\varepsilon}(\omega))$. But
$$v(p(R(y))) = v(p(p(y) + \frac{k}{2h+\varepsilon} b(y) \nabla b(y))) = v(p(y))$$
since ∇b, b and p are defined in $S_{2h+\varepsilon}(\omega)$. This implies that
$$v \circ p = (v \circ p) \circ R \in W^{1,p}(S_{2h+\varepsilon}(\omega)).$$
The case $p = \infty$ follows from that argument and the proof of Theorem 4.5 (iii). □

We first make sense of $\nabla_\Gamma f$ for Lipschitz continuous functions f.

THEOREM 4.6. *Let ω be a bounded open subset of Γ which satisfies the assumptions of Definition 4.1. Consider a function*

(4.11) $\quad F : S_h(\omega) \to \mathbf{R}$ *such that* $F \circ p \in C^{0,1}(S_h(\omega))$ *(resp. $C^{0,1}(\overline{S_h(\omega)})$).*

(i) *If $\nabla(F \circ p)(x)$ and $D^2 b(x)$ exist at $x \in S_h(\omega)$, then $\nabla(F \circ p)(p(x))$ exists,*

(4.12) $\quad \nabla(F \circ p)(p(x)) = [I - b(x) D^2 b(x)]^{-1} \nabla(F \circ p)(x)$

(4.13) $\quad 0 = \nabla(F \circ p)(p(x)) \cdot \nabla b(x) = (\nabla(F \circ p) \cdot \nabla b)(p(x))$

The function $f \stackrel{\text{def}}{=} F|_\omega$ belongs to $C^{0,1}(\omega)$ (resp. $C^{0,1}(\overline{\omega})$). Moreover $\nabla(F \circ p)$ exists H_{N-1} almost everywhere in ω and

(4.14) $\quad \nabla_\Gamma f \stackrel{\text{def}}{=} \nabla(F \circ p)|_\omega \in L^\infty(\omega)^N$ *and* $\nabla_\Gamma f \cdot n = 0.$

(ii) *If $F \in C^{0,1}(S_h(\omega))$ and $\nabla(F \circ p)(x)$ and $D^2 b(x)$ exist at $x \in S_h(\omega)$, then $(P\nabla F)(p(x))$ exists and*

(4.15) $\quad (P\nabla F)(p(x)) = [I - b(x) D^2 b(x)]^{-1} \nabla(F \circ p)(x).$

Moreover $F \circ p = f \circ p$ and

(4.16) $\quad \nabla_\Gamma f = (P\nabla F)|_\omega$ *and* $(P\nabla F) \circ p = \nabla_\Gamma f \circ p$

(4.17) $\quad P\nabla(F \circ p) = \nabla(F \circ p) = [I - b D^2 b](P\nabla F) \circ p$

(4.18) $\quad P\nabla(f \circ p) = \nabla(f \circ p) = [I - b D^2 b] \nabla_\Gamma f \circ p.$

(iii) *For any $g \in C^{0,1}(\omega)$, $\nabla(g \circ p)$ exists H_{N-1} a. e. in ω and*

(4.19) $\quad \nabla_\Gamma g \stackrel{\text{def}}{=} \nabla(g \circ p)|_\omega \in L^\infty(\omega)^N$

(4.20) $\quad \nabla_\Gamma g \cdot n = 0$ *and* $\nabla_\Gamma g = P\nabla(g \circ p)|_\omega$

(4.21) $\quad P\nabla(g \circ p) = \nabla(g \circ p) = [I - b D^2 b] \nabla_\Gamma g \circ p.$

In particular

(4.22) $\quad C^{0,1}(\omega) \subset W^{1,\infty}_{\text{loc}}(\omega)$ *and* $C^{0,1}(\overline{\omega}) \subset W^{1,\infty}(\omega).$

PROOF. The proof is similar to the one of Theorem 4.4. (i) By assumption $D^2 b(x)$ exists and necessarily

$$Dp(x) = I - \nabla b(x)^* \nabla b(x) - b(x) D^2 b(x) \text{ exists.}$$

By assumption the directional derivative of $G = F \circ p$ at x in the direction v

$$dG(x; v) \stackrel{\text{def}}{=} \frac{1}{t}[G(x + tv) - G(x)] = \nabla G(x) \cdot v$$

exists. So does $d(G \circ p)(x; v)$ since $G \circ p = (F \circ p) \circ p = F \circ p = G$ and $d(G \circ p)(x; v) = dG(x; v)$. Consider the differential quotient for $G \circ p$

$$q_t \stackrel{\text{def}}{=} \frac{1}{t}[G(p(t + tv)) - G(p(x))].$$

Then

$$q_t = \frac{1}{t}\left[G(p(x) + t\frac{p(x + tv) - p(x)}{t}) - G(p(x))\right]$$

$$= \frac{1}{t}[G(p(x) + t dp(x; v)) - G(p(x))]$$

$$+ \frac{1}{t}\left[G(p(x) + t\frac{p(x + tv) - p(x)}{t}) - G(p(x) + t dp(x; v))\right].$$

The second term on the right-hand side of the last identity goes to zero as t goes to zero. By Lipschitz continuity of G in $p(x)$ it is bounded by

$$\frac{c}{t}\left|p(x) + t\frac{p(x+tv) - p(x)}{t} - p(x) - tdp(x;v)\right|$$

$$\leq c\left|\frac{p(x+tv) - p(x)}{t} - dp(x;v)\right| \to 0 \text{ as } t \searrow 0$$

since $Dp(x)$ and hence $dp(x;v)$ exists. Therefore for all v

$$\boxed{dG(p(x); dp(x;v)) = \nabla G(x) \cdot v}$$

But

$$dp(x;v) = Dp(x)v = [I - b(x)D^2b(x)]v - \nabla b(x)\,{}^*\nabla b(x)v.$$

Therefore for tangential directions v_Γ, $Dp(x)v_\Gamma = [I - b(x)D^2b(x)]v_\Gamma$ and

$$dG(p(x); [I - b(x)D^2b(x)]v_\Gamma) = \nabla G(x) \cdot v_\Gamma.$$

Now $I - b(x)D^2b(x)]$ is invertible. It maps $\nabla b(x)$ onto $\nabla b(x)$ and tangential vectors onto other tangential vectors. So for all tangential vectors v_Γ

$$\boxed{dG(p(x); v_\Gamma) = \nabla G(x) \cdot [I - b(x)D^2b(x)]^{-1}v_\Gamma}$$

This characterizes $dG(p(x); v)$ for tangential vectors v's. For general vectors v

$$\frac{1}{t}[G(p(x+tv)) - G(p(x))] = \frac{1}{t}[G(p(x+tv)) - G(p(x+tv_\Gamma))]$$
$$+ \frac{1}{t}[G(p(x+tv_\Gamma)) - G(p(x))].$$

The limit of the second term is $dG(p(x); v_\Gamma)$ where v_Γ is the tangential component of v. As for the first term

$$T_1 \stackrel{\text{def}}{=} \frac{1}{t}[G(p(x+tv)) - G(p(x+tv_\Gamma))]$$
$$= \frac{1}{t}[G(x+tv) - G(x)] - \frac{1}{t}[G(x+tv_\Gamma) - G(x)]$$

since $G = G \circ p$. But $\nabla G(x)$ exists and as t goes to zero

$$T_1 \to \nabla G(x) \cdot v - \nabla G(x) \cdot v_\Gamma = \nabla G(x) \cdot (v - v_\Gamma) = \nabla G(x) \cdot (v_n \nabla b(x)) = 0,$$

where v_n is the normal component of v since $\nabla G(x) \cdot \nabla b(x) = 0$. This arises from the fact that $p(x + t\nabla b(x)) = p(x)$ and

$$G(x + t\nabla b(x)) - G(x) = G(p(x + t\nabla b(x))) - G(p(x)) = 0.$$

Finally for any $v \in \mathbf{R}^N$

$$dG(p(x); v) = dG(p(x); v_\Gamma) = \nabla G(x) \cdot [I - b(x)D^2b(x)]^{-1}v_\Gamma$$
$$= \nabla G(x) \cdot [I - b(x)D^2b(x)]^{-1}v$$

since $[I - b(x)D^2b(x)]^{-1}\nabla b(x) = \nabla b(x)$ and $\nabla G(x) \cdot \nabla b(x) = 0$. In conclusion

$$\nabla G(p(x)) \,\exists\, \text{ and } \boxed{\nabla G(p(x)) = [I - b(x)D^2b(x)]^{-1}\nabla G(x)}$$

The restriction $f = (F \circ p)|_\omega = F|_\omega$ of a $C^{0,1}$ function to ω belongs to $C^{0,1}(\omega)$. Since both $G = F \circ p$ and ∇b belongs to $C^{0,1}(S_h(\omega))$, then, by Rademacher's theorem, ∇G and $D^2 b$ exist almost everywhere. So the set of points where either ∇G or $D^2 b$

does not exist is of zero measure and we are back to the conditions of part (i) m_N a.e. in $S_h(\omega)$. In such points
$$\nabla G(p(x)) = [I - b(x)D^2 b(x)]^{-1}\nabla G(x)$$
$$\nabla_\Gamma f = \nabla G|_\omega \text{ and } \nabla_\Gamma f \cdot n = 0 \Rightarrow \nabla_\Gamma f = P\nabla G|_\omega.$$

To show that ∇G exists H_{N-1} almost everywhere we proceed by contradiction. If there exists $A \subset \omega$ such that $H_{N-1}(A) > 0$, where $\nabla G(x)$ does not exists, then for all $y \in S_h(\omega)$ such that $p(x) = y$, $\nabla G(y)$ does not exist. Otherwise this would contradict part (i). Therefore ∇G does not exist in
$$S_h(A) \stackrel{\text{def}}{=} \{x \in S_h(\omega) : p(x) \in A\} = \{x \in \mathbf{R}^N : |b(x)| < h, p(x) \in A\}.$$
But we have seen in the proof of Theorem 4.4 that
$$H_{N-1}(A) > 0 \Rightarrow m_N(S_h(A)) > 0.$$
This contradicts the fact that, for the Lipschitz continuous function G, ∇G exists m_N almost everywhere in $S_h(\omega)$ by Rademacher's Theorem. To show that $\nabla_\Gamma f$ is H_{N-1} measurable, it is sufficient to show that for all open $V \subset \mathbf{R}^N$, $(\nabla_\Gamma f)^{-1}(V)$ is H_{N-1} measurable. This follows from the identity of part (i)
$$\nabla_\Gamma f \circ p = \nabla G \circ p = [I - bD^2 b]^{-1}\nabla G$$
where the right-hand side is m_N-measurable in $S_h(\omega)$. Then $[\nabla G \circ p]^{-1}(V) = p^{-1}((\nabla_\Gamma f)^{-1}(V))$ is m_N measurable. But by Theorem 4.1 (ii), p maps m_N measurable sets onto H_{N-1} measurable subsets of Γ. So
$$(\nabla_\Gamma f)^{-1}(V) = p([\nabla G \circ p]^{-1}(V))$$
is H_{N-1} measurable. Finally $\nabla_\Gamma f \in L^\infty(\omega)^N$ since $\nabla_\Gamma f$ is uniformly bounded a.e.

(ii) We proceed as in part (i) and recall that $\nabla(F \circ p)(x)$ exists, and that $\nabla(F \circ p)(x) = P(x)\nabla(F \circ p)(x)$. Consider the differential quotient
$$q_t \stackrel{\text{def}}{=} \frac{1}{t}[F(p(t+tv)) - F(p(x))] \quad \lim_{t \searrow 0} q_t = \nabla(F \circ p)(x) \cdot v.$$
Then
$$q_t = \frac{1}{t}\left[F(p(x) + t\frac{p(x+tv) - p(x)}{t}) - F(p(x))\right]$$
$$= \frac{1}{t}[F(p(x) + tdp(x;v)) - F(p(x))]$$
$$+ \frac{1}{t}\left[F(p(x) + t\frac{p(x+tv) - p(x)}{t}) - GF(p(x) + tdp(x;v))\right].$$

The second term on the right-hand side of the last identity goes to zero as t goes to zero. By Lipschitz continuity of F in $p(x)$ it is bounded by
$$\frac{c}{t}\left|p(x) + t\frac{p(x+tv) - p(x)}{t} - p(x) - tdp(x;v)\right|$$
$$\leq c\left|\frac{p(x+tv) - p(x)}{t} - dp(x;v)\right| \to 0 \text{ as } t \searrow 0$$
since $Dp(x)$ and hence $dp(x;v)$ exists. Therefore for all v
$$\boxed{dF(p(x); dp(x;v)) = \nabla(F \circ p)(x) \cdot v}$$

But $dp(x;v) = Dp(x)v = [I - b(x)D^2b(x)]v - \nabla b(x)\,^*\nabla b(x)v$. For all directions v, $P(x)v$ is tangential,

$$dp(x; P(x)v) = Dp(x)P(x)v$$
$$= [I - b(x)D^2b(x)]P(x)v = P(x)[I - b(x)D^2b(x)]v$$
$$dF(p(x); P(x)[I - b(x)D^2b(x)]v) = dF(p(x); dp(x; P(x)v))$$
$$= \nabla(F \circ p)(x) \cdot P(x)v = \nabla(F \circ p)(x) \cdot v.$$

But $I - b(x)D^2b(x)]$ is invertible and for all directions v

$$dF(p(x); P(x)v) = \nabla(F \circ p)(x) \cdot [I - b(x)D^2b(x)]^{-1}v$$

Using the fact that $P(x) = P(p(x))$

$$\boxed{dF(p(x); P(p(x))v) = [I - b(x)D^2b(x)]^{-1}\nabla(F \circ p)(x) \cdot v}$$

which means that $P\nabla F$ is defined at $p(x)$ and that

$$(P\nabla F)(p(x)) = [I - b(x)D^2b(x)]^{-1}\nabla(F \circ p)(x).$$

Another way to prove the same result is to use the approximation of Lipschitz functions by C^1 functions (cf [**11**], Thm 1, p. 251). Here $F \in C^{0,1}(S_h(\omega))$ and $p \in C^{0,1}(S_h(\omega))^N$ imply that $G = F \circ p \in C^{0,1}(S_h(\omega))$. There exist $\{F_n\} \subset C^1(S_h(\omega))$ and $\{p_n\} \subset C^1(S_h(\omega))^N$ such that the measure $m_N(N_n)$ of the set

$$N_n = \{x \in S_h(\omega) : F_n(x) \neq F(x), p_n(x) \neq p(x)$$
$$\nabla F_n(x) \neq \nabla F(x), Dp_n(x) \neq Dp(x)\}$$

is less than $1/n$ and there exists $c(N) > 0$ such that

$$|\nabla F_n(x)| \leq c(N)\,\text{Lip}(F) \text{ and } |\nabla p_n(x)| \leq c(N)\,\text{Lip}(p).$$

Therefore

$$\nabla(F_n \circ p_n) = {}^*Dp_n(x)\nabla F_n \circ p_n$$
$$\Rightarrow \nabla(F \circ p) = {}^*Dp(x)\nabla F \circ p \text{ in } S_h(\omega) \setminus N_n$$
$$\Rightarrow \nabla(F \circ p) = {}^*Dp(x)\nabla F \circ p \text{ in } S_h(\omega) \setminus N, \quad N \stackrel{\text{def}}{=} \cap_n N_n$$
$$\Rightarrow \nabla(F \circ p) = {}^*Dp(x)\nabla F \circ p \text{ a.e. in } S_h(\omega)$$

since $m_N(N) = 0$. But for all v

$$\nabla(F \circ p)(x) \cdot v = \nabla(F \circ p)(x) \cdot P(x)v$$
$$= {}^*Dp(x)\nabla F(p(x)) \cdot P(x)v$$
$$= \nabla F(p(x)) \cdot Dp(x)P(x)v$$
$$= \nabla F(p(x)) \cdot [I - b(x)D^2b(x)]P(x)v$$
$$= \nabla F(p(x)) \cdot P(x)[I - b(x)D^2b(x)]v$$
$$= \nabla F(p(x)) \cdot P(p(x))[I - b(x)D^2b(x)]v$$
$$= (P\nabla F)(p(x)) \cdot [I - b(x)D^2b(x)]v$$
$$\Rightarrow \nabla(F \circ p)(x) = [I - b(x)D^2b(x)](P\nabla F)(p(x)),$$

since $P(x) = P(p(x))$ and
$$Dp(x)P(x) = [I - b(x)D^2b(x)]P(x)$$
$$= P(x)[I - b(x)D^2b(x)] = P(p(x))[I - b(x)D^2b(x)].$$
Moreover since $[I - b(x)D^2b(x)]$ is invertible
$$(P\nabla F)(p(x)) = [I - b(x)D^2b(x)]^{-1}\nabla(F \circ p)(x)$$
$$\Rightarrow (P\nabla F)|_\omega = \nabla_\Gamma f \in L^\infty(\omega)^N.$$

(iii) This follows from part (ii) for the function $F = g \circ p$ which is $C^{0,1}(S_h(\omega))$ as the composition of two $C^{0,1}$ functions. For any $g \in C^{0,1}(\overline{\omega})$, $g \circ p \in C^{0,1}(\overline{S_h(\omega)})$ and by Rademacher's theorem $g \circ p \in W^{1,\infty}(S_h(\omega))$. Therefore, by definition of $W^{1,\infty}(\omega)$, $C^{0,1}(\overline{\omega}) \subset W^{1,\infty}(\omega)$. Similarly for any $g \in C^{0,1}(\omega)$, $g \circ p \in C^{0,1}(S_h(\omega))$ and by Rademacher's theorem $g \circ p \in W^{1,\infty}_{\text{loc}}(S_h(\omega))$ and $C^{0,1}(\omega) \subset W^{1,\infty}_{\text{loc}}(\omega)$. □

We have seen in Theorem 4.6 (iii) that we can associate with each $f \in C^{0,1}(\overline{\omega})$ a *tangential gradient*
$$\nabla_\Gamma f \stackrel{\text{def}}{=} \nabla(f \circ p)|_\omega \in L^\infty(\omega)^N \text{ and } f \in W^{1,\infty}(\omega).$$
In view of the equivalence of norms in $L^p(\omega)$ and $L^p(S_h(\omega))$ for $p \geq 1$ in Theorem 4.5, there exist $c > 0$ and $\bar{h} > 0$ such that for all $0 < h \leq \bar{h}$

(4.23)
$$(1 - ch^2)\int_\omega |\nabla_\Gamma f|^p d\Gamma \leq \frac{1}{2h}\int_{S_h(\omega)} |\nabla_\Gamma f \circ p|^p dx$$
$$\leq (1 + ch^2)\int_\omega |\nabla_\Gamma f|^p d\Gamma$$

and from the fact that $\nabla(f \circ p) = [I - bD^2b]\nabla_\Gamma f \circ p$ for all p, $1 \leq p < \infty$,

(4.24)
$$\boxed{\begin{aligned}\lim_{h \searrow 0} \frac{1}{2h}\int_{S_h(\omega)} |[I - bD^2b]^{-1}\nabla(f \circ p)|^p dx &= \int_\omega |\nabla_\Gamma f|^p d\Gamma \\ \|[I - bD^2b]^{-1}\nabla(f \circ p)\|_{L^\infty(S_h(\omega))} &= \|\nabla_\Gamma f\|_{L^\infty(\omega)}\end{aligned}}$$

Since $I - bD^2b$ is bounded and goes to I in the L^∞ norm, there exists another constant $c > 0$ such that, as h goes to zero, we also have

(4.25) $$(1 - ch)\int_\omega |\nabla_\Gamma f|^p d\Gamma \leq \frac{1}{2h}\int_{S_h(\omega)} |\nabla(f \circ p)|^p dx \leq (1 + ch)\int_\omega |\nabla_\Gamma f|^p d\Gamma$$

(4.26)
$$\boxed{\begin{aligned}\forall p, 1 \leq p < \infty, \quad \lim_{h \searrow 0} \frac{1}{2h}\int_{S_h(\omega)} |\nabla(f \circ p)|^p dx &= \int_\omega |\nabla_\Gamma f|^p d\Gamma \\ \lim_{h \searrow 0} \|\nabla(f \circ p)\|_{L^\infty(S_h(\omega))} &= \|\nabla_\Gamma f\|_{L^\infty(\omega)}\end{aligned}}$$

The next theorem makes sense of $\nabla_\Gamma f$ for any f in $W^{1,p}(\omega)$ and $1 \leq p < \infty$.

THEOREM 4.7. *Let ω be a bounded open subset of Γ satisfying the assumptions of Definition 4.1 and $1 \leq p \leq \infty$.*

(i) *The space $W^{1,p}(\omega)$ is complete when endowed with the norm*

(4.27) $$n_h(f) \stackrel{\text{def}}{=} \|f \circ p\|_{W^{1,p}(S_h(\omega))}.$$

(ii) *For each $f \in W^{1,p}(\omega)$ there exists a unique $\nabla_\Gamma f \in L^p(\omega)^N$ such that*

(4.28) $$\nabla(f \circ p) = [I - bD^2b]\nabla_\Gamma f \circ p.$$

(iii) The space $W^{1,p}(\omega)$ is also complete for the norm

(4.29) $$\boxed{\|f\|_{W^{1,p}(\omega)} \stackrel{\text{def}}{=} \{\|f\|^p_{L^p(\omega)} + \|\nabla_\Gamma f\|^p_{L^p(\omega)}\}^{1/p}}$$

which is equivalent to $n_h(f)$ and identities (4.24) and (4.26) hold.

(iv) For $1 \leq p < \infty$, the space $C^{0,1}(\overline{\omega})$ is dense in $W^{1,p}(\omega)$ and the space

(4.30) $$C^\infty(|\omega) \stackrel{\text{def}}{=} \{F|_\omega : \forall F \in C^\infty(S_{2h}(\omega))\}$$

is dense in $W^{1,p}(\omega)$.

(v) There exist $c > 0$ and $\bar{h} > 0$ such that for all $0 < h \leq \bar{h}$ and $f \in W^{1,p}(\omega)$

(4.31) $$\boxed{\begin{aligned}(1-ch^2)\int_\omega |\nabla_\Gamma f|^p d\Gamma &\leq \frac{1}{2h}\int_{S_h(\omega)} |\nabla_\Gamma f \circ p|^p dx \\ &\leq (1+ch^2)\int_\omega |\nabla_\Gamma f|^p d\Gamma\end{aligned}}$$

(4.32) $$\boxed{(1-ch)\int_\omega |\nabla_\Gamma f|^p d\Gamma \leq \frac{1}{2h}\int_{S_h(\omega)} |\nabla(f \circ p)|^p dx \leq (1+ch)\int_\omega |\nabla_\Gamma f|^p d\Gamma}$$

For p, $1 \leq p < \infty$,

(4.33) $$\lim_{h \searrow 0} \frac{1}{2h}\int_{S_h(\omega)} |[I-bD^2b]^{-1}\nabla(f \circ p)|^p dx = \int_\omega |\nabla_\Gamma f|^p d\Gamma$$
$$\lim_{h \searrow 0} \frac{1}{2h}\int_{S_h(\omega)} |\nabla(f \circ p)|^p dx = \int_\omega |\nabla_\Gamma f|^p d\Gamma$$

and $p = \infty$

(4.34) $$\|[I-bD^2b]^{-1}\nabla(f \circ p)\|_{L^\infty(S_h(\omega))} = \|\nabla_\Gamma f\|_{L^\infty(\omega)}$$
$$\lim_{h \searrow 0} \|\nabla(f \circ p)\|_{L^\infty(S_h(\omega))} = \|\nabla_\Gamma f\|_{L^\infty(\omega)}$$

Moreover as h goes to zero

(4.35) $$\frac{1}{2h}\int_{-h}^h \nabla(f \circ p) \circ T_z\, dz \to \nabla_\Gamma f \text{ in } L^p(\omega)^N$$

(4.36) $$\frac{1}{2h}\int_{S_h(\omega)} f \circ p\, dx \to \int_\omega f\, d\Gamma, \quad \frac{1}{2h}\int_{S_h(\omega)} \nabla(f \circ p)\, dx \to \int_\omega \nabla_\Gamma f\, d\Gamma.$$

PROOF. (i) Consider a Cauchy sequence $\{f_n\}$ in $W^{1,p}(\omega)$. By definition of n_h, $\{f_n \circ p\}$ is also Cauchy in $W^{1,p}(S_h(\omega))$ and there exists $F \in W^{1,p}(S_h(\omega))$ such that

$$f_n \circ p \to F \text{ in } W^{1,p}(S_h(\omega)) \Rightarrow f_n \circ p \to F \text{ in } L^p(S_h(\omega)).$$

But $\{f_n \circ p\} \subset L^p(S_h(\omega))$ and by Theorem 4.5 (v) there exists $f \in L^p(\omega)$ such that $f_n \circ p \to f \circ p$ in $L^p(S_h(\omega))$. Hence $F = f \circ p$ and $f_n \circ p \to f \circ p$ in $W^{1,p}(S_h(\omega))$. Therefore $W^{1,p}(\omega)$ is closed for the norm n_h.

(ii) By Theorem 4.6 (iii) we have already shown the existence of $\nabla_\Gamma f \in L^\infty(\omega)^N$ for elements f of $C^{0,1}(\overline{\omega})$. For $1 \leq p < \infty$ we proceed by density. Indeed given $f \in L^p(\omega)$ such that $f \circ p \in W^{1,p}(S_h(\omega))$, then $f \circ p \in W^{1,p}(S_{2h+\varepsilon}(\omega))$ by Lemma 4.1. Let $F = f \circ p\, \psi$ where $\psi \in C_c^\infty(\mathbf{R}^N)$ is as defined in Theorem 2.2 of § 2. Then $F \in W^{1,p}(\mathbf{R}^N)$. There exists a sequence $\{F_n\} \subset C_c^\infty(\mathbf{R}^N)$ such that $F_n \to F$ in $W^{1,p}(\mathbf{R}^N)$. From Theorem 2.2 of § 2 for all $|z| < h$, the map $T_z : \mathbf{R}^N \to \mathbf{R}^N$ is a bijection, T_z and T_z^{-1} are Lipschitz continuous, and $F_n \circ T_z \to F \circ T_z$ in $W^{1,p}(\mathbf{R}^N)$

$$G_n \stackrel{\text{def}}{=} [I - bD^2b]^{-1}P\nabla F_n \to [I - bD^2b]^{-1}P\nabla F \text{ in } L^p(S_{2h+\varepsilon}(\omega))^N$$

$$G_n \circ T_z \to \{[I - bD^2b]^{-1}P\nabla F\} \circ T_z \text{ in } L^p(S_h(\omega))^N.$$

Moreover on $S_h(\omega)$, $F \circ T_z = f \circ p \circ T_z = f \circ p$, $[I - bD^2b]^{-1}P\nabla F = [I - bD^2b]^{-1}\nabla(f \circ p)$, and by Theorem 4.4 (i)

$$\begin{aligned}\{[I - bD^2b]^{-1}\nabla(f \circ p)\} \circ T_z &= [I - bD^2b]^{-1} \circ T_z \nabla(f \circ p) \circ T_z \\ &= [I - bD^2b]^{-1} \circ T_z [I + zD^2b]^{-1}\nabla(f \circ p \circ T_z) \\ &= [I - bD^2b]^{-1}\nabla(f \circ p \circ T_z) = [I - bD^2b]^{-1}\nabla(f \circ p).\end{aligned}$$

Hence for $|z| < h$, $F_n \circ T_z \to f \circ p$ in $W^{1,p}(S_h(\omega))$. Define

$$\boxed{\overline{F}_n \stackrel{\text{def}}{=} \frac{1}{2h} \int_{-h}^{h} F_n \circ T_z \, dz} \qquad \boxed{\overline{f}_n \stackrel{\text{def}}{=} \frac{1}{2h} \int_{-h}^{h} (F_n \circ T)(\cdot, z) \, dz}$$

$$\Rightarrow \{\overline{F}_n\} \subset C^{0,1}(\overline{S_h(\omega)}), \quad \overline{F}_n|_\omega = \overline{f}_n, \quad \overline{F}_n \circ p = \overline{f}_n \circ p.$$

By Theorem 4.6 (iii) there exists $\nabla_\Gamma \overline{f}_n \in L^\infty(\omega)^N$ such that

(4.37) $$\nabla(\overline{f}_n \circ p) = [I - bD^2b]\nabla_\Gamma \overline{f}_n \circ p.$$

Define

(4.38) $$\overline{g}_n \stackrel{\text{def}}{=} \frac{1}{2h} \int_{-h}^{h} (G_n \circ T)(\cdot, z) \, dz \in L^\infty(\omega)^N.$$

We wish to show that $\overline{g}_n = \nabla_\Gamma \overline{f}_n$. The function $F_n \circ T_z$ belongs to $C^{0,1}(\overline{S_h(\omega)})$ and by Theorem 4.6 (ii)

$$\nabla(F_n \circ T_z \circ p) = [I - bD^2b](P\nabla(F_n \circ T_z)) \circ p.$$

Since the map T_z is bijective and T_z and T_z^{-1} are Lipschitz continuous

$$\nabla(F_n \circ T_z) = {}^*DT_z \nabla F_n \circ T_z = [I + zD^2b]\nabla F_n \circ T_z.$$

From Theorem 4.4 (i) and since $D^2b \circ p$ is well-defined and $b \circ p = 0$

$$I + zD^2b = [I - bD^2b][I - bD^2b]^{-1} \circ T_z$$
$$P[I - bD^2b] = P[I - bD^2b] \text{ and } P = P \circ T_z$$
$$\nabla(F_n \circ T_z) = [I - bD^2b]\{[I - bD^2b]^{-1}\nabla F_n\} \circ T_z$$
$$\begin{aligned}P\nabla(F_n \circ T_z) &= P[I - bD^2b]\{[I - bD^2b]^{-1}\nabla F_n\} \circ T_z \\ &= [I - bD^2b]P\{[I - bD^2b]^{-1}\nabla F_n\} \circ T_z \\ &= [I - bD^2b]P \circ T_z\{[I - bD^2b]^{-1}\nabla F_n\} \circ T_z \\ &= [I - bD^2b]\{P[I - bD^2b]^{-1}\nabla F_n\} \circ T_z \\ &= [I - bD^2b]\{[I - bD^2b]^{-1}P\nabla F_n\} \circ T_z = [I - bD^2b]G_n \circ T_z\end{aligned}$$
$$(P\nabla(F_n \circ T_z)) \circ p = G_n \circ T_z \circ p \quad \Rightarrow \quad \nabla(F_n \circ T_z \circ p) = [I - bD^2b]G_n \circ T_z \circ p.$$

Now take the z-average of both side of the last identity

$$\frac{1}{2h}\int_{-h}^{h}\nabla(F_n\circ T_z\circ p)\,dz = \nabla\left(\left[\frac{1}{2h}\int_{-h}^{h}F_n\circ T_z\,dz\right]\circ p\right)$$
$$= \nabla(\overline{F}_n\circ p) = \nabla(\overline{f}_n\circ p)$$

and from (4.38)

$$\nabla(\overline{F}_n\circ p) = \frac{1}{2h}\int_{-h}^{h}[I - bD^2b]G_n\circ T_z\circ p\,dz$$
$$= [I - bD^2b]\frac{1}{2h}\int_{-h}^{h}G_n\circ T_z\,dz\circ p$$
$$= [I - bD^2b]\frac{1}{2h}\int_{-h}^{h}(G_n\circ T)(\cdot,z)\,dz\circ p = [I - bD^2b]\overline{g}_n\circ p$$

by Theorem 4.5 (i). But $\overline{F}_n\circ p = \overline{f}_n\circ p$ and from (4.37)

$$\nabla(\overline{F}_n\circ p) = \nabla(\overline{f}_n\circ p) = [I - bD^2b]\nabla_\Gamma\overline{f}_n\circ p \quad\Rightarrow\quad \boxed{\overline{g}_n = \nabla_\Gamma\overline{f}_n}$$

Since $\{F_n\}$ is Cauchy in $W^{1,p}(S_h(\omega))$, so is $\{G_n\}$ in $L^p(S_h(\omega))^N$. By continuity of the averaging map of Theorem 4.5 (i), the sequences $\{\overline{f}_n\}$ and $\{\overline{g}_n = \nabla_\Gamma\overline{f}_n\}$ are Cauchy in $L^p(S\omega)$ and $L^p(S\omega)^N$. So there exists $g \in L^p(S\omega)^N$ such that

$$\overline{f}_n \to \frac{1}{2h}\int_{-h}^{h}F\circ T_z\,dz = f \text{ in } L^p(\omega) \text{ and } \overline{g}_n = \nabla_\Gamma\overline{f}_n \to g \text{ in } L^p(\omega)^N.$$

By Theorem 4.5 (i)

$$\overline{f}_n\circ p \to f\circ p \text{ in } L^p(\omega) \text{ and } \overline{g}_n\circ p \to g\circ p \text{ in } L^p(\omega)^N$$
$$\nabla(\overline{f}_n\circ p) = [I - bD^2b]\overline{g}_n\circ p \to [I - bD^2b]g\circ p \text{ in } L^p(\omega)^N$$

and necessarily

$$\overline{f}_n\circ p \to f\circ p \text{ in } W^{1,p}(\omega) \quad\Rightarrow\quad \nabla(f\circ p) = [I - bD^2b]g\circ p.$$

By defining

(4.39)
$$\boxed{\nabla_\Gamma f \stackrel{\text{def}}{=} g \in L^p(\omega)^N}$$

we get the required identities. For $p = \infty$ we further show that $g \in L^\infty(\omega)^N$.

(iii) Recall that from the equivalence of L^p norms in Theorem 4.5: there exist $c > 0$ and $\overline{h} > 0$ such that for al $0 < h \le \overline{h}$

$$(1 - ch^2)\int_\omega |f|^p + |\nabla_\Gamma f|^p\,d\Gamma \le \frac{1}{2h}\int_{S_h(\omega)} |f\circ p|^p + |\nabla_\Gamma f\circ p|^p\,dx$$
$$\le (1 + ch^2)\int_\omega |f|^p + |\nabla_\Gamma f|^p\,d\Gamma.$$

But $\nabla_\Gamma f\circ p = [I - bD^2b]^{-1}\nabla(f\circ p)$ and since D^2b is bounded a.e. in $S_h(\omega)$ by a constant independent of h, there exists $c > 0$ such that

$$(1 - ch)\|\nabla(f\circ p)\|_{L^p(S_h(\omega))} \le \|[I - bD^2b]^{-1}\nabla(f\circ p)\|_{L^p(S_h(\omega))}$$
$$\le (1 + ch)\|\nabla(f\circ p)\|_{L^p(S_h(\omega))}.$$

Therefore there exists $c' > 0$ such that as h goes to zero

$$(1 - c'h) \int_\omega |f|^p + |\nabla_\Gamma f|^p d\Gamma \le \frac{1}{2h} \int_{S_h(\omega)} |f \circ p|^p + |\nabla(f \circ p)|^p dx$$

$$\le (1 + c'h)) \int_\omega |f|^p + |\nabla_\Gamma f|^p d\Gamma.$$

and

$$\lim_{h \searrow 0} \frac{1}{2h} \int_{S_h(\omega)} |f \circ p|^p + |\nabla(f \circ p)|^p dx = \int_\omega |f|^p + |\nabla_\Gamma f|^p d\Gamma.$$

Therefore we have the equivalence of the n_h and $W^{1,p}(\omega)$ norms. Since $W^{1,p}(\omega)$ was complete for the $\|f \circ p\|_{W^{1,p}(\omega)}$ norm, it is also complete for the $W^{1,p}(\omega)$ norm.

(iv) The density follows from parts (ii) and (iii).

(v) By the density of $C^{0,1}(\overline{\omega})$ in $W^{1,p}(\omega)$ of part (iv), inequalities (4.31) and (4.32) follow from inequalities (4.23) and (4.25) and the limits (4.33) and (4.34) from the limits (4.24) and (4.26). Finally consider the difference of the average of $\nabla(f \circ p)$ and $\nabla_\Gamma f$ in $L^p(\omega)^N$

$$\frac{1}{2h} \int_{-h}^{h} \nabla(f \circ p) \circ T_z \, dz - \nabla_\Gamma f = \frac{1}{2h} \int_{-h}^{h} ([I - bD^2 b] \nabla_\Gamma f \circ p) \circ T_z \, dz - \nabla_\Gamma f$$

$$= \frac{1}{2h} \int_{-h}^{h} [I + zD^2 b]^{-1} \nabla_\Gamma f \, dz - \nabla_\Gamma f = \frac{1}{2h} \int_{-h}^{h} \left\{ [I + zD^2 b]^{-1} - I \right\} dz \, \nabla_\Gamma f$$

$$\Rightarrow \int_\omega \left| \frac{1}{2h} \int_{-h}^{h} \nabla(f \circ p) \circ T_z \, dz - \nabla_\Gamma f \right|^p d\Gamma$$

$$\le \int_\omega \left| \frac{1}{2h} \int_{-h}^{h} \left\{ [I + zD^2 b]^{-1} - I \right\} dz \right| |\nabla_\Gamma f|^p d\Gamma$$

$$\le \|[I + zD^2 b]^{-1} - I\|_{L^\infty(\omega)} \|\nabla_\Gamma f\|^p_{L^p(\omega)}$$

and $\|[I + zD^2 b]^{-1} - I\|_{L^\infty(\omega)}$ goes to zero as h goes to zero. The other limits are obvious by using similar techniques. $\square$

4.3. Parametric definitions of tangential Sobolev spaces. A special case of the intrinsic definition of Sobolev spaces is when Γ is an $(N - 1)$ dimensional hyperplane in $\mathbf{R}^N$. More precisely consider the half-space

(4.40) $$H \stackrel{\text{def}}{=} \{\zeta = (\zeta', \zeta_N) : \forall \zeta' \in \mathbf{R}^{N-1}, \zeta_N > 0\}$$

(4.41) $$H_0 \stackrel{\text{def}}{=} \partial H = \{\zeta = (\zeta', \zeta_N) : \forall \zeta' \in \mathbf{R}^{N-1}, \zeta_N = 0\}$$

and its associated oriented distance function

(4.42) $$b_H(\zeta) = b_H(\zeta', \zeta_N) = -\zeta_N$$

which is a C^∞ function. It is easy to check that

(4.43) $$\nabla b_H = -\begin{bmatrix} 0 \\ 1 \end{bmatrix}, \; D^2 b_H = 0, \; p_{H_0}(\zeta) = \begin{bmatrix} \zeta' \\ 0 \end{bmatrix}, \; P_{H_0}(\zeta) = \begin{bmatrix} I_{N-1} & 0 \\ 0 & 0 \end{bmatrix}$$

The tangential derivative of a function $f : B_0 \subset H_0 \to \mathbf{R}$ is given by

$$\nabla_{H_0} f = \nabla(f \circ p_{H_0})|_{H_0}.$$

But $f \circ p_{H_0}(\zeta', \zeta_N) = f(\zeta', 0)$ and we can associate with $f(\zeta', 0)$ the function
$$f'(\zeta') \stackrel{\text{def}}{=} f(\zeta', 0) : B' \subset \mathbf{R}^{N-1} \to \mathbf{R}$$
where $B' = \{\zeta' \in \mathbf{R}^{N-1} : |\zeta'| < 1\}$. Then the tangential gradient reduces to the usual gradient of f' in $\mathbf{R}^{N-1}$
$$\nabla_{H_0} f(\zeta', 0) = \begin{bmatrix} \nabla f'(\zeta') \\ 0 \end{bmatrix}.$$
and we have the following equivalence for any p, $1 \leq p \leq \infty$:
$$f \in W^{1,p}(B_0) \iff f' \in W^{1,p}(B').$$

Assume that Γ satisfies the assumptions of Definition 4.1. For each $x \in \Gamma$, there exists an open neighborhood $U(x)$ of x and
$$\exists f_x \in C^{1,1}(U(x), B) \text{ such that } h_x = f_x^{-1} \in C^{1,1}(B, U(x))$$
such that
(4.44) $\quad \text{int } \Omega \cap U(x) = h_x(B_+) \quad \text{and} \quad \Gamma_x \stackrel{\text{def}}{=} \partial\Omega \cap U(x) = h_x(B_0),$

where B is the open unit ball with center at 0 in $\mathbf{R}^N$. For each $x \in \omega$, associate with a function w on ω the functions

(4.45) $\quad W_x \stackrel{\text{def}}{=} w \circ f_x^{-1} : B_0 \cap f_x(\omega) \to \mathbf{R}$

(4.46) $\quad \zeta' \to W_x'(\zeta') \stackrel{\text{def}}{=} (w \circ f_x^{-1})(\zeta', 0) : B_x' \to \mathbf{R}$

where
$$B_x' \stackrel{\text{def}}{=} \{\zeta' \in \mathbf{R}^{N-1} : (\zeta', 0) \in B_0 \cap f_x(\omega)\}.$$

DEFINITION 4.2 (Parametric definition of Sobolev spaces). Let ω be a bounded open subset of Γ which satisfies the assumptions of Definition 4.1.

(4.47) $\quad \forall p, 1 \leq p \leq \infty, \quad W_f^{1,p}(\omega) \stackrel{\text{def}}{=} \{w : \omega \to \mathbf{R} : \forall x \in \omega, W_x' \in W^{1,p}(B_x')\}.$

Of course there is a direct relationship between the derivatives of the functions $w : \omega \to \mathbf{R}$ and the set of functions $W_x' : B_x' \to \mathbf{R}$ indexed by $x \in \omega$.

THEOREM 4.8. *Let ω be a bounded open subset f Γ satisfying the assumptions of Definition 4.1. Given p, $1 \leq p \leq \infty$,*

(4.48) $\quad W_f^{1,p}(\omega) = W^{1,p}(\omega).$

Moreover on $\Gamma_x \cap \omega$

(4.49) $\quad \nabla_\Gamma w = {}^*D_\Gamma f_x \nabla_{H_0} W_x \circ f_x = P {}^*Df_x \nabla_{H_0} W_x \circ f_x$

(4.50) $\quad \nabla_\Gamma w(y) = P(y) {}^*Df_x(y) \begin{bmatrix} \nabla W_x'((f_x(y))') \\ 0 \end{bmatrix}$

and on $B_0 \cap f_x(\omega)$

(4.51) $\quad \nabla_{H_0} W_x = {}^*D_{H_0} f_x^{-1} \nabla_\Gamma w \circ f_x^{-1} = P_{H_0} {}^*Df_x^{-1} \nabla_\Gamma w \circ f_x^{-1}$

(4.52) $\quad \begin{bmatrix} \nabla W_x'(\zeta') \\ 0 \end{bmatrix} = P_{H_0} {}^*Df_x^{-1}(\zeta', 0) \nabla_\Gamma w(f_x^{-1}(\zeta', 0)).$

COROLLARY 4.8.1. *Given two $C^{1,1}$ sets Ω_1 and Ω_2, $\Gamma_1 = \partial\Omega_1$, $\Gamma_2 = \partial\Omega_2$, assume that $g : U(\Gamma_1) \to U(\Gamma_2)$ is a $C^{1,1}$-diffeomorphism such that*

$$g(\Gamma_1) = \Gamma_2 \quad g(\Omega_1 \cap U(\Gamma_1)) = \Omega_2 \cap U(\Gamma_2)$$

for some neighborhoods $U(\Gamma_i)$ of Γ_i, $i = 1, 2$. Then for any open $\omega_1 \subset \Gamma_1$ and $\omega_2 = g(\omega_1)$

$$\forall w_1 \in W^{1,p}(\omega_1), \quad \nabla_{\Gamma_2}(w_1 \circ g) = {}^*D_{\Gamma_2}g \nabla_{\Gamma_1} w_1 \circ g$$

$$\forall w_2 \in W^{1,p}(\omega_2), \quad \nabla_{\Gamma_1}(w_2 \circ g^{-1}) = {}^*D_{\Gamma_1}g^{-1}\nabla_{\Gamma_2} w_2 \circ g^{-1}$$

In particular

$$^*D_{\Gamma_2}g(x) : T_{\Gamma_1}(x) \to T_{\Gamma_2}(g(x)) \quad \text{and} \quad {}^*D_{\Gamma_1}g^{-1}(y) : T_{\Gamma_2}(y) \to T_{\Gamma_1}(g^{-1}(y))$$

$$^*D_{\Gamma_2}g\, ({}^*D_{\Gamma_1}g^{-1}) \circ g = I_{T_{\Gamma_2}\circ g} \quad \text{and} \quad {}^*D_{\Gamma_1}g^{-1}\, ({}^*D_{\Gamma_2}g) \circ g^{-1} = I_{T_{\Gamma_1}}$$

where $I_{\Gamma_i}(x)$ is the identity transformation of $T_{\Gamma_i}(x)$.

4.4. Intermediary Sobolev spaces. Because of the connection between tangential Sobolev spaces in ω and their counterpart in $S_h(\omega)$, we can define Sobolev spaces of functions with s-derivatives, $0 < s < 1$ by direct interpolation.

DEFINITION 4.3. Assume that the conditions of Definition 4.1 are satisfied. For any real s, $0 < s < 1$, and $p \geq 1$,

(4.53) $\quad W^{s,p}(\omega) \stackrel{\text{def}}{=} \{f : \omega \to \mathbf{R} : f \circ p \in W^{s,p}(S_h(\omega))\}, \quad H^s(\omega) \stackrel{\text{def}}{=} W^{s,2}(\omega).$

For instance it is easy to check that for $p = 2$ we can interpolate between spaces on both sides of the definition

$$H^s(\omega) = [H^1(\omega), L^2(\omega)]_{1-s} = \{f : \omega \to \mathbf{R} : f \circ p \in H^s(S_h(\omega))\}.$$

4.5. Smoothness of the boundary γ and extension operators.

4.5.1. Smoothness of the boundary γ. The next issue is the smoothness of the boundary γ of the domain ω in Γ. It is natural to characterize it by the smoothness of the *normal set* generated by the flow of the gradient of b through γ in a small neighborhood of ω. To be more precise recall that for a domain Ω of class $C^{1,1}$ in each point x of its boundary Γ there exists a neighborhood $U(x)$ where $b \in C^{1,1}(U(x))$. So for a bounded open domain ω in Γ there exists a $h > 0$ such that b_Ω is $C^{1,1}$ in the set

(4.54) $\quad S_h(\omega) \stackrel{\text{def}}{=} \{x \in \mathbf{R}^N : |b_\Omega(x)| < h \text{ and } p(x) \in \omega\}$

generated by the normal flow through ω. If γ is the (relative) boundary of ω in Γ, the *lateral boundary* of $S_h(\omega)$

(4.55) $\quad \Sigma_h(\gamma) \stackrel{\text{def}}{=} \{x \in \mathbf{R}^N : |b_\Omega(x)| < h \text{ and } p(x) \in \gamma\}$

is a submanifold of $\mathbf{R}^N$ of codimension one *normal* to Γ. The smoothness of the boundary γ of ω in Γ will be characterized by the smoothness of the lateral boundary $\Sigma_h(\gamma)$ near γ.

DEFINITION 4.4. Let ω be a bounded open subset of Γ which satisfies the assumptions of Definition 4.1.

(i) Given an integer $k \geq 1$ and a real number $0 \leq \lambda \leq 1$, the boundary γ is $C^{k,\lambda}$ if there exist $h > 0$ and $0 < h' \leq h$ such that the piece $\Sigma_{h'}(\gamma)$ of the lateral boundary of $S_h(\omega)$ is $C^{k,\lambda}$.

(ii) The boundary γ is Lipschitzian if there exist $h > 0$ and $0 < h' \leq h$ such that $\Sigma_{h'}(\gamma)$ is Lipschitzian.

(iii) The boundary γ has the cone property if there exist $h > 0$ and $0 < h' \leq h$ such that $\Sigma_{h'}(\gamma)$ is has the cone property.

(iv) The boundary γ has the *uniform cone property* if there exist $h > 0$, $0 < h' \leq h$, $\lambda > 0$, $0 < \beta \leq \pi/2$, $\rho > 0$ such that

$$\forall x \in \Sigma_{h'}(\gamma), \quad \exists \zeta_x \in \mathbf{R}^N, \quad |\zeta_x| = 1 \text{ such that}$$
$$\forall y \in B(x,\rho) \cap \overline{S}_h(\omega), \quad y + C(\lambda,\beta,\zeta_x) \subset S_h(\omega)$$

where $C(\lambda,\beta,\zeta)$ is the open cone of vertex 0, height λ, aperture β, and direction ζ defined as

$$C(\lambda,\beta,\zeta) \stackrel{\text{def}}{=} \left\{ y \in \mathbf{R}^N : \frac{1}{\tan\beta}|P(y)| < y \cdot \zeta < \lambda \right\}$$

and P is the orthogonal projection onto the hyperplane orthogonal to ζ.

(v) The domain ω is connected if there exists h', $0 < h' < h$, such that $S_{h'}(\omega)$ is connected.

The definitions correspond to the usual ones in $\mathbf{R}^N$. For instance condition (i) is equivalent to say that the oriented distance function $b_{S_h(\omega)}$ associated with the set $S_h(\omega)$ has the required smoothness in a neighborhood of $\Sigma_{h'}(\gamma)$.

DEFINITION 4.5. Let ω be a bounded open subset of Γ which satisfies the assumptions of Definition 4.1. Further assume that ω is connected with a Lipschitzian boundary γ, and that γ_0 is an $(N-2)$-Hausdorff measurable subset of γ such that $H_{N-2}(\gamma_0) > 0$. Given $1 \leq p \leq \infty$, define

$$W^{1,p}_{\gamma_0}(\omega) \stackrel{\text{def}}{=} \{ f : \omega \to \mathbf{R} : f \circ p \in W^{1,p}_{\Sigma_h(\gamma_0)}(S_h(\omega)) \}$$

where

$$W^{1,p}_{\Sigma_h(\gamma_0)}(S_h(\omega)) \stackrel{\text{def}}{=} \{ F \in W^{1,p}(S_h(\omega)) : F|_{\Sigma_h(\gamma_0)} = 0 \}$$
$$\Sigma_h(\gamma_0) = \{ x : |b(x)| < h \text{ and } p(x) \in \gamma_0 \}.$$

When a domain ω is bounded and connected with a Lipschitzian boundary γ in Γ and Γ is of class C^2, there exists a sufficiently small $h > 0$ such that $S_h(\omega)$ is a bounded connected domain in $\mathbf{R}^N$ with a Lipschitzian boundary. So for $f \in C^1(\omega)$ such that $\nabla_\Gamma f = 0$ in ω

$$\nabla(f \circ p) = [I - bD^2b]\nabla_\Gamma f \circ p = 0 \text{ in } S_h(\omega)$$

and since $S_h(\omega)$ is a bounded connected open domain with a Lipschitzian boundary, there exists a constant such that $f \circ p = c$ in $S_h(\omega)$ and a fortiori $f = c$ in ω which is the analogue of what we have in $\mathbf{R}^N$.

4.5.2. *Extension of $W^{1,p}(\omega)$-functions to Γ.* Paralleling [13] (§3.6 and §3.7, Thms 3.9 and 3.10, Chap. 2), we give for the submanifold Γ the equivalent of the methods of Nikolskij-Babič and Calderon-Nečas.

THEOREM 4.9. *Assume that Γ satisfies the assumptions of Theorem 2.2 in § 2. Let ω be a bounded open subset of Γ satisfying the conditions of Definition 4.1. Further assume that ω is connected with Lipschitzian boundary γ.*

(i) *Given p, $1 \leq p < \infty$, the Nikolskij-Babič extension operator*

$$P_\Gamma : W^{1,p}(\omega) \to W^{1,p}(\mathbf{R}^N) \quad \text{exists.}$$

(ii) *Given p, $1 < p < \infty$, the Calderon-Nečas extension operator*
$$P_\Gamma : W^{1,p}(\omega) \to W^{1,p}(\mathbf{R}^N) \quad \text{exists and } (P_\Gamma(f)|_\omega = f.$$

5. Tangential Korn's inequalities

The tangential versions of Korn's inequalities are central in the theory of shells. With our intrinsic definition of tangential Sobolev spaces, they almost come as a corollary to their analogues in the Euclidean N-dimensional space.

5.0.3. *First Korn's inequality.* Recall Korn's inequality for a bounded open Lipschitzian domain Ω in $\mathbf{R}^N$. There exists a constant $c(\Omega) > 0$ such that
$$\forall V \in \{V \in L^2(\Omega) : \|\varepsilon(V)\|_{L^2(\Omega)} < \infty\}$$
(5.1) $\quad \int_\Omega \|D(V)\|^2 \, dx \leq c(\Omega) \int_\Omega \|\varepsilon(V)\|^2 + \|V\|^2 \, dx, \quad \varepsilon(V) \stackrel{\text{def}}{=} \frac{1}{2}\{D(V) + {}^*D(V)\}.$

THEOREM 5.1. *Let ω be a bounded open subset of Γ satisfying the assumptions of Definition 4.1. Further assume that ω is Lipschitzian if the boundary γ of ω is not non-empty.*

(i) *There exists a constant $c(\omega) > 0$ such that*
$$\forall v \in E(\omega) = \{v \in L^2(\omega)^N : \varepsilon_\Gamma(v) \in L^2(\omega)^{N\times N}\}$$
(5.2) $\quad \int_\omega \|D_\Gamma(v)\|^2 \, d\Gamma \leq c(\omega)^2 \int_\omega |v|^2 + \|\varepsilon_\Gamma(v)\|^2 \, d\Gamma$

and $E(\omega) = H^1(\omega)^N$.

(ii) *There exists a constant $c(\omega) > 0$ such that*
$$\forall v \in E_\Gamma(\omega) = \{v \in L^2(\omega)^N : v_n = v \cdot n = 0 \text{ and } \varepsilon_\Gamma^P(v_\Gamma) \in L^2(\omega)^{N\times N}\}$$
(5.3) $\quad \begin{aligned} \int_\omega \|D_\Gamma(v_\Gamma)\|^2 \, d\Gamma &\leq c(\omega)^2 \int_\omega |v_\Gamma|^2 + \|\varepsilon_\Gamma^P(v_\Gamma)\|^2 \, d\Gamma \\ \int_\omega \|D_\Gamma^P(v_\Gamma)\|^2 \, d\Gamma &\leq c(\omega)^2 \int_\omega |v_\Gamma|^2 + \|\varepsilon_\Gamma^P(v_\Gamma)\|^2 \, d\Gamma \end{aligned}$

and $E_\Gamma(\omega) = \{v \in H^1(\omega)^N : v_n = v \cdot n = 0 \text{ on } \omega\}$. In particular
(5.4) $\quad \left\{\|v\|_{L^2(\omega)}^2 + \|\varepsilon_\Gamma^P(v)\|_{L^2(\omega)}^2\right\}^{1/2} \text{ and } \left\{\|v\|_{L^2(\omega)}^2 + \|D_\Gamma^P(v)\|_{L^2(\omega)}^2\right\}^{1/2}$

are equivalent norms on the space
(5.5) $\quad H_t^1(\omega)^N \stackrel{\text{def}}{=} \{v \in H^1(\omega)^N : v \cdot n = 0\}$

endowed with the norm
(5.6) $\quad \left\{\|v_n\|_{L^2(\omega)}^2 + \|v_\Gamma\|_{L^2(\omega)}^2 + \|D_\Gamma(v_\Gamma)\|_{L^2(\omega)}^2\right\}^{1/2}.$

PROOF. (i) If $\omega = \Gamma$ then $S_h(\omega)$ is Lipschitzian. If ω has a boundary, then by assumption on ω and the definition of a Lipschitzian domain ω in Γ, we can choose h sufficiently small to make the lateral boundary $\Sigma_h(\gamma)$ and hence $S_h(\omega)$ Lipschitzian. The main idea of the proof is to apply Korn's inequality in $S_h(\omega)$,
$$\exists c(h) > 0, \quad \forall V \in H^1(S_h(\omega))^N$$
(5.7) $\quad \int_{S_h(\omega)} \|DV\|^2 \, dx \leq c(h) \int_{S_h(\omega)} \|\varepsilon(V)\|^2 + \|V\|^2 \, dx,$

to the function
$$V = \begin{cases} \dfrac{k^2 - b^2}{k^2} v \circ p, & \text{in } S_k(\omega) \\ 0, & \text{in } S_h(\omega) \setminus S_k(\omega) \end{cases}$$

for vector functions $v : \omega \to \mathbf{R}^N$ and k, $0 < k \leq h$. The function V belongs to $H^1(S_h(\omega))^N$, has support in $S_k(\omega)$, and in $S_k(\omega)$

$$DV = \frac{k^2 - b^2}{k^2} D(v \circ p) - \frac{2b}{k^2} v \circ p \, {}^*\nabla b$$

$$\varepsilon(V) = \frac{k^2 - b^2}{k^2} \varepsilon(v \circ p) - \frac{b}{k^2} \left[v \circ p \, {}^*\nabla b + \nabla b \, {}^*(v \circ p) \right],$$

The final result is obtained through estimates for k sufficiently small. It will be convenient to use the respective notation $\| \ \|_k$ and $\| \ \|_h$ for the L^2-norms on $S_k(\omega)$ and $S_h(\omega)$, and $\| \ \|$ for the L^2-norm on ω. Observe that for k, $0 < k \leq h$

$$\frac{1}{2} \|v \circ p\|_{k/2} \leq \|V\|_{k/2} \leq \|v \circ p\|_{k/2}$$

$$\frac{1}{2} \|\varepsilon(v \circ p)\|_{k/2} - \frac{1}{k} \|v \circ p\|_{k/2} \leq \|\varepsilon(V)\|_{k/2} \leq \|\varepsilon(v \circ p)\|_{k/2} + \frac{1}{k} \|v \circ p\|_{k/2}$$

and by Korn's inequality in $S_{k/2}(\omega)$ and definition of $H^1(\omega)$ in Definition 4.1 and property (4.10) in Lemma 4.1, $v \in H^1(\omega)^N$.

From Theorem 4.7 (v) and Theorem 4.5 (ii), there exists $\bar{h}$, $0 < \bar{h} \leq h$, such that for all k, $0 < k \leq \bar{h}$,

$$(5.8) \qquad \frac{1}{2} \|D_\Gamma(v)\|^2 \leq \frac{1}{2k} \int_{S_{2k}(\omega)} \|D(v \circ p)\|^2 \, dx \leq 2 \|D_\Gamma(v)\|^2$$

$$(5.9) \qquad \frac{1}{2} \|v\|^2 \leq \frac{1}{2k} \int_{S_{2k}(\omega)} \|v \circ p\|^2 \, dx \leq 2 \|v\|^2.$$

Also

$$\varepsilon(v \circ p) = \varepsilon_\Gamma(v) \circ p - \frac{b}{2} \left[D_\Gamma v \circ p \, D^2 b + D^2 b \, {}^*D_\Gamma v \circ p \right].$$

But the elements of $D^2 b$ belong to $L^\infty(S_{2h+2\varepsilon}(\omega))$ and Theorem 4.5 (ii) extends to matrix functions $\varepsilon_\Gamma(v)$. Therefore there exists a constant $c > 0$ such that

$$(5.10) \qquad \frac{1}{2k} \int_{S_{2k}(\omega)} \|\varepsilon(v \circ p)\|^2 \, dx \leq 2 \|\varepsilon_\Gamma(v)\|^2 + ck^2 \|D_\Gamma(v)\|^2.$$

Moreover

$$(5.11) \qquad \|V\|_h = \|V\|_k = \left\| \frac{k^2 - b^2}{k^2} v \circ p \right\|_k \leq \|v \circ p\|_k$$

$$(5.12) \qquad \|DV\|_h = \|DV\|_k \leq \left\| \frac{k^2 - b^2}{k^2} D(v \circ p) \right\|_k + \left\| \frac{2b}{k^2} v \circ p \, {}^*\nabla b \right\|_k$$
$$\leq \|D(v \circ p)\|_k + \frac{2}{k} \|v \circ p\|_k$$

$$(5.13) \qquad \|\varepsilon(V)\|_h = \|\varepsilon(V)\|_k \leq \|\varepsilon(v \circ p)\|_k + \frac{2}{k} \|v \circ p\|_k$$

In addition

$$\left\|\frac{k^2-b^2}{k^2}D(v\circ p)\right\|_k^2 = \int_{S_k(\omega)}\left|\frac{k^2-b^2}{k^2}\right|^2\|D(v\circ p)\|^2\,dx$$

$$= \int_\omega\int_{-k}^k j_z\left|\frac{k^2-z^2}{k^2}\right|^2\|D_\Gamma(v)\|^2\|[I+zD^2b]^{-1}\|^2\,dz d\Gamma$$

$$\geq \frac{1}{2}\int_\omega\int_{-k}^k\left|\frac{k^2-z^2}{k^2}\right|^2\|D_\Gamma(v)\|^2\,dz d\Gamma \geq \frac{8}{15}k\|D_\Gamma(v)\|^2$$

(5.14) $$\Rightarrow \left\|\frac{k^2-b^2}{k^2}D(v\circ p)\right\|_k^2 \geq \frac{8}{15}k\|D_\Gamma(v)\|^2$$

for k, $0 < k \leq \bar{h}$, and hence

(5.15) $$\frac{8}{15}k\|D_\Gamma(v)\|^2 \leq 2\left(\|DV\|_h^2 + \left\|\frac{2b}{k^2}v\circ p\,{}^*\nabla b\right\|_k^2\right)$$

$$\leq 2\left(\|DV\|_h^2 + \frac{4}{k^2}\|v\circ p\|_k^2\right).$$

Now uses Korn's inequality

$$\|DV\|_h^2 \leq c(h)\left(\|\varepsilon(V)\|_h^2 + \|V\|_h^2\right) \leq c(h)\left(\|\varepsilon(V)\|_k^2 + \|V\|_k^2\right)$$

$$\leq c(h)\left(2\|\varepsilon(v\circ p)\|_k^2 + \frac{4}{k}\|v\circ p\|_k^2 + \|v\circ p\|_k^2\right)$$

$$\leq c(h)\left(2\|\varepsilon(v\circ p)\|_k^2 + \left(\frac{4}{k}+1\right)\|v\circ p\|_k^2\right)$$

from (5.11) and (5.13). Now use (5.10) and (5.9) in the last inequality

$$\|DV\|_h^2 \leq 2c(h)\|\varepsilon(v\circ p)\|_k^2 + c(h)\left(\frac{4}{k}+1\right)\|v\circ p\|_k^2$$

$$\frac{1}{2k}\|DV\|_h^2 \leq 2c(h)\frac{1}{2k}\|\varepsilon(v\circ p)\|_k^2 + c(h)\left(\frac{4}{k}+1\right)\frac{1}{2k}\|v\circ p\|_k^2$$

$$\leq 2c(h)\left(2\|\varepsilon_\Gamma(v)\|^2 + ck^2\|D_\Gamma(v)\|^2\right) + 2c(h)\left(\frac{4}{k}+1\right)\|v\|^2$$

for k, $0 < k \leq \bar{h}$. By combining this last inequality and (5.15), we finally get

$$\frac{2}{15}\|D_\Gamma(v)\|^2 \leq \frac{1}{2k}\left(\|DV\|_h^2 + \frac{4}{k^2}\|v\circ p\|_k^2\right) \leq \frac{1}{2k}\|DV\|_h^2 + \frac{8}{k^2}\|v\|^2$$

$$\leq 4c(h)\|\varepsilon_\Gamma(v)\|^2 + 2c(h)\,ck^2\|D_\Gamma(v)\|^2 + \left(2c(h)\left(\frac{4}{k}+1\right)+\frac{8}{k^2}\right)\|v\|^2$$

and

$$\left(\frac{2}{15} - 2c(h)\,ck^2\right)\|D_\Gamma(v)\|^2 \leq 4c(h)\|\varepsilon_\Gamma(v)\|^2 + \left(2c(h)\left(\frac{4}{k}+1\right)+\frac{8}{k^2}\right)\|v\|^2$$

Now chose k, $0 < k \leq \inf\{\bar{h},h\}$, such that

$$\frac{1}{15} \leq \frac{2}{15} - 2c(h)\,ck^2 \quad \Rightarrow \quad k^2 \leq \frac{1}{30\,c(h)\,c}$$

and this yields the tangential Korn's inequality.

(ii) For tangential vector functions

$$D_\Gamma(v_\Gamma) = D_\Gamma^P(v_\Gamma) + n\,{}^*n\,D_\Gamma(v_\Gamma) = D_\Gamma^P(v_\Gamma) - n\,{}^*(D^2 b v_\Gamma)$$

$$\Rightarrow \varepsilon_\Gamma(v_\Gamma) = \varepsilon_\Gamma^P(v_\Gamma) - \frac{1}{2}\left[n\,{}^*(D^2 b v_\Gamma) + D^2 b v_\Gamma\,{}^*n\right].$$

Therefore since the elements of $D^2 b$ belong to $L^\infty(S_{2h+\varepsilon}(\omega))$

$$\|\varepsilon_\Gamma(v_\Gamma)\| \leq \|\varepsilon_\Gamma^P(v_\Gamma)\| + c\,\|v_\Gamma\|$$

and from part (i) we get

$$\|D_\Gamma(v_\Gamma)\|^2 \leq c\left[2\|\varepsilon_\Gamma(v_\Gamma)\|^2 + 2c^2\|v_\Gamma\|^2 + \|v_\Gamma\|^2\right] \leq c'\left[\|\varepsilon_\Gamma(v_\Gamma)\|^2 + \|v_\Gamma\|^2\right]$$

and necessarily $\|D_\Gamma^P(v_\Gamma)\|^2 \leq c''\left[\|\varepsilon_\Gamma(v_\Gamma)\|^2 + \|v_\Gamma\|^2\right]$. From this we readily obtain the equivalences of norms. □

5.0.4. *Second Korn's inequality and finite dimensionality of* $\ker \varepsilon_\Gamma$. The second Korn's inequality is an equivalence of norms relationship which follows from the fact that the kernel of ε_Γ (resp. ε_Γ^P) is finite dimensional. So $\|\varepsilon_\Gamma(v)\|$ (resp. $\|\varepsilon_\Gamma^P(v)\|$) becomes a norm which is equivalent to the usual quotient norm on $H^1(\omega)^N/\ker \varepsilon_\Gamma$ (resp. $H_t^1(\omega)^N/\ker \varepsilon_\Gamma^P$). Because of the finite dimensionality of the kernel the H^1-norm is equivalent to the norm $\|\varepsilon_\Gamma(v)\|$ (resp. $\|\varepsilon_\Gamma^P(v)\|$) plus any norm on the H^1-projection onto $\ker \varepsilon_\Gamma$ (resp. $\ker \varepsilon_\Gamma^P$). For instance this property can be used to obtain Poincaré's inequalities. The finite dimensionality of the kernels is a consequence of the following lemma.

LEMMA 5.1. *Let V and H be two real Hilbert spaces with continuous compact injection of V into H and identify the elements of the dual H' of H with those of H. Assume that $A : V \to V'$ is a linear continuous symmetrical operator which is V-H coercive, that is there exist λ and $\alpha > 0$ such that*

$$\forall v \in V, \quad \lambda\,|v|_H^2 + \langle Av, v\rangle_{V' \times V} \geq \alpha\|v\|_V^2.$$

Then $\ker A \stackrel{\text{def}}{=} \{v \in V : Av = 0\}$ *is a finite dimensional subspace of V and*

(5.16) $$\exists c > 0,\ \forall v \in \ker A, \quad \|v\|_V \leq c\,|v|_H.$$

THEOREM 5.2. *Let ω be a bounded open subset $f\,\Gamma$ satisfying the assumptions of Definition 4.1. Further assume that ω is Lipschitzian if ω has a non-empty boundary γ.*

(i) *For the operator*

$$\langle Au, v\rangle \stackrel{\text{def}}{=} \int_\omega \varepsilon_\Gamma(u) \cdot\cdot\, \varepsilon_\Gamma(v)\,d\Gamma, \quad \forall u, v \in H^1(\omega)^N$$

the assumptions of Lemma 5.1 are satisfied and

$$\ker \varepsilon_\Gamma \stackrel{\text{def}}{=} \left\{v \in H^1(\omega)^N : \varepsilon_\Gamma(v) = 0\right\}$$

is finite dimensional. Moreover if $\pi_{\ker \varepsilon_\Gamma}(v)$ denotes the H^1-projection of v onto $\ker \varepsilon_\Gamma$

$$\left\{\|\varepsilon_\Gamma(v)\|_{L^2(\omega)}^2 + \|\pi_{\ker \varepsilon_\Gamma}(v)\|_{L^2(\omega)}^2\right\}^{1/2}$$

is an equivalent norm on $H^1(\omega)^N$ and $\|\varepsilon_\Gamma(v)\|$ is a norm on the quotient space $H^1(\omega)^N/\ker \varepsilon_\Gamma$ which is equivalent to the natural quotient norm.

(ii) *For the operator*
$$\langle A_\Gamma u, v\rangle \stackrel{\text{def}}{=} \int_\omega \varepsilon_\Gamma^P(u) \cdot \cdot \varepsilon_\Gamma^P(v)\, d\Gamma$$
$$\forall u, v \in H_t^1(\omega)^N \stackrel{\text{def}}{=} \{v \in H^1(\omega)^N : v \cdot n = 0\}$$
the conditions of Lemma 5.1 are satisfied and
$$\ker \varepsilon_\Gamma^P \stackrel{\text{def}}{=} \{v \in H^1(\omega)^N : \varepsilon_\Gamma^P(v) = 0 \text{ and } v \cdot n = 0\}$$
is finite dimensional. Moreover if $\pi_{\ker \varepsilon_\Gamma^P}(v)$ *denotes the* H^1-*projection of* v *onto* $\ker \varepsilon_\Gamma^P$
$$\left\{\|\varepsilon_\Gamma^P(v)\|^2 + \|\pi_{\ker \varepsilon_\Gamma^P}(v)\|^2\right\}^{1/2}$$
is an equivalent norm on $H_t^1(\omega)^N$ *and* $\|\varepsilon_\Gamma^P(v)\|$ *is a norm on the quotient space* $H_t^1(\omega)^N / \ker \varepsilon_\Gamma^P$ *which is equivalent to the natural quotient norm.*

PROOF. In view of Korn's inequality (5.2) of Theorem 5.1, we have the equivalence of the norms
$$\left\{\int_\omega |v|^2 + \|D_\Gamma(v)\|^2\, d\Gamma\right\} \text{ and } \left\{\int_\omega |v|^2 + \|\varepsilon_\Gamma(v)\|^2\, d\Gamma\right\}$$
on $H^1(\omega)^N$. Hence the conditions of the lemma are satisfied with $\lambda = 1$ for the operator A and
$$Av = 0 \Rightarrow \|\varepsilon_\Gamma(v)\|^2 = \langle Av, v\rangle = 0 \text{ and } v \in \ker \varepsilon_\Gamma$$
and the converse is obvious. The equivalence of norms now follows from the fact that all norms are equivalent on a finite dimensional subspace and the continuity of the H^1-projection. The proof is similar for A_Γ. □

References

[1] M.C. Delfour, *Intrinsic Differential Geometric Methods in the Asymptotic Analysis of Linear Thin Shells*, in "Boundaries, interfaces and transitions", M. Delfour, ed., pp. 19-90, CRM Proc. Lecture Notes, vol. 13, AMS Publications, Providence, R.I. 1998.

[2] ———, *Intrinsic $P(2,1)$ thin shell model and Naghdi's models without a priori assumption on the stress tensor*, in Proc International Conference on Optimal Control of Partial Differential Equations, K.H. Hoffmann, G. Leugering, F. Tröltzsch, eds., pp. 99-113, Int. Ser. of Numerical Mathematics, Vol. 133, Birkhäuser Verlag, Basel 1999.

[3] ———, *Membrane shell equation: characterization of the space of solutions*, in "Control of Distributed Parameter and Stochastic Systems", Shuping Chen, Xunjing Li, Jiongmin Yong, Xun Yu Zhou, eds., pp 21-29, Chapman and Hall, 1999.

[4] ———, *Characterization of the space of solutions of the membrane shell equation for arbitrary $C^{1,1}$ midsurfaces*, Control and Cybernetics 28 (3) (1999), pp 481-501.

[5] M.C. Delfour and J.-P. Zolésio, *Shape analysis via distance functions*, J. Functional Analysis **123** (1994), 129–201.

[6] ———, *A boundary differential equation for thin shells*, J. Differential Equations **119** (1995), 426–449.

[7] ———, *Tangential differential equations for dynamical thin/shallow shells*, J. Differential Equations 128 (1996), 125-167.

[8] ———, *Differential equations for linear shells: comparison between intrinsic and classical models*, in "Advances in the Mathematical Sciences - CRM's 25 years", (Luc Vinet, ed.), pp. 42–124, CRM Proc. Lecture Notes, vol. 11, Amer. Math. Soc., Providence, RI, 1997.

[9] ———, *Shape analysis via distance functions: local theory*, in "Boundaries, interfaces and transitions", M. Delfour, ed., pp. 91–123, CRM Proc. Lecture Notes, Vol. 13, Amer. Math. Soc., Providence, RI, 1998.

[10] _____, *On the design and control of systems governed by differential equations on submanifolds*, Control and Cybernetics 25 (1996), 497–514.

[11] L.C. Evans and R.F. Gariepy, *Measure theory and the properties of functions*, CRC Press, Inc, Boca Raton, Fla 1992.

[12] H. Federer, *Curvature measures*, Trans. Amer. Math. Soc. **93** (1959), 418–419.

[13] J. Nečas, *Les méthodes directes en théorie des équations elliptiques*, Masson, Paris and Academia, Prague, 1967.

CENTRE DE RECHERCHES MATHÉMATIQUES ET DÉPARTEMENT DE MATHÉMATIQUES ET DE STATISTIQUE, UNIVERSITÉ DE MONTRÉAL, C. P. 6128, SUCC. CENTRE-VILLE, MONTRÉAL QC, CANADA H3C 3J7

E-mail address: `delfour@CRM.UMontreal.CA`

Carleman Estimates with Two Large Parameters and Applications

Matthias M. Eller and Victor Isakov

ABSTRACT. We derive Carleman estimates with two large parameters for general elliptic, parabolic, and some hyperbolic partial differential operators of second order with variable coefficients. As an application we prove unique continuation theorems for two systems of equations of thermo-elasticity.

1. Introduction and main result

In recent years Carleman estimates have been used to prove unique continuation for certain systems of partial differential equations. In the paper [**E-I-N-T**] uniqueness results for principally scalar systems are proved. Moreover, it is shown that Maxwell's system and the equations of classical elasticity can be represented as principally scalar systems. The proof of these uniqueness results is based on Carleman estimates for second order operators, in particular on results by L. Hörmander [**H**] and D. Tataru [**T**].

On the other hand there are systems were the principal part can not be diagonalized. One example is the thermoelastic systems which describes the dynamics of a thin elastic plate taking thermal forces into account [**L**]. V. Isakov [**I.3**] managed to prove a uniqueness result for this system in the case of constant coefficients in the principal part. Starting from his paper [**I.1**] he developed Carleman estimates with two large parameters for the Laplace and heat operator and for the wave operator with a constant speed of propagation.

We will study those Carleman estimates for basic second order operators with variable coefficients. In fact, we consider general second order elliptic and parabolic operators with different principal parts and the wave operator with a variable speed of propagation. This increases the technical difficulty of the proofs substantially. Moreover, even for isotropic wave operators we have to impose additional conditions on the speed of propagation since otherwise pseudo-convexity fails and non-uniqueness can occur. At the end we give applications towards two thermoelastic systems. These Carleman estimates will enable us give uniqueness theorems for these system with variable coefficients, even in the principal part. This is an improvement over the paper [**I.3**].

1991 *Mathematics Subject Classification.* Primary 35.

The work of Victor Isakov was in part supported by the NSF grant DMS-9803397.

© 2000 American Mathematical Society

At this point we emphasize that the uniqueness theorems discussed in the papers above are of local or semi-global nature. In order to obtain global uniqueness theorems, i.e. uniqueness theorems for certain Cauchy problems with boundary data given on the lateral surface one needs statements on forward and backwards uniqueness of these problems. For scalar equations such as the wave equation and the heat equation backwards uniqueness properties are well established; however, for certain systems this question is much harder. In particular, for the thermoelastic system a positive answer was given only recently [**L-R-T**]. Here, we will not touch the issue of global uniqueness; we will only give uniqueness theorems of semi-global nature, see also remark 3.4.

We introduce the following notations. Let $x \in \mathbf{R}^{n+1}$ and we distinguish between time and space variable, i.e. $x = (t, x')$ with $t = x_0$ and $x' = (x'', x_n)$. Moreover, let Ω denote a bounded connected domain in $\mathbf{R}^n$ and let $Q = (0, T) \times \Omega$ for some $T > 0$. By Γ we denote a subset of the boundary of Ω, i.e. $\Gamma \subset \partial\Omega$. Moreover, $D_j = -i\partial_j$ for $j = 0, .., n$, $\partial_t = \partial_0$, $\nabla = (\partial_t, \partial_1, ..., \partial_n)$ and $\nabla' = (\partial_1, \partial_2, ..., \partial_n)$. By $B(x, \delta)$ we denote the ball of radius δ with center at x and by $L^2(S)$ we denote the space of square integrable functions on S with the norm

$$\|u\|^2_{L^2(S)} = \int_S |u|^2 \qquad u \in L^2(S)$$

and we agree that in the case $S = \mathbf{R}^{n+1}$ we will drop the subscripts in the norm and in integrals.

Regarding the geometry of Ω and the choice of Γ we will distinguish two cases.

- (i): We have $0 \in \Omega$ and set $\Gamma = \partial\Omega$.
- (ii): For some $h > 0$ and $r > 0$ we have $\Omega \subset \{-h < x_n < 0, |x''| < r\}$ and we set $\Gamma = \partial\Omega \cap \{x_n < 0\}$

For a discussion of the geometry of these two domains we refer to V. Isakov's book, [**I.2**] section 3.4..

The idea behind Carleman estimates is the following. Given a function $\psi(x)$ with strongly pseudo-convex level surfaces with respect to the linear partial differential operator $P(x, D)$ of order m in Q, one can obtain estimates of the form

$$(1.1) \qquad \sum_{|\alpha| \leq m-1} \tau^{m-|\alpha|-\frac{1}{2}} \|e^{\tau\varphi} D^\alpha u\| \leq C(\lambda) \|e^{\tau\varphi} P(x,D)u\| \qquad \tau \geq \tau_0$$

for $u \in C_0^\infty(Q)$ where $\varphi = e^{\lambda\psi}$ and $\lambda \geq \lambda_0$, see [**H**], chapter 8. We will show that for certain second order operators one can keep track of the parameter λ and make it appear in the estimate (1.1) explicitly. Observe that in our estimates C will not depend on λ. Moreover, we will choose ψ to be a function which has strongly pseudo-convex level surfaces with respect to basic second order operators.

More specifically we will work with

$$(1.2) \qquad \psi(x) = -\theta^2 \left(t - \frac{T}{2}\right)^2 + |(x-a)'|^2 - \rho$$

where $\theta, \rho \in \mathbf{R}_+ = \{t \in \mathbf{R} : t \geq 0\}$ and $a \in \mathbf{R}^n$ are constants which will be determined later. Let

$$Q_\varepsilon = Q \cap \{\psi(x) > \varepsilon\}.$$

In order to formulate our results we set

(1.3) $$\varphi(x) = e^{\frac{\lambda}{2}\psi(x)}$$

where λ is a large positive parameter.

By C we will denote generic constants depending only on the differential operator, the domain Q, and on ψ. Any additional dependence will be specified. We will agree that the next constant C is not less than the previous one. We let $\sigma = \lambda\tau\varphi$.

We start with the Carleman estimate for a second order elliptic operator

(1.4) $$A(x,D) = \sum_{j,k=1}^{n} a_{jk}(x) D_j D_k .$$

The matrix (a_{jk}) is symmetric and real-valued and the operator is assumed to be elliptic, i.e. there exists a positive constant ϵ_a such that

(1.5) $$\sum_{j,k=1}^{n} a_{jk}(x)\xi_j\xi_k \geq \epsilon_a |\xi'|^2 \qquad x \in Q, \text{ for all } \xi \neq 0$$

and the constant ϵ_a does not depend on x. We like to emphasize that this operator is elliptic only with respect to the space variables. Taking the time variable t into account, this operator will not be elliptic since it does not have any derivatives with respect to t.

For this operator we will prove the following result.

THEOREM 1.1. *Assume that $a_{jk}(x) \in C^1(\overline{Q})$ and that the ellipticity condition (1.5) is satisfied.*

Then there exist constants $C = C(\varepsilon), C_0 = C_0(\varepsilon, \lambda)$ such that

(1.6) $$\sqrt{\lambda}\|\sigma^{\frac{3}{2}-|\alpha|} e^{\tau\varphi}\partial^\alpha u\| \leq C\|e^{\tau\varphi} A(x,D) u\|$$

for $u \in C_0^\infty(Q_\varepsilon)$ provided $|\alpha| \leq 2$, $\alpha_0 = 0$ and $\lambda \geq C$ and $\tau \geq C_0$.

Secondly, we consider the second order parabolic operator

(1.7) $$H(x,D) = \partial_t + A(x,D)$$

where the second part of the right hand side is the elliptic operator (1.4). We will prove the following result.

THEOREM 1.2. *Assume that $a_{jk}(x) \in C^1(\overline{Q})$ and that the ellipticity condition (1.5) is satisfied.*

Then there exist constants $C = C(\varepsilon), C_0 = C_0(\varepsilon, \lambda)$ such that

(1.8) $$\sqrt{\lambda}\|\sigma^{\frac{3}{2}-|\alpha|} e^{\tau\varphi}\partial^\alpha u\| \leq C\|e^{\tau\varphi} H(x,D) u\|$$

for $u \in C_0^\infty(Q_\varepsilon)$ provided $|\alpha| \leq 2$ with $\alpha_0 = 0$ or $|\alpha| \leq 1$ and $\lambda \geq C$ and $\tau \geq C_0$.

Next we are concerned with the hyperbolic operator

$$\Box_\gamma = -\gamma(x)\partial_t^2 + \Delta$$

where Δ is the Laplacian with respect to the space variables.

THEOREM 1.3. *Assume that $\gamma(x) \in C^1(\overline{Q})$ and that*

(1.9) $$0 < \gamma(x)\theta^2 < 1 \qquad \text{for all } x \in \overline{Q}$$

Moreover, assume that

$$\theta^2 \left(2\gamma(x)^2 + \partial_0\gamma(x)\gamma(x)(t-T/2) + 2|(t-T/2)\nabla'\gamma(x)|\sqrt{\gamma(x)}\right)$$
(1.10) $\qquad < 2\gamma(x) + \nabla'\gamma(x)\cdot(x-a)'$

for all $x \in \overline{Q}$.

Then there exist constants $C = C(\varepsilon), C_0 = C_0(\varepsilon, \lambda)$ such that

(1.11) $\qquad \|\sigma^{\frac{3}{2}-|\alpha|}e^{\tau\varphi}\partial^\alpha u\| \leq C\|e^{\tau\varphi}\Box_\gamma u\|$

for all $u \in C_0^\infty(\Omega_\varepsilon)$, provided $|\alpha| \leq 1$ and $\lambda \geq C$ and $\tau \geq C_0$.

In the following section we will prove these theorems. Our proofs will rely on differential quadratic forms, a technique used by Hörmander [**H**], chapter 8. However, we will make many computations explicitly using $\varphi(x)$ given in (1.3) as the strongly pseudo-convex function.

2. Proof of the theorems

For the proofs of all three theorems we need to discuss the strongly pseudo-convex function (1.3) and differential quadratic forms. For a motivation of this approach we refer to Hörmander's book, [**H**], chapter 8.

For future reference we state formulas for the derivatives of $\varphi(x)$.

(2.1)
$$\begin{aligned}
\partial_0\varphi &= -\lambda\theta^2\left(t-\frac{T}{2}\right)\varphi \\
\partial_0^2\varphi &= \lambda^2\theta^4\left(t-\frac{T}{2}\right)^2\varphi - \lambda\theta^2\varphi \\
\partial_j\varphi &= \lambda(x-a)_j\varphi \qquad j=1,..,n \\
\partial_j\partial_k\varphi &= \lambda^2(x-a)_j(x-a)_k\varphi + \delta_{jk}\lambda\varphi \qquad j,k=1,..,n \\
\partial_0\partial_j\varphi &= -\lambda^2(x-a)_j\theta^2\left(t-\frac{T}{2}\right)\varphi \qquad j=1,..,n
\end{aligned}$$

The choice of the parameters a and ρ depends on the geometry of the domain Ω as well. According to the two cases stated in the introduction we set
- (i): $a = 0, \rho = 0$
- (ii): $a'' = 0, a_n \geq 0$ and $\rho = a_n^2 + r^2$

We note that

(2.2) $\qquad \varphi(x) > 1 \qquad x \in Q_\varepsilon$

and that there exist two constant $c_1(\varepsilon)$ and c_2 such that

(2.3) $\qquad c_1 < |(x-a)'| < c_2 \qquad x \in Q_\varepsilon$

The dual variable of $x = (t, x')$ will be denoted by $\xi = (\xi_0, \xi')$.

Next we introduce differential quadratic forms. Given a linear partial differential operator

$$P(x,D) = \sum_{|\alpha|\leq m} a_\alpha(x)D^\alpha$$

of order m we denote its principal symbol by $p(x,\xi)$, i.e.

$$p(x,\xi) = \sum_{|\alpha|=m} a_\alpha(x)\xi^\alpha \ .$$

We make the following assumptions.
- $a_\alpha \in C^1(\overline{Q})$
- a_α is real-valued for $|\alpha| = m$.

Moreover, we introduce the symbol

(2.4)
$$p(x, \xi + i\tau \nabla\varphi(x))$$

and the operator with this symbol will be denoted by

$$P^\bullet(x, D + i\tau \nabla\varphi(x)).$$

We like to point out that the operators $P^\bullet(x, D + i\tau\nabla\varphi(x))$ and $P(x, D + i\tau\nabla\varphi(x))$ do not coincide, however their principal part in (D, τ) is the same.

In the following we will use the abbreviation $\zeta = \xi + i\tau \nabla\varphi(x)$. Furthermore

$$p^{(k)}(x, \zeta) = \frac{\partial p(x, \zeta)}{\partial \zeta_k}$$

and

$$p_{(k)}(x, \zeta) = \frac{\partial p(x, \zeta)}{\partial x_k}.$$

Based on the operator $P(x, D)$ we introduce the differential quadratic form

$$\mathcal{F}(x, D, \overline{D})v\overline{v} = |P^\bullet(x, D + i\tau\nabla\varphi(x))v|^2 - |\overline{P^\bullet}(x, D - i\tau\nabla\varphi(x))v|^2.$$

This differential quadratic form is of order $(2m - 1, m)$ since the coefficients of the principal part are real valued.

By lemma 8.2.2 in [**H**] there exists differential quadratic form $\mathcal{G}(x, D, \overline{D})$ of order $(2m - 2, m - 1)$ such that

(2.5)
$$\int_Q \mathcal{G}(x, D, \overline{D})v\overline{v} = \int_Q \mathcal{F}(x, D, \overline{D})v\overline{v} \leq \int_Q |P^\bullet(x, D + i\tau\nabla\varphi(x))v|^2.$$

The same lemma in Hörmander's book gives a rule how to compute the symbol of $\mathcal{G}$. We obtain

(2.6)
$$\begin{aligned}\mathcal{G}(x, \xi, \xi) &= 2\tau \sum_{j,k=0}^n p^{(j)}(x, \zeta)\overline{p^{(k)}(x, \zeta)}\partial^2_{jk}\varphi(x) \\ &\quad + 2\mathrm{Im} \sum_{k=0}^n p_{(k)}(x, \zeta)\overline{p^{(k)}(x, \zeta)} \\ &\quad + 2\mathrm{Im} \sum_{k=0}^n p(x, \zeta)\left(\overline{p^{(k)}_{(k)}(x, \zeta)} + i\tau \sum_{j=0}^n \overline{p^{(k,j)}(x, \zeta)}\partial_j\partial_k\varphi(x)\right).\end{aligned}$$

2.1. Proof of theorem 1.1. We denote the symbol of the operator (1.4) by a and start with computing the symbol of the differential quadratic form $\mathcal{G}$ on the characteristic set of a, i.e. on the set

$$\{(x, \zeta) : a(x, \zeta) = 0\}.$$

Using formulas (2.6) and (2.1) and the abbreviation $\sigma = \lambda\tau\varphi$ we have

$$\begin{aligned}\mathcal{G}(x,\xi,\zeta) &= 2\tau \sum_{j,k=1}^{n} a^{(j)}(x,\zeta)\overline{a^{(k)}(x,\zeta)}\partial_j\partial_k\varphi(x) \\ &\quad + 2\operatorname{Im}\sum_{k=1}^{n} a_{(k)}(x,\zeta)\overline{a^{(k)}(x,\zeta)} \\ &= 8\lambda\sigma(x)\left(\sum_{j,l=1}^{n} a_{jl}(x)\zeta_l(x-a)_j\right)\left(\sum_{k,m=1}^{n} a_{km}(x)\overline{\zeta}_m(x-a)_k\right) \\ &\quad + 8\sigma(x)\sum_{j,m,l=1}^{n} a_{jl}(x)a_{km}(x)\zeta_m\overline{\zeta}_l \\ &\quad + 4\operatorname{Im}\sum_{j,k,l,m=1}^{n} \partial_k a_{lm}(x)a_{kj}(x)\zeta_l\zeta_m\overline{\zeta}_j \\ &\geq 8\lambda\sigma(x)\left|\sum_{j,l=1}^{n} a_{jl}(x)(\xi_l+i\sigma(x)(x-a)_l)(x-a)_j\right|^2 \\ &\quad - C\sigma|\zeta|^2 - C|\zeta|^3 \:.\end{aligned}$$

Keeping only the imaginary part of the first term and using the ellipticity condition (1.5) we arrive at the inequality

$$(2.7) \qquad \mathcal{G}(x;\xi,\zeta) \geq C^{-1}\lambda\sigma^3 - C\sigma|\zeta|^2 - C|\zeta|^3$$

In order to bound the last two terms by the first one we consider the characteristic equation

$$\begin{aligned}0 = a(x,\zeta) &= \sum_{j,k=1}^{n} a_{jk}(x)(\xi_j+i\sigma(x)(x-a)_j)(\xi_k+i\sigma(x)(x-a)_k) \\ &= \sum_{j,k=1}^{n} a_{jk}(x)\xi_j\xi_k - \sigma(x)^2\sum_{j,k} a_{jk}(x)(x-a)_j(x-a)_k \\ &\quad + 2i\sigma(x)\sum_{j,k} a_{jk}(x)\xi_j(x-a)_k \:.\end{aligned}$$

Considering only the real part we have

$$\sum_{j,k} a_{jk}(x)\xi_j\xi_k = \sigma(x)^2\sum_{j,k} a_{jk}(x)(x-a)_j(x-a)_k \:.$$

which implies

$$|\xi'|^2 \leq C\sigma(x)^2$$

by ellipticity and continuity.

Continuing the estimate of the differential quadratic form we obtain from (2.7) that

$$\mathcal{G}(x,\xi,\zeta) \geq C^{-1}\lambda\sigma(x)^3 - C\sigma(x)^3 \:.$$

Choosing $\lambda \geq C_0 = 2C^2$ we obtain

$$\mathcal{G}(x,\xi,\zeta) \geq C^{-1}\lambda\sigma(x)^3 \:.$$

Combining with the above bounds on ξ we will have

$$\tau\varphi(x)\mathcal{G}(x,\xi,\xi) \geq C^{-1}\sigma(x)^4 \geq C^{-1}|\zeta'|^4$$

Here we emphasize that C will depend on ε.

By continuity of the differential quadratic form $\mathcal{G}$ and homogeneity of the symbol $a(x,\zeta)$ we obtain

$$C^{-1}|\zeta'|^4 \leq \tau\varphi(x)\mathcal{G}(x,\xi,\xi) + C|a(x,\zeta)|^2 \ .$$

Now we fix $x = x_0 = (t_0, x'_0) \in Q_\varepsilon$.

Multiplying this inequality with $|\hat{v}|^2$, the square of the Fourier transform of $v \in C_0^\infty(Q_\varepsilon)$ and integrating over the whole space in the ξ'-variable we obtain after using Parseval's identity

(2.8)
$$\begin{aligned}C^{-1}\sum_{\substack{|\alpha|\leq 2\\ \alpha_0=0}}\int_{\{t=t_0\}}\sigma(x_0)^{4-2|\alpha|}|D^\alpha v|^2 &\leq \tau\varphi(x_0)\int_{\{t=t_0\}}\mathcal{G}(x_0,D,\overline{D})v\overline{v}\\ &+ C\int_{\{t=t_0\}}|A^\bullet(x_0, D+i\tau\nabla\varphi(x_0))v|^2 \ .\end{aligned}$$

Because of formula (2.6) we have

(2.9) $\quad \mathcal{G}(x_0,D,\overline{D})v\overline{v} \leq \mathcal{G}(x,D,\overline{D})v\overline{v} + \omega_1(\delta,\lambda)\sum_{\substack{|\alpha|\leq 1\\ \alpha_0=0}}(\tau\varphi(x_0))^{3-2|\alpha|}|D^\alpha v|^2$

for $v \in C_0^\infty(Q_\varepsilon \cap B(x_0,\delta))$ and $\omega_1(\delta,\lambda) \to 0$ for $\delta \to 0$ and fixed λ. Moreover, $|A^\bullet(x, D+i\tau\nabla\varphi(x))|^2$ is a differential quadratic form of order $(4,2)$ in (D,τ), so we have

$$\begin{aligned}\int_{\{t=t_0\}}|A^\bullet(x_0, D+i\tau\nabla\varphi(x_0))v|^2 &\leq \int_{\{t=t_0\}}|A^\bullet(t_0, x', D+i\tau\nabla\varphi(t_0,x'))v|^2\\ &+ \omega_2(\delta,\lambda)\sum_{\substack{|\alpha|\leq 2\\ \alpha_0=0}}(\tau\varphi(x_0))^{4-2|\alpha|}\int_{\{t=t_0\}}|D^\alpha v|^2\end{aligned}$$

(2.10)

for $v \in C_0^\infty(Q_\varepsilon \cap B(x_0,\delta))$ and $\omega_2(\delta,\lambda) \to 0$ for $\delta \to 0$.

Given $\lambda \geq C_0$ and choosing $\delta > 0$ such that C in formula (2.8) satisfies

$$C^{-1} > \frac{\omega_1(\delta,\lambda) + C\omega_2(\delta,\lambda)}{2}$$

we can use the two estimates (2.9) and (2.10) and obtain the estimate (2.8) with x_0 replaced by (t_0, x') in the integrals in the right hand side. Dividing by $\tau\varphi(x_0)$ and observing that

$$\frac{1}{\tau\varphi(x_0)} < 1 \qquad \text{for } \tau \geq 1$$

we obtain

$$C^{-1}\lambda \sum_{\substack{|\alpha|\leq 2 \\ \alpha_0=0}} \int_{\{t=t_0\}} \sigma(x_0)^{3-2|\alpha|}|D^\alpha v|^2 \leq \int_{\{t=t_0\}} \mathcal{G}(t_0, x', D, \overline{D})v\overline{v}$$
$$+ C \int_{\{t=t_0\}} |A^\bullet(t_0, x', D + i\tau\nabla\varphi(t_0, x'))v|^2$$

for $v \in C_0^\infty(Q_\varepsilon \cap B(x_0, \delta))$ provided $\tau \geq 1$. Next we use

$$\sigma(x_0) = \sigma(t_0, x')(1 + \omega_3(\delta, \lambda)) \quad \text{for } (t_0, x') \in B(x_0, \delta)$$

where $\omega_3(\delta, \lambda) \to 0$ for $\delta \to 0$ when λ is fixed. This leads to

$$C^{-1}\lambda \sum_{\substack{|\alpha|\leq 2 \\ \alpha_0=0}} \int_{\{t=t_0\}} \sigma(t_0, x')^{3-2|\alpha|}|D^\alpha v|^2 \leq \int_{\{t=t_0\}} \mathcal{G}(t_0, x', D, \overline{D})v\overline{v}$$
$$+ C \int_{\{t=t_0\}} |A^\bullet(t_0, x', D + i\tau\nabla\varphi(t_0, x'))v|^2$$

for $v \in C_0^\infty(Q_\varepsilon \cap B(x_0, \delta))$ provided $\delta > 0$ is small enough. Integrating in time from 0 to T we have

$$C^{-1}\lambda \sum_{\substack{|\alpha|\leq 2 \\ \alpha_0=0}} \int \sigma^{3-2|\alpha|}|D^\alpha v|^2 \leq \int \mathcal{G}(x, D, \overline{D})v\overline{v}$$
$$+ C \int |A^\bullet(x, D + i\tau\nabla\varphi(x))v|^2$$

and applying (2.5) we arrive at

$$C^{-1}\lambda \sum_{\substack{|\alpha|\leq 2 \\ \alpha_0=0}} \int \sigma^{3-2|\alpha|}|D^\alpha v|^2 \leq C \int |A^\bullet(x, D + i\tau\nabla\varphi(x))v|^2$$

provided $\lambda \geq \lambda_0$ and $\tau \geq 1$. Since the difference between the two operators $A^\bullet(x, D + i\tau\nabla\varphi(x))$ and $A(x, D + i\tau\nabla\varphi(x))$ consists only of lower order terms we can replace $A^\bullet$ by A in the estimate above when we choose τ sufficiently large. Setting $u = e^{-\tau\varphi}v$ we have

$$A(x, D + i\tau\nabla\varphi(x))v = e^{\tau\varphi}A(x, D)u$$

and the proof is finished for $u \in C_0^\infty(Q_\varepsilon \cap B(x_0, \delta))$. Finally, we prove the theorem by using a partition of unity. This step requires that we choose τ large. Since the number of sets in the this process depends on $\delta(\lambda)$, the lower bound for the parameter τ will depend on λ.

2.2. Proof of theorem 1.2. The proof is similar to the proof of theorem 1.1. At first we will compute the symbol of the differential quadratic form $\mathcal{G}$ on the characteristic set. Note that $h^{(0)}(x, \zeta) = i$ and that $h_{(0)}(x, \zeta) = 0$. As in the proof of Theorem 1.1,

$$\begin{aligned}
\mathcal{G}(x,\xi,\xi) &= 2\tau \sum_{j,k=0}^{n} h^{(j)}(x,\zeta)\overline{h^{(k)}(x,\zeta)}\partial_j\partial_k\varphi(x) \\
&\quad + 2\,\mathrm{Im} \sum_{k=0}^{n} h_{(k)}(x,\zeta)\overline{h^{(k)}(x,\zeta)} \\
&= 2\tau\left(\lambda^2\theta^4(t-T/2)^2 - \lambda\theta^2\right)\varphi \\
&\quad + i\sum_{j,k=1}^{n} 2a_{jk}(x)\overline{\zeta}_j(-\lambda^2)(x-a)_k\theta^2\left(t-\frac{T}{2}\right)\varphi(x) \\
&\quad - i\sum_{j,k=1}^{n} 2a_{jk}\zeta_j(-\lambda^2)(x-a)_k\theta^2\left(t-\frac{T}{2}\right)\varphi(x) \bigg) \\
&\quad + 8\lambda\sigma(x)\sum_{j,k,l,m=1}^{n} a_{jl}(x)a_{km}(x)\zeta_l\overline{\zeta}_m(x-a)_j(x-a)_k + \\
&\quad 8\sigma(x)\sum_{j,l,m=1}^{n} a_{jl}(x)a_{jm}(x)\zeta_l\overline{\zeta}_m \\
&\quad + 4\,\mathrm{Im}\sum_{k=1}^{n}\left(\sum_{l,m=1}^{n}\partial_k a_{lm}(x)\zeta_l\zeta_m\right)\left(\sum_j a_{jk}\overline{\zeta}_j\right) \\
&\geq -C\lambda\sigma - C\lambda\sigma|\zeta'| \\
&\quad + 8\lambda\sigma(x)\left|\sum_{j,l=1}^{n} a_{jl}(x)(\xi_l + i\sigma(x)(x-a)_l)(x-a)_j\right|^2 - \\
&\quad C\sigma|\zeta'|^2 - C|\zeta'|^3\,.
\end{aligned}$$

The characteristic equation $h(x,\zeta) = 0$ leads to the identities

$$\sum_{j,k=1}^{n} a_{jk}(x)\xi_j\xi_k = \sigma(x)^2 \sum_{j,k=1}^{n} a_{jk}(x)(x-a)_j(x-a)_k - \sigma(x)\theta^2\left(t-\frac{T}{2}\right)$$

and

$$\xi_0 = -2\sigma\sum_{j,k=1}^{n} a_{jk}(x)\xi_j(x-a)_k\,.$$

The first identity and ellipticity yield

(2.11) $$|\xi'|^2 \leq C\sigma(x)^2\,.$$

The second identity gives

(2.12) $$|\xi_0| \leq C\sigma(x)^2$$

in connection with the estimate (2.11).

Now we estimate the differential quadratic form and obtain

$$\mathcal{G}(x,\xi,\xi) \geq -C\lambda\sigma - C\lambda\sigma^2 + C^{-1}\lambda\sigma^3 - C\sigma^3\,.$$

As in the previous subsection we obtain
$$\mathcal{G}(x,\xi,\xi) \geq C^{-1}\lambda\sigma^3$$
for $\lambda \geq C$. From this and from 2.11, 2.12 we conclude that
$$\tau\varphi(x)\mathcal{G}(x,\xi,\xi) \geq C^{-1}(|\zeta'|^4 + |\zeta_0|^2)$$
on the characteristic set $h(x,\zeta) = 0$. The proof of this theorem is finished by the same procedure as the proof of theorem 1.1. The only difference is that we will integrate over all variables in the Fourier space before applying Parseval's identity. We like to emphasize that the analogues of formulas (2.9) and (2.10) are valid for the parabolic operator as well. This follows from the fact that the coefficient in front of the time derivative is a constant and from a careful analysis of (2.6).

COROLLARY 2.1. *Let all the assumptions of theorem 1.2 be satisfied. Then there exist constants* $C = C(\varepsilon), C_0 = C_0(\varepsilon, \lambda)$ *such that*

(2.13) $$\sqrt{\lambda}\|\sigma^{2-|\alpha|}e^{\tau\varphi}\partial^\alpha v\| \leq C\|\sigma^{\frac{1}{2}}e^{\tau\varphi}H(x,D)v\|$$

for $v \in C_0^\infty(Q_\varepsilon)$ *provided* $|\alpha| \leq 2$ *with* $\alpha_0 = 0$ *or* $|\alpha| \leq 1$ *and* $\lambda > C$ *and* $\tau \geq C_0$.

PROOF. We use theorem 1.2 with $u = \sigma^{1/2}v$. Furthermore, let α be a multi index subject to the same constraints as in the corollary. From the product rule of differentiation we obtain
$$\partial^\alpha(\sigma^{\frac{1}{2}}v) = \sigma^{\frac{1}{2}}\partial^\alpha v + \tau^{\frac{1}{2}}b^{(\alpha)}(x,D)v$$
where $b^{(\alpha)}$ is a first operator in space for $|\alpha| = 2$ and a operator of order zero for $|\alpha| = 1$ with coefficients depending only on λ. Similarly, we have
$$H(x,D)(\sigma^{\frac{1}{2}}v) = \sigma^{\frac{1}{2}}H(x,D)v + \tau^{\frac{1}{2}}b(x,D)v$$
where b is an first order operator in space with coefficients depending only on λ. By means of the triangle inequality we get
$$\sqrt{\lambda}\sum_{\substack{|\alpha|\leq 1 \\ \alpha_0=0}}\|\sigma^{2-|\alpha|}e^{\tau\varphi}\partial^\alpha v\| - C(\lambda)\tau^{\frac{1}{2}}(\|\sigma^{-\frac{1}{2}}e^{\tau\varphi}\nabla v\| + \|\sigma^{\frac{1}{2}}e^{\tau\varphi}v\|)$$
$$\leq C\|\sigma^{\frac{1}{2}}H(x,D)v\| + C(\lambda)\tau^{\frac{1}{2}}\|\sigma^{\frac{1}{2}}e^{\tau\varphi}v\|.$$

Choosing τ large enough we can control the last two term in the left hand side and the second term on the right hand side by the first term in the left hand side and we obtain the desired conclusion.

2.3. Proof of theorem 1.3. For the proof we will denote the symbol of the hyperbolic operator $\Box_\gamma$ by $p(x,\zeta)$, i.e.
$$p(x,\zeta) = \gamma(x)\zeta_0^2 - \zeta'\cdot\zeta'.$$
We compute the differential quadratic form on the characteristic set. Substituting $\zeta = \xi + i\tau\nabla\varphi(x)$ in the hyperbolic operator and using (2.1) we obtain two characteristic equations (one for the real part, one for the imaginary part)
$$\gamma(x)\xi_0^2 - |\xi'|^2 - \gamma(x)\sigma(x)^2\theta^4\left(t-\frac{T}{2}\right)^2 + \sigma(x)^2|(x-a)'|^2 = 0$$
$$\gamma(x)\xi_0\sigma(x)\theta^2\left(t-\frac{T}{2}\right) + \sigma(x)\xi'\cdot(x-a)' = 0.$$

Setting $s = t - T/2$ we obtain

$$\begin{aligned}
\mathcal{G}(x,\xi,\xi) &= 2\tau \left(4\gamma(x)^2|\zeta_0|^2\right)\left(\lambda^2\theta^4 s^2\varphi(x) - \lambda\theta^2\varphi(x)\right) \\
&\quad - \sum_{k=1}^{n} 4\gamma(x)(\zeta_k\overline{\zeta}_0 + \zeta_0\overline{\zeta}_k)\left(-\lambda^2(x-a)_k\theta^2 s\varphi(x)\right) \\
&\quad + \sum_{j,k=1}^{n} 4\left(\lambda^2(x-a)_j(x-a)_k\varphi(x) + \delta_{jk}\lambda\varphi(x)\right)\zeta_j\overline{\zeta}_k \\
&\quad + 2\operatorname{Im}\left(2\partial_0\gamma(x)\gamma(x)\zeta_0^2\overline{\zeta}_0 - \sum_{k=1}^{n} 2\partial_k\gamma\zeta_0^2\overline{\zeta}_k\right) \\
&= 8\sigma(x)\left[-\gamma(x)^2\theta^2|\zeta_0|^2 + |\zeta'|^2 \right.\\
&\quad \left. + \lambda\left(\theta^2 s\gamma(x)\zeta_0 + (x-a)'\cdot\zeta'\right)\left(\theta^2 s\gamma(x)\overline{\zeta}_0 + (x-a)'\cdot\overline{\zeta}'\right)\right] \\
&\quad + 2\operatorname{Im}\left[\partial_0\gamma(x)2\gamma(x)\left(\xi_0^2 - 2i\sigma(x)\theta^2\xi_0 s - \sigma(x)^2\theta^4 s^2\right)\right.\\
&\quad \times \left(\xi_0 + i\sigma(x)\theta^2 s\right) \\
&\quad - 2\sum_{k=1}^{n} \partial_k\gamma(x)\left(\xi_0^2 - 2i\xi_0\sigma(x)\theta^2 s - \sigma(x)^2\theta^4 s^2\right) \\
&\quad \left. \times \left(\xi_k - i\sigma(x)(x-a)_k\right)\right].
\end{aligned}$$

Calculating the imaginary part we will have

$$\begin{aligned}
\mathcal{G}(x,\xi,\xi) &= 8\sigma(x)\left[-\gamma(x)^2\theta^2|\zeta_0|^2 + |\zeta'|^2 \right.\\
&\quad \left. + \lambda\left|i\left(-\gamma(x)\theta^4\sigma(x)s^2 + \sigma(x)|(x-a)'|^2\right)\right|^2\right] \\
&\quad + 4\partial_0\gamma(x)\gamma(x)\sigma(x)\theta^2\left(\xi_0^2 s - \sigma(x)^2\theta^4 s^3 - 2\xi_0^2 s\right) \\
&\quad + 4\sigma(x)\sum_{k=1}^{n} \partial_k\gamma(x)\left(\xi_0^2(x-a)_k - \sigma(x)^2\theta^4 s^2(x-a)_k + 2\xi_0\xi_k\theta^2 s\right) \\
&= 8\sigma(x)\left[-\gamma(x)^2\theta^2|\zeta_0|^2 + |\zeta'|^2 + \lambda\sigma(x)^2\left|-\gamma(x)\theta^4 s^2 + |(x-a)'|^2\right|^2\right] \\
&\quad - 4\partial_0\gamma(x)\gamma(x)\sigma(x)\theta^2 s|\zeta_0|^2 \\
&\quad + 4\sigma(x)\left(\operatorname{Re}\zeta_0^2\nabla'\gamma(x)\cdot(x-a)' + 2\xi_0\theta^2 s\nabla'\gamma(x)\cdot\xi'\right).
\end{aligned}$$

Summing up we have

$$\sigma^{-1}(x)\mathcal{G}(x,\xi,\xi) = \mathcal{H}_1(x,\xi,\xi) + \mathcal{H}_2(x,\xi,\xi)$$

where

$$\begin{aligned}
\mathcal{H}_1(x,\xi,\xi) &= 8\left[-\gamma(x)^2\theta^2|\zeta_0|^2 + |\zeta'|^2 - \frac{1}{2}\partial_0\gamma(x)\gamma(x)\theta^2 s|\zeta_0|^2\right.\\
&\quad \left. + \frac{1}{2}\operatorname{Re}\zeta_0^2\nabla'\gamma(x)\cdot(x-a)' + \xi_0\theta^2 s\nabla'\gamma(x)\cdot\xi'\right]
\end{aligned}$$

and

$$\mathcal{H}_2(x,\xi,\xi) = 8\lambda\sigma(x)^2\left|-\gamma(x)\theta^4 s^2 + |(x-a)'|^2\right|^2.$$

Notice that the right hand side in both quadratic forms is homogeneous of degree 2 in ζ. This shows that it will be sufficient to work only with $|\zeta| = 1$.

First we estimate $\mathcal{H}_2$. Because of assumption (1.9) we have

$$\mathcal{H}_2(x,\xi,\xi) \geq 8\lambda\sigma(x)^2\varepsilon^2$$

For the estimate of $\mathcal{H}_1$ we consider at first the case $\tau = 0$. Then the characteristic equation becomes $\gamma(x)\xi_0^2 = |\xi'|^2$ and consequently

$$\xi_0^2 = \frac{1}{1+\gamma(x)} \quad \text{and} \quad |\xi'|^2 = \frac{\gamma(x)}{1+\gamma(x)}.$$

This leads to

$$\begin{aligned}
\mathcal{H}_1(x,\xi,\xi) &= \frac{4}{1+\gamma(x)}\left[-2\gamma(x)^2\theta^2 + 2\gamma(x) - \partial_0\gamma(x)\gamma(x)\theta^2 s\right.\\
&\quad \left.+\nabla'\gamma(x)\cdot(x-a)' + 2(\xi_0\theta^2 s\nabla'\gamma(x)\cdot\xi')(1+\gamma(x))\right]\\
&\geq \frac{4}{1+\gamma(x)}\left[-2\gamma(x)^2\theta^2 + 2\gamma(x) - \partial_0\gamma(x)\gamma(x)\theta^2 s\right.\\
&\quad \left.+\nabla'\gamma(x)\cdot(x-a)' - 2\theta^2|s\nabla'\gamma(x)|\sqrt{\gamma(x)}\right] \geq \epsilon
\end{aligned}$$

where we made use of the characteristic equations and assumption (1.10). The same estimate is valid for small τ by continuity. This shows that

$$\mathcal{H}_1(x,\xi,\xi) + \mathcal{H}_2(x,\xi,\xi) \geq \mathcal{H}_1(x,\xi,\xi) \geq \epsilon$$

on the characteristic set.

If τ is large and $|\zeta| = 1$ we have $\mathcal{H}_1 > -C$ and choosing λ large we can guarantee

$$\mathcal{H}_1(x,\xi,\xi) + \mathcal{H}_2(x,\xi,\xi) > 0$$

on the characteristic set in that case.

By homogeneity we obtain the following estimate

$$C^{-1}|\zeta|^2 \leq \sigma(x)^{-1}\mathcal{G}(x,\xi,\xi) + \frac{|p(x,\zeta)|^2}{|\zeta|^2}.$$

Now we fix $x = x_0 \in Q_\varepsilon$ and multiply this estimate by $|\hat{v}(\xi)|^2$ for $v \in C_0^\infty(\Omega_\varepsilon)$ and integrate and arrive at

$$C^{-1}\sum_{|\alpha|\leq 1}\sigma(x_0)^{2-2|\alpha|}\int |D^\alpha v|^2 \leq \frac{1}{\sigma(x_0)}\int \mathcal{G}(x_0, D, \overline{D})v\overline{v}$$
$$+|||P(x_0, D+i\tau\nabla\varphi(x_0))v|||^2$$

where

(2.14) $$|||w|||^2 = \int \frac{|\hat{w}(\xi)|^2}{|\xi|^2 + \tau^2|\nabla\varphi(x_0)|^2}d\xi.$$

Using formula (2.6) we obtain

$$\mathcal{G}(x_0, D, \overline{D})v\overline{v} \leq \mathcal{G}(x, D, \overline{D})v\overline{v} + \omega_1(\delta,\lambda)\sum_{|\alpha|\leq 1}(\tau\varphi(x_0))^{3-2|\alpha|}|D^\alpha v|^2$$

for $v \in C_0^\infty(Q_\varepsilon \cap B(x_0, \delta))$ where $\omega_1(\delta, \lambda) \to 0$ for λ fixed and $\delta \to 0$. Together with lemma 4.1 from the appendix we get

$$C^{-1} \sum_{|\alpha| \leq 1} \sigma(x_0)^{2-2|\alpha|} \int |D^\alpha v|^2 \leq \frac{1}{\sigma(x_0)} \int \mathcal{G}(x, D, \overline{D}) v \overline{v}$$

$$+ \tau^{-2} \| P^\bullet(x, D + i\tau \nabla \varphi(x)) v \|^2 + \omega(\delta, \lambda) \sum_{|\alpha| \leq 1} \sigma(x_0)^{2-2|\alpha|} \int |D^\alpha v|^2$$

for $v \in C_0^\infty(Q_\varepsilon \cap B(x_0, \delta))$ where $\omega(\delta, \lambda) \to 0$ for λ fixed and $\delta \to 0$. Choosing $\delta > 0$ small enough such that $C > \omega(\delta, \lambda)/2$ we can cancel the last term on the right hand side by one half of the left hand side and obtain

$$C^{-1} \sum_{|\alpha| \leq 1} \sigma(x_0)^{2-2|\alpha|} \int |D^\alpha v|^2 \leq \frac{1}{\sigma(x_0)} \| P^\bullet(x, D + i\tau \nabla \varphi(x)) v \|^2$$

$$+ \tau^{-2} \| P^\bullet(x, D + i\tau \nabla \varphi(x) v \|^2$$

where we used formula (2.5) as well. Choosing τ large such that $\sigma(x_0)^{-1} \tau^{-2}$ we can combine the two terms in the right hand side into one term and obtain

$$C^{-1} \sum_{|\alpha| \leq 1} \sigma(x_0)^{3-2|\alpha|} \int |D^\alpha v|^2 \leq \| P^\bullet(x, D + i\tau \nabla \varphi(x)) v \|^2$$

with a new constant C.

¿From here we complete the proof as in theorem 1.1. □

3. Applications to thermoelasticity

In this section we apply Carleman estimates to obtain unqueness of the continuation results for two imporatant systems of thermo-elasticity.

The proof of the first uniqueness theorem needs the following

LEMMA 3.1. *Let the operators $A(x, D)$ and $\Box_\gamma$ satisfy the assumptions of theorem 1.1 and of theorem 1.3.*

Then there exist constants $C = C(\varepsilon), C_0 = C_0(\varepsilon, \lambda)$ such that

$$\lambda \left(\| \sigma^{3-|\alpha|} e^{\tau\varphi} \partial^\alpha w \| + \| \sigma^{2-|\alpha|} e^{\tau\varphi} \partial^\alpha \nabla' w \| + \| \sigma^{\frac{3}{2}-|\beta|} e^{\tau\varphi} \partial^\beta A(x, D) w \| \right.$$

(3.1) $$\left. + \| \sigma^{\frac{3}{2}} e^{\tau\varphi} \partial_t^2 w \| + \| \sigma^{\frac{1}{2}} e^{\tau\varphi} \partial_t^2 \nabla' w \| \right) \leq C \| e^{\tau\varphi} \Box_\gamma A(x, D) w \|$$

for $w \in C_0^\infty(Q_\varepsilon)$ provided $|\alpha| \leq 2, \alpha_0 = 0, |\beta| \leq 1, \lambda \geq C$, and $\tau \geq C_0$.

PROOF. The proof is based on a combination of theorem 1.1 and theorem 1.3. For this proof α will denote a multi index with $|\alpha| \leq 2$ and $\alpha_0 = 0$ whereas β is a multi-index such that $|\beta| \leq 1$.

For large τ it follows from (1.6) with $u = \sigma^{3/2} w$ that

$$\sqrt{\lambda} \| \sigma^{3-|\alpha|} e^{\tau\varphi} \partial^\alpha w \| \leq C \| \sigma^{\frac{3}{2}} e^{\tau\varphi} A(x, D) w \| .$$

This can be proved like corollary 2.1. A straightforward application of (1.11) leads to

(3.2) $$\sqrt{\lambda} \| \sigma^{3-|\alpha|} e^{\tau\varphi} \partial^\alpha w \| \leq C \| e^{\tau\varphi} \Box_\gamma A(x, D) w \| .$$

Similarly we obtain

$$\sqrt{\lambda}\|\sigma^{2-|\alpha|}e^{\tau\varphi}\partial^\alpha\nabla' w\| \le C\|e^{\tau\varphi}\Box_\gamma A(x,D)w\| \tag{3.3}$$

provided τ is sufficiently large. Next, applying (1.11) to $u = A(x,D)w$ we derive

$$\|\sigma^{\frac{3}{2}-|\beta|}e^{\tau\varphi}\partial^\beta A(x,D)w\| \le C\|e^{\tau\varphi}\Box_\gamma A(x,D)w\| . \tag{3.4}$$

Next we need to find estimates for the time derivative. At first we notice that

$$\|\sigma^{\frac{3}{2}}e^{\tau\varphi}\partial_t^2 w\| \le C(\|\sigma^{\frac{3}{2}}e^{\tau\varphi}\Box_\gamma w\| + \sum_\alpha \|\sigma^{\frac{3}{2}}e^{\tau\varphi}\partial^\alpha w\|) \tag{3.5}$$

and that

$$\|\sigma^{\frac{1}{2}}e^{\tau\varphi}\nabla'\partial_t^2 w\| \le C(\|\sigma^{\frac{1}{2}}e^{\tau\varphi}\nabla'\Box_\gamma w\| + \sum_\alpha \|\sigma^{\frac{1}{2}}e^{\tau\varphi}\nabla'\partial^\alpha w\|) . \tag{3.6}$$

Furthermore, an application of (1.6) with $u = \Box_\gamma w$ provides

$$\sqrt{\lambda}\|\sigma^{\frac{3}{2}-|\alpha|}e^{\tau\varphi}\partial^\alpha \Box_\gamma w\| \le C\|e^{\tau\varphi}A(x,D)\Box_\gamma w\|$$

Comparing the right hand side with the desired estimate (3.1) it is clear that we have to investigate the commutator of the two operators $A(x,D)$ and $\Box_\gamma$. A careful analysis gives

$$\|e^{\tau\varphi}A(x,D)\Box_\gamma w\| \le \|e^{\tau\varphi}\Box_\gamma A(x,D)w\| + C\left(\sum_\alpha \|e^{\tau\varphi}\partial^\alpha w\| \right.$$
$$\left. + \sum_\alpha \|e^{\tau\varphi}\partial^\alpha \nabla' w\| + \|e^{\tau\varphi}\partial_t^2 w\| + \|e^{\tau\varphi}\partial_t^2 \nabla' w\|\right) .$$

In view of (3.5) and (3.6) we get

$$\sqrt{\lambda}\left(\|\sigma^{\frac{3}{2}}e^{\tau\varphi}\partial_t^2 w\| + \|\sigma^{\frac{1}{2}}e^{\tau\varphi}\nabla'\partial_t^2 w\|\right) \le \|e^{\tau\varphi}\Box_\gamma A(x,D)w\|$$
$$+C\left(\sum_\alpha \|e^{\tau\varphi}\partial^\alpha w\| + \sum_\alpha \|e^{\tau\varphi}\partial^\alpha \nabla' w\| + \|e^{\tau\varphi}\partial_t^2 w\| + \|e^{\tau\varphi}\partial_t^2 \nabla' w\|\right) .$$

For large τ the last two terms are controlled by the right hand side and the second and third term on the right hand side can be estimated using (3.2) and (3.3). $\square$

The first thermoelasticity system describes the dynamics of a thin elastic plate $\Omega \subset \mathbf{R}^2$ subject to thermal effects. Let w denote the vertical displacement and v the temperature. We consider the following system of partial differential equations

$$\begin{aligned} \partial_t^2 w + \Box_\gamma A_1(x,D)w + A_3(x,D)w + A_4(x,D)v &= 0 \\ \partial_t v - A_2(x,D)v + A_1(x,D)\partial_t w &= 0 \quad x \in Q . \end{aligned} \tag{3.7}$$

Here $A_j(x,D), j = 1, 2$ are linear second order elliptic operators as introduced in (1.4) and A_3, A_4 are linear second order with $L^\infty(Q)$-coefficients. Moreover, A_4 does not involve $\partial_t^2 v$. In this sense the first differential equation in (3.7) is a generalization of the Kirchhoff plate equation.

The second system describes thermal and elastic effects in a three dimensional body $\Omega \subset \mathbf{R}^3$. We introduce the displacement vector $\mathbf{u} = (u_1, u_2, u_3)^T$ and the temperature v and consider the following system

(3.8) $$\begin{aligned} \rho\partial_t^2 \mathbf{u} - \mu(\Delta + \nabla' div)\mathbf{u} - \nabla'(\lambda div\mathbf{u}) \\ -(\nabla'\mathbf{u} + \nabla'\mathbf{u}^T)\nabla\mu + \mathbf{B}(x,D)(\mathbf{u},v) &= 0 \\ \partial_t v - A(x,D)v + B_1 div\partial_t \mathbf{u} &= 0 . \end{aligned}$$

for $x \in Q$ where $\mathbf{B}$ is the (matrix) linear partial differential operator of first order with respect to v and of zero order with respect to $\mathbf{u}$ with $C^1(\overline{Q})$-coefficients, and $B_1 \in L^\infty(Q)$. Furthermore, ρ denotes the density and μ and λ are the Lame parameter and we assume in the following $\rho > 0$, $\mu > 0$ and $2\mu + \lambda > 0$ for all $x \in \overline{Q}$.

In our uniqueness results we refer to the two types of the domain Ω stated in the introduction.

THEOREM 3.2. *Assume that* $\gamma \in C^1(\overline{Q})$, $\theta \in \mathbf{R}_+$ *and* $T > 0$ *satisfy*

(3.9) $$c < \gamma(x)\theta^2 < 1 \qquad x \in Q$$

for some positive constant c and that

(3.10) $$\theta^2 \left(2\gamma(x)^2 + \partial_0\gamma(x)\gamma(x)(t - T/2) + 2|(t - T/2)\nabla'\gamma(x)|\sqrt{\gamma(x)} \right) \\ < 2\gamma(x) + \nabla'\gamma(x) \cdot (x - a)'$$

for $x \in \overline{Q}$ and some vector a. Furthermore, assume that $(w, v) \in H^3(Q) \times H^1(Q)$ is a solution of (3.7) and has vanishing Cauchy data on $\Gamma \times (0, T)$, i.e.

$$\partial_\nu^j w = \partial_\nu^k v = 0 \qquad on \; \Sigma = (0,T) \times \Gamma \qquad j = 0,1,2,3, \quad k = 0,1 .$$

Then

$$(w, v) \equiv 0 \qquad in \; Q_0$$

provided $\theta > 2\sup_{x' \in \Omega} |x'|/T$ in case (i) and $\theta^2 > 4h(h + 2a_n)/T^2$ in case(ii).

PROOF. We start by extending the domain Q. In case (i) we set $Q^* = (0, T) \times \mathbf{R}^2$ and in case (ii) we set $Q^* = (0, T) \times \mathbf{R} \times \{x_2 : x_2 > 0\}$. We extend the functions (w, v) to functions $(w^*, v^*) \in H^3(Q^*) \times H^1(Q^*)$ by setting them zero on $Q^* \setminus Q$. This can be done since (w, v) have zero Cauchy data on Σ. Moreover, (w^*, v^*) is a solution to the system (3.7) in Q^*.

Now we define a cutoff function $\chi \in C_0^\infty(Q^*)$ such that

$$\chi = \begin{cases} 1 & \text{for } x \in Q_{2\varepsilon} \\ 0 & \text{for } x \in Q \setminus Q_\varepsilon . \end{cases}$$

This can be done since the condition $\theta > 2\sup_{x' \in \Omega} |x'|/T$ in case (i) and $\theta^2 > 4h(h + 2a_n)/T^2$ guarantee that $Q_\varepsilon \cap \partial Q \subset \Sigma$.

We set $w^0 = \chi w^*$ and $v^0 = \chi v^*$ and observe that these two functions satisfy the following system

(3.11) $$\begin{aligned} \Box_\gamma A_1(x,D)w^0) &= -A_2(x,D)v^0 - \partial_t^2 w^0 + F_1 \\ \partial_t v^0 + A_3(x,D)v^0 &= A_1(x,D)\partial_t w^0 + F_2 \end{aligned}$$

for $x \in Q^*$ where $F_j \in L_2(Q^*)$ with $supp \, F_j \subset Q_\varepsilon \setminus Q_{2\varepsilon}$. This follows from the product rule of differentiation and from the regularity of w^* and v^*.

Next we apply lemma 3.1 to w^0 and corollary 2.1 to v^0. In view of (3.11) we obtain

$$\lambda \left(\sum_{\substack{|\alpha|\leq 2 \\ \alpha_0=0}} \|\sigma^{3-|\alpha|}e^{\tau\varphi}\partial^\alpha w^0\| + \sum_{\substack{|\alpha|\leq 2 \\ \alpha_0=0}} \|\sigma^{2-|\alpha|}e^{\tau\varphi}\partial^\alpha \nabla w^0\| \right.$$

$$\left. + \sum_{|\beta|\leq 1} \|\sigma^{\frac{3}{2}-|\beta|}e^{\tau\varphi}\partial^\beta A_1(x,D)w^0\| + \|\sigma^{\frac{3}{2}}e^{\tau\varphi}\partial_t^2 w^0\| + \|\sigma^{\frac{1}{2}}e^{\tau\varphi}\partial_t^2 \nabla w^0\| \right)$$

$$(3.12) \quad \leq C\|e^{\tau\varphi}A_2(x,D)v^0\| + \|e^{\tau\varphi}\partial_t^2 w^0\| + \|e^{\tau\varphi}F_1\|$$

and

$$\sqrt{\lambda}\|\sigma^{2-|\alpha|}e^{\tau\varphi}\partial^\alpha v^0\| \leq C\|\sigma^{\frac{1}{2}}e^{\tau\varphi}A_1(x,D)\partial_t w^0\| + \|\sigma^{\frac{1}{2}}e^{\tau\varphi}F_2\|.$$

In the first term on the right hand side of the second estimate we can commute the operators $A_1(x,D)$ and ∂_t at the expense of lower order terms

$$\sqrt{\lambda}\|\sigma^{2-|\alpha|}e^{\tau\varphi}\partial^\alpha v^0\| \leq C\|\sigma^{\frac{1}{2}}e^{\tau\varphi}\partial_t A_1(x,D)w^0\| + \sum_{|\alpha|\leq 2}\|\sigma^{\frac{1}{2}}e^{\tau\varphi}\partial^\alpha w^0\| + \|\sigma^{\frac{1}{2}}e^{\tau\varphi}F_2\|.$$

Now we add the last estimate and (3.12) together and observe that with $\lambda = 4C^2$ we will be able to cancel the terms in the right hand side involving w^0 and v^0 against $1/2$ of the corresponding terms in the left hand side and obtain

$$\tau^2\|e^{\tau\varphi}v^0\| + \tau^3\|e^{\tau\varphi}w^0\| \leq C(\|e^{\tau\varphi}F_1\| + \tau^{\frac{1}{2}}\|e^{\tau\varphi}F_2\|).$$

Next we shrink the domain of integration on the left hand side from Q_ε to $Q_{2\varepsilon}$ and observe that

$$\inf_{x\in Q_{2\varepsilon}} \varphi(x) = \sup_{x\in Q_\varepsilon \setminus Q_{2\varepsilon}} \varphi(x).$$

Hence, keeping in mind the regularity and the support of F_1 and F_2 we arrive at

$$\tau^2\|w^0\|_{L_2(Q_{2\varepsilon})} + \tau^3\|v^0\|_{L_2(Q_{2\varepsilon})} \leq C\tau^{\frac{1}{2}}M.$$

Dividing, by τ and letting $\tau \to \infty$ we conclude

$$w^0 = v^0 = 0 \text{ for } x \in Q_{2\varepsilon}.$$

Since $w^0 = w$ and $v^0 = v$ on $Q_{2\varepsilon}$ we obtain the same conclusion for w and v. Finally, we can choose ε as small as we wish and the theorem is proved. □

The next theorem is about the thermo-elastic system (3.8)

THEOREM 3.3. *Assume that $\rho \in C^1(\overline{Q})$ and $\lambda, \mu \in C^2(\overline{Q})$. Moreover, assume that that the conditions (3.9) and (3.10) are satisfied for*

$$\gamma = \frac{\rho}{\mu} \text{ and } \gamma = \frac{\rho}{2\mu + \lambda}.$$

Furthermore, assume that $(\mathbf{u}, v) \in H^1(Q)^3 \times H^1(Q)$ is a solution of the system (3.8) and has vanishing Cauchy data on $\Sigma = (0,T) \times \Gamma$, i.e.

$$\partial_\nu^j \mathbf{u} = \partial_\nu^j v = 0 \qquad \text{on } \Sigma \qquad j = 0, 1.$$

Then

$$(\mathbf{u}, v) \equiv 0 \qquad \text{in } Q_0$$

provided $\theta > 2\sup_{x' \in \Omega} |x'|/T$ in case (i) and $\theta^2 > 4h(h + 2a_n)/T^2$ in case (ii).

REMARK 3.4. In case (i) the domain Q_0 contains $\{T/2\} \times \Omega$ and hence by uniqueness for well-posed initial boundary value problem for thermoelasticity we have $\mathbf{u} = 0$ and $v = 0$ on $(T/2, T) \times \Omega$.

In case (ii), $Q_0 \cap \{t = t_1\}$ is never Ω while $Q_0 \cap \{t = T/2\}$ approaches $\{T/2\} \times \Omega$ as $T \longrightarrow \infty$ (see [**I.2**], p.61).

To prove this theorem we need a minor modification of Lemma 5.1 in ([**E-I-N-T**]) which is claiming that the first three equations of (3.8) with respect to $\mathbf{u}$ can be principally diagonalized.

LEMMA 3.5. *Let* $u_4 = div\mathbf{u}, \mathbf{u}_5 = curl\mathbf{u}$. *If* $\rho \in C^1(\overline{Q}), \lambda, \mu \in C^2(\overline{Q})$ *and*

(3.13) $\quad \rho\partial_t^2\mathbf{u} - \mu(\Delta + \nabla'div)\mathbf{u} - \nabla'(\lambda div\mathbf{u}) - (\nabla'\mathbf{u} + \nabla'\mathbf{u}^T)\nabla'\mu = \mathbf{f} \qquad$ *in* Q

then

(3.14)
$$\frac{\rho}{\mu}\partial_t^2\mathbf{u} - \Delta\mathbf{u} + \mathbf{A}_{1;1}(\mathbf{u}, u_4) = \frac{\mathbf{f}}{\mu}$$
$$\frac{\rho}{2\mu + \lambda}\partial_t^2 u_4 - \Delta u_4 + \mathbf{A}_{2;1}(\mathbf{u}, u_4, \mathbf{u}_5) = \frac{\rho}{2\mu + \lambda}div\left(\frac{\mathbf{f}}{\rho}\right)$$
$$\frac{\rho}{\mu}\partial_t^2\mathbf{u}_5 - \Delta\mathbf{u}_5 + \mathbf{A}_{3;1}(\mathbf{u}, u_4, \mathbf{u}_5) = \frac{\rho}{\mu}curl\left(\frac{\mathbf{f}}{\rho}\right)$$

in Q *where* $\mathbf{A}_{j;1}$ *are linear (matrix) partial differential operators of first order with the coefficients in* $L^\infty(Q)$.

PROOF. Dividing the equations (3.13) by ρ, applying the divergence with the use of the identity $div(g\mathbf{v}) = \nabla'g \cdot \mathbf{v} + gdiv\mathbf{v}$, and rearranging terms we obtain

$$\partial_t^2 u_4 - \frac{2\mu + \lambda}{\rho}\Delta u_4 - \left(\nabla'\frac{\mu}{\rho} + \frac{2\nabla'\lambda}{\rho} + \lambda\nabla'\frac{1}{\rho} + \frac{\nabla'\mu}{\rho}\right) \cdot \nabla' u_4 - div\left(\frac{\nabla'\lambda}{\rho}\right)u_4$$
$$- \left(\nabla'\frac{\mu}{\rho} + \frac{\nabla'\mu}{\rho}\right) \cdot \Delta\mathbf{u} - \sum_{j=1}^{3}\partial_j\left(\frac{\nabla'\mu}{\rho}\right) \cdot (\nabla'u_j + \partial_j\mathbf{u}) = div\frac{\mathbf{f}}{\rho}.$$

Using the identity $\Delta\mathbf{u} = \nabla'div\mathbf{u} - curlcurl\mathbf{u}$ we obtain the equation (3.14) for u_4.

Similarly, dividing the equation (3.13) by ρ, applying the *curl* and using the identity $curl(g\mathbf{u}) = gcurl\mathbf{u} + \nabla'g \times \mathbf{u}$ gives

$$\partial_t^2\mathbf{u}_5 - \frac{\mu}{\rho}\Delta\mathbf{u}_5 - \nabla'\frac{\mu}{\rho} \times (\Delta\mathbf{u} + \nabla'u_4) - \nabla'\frac{1}{\rho} \times \nabla'(\lambda u_4)$$
$$- \sum_{j=1}^{3}\left(\frac{\partial_j\mu}{\rho}\partial_j\mathbf{u}_5 + \nabla'\frac{\partial_j\mu}{\rho} \times (\partial_j\mathbf{u} + \nabla'u_j)\right) = curl'\frac{\mathbf{f}}{\rho}.$$

Utilizing the above identity for $\Delta\mathbf{u}$ we arrive at the conclusion of Lemma 3.4. □

Proof of theorem 3.3: Applying Lemma 3.4 with $\mathbf{f} = -\mathbf{B}(\mathbf{u}, v)$ we conclude from (3.8) that the vector-function $\mathbf{U} = (\mathbf{u}, u_4, \mathbf{u}_5)$ satisfies the following principally diagonal system of the differential equations

(3.15) $\qquad\qquad \mathbf{D}\partial_t^2\mathbf{U} - \Delta\mathbf{U} + \mathbf{A}_1(\mathbf{U}) = \mathbf{A}_2(v) \qquad$ in Q

where $\mathbf{D}$ is the diagonal 7×7-matrix with the diagonal entries $\rho\mu^{-1}, \rho\mu^{-1}, \rho\mu^{-1},$ $\rho(2\mu+\lambda)^{-1}, \rho\mu^{-1}, \rho\mu^{-1}, \rho\mu^{-1}$, $\mathbf{A}_1$ is a 7×7-matrix linear differential operator of the first order with the coefficients in $L^\infty(Q)$ and $\mathbf{A}_2$ is a (matrix) second order linear partial differential operator (with respect to x_1, x_2, x_3) with $L^\infty(Q)$-coefficients.

Introducing a cut-off function $\chi \in C^\infty(\mathbf{R}^4)$, $\chi = 1$ on $Q_{2\varepsilon}$, $\chi = 0$ on $Q \setminus Q_\varepsilon$, letting $\mathbf{U}^0 = \chi \mathbf{U}, v^0 = \chi v$, and using Leibniz' formula we derive from (3.15),(3.8) that

$$\mathbf{D}\partial_t^2 \mathbf{U}^0 - \Delta \mathbf{U}^0 = \chi \mathbf{A}_2(v^0) + \mathbf{A}_2^\bullet(\mathbf{U}, v)$$

(3.16) $$\partial_t v^0 - A v^0 = \mathbf{A}_1^\bullet(\mathbf{U}, v) \quad \text{in } Q$$

where $\mathbf{A}_k^\bullet$ are matrix linear partial differential operators with the coefficients from $L^\infty(Q)$. In addition, $\mathbf{A}_2^\bullet$ is of second order and does not involve second derivatives of $\mathbf{U}$ and $\partial_t^2 v$, and $\mathbf{A}_1^\bullet$ is of first order and does not involve $\partial_t v$.

Applying Theorem 1.3 to each of the first seven equations of (3.16) and summing the results we obtain

$$\sum_{|\alpha|\leq 1} \|\sigma^{3/2-|\alpha|} e^{\tau\varphi} \partial^\alpha \mathbf{U}^0\| \leq C \left(\sum_{|\alpha|\leq 1} \|e^{\tau\varphi} \partial^\alpha \mathbf{U}\| + \sum_{\substack{|\beta|\leq 2 \\ \beta_0 \leq 1}} \|e^{\tau\varphi} \partial^\beta v^0\| + \sum_{\substack{|\alpha|\leq 1 \\ \alpha_0 \leq 0}} \|e^{\tau\varphi} \partial^\alpha v\| \right).$$

Similarly from Corollary 2.1 and from the last equation in (3.16) we have

$$\lambda^{1/2} \sum_{\substack{|\beta|\leq 2 \\ \beta_0 \leq 1}} \|\sigma^{2-|\alpha|} e^{\tau\varphi} \partial^\alpha v^0\| \leq C \left(\sum_{|\alpha|\leq 1} \|e^{\tau\varphi} \sigma^{1/2} \partial^\alpha \mathbf{U}\| + \sum_{\substack{|\beta|\leq 1 \\ \beta_0=0}} \|\sigma^{1/2} e^{\tau\varphi} \partial^\beta v\| \right)$$

After that the proof proceeds exactly as the proof of theorem 3.1. $\square$

4. Appendix

Here we prove the estimate needed in the proof of theorem 1.3.

LEMMA 4.1. *Consider the hyperbolic operator $\square_\gamma = P(x, D)$ subject to the same restriction as in theorem 1.3. Then*

$$|||P(x_0, D + i\tau \nabla \varphi(x_0))v|||^2 \leq 2\tau^{-2} \|P^\bullet(x_0, D + i\tau \nabla \varphi(x))v\|^2 + \omega(\delta, \lambda) \sum_{|\alpha|\leq 1} \sigma(x_0)^{2-2|\alpha|} |D^\alpha v|^2$$

for $v \in C^\infty(Q_\varepsilon \cap B(x_0, \delta))$ where $\omega(\delta, \lambda) \to 0$ for $\delta \to 0$ and λ fixed. The norm $||| \cdot |||$ was introduced in (2.14).

PROOF. We start with the triangle inequality

$$|||P^\bullet(x, D + i\tau\nabla\varphi(x))v - P(x_0, D + i\tau\nabla\varphi(x_0))v|||$$
$$\leq |||P^\bullet(x, D + i\tau\nabla\varphi(x))v - P(x, D + i\tau\nabla\varphi(x_0))v|||$$
$$+ |||P(x, D + i\tau\nabla\varphi(x_0))v - P(x_0, D + i\tau\nabla\varphi(x_0))v|||.$$

We will make two separate considerations for the two terms on the right hand side. For the first one we use the fact that

$$P^\bullet(x, D + i\tau\nabla\varphi(x))v - P(x, D + i\tau\nabla\varphi(x_0))v$$
$$= +\gamma(x)\left(2i\tau(\partial_0\varphi(x) - \partial_0\varphi(x_0))D_0 - \tau^2(\partial_0\varphi(x))^2 + \tau^2(\partial_0\varphi(x_0))^2\right)v$$
$$- \sum_{j=1}^{n}\left(2i\tau(\partial_j\varphi(x) - \partial_j\varphi(x_0))D_0 - \tau^2(\partial_j\varphi(x))^2 + \tau^2(\partial_j\varphi(x_0))^2\right)v$$

which allows us to estimate

$$|||P^\bullet(x, D + i\tau\nabla\varphi(x))v - P(x, D + i\tau\nabla\varphi(x_0))v|||^2$$
$$\leq C\tau^{-2}\|P^\bullet(x, D + i\tau\nabla\varphi(x))v - P(x, D + i\tau\nabla\varphi(x_0))v\|^2$$
$$\leq C\tau^{-2}\left(\tau^2 \sup_x |\nabla\varphi(x) - \nabla\varphi(x_0)|^2 \|\nabla v\|^2 + \tau^4 \sup_x |\varphi(x) - \varphi(x_0)|^2 \|v\|^2\right)$$
$$\leq \omega(\lambda,\delta) \sum_{|\alpha|\leq 1} \tau^{2-2|\alpha|}\|D^\alpha v\|^2.$$

For the second part we observe that

$$|||P(x, D + i\tau\nabla\varphi(x_0))v - P(x_0, D + i\tau\nabla\varphi(x_0))v|||$$
$$= |||(\gamma(x) - \gamma(x_0))(D_0 + i\tau\partial_0\varphi(x_0))^2 v|||^2$$
$$\leq C\|\gamma(x) - \gamma(x_0)\|^2_{C^1_\tau(B(x_0,\delta))} \int \frac{|(\xi_0 + i\tau\partial_0\varphi(x_0))^2|^2 |\hat{v}(\xi)|^2}{|\xi|^2 + \tau^2|\nabla\varphi(x_0)|^2} d\xi$$
$$\leq C\|\gamma(x) - \gamma(x_0)\|^2_{C^1_\tau(B(x_0,\delta))} \int |\xi_0 + i\tau\partial_0\varphi(x_0)|^2 |\hat{v}(\xi)|^2 d\xi$$
$$\leq C\|\gamma(x) - \gamma(x_0)\|^2_{C^1_\tau(B(x_0,\delta))} \sum_{|\alpha|\leq 1} (\sigma(x_0)^{2-2|\alpha|}|D^\alpha v|^2$$

where we used lemma 2.1 in [**T**] (see also lemma 2.1 in [**E**]) and

$$\|w\|_{C^1_\tau} = \sum_{|\alpha|\leq 1} \tau^{-|\alpha|} \sup |D^\alpha w|.$$

Putting these two parts together we have

$$|||P^\bullet(x, D + i\tau\nabla\varphi(x))v - P(x, D + i\tau\nabla\varphi(x_0))v|||^2$$
$$\leq \omega(\lambda,\delta) \sum_{|\alpha|\leq 1} \tau^{2-2|\alpha|}\|D^\alpha v\|^2$$

and the lemma follows from the triangle inequality and the estimate

$$|||w|||^2 \leq C\tau^{-2}\|w\|.$$

$\square$

References

[E] M. Eller *Uniqueness of continuation theorems* Direct and inverse problems of mathematical physics, Kluwer Academic, Dordrecht/Norwell, MA pp.53-106 (1999)

[E-I-N-T] M. Eller, V. Isakov, G. Nakamura and D.Tataru *Uniqueness and stability in the Cauchy problem for Maxwell's and the elasticity system*, Nonlinear partial differential equations, Vol. 16, College de France Seminar, Chapman & Hall, London/New York (2000)

[H] L. Hörmander *Linear partial differential operators* Springer Verlag, New York (1963)

[I.1] V. Isakov *A nonhyperbolic Cauchy problem for $\square_b\square_c$ and its applications to elasticity theory*, Comm. Pure Appl. Math. **39** pp. 747-769 (1986)

[I.2] V. Isakov *Inverse problems in partial differential equations* Springer Verlag, New York (1997)
[I.3] V. Isakov *On the uniqueness of the continuation for a thermoelasticity system* preprint 1998, 17 pp.
[L] J. Lagnese *Boundary stabilization for thin plates* SIAM, Philadelphia (1989)
[L-R-T] I. Lasiecka, M. Renardy and R. Triggiani *Backwards uniqueness for thermoelstic plates with rotational forces*, Semigroup forum, to appear
[T] D. Tataru *Unique continuation for solutions to PDE's; between Hörmander's and Holmgren's theorem* Comm. Part. Diff. Equations **20** pp. 855-884 (1995)

DEPARTMENT OF MATHEMATICS, GEORGETOWN UNIVERSITY, WASHINGTON, D.C. 20057

DEPARTMENT OF MATHEMATICS, WICHITA STATE UNIVERSITY, WICHITA, K.S. 67260
E-mail address: isakov@twsuvm.uc.twsu.edu

ON THE PRESCRIBED SCALAR CURVATURE PROBLEM ON COMPACT MANIFOLDS WITH BOUNDARY

JOSÉ F. ESCOBAR

Let (M^n, g) be an n-dimensional compact Riemannian manifold with boundary, with $n \geq 3$. In this paper we address the problem of prescribing the scalar curvature in a compact manifold with boundary.

Let $\tilde{g} = u^{\frac{4}{n-2}} g$ be a metric conformally related to the metric g. The relation between the scalar curvature of the metric g, R_g, and the scalar curvature of the metric $\tilde{g}$, $R_{\tilde{g}}$, is given by

$$(1) \qquad R_{\tilde{g}} = -\frac{4(n-1)}{n-2} \frac{Lu}{u^{\frac{n+2}{n-2}}}, \quad L = \Delta_g - \frac{(n-2)}{4(n-1)} R_g,$$

where Δ_g is the Laplacian calculated with respect to the metric g. The relation between the mean curvature of ∂M with respect to the metric g, h_g, and the mean curvature of ∂M with respect to the metric $\tilde{g}$, $h_{\tilde{g}}$, is given by the equation

$$(2) \qquad h_{\tilde{g}} = \frac{2}{n-2} \frac{Bu}{u^{\frac{n}{n-2}}}, \quad B = \frac{\partial}{\partial \eta} + \frac{n-2}{2} h_g,$$

where $\frac{\partial}{\partial \eta}$ is the outward normal derivative calculated with respect to the metric g.

In our previous papers [E1] and [E2] we pointed out that the sign of the first eigenvalue, ν_1, for the problem

$$(3) \qquad \Delta_g u - \frac{(n-2)}{4(n-1)} R_g u = 0 \quad \text{in} \quad M,$$

$$\frac{\partial u}{\partial \eta} + \frac{(n-2)}{2} h_g u = \nu_1 u \quad \text{on} \quad \partial M,$$

is a conformal invariant.

1991 *Mathematics Subject Classification.* Primary 53C21; Secondary 58E11, 35J60.
Research supported by NSF Grant 9704482.

Compact Riemannian manifolds can be divided into four groups according to the sign of the first eigenvalue of the above problem when it is finite or when the first eigenvalue of the above problem is $-\infty$.

If the first eigenvalue of problem (3) is finite we can change the metric g with the metric $g_1 = \varphi_1^{\frac{4}{n-2}} g$, where φ_1 is the first eigenfunction for problem (3). The transformation law (1) implies that $R_{g_1} = 0$. Given a smooth function K on M, such a function is the scalar curvature of a metric conformally equivalent to g_1 (and hence to the metric g), provided that there exists a smooth positive function u satisfying the equation

$$(4) \qquad \Delta_{g_1} u + \frac{n-2}{4(n-1)} K u^{\frac{n+2}{n-2}} = 0 \quad \text{in} \quad M.$$

In this paper we study the above equation under Dirichlet boundary condition and under a mixed boundary condition.

Since the prescribed scalar curvature problem is a conformal invariant one, the discussion above implies that Theorem 1 and Theorem 2 below apply to compact manifolds with boundary (M^n, g) for which the first eigenvalue ν_1 in problem (3) is finite. Since those manifolds admit a scalar flat metric we assume from now on that K is a smooth function on M that does not vanish everywhere.

In order to solve the equation (4) on any compact Riemaniann manifold with boundary we consider the Dirichlet integral of u

$$D(u) = \int_M |\nabla u|^2$$

for $u \in H^1(M)$, the Sobolev space of L^2 functions on M with first derivatives in L^2. We define

$$C_{\alpha,s} = \left\{ u \in H^1(M) : \int_M K|u|^\alpha = s \right\},$$

where $1 < \alpha \leq \frac{2n}{n-2}$. The condition that K is positive guarantees that the set $C_{\alpha,s}$ is not empty for s positive.

The main point of our next Theorem is to estimate the Lagrange multiplier for the associate variational problem. Cafarelli and Spruk in [C-S] studied the equation (5) below on domains in the Euclidean space with $K = 1$. Here we extend their result to Riemannian manifolds and K a smooth non negative function.

Theorem 1. *Let M be a compact Riemannian manifold with boundary. Given smooth functions $K \geq 0$ on M and $\varphi > 0$ on ∂M, and $s > \int_M K f^{\frac{2n}{n-2}}$, where f is the harmonic extension of φ, there is a smooth positive minimizer of $D(u)$ in the constraint set $C_{\frac{2n}{n-2},s}$ satisfying that*

$$(5) \qquad \Delta u + \lambda K u^{\frac{n+2}{n-2}} = 0 \quad \text{in} \quad M,$$
$$u = \varphi \quad \text{on} \quad \partial M,$$

where $0 < \lambda < C(n,K,\varphi)(s - \int_M Kf^{\frac{2n}{n-2}})$. Moreover $\lambda \to 0$ when $s \to \infty$.

Proof. To show the existence of a smooth solution to problem (5) we define the constraint set
$$C_{\alpha,\varphi} = \left\{ u \in C_{\alpha,s} : u - f = 0 \text{ a.e. on } \partial M \right\}.$$
Observe that the set
$$E_\Lambda = \{ u \in C_{\alpha,\varphi} : D(u) \leq \Lambda \}$$
is bounded in $H^1(M)$. The direct method gives us a minimizing function u_α for the functional D in $C_{\alpha,\varphi}$ for any $\alpha \in (1, \frac{2n}{n-2})$. Replacing u by its absolute value $|u|$ we may assume $u \geq 0$. Standard regularity theory shows that u_α is a smooth function and satisfies the equation
$$\Delta u_\alpha + \lambda_\alpha K u^{\alpha-1} \, 0 \quad \text{in} \quad M,$$
$$u_\alpha = \varphi \quad \text{on} \quad \partial M,$$
where λ_α is a Lagrange multiplier. Observe that $\lambda_\alpha > 0$ (see the argument below for the case $\alpha = \frac{2n}{n-2}$), and that $u_\alpha > 0$ in $\overline{M}$ by the maximum principle. The numbers λ_α are bounded and $\lambda_\alpha \to \lambda_{\frac{2n}{n-2}}$ when $\alpha \to \frac{2n}{n-2}$ with $\alpha < \frac{2n}{n-2}$.

Thus the sequence $\{u_\alpha\}$, $\alpha \in (1, \frac{2n}{n-2})$, is bounded in $H^1(M)$. There exists a subsequence of $\{u_\alpha\}$, say $\{u_{\alpha_i}\}$, such that $u_{\alpha_i} \rightharpoonup u_0 \in H^1(M)$ when $\alpha_i \to \frac{2n}{n-2}$. The function u_0 is an H^1-weak solution of the equation
$$\Delta u_0 + \lambda_{\frac{n+2}{n-2}} K u_0^{\frac{n+2}{n-2}} = 0 \quad \text{in} \quad M,$$
$$u_0 = \varphi \quad \text{on} \quad \partial M.$$

From the Sobolev trace inequality we have
$$\int_{\partial M} \varphi^2 = \int_{\partial M} u_\alpha^2 \leq C \|u_\alpha\|^2_{H^1(M)}.$$

Thus $\|u_\alpha\|^2_{H_1} \geq C$ for all $\alpha \in (1, \frac{2n}{n-2})$. By compactness of the embedding of $L^2(\partial M) \hookrightarrow H^1(M)$ we have that the same bound holds for the limiting function u_0, $\|u_0\|^2_{H_1} \geq C > 0$. Hence $u_0 \geq 0$ and $u_0 \neq 0$. It is well known that u_0 is a smooth solution and by the strong maximum principle $u_0 > 0$ on $\overline{M}$.

Now we estimate λ in equation (5) in terms of $s - \int_M Kf^{\frac{2n}{n-2}}$.

Let u be a solution of (5). Multiplying equation (5) by $u - f$ and integrating by parts we obtain

(6) $$\int_M |\nabla u|^2 - \int_M \nabla u \cdot \nabla f = \lambda \left(\int_M K u^{\frac{2n}{n-2}} - \int_M K u^{\frac{n+2}{n-2}} f \right)$$
$$= \lambda \left(s - \int_M K u^{\frac{n+2}{n-2}} f \right).$$

Hölder's inequality implies that

$$\int_M Ku^{\frac{n+2}{n-2}} f \leq \left(\int_M Kf^{\frac{2n}{n-2}}\right)^{\frac{n-2}{2n}} \left(\int_M Ku^{\frac{2n}{n-2}}\right)^{\frac{n+2}{2n}} \leq s.$$

Observe that since f is harmonic and $u = f$ on ∂M

$$\int_M \nabla u \cdot \nabla f = \int_M |\nabla f|^2$$

and hence

$$\int_M |\nabla u|^2 - \int_M \nabla u \cdot \nabla f = \int_M |\nabla(u-f)|^2 > 0.$$

Using this in the equality (6) we obtain

(7) $$0 < \int_M |\nabla(u-f)|^2 = \lambda\left(s - \int_M Ku^{\frac{n+2}{n-2}} f\right).$$

The inequality (7) implies that $\lambda > 0$ and thus $u > f$ by the maximum principle. The Mean Value Theorem implies

$$Ku^{\frac{2n}{n-2}} - Kf^{\frac{2n}{n-2}} \leq \frac{2n}{n-2} Ku^{\frac{n+2}{n-2}}(u-f).$$

Thus

$$\int_M |\nabla(u-f)|^2 \geq \lambda\left(\frac{n-2}{2n}\right)\left(s - \int_M Kf^{\frac{2n}{n-2}}\right)$$

or, equivalently,

(8) $$\lambda \leq \frac{2n}{n-2} \frac{\int_M |\nabla(u-f)|^2}{\left(s - \int_M Kf^{\frac{2n}{n-2}}\right)}.$$

To estimate

$$\int_M |\nabla(u-f)|^2 = \int_M |\nabla u|^2 - \int_M |\nabla f|^2$$

we let $\psi \geq 0$ be a fixed nontrivial function $\psi \in C_0^\infty(\Omega)$ and $v = f + \gamma\psi$. Then there exists a unique $\gamma > 0$ such that $\int_M Kv^{\frac{2n}{n-2}} = s$, so that $v \in \mathcal{C}_{\frac{2n}{n-2},s}$. To estimate γ we observe that

$$s = \int_M K(f + \gamma\psi)^{\frac{2n}{n-2}} \geq \int_M Kf^{\frac{2n}{n-2}} + \frac{2n}{n-2}\gamma \int_M Kf^{\frac{n+2}{n-2}}\psi.$$

Thus

$$\gamma \leq \frac{n-2}{2n} \frac{\left(s - \int_M Kf^{\frac{2n}{n-2}}\right)}{\int_M Kf^{\frac{n+2}{n-2}}\psi},$$

or, equivalently,
$$\gamma \leq C\left(s - \int_M Kf^{\frac{2n}{n-2}}\right),$$
where

(9) $$C = C(n, K, f) = \frac{n-2}{2n \int_M Kf^{\frac{n+2}{n-2}} \psi}.$$

By minimality of the function u we have
$$\int_M |\nabla u|^2 \leq \int_M |\nabla v|^2 = \int_M |\nabla f|^2 + \gamma^2 \int_M |\nabla \psi|^2.$$

Therefore
$$\int_M |\nabla u|^2 - \int_M |\nabla f|^2 \leq C(n, K, f)^2 \left(s - \int_M Kf^{\frac{2n}{n-2}}\right)^2$$

and hence
$$\lambda \leq C(n, K, f)^2 \left(s - \int_M Kf^{\frac{2n}{n-2}}\right),$$

where $C(n, K, f)$ is given by equality (9). To estimate λ when s is large let $v = (1-a)s^{\frac{n-2}{2n}} u_2 + f$, where u_2 satisfies that $\int_M K|u_2|^{\frac{2n}{n-2}} = 1$, $u_2 \in H_0^1(M)$ and a is choosen so that $v \in C_{\frac{n+2}{n-2},s}$. Then

$$\int_M |\nabla v|^2 = (1-a)^2 s^{\frac{n-2}{n}} \int_M |\nabla u_2|^2 + \int_M |\nabla f|^2.$$

Since $\int_M |\nabla u|^2 \leq \int_M |\nabla v|^2$, using inequality (8) we get

$$\lambda \leq \frac{2n}{n-2} \frac{(1-a)^2 s^{\frac{n-2}{n}} \int_M |\nabla(u_2)|^2}{(s - \int_M Kf^{\frac{2n}{n-2}})} \leq cs^{-\frac{2}{n}}.$$

From this inequality follows the last statement of our Theorem.

Remark. The existence of a smooth positive solution to equation (5) when $\int_M K > 0$ can be proved using the same method as in Theorem 1. The only difficult part is to prove that a minimizing sequence is bounded in $H^1(M)$. To do that, we use the same method as in the proof of Theorem 2 below. We used the nonnegativity of the function K to prove the estimates on the lagrange multiplier λ. If the function K is nonpositive then one can prove the existence of a smooth positive solution to problem (5) quite easily because the associated functional is convex.

In the next Theorem we consider the case when the prescribed scalar curvature function K satisfies

i) K changes sign,

ii) $\int_M K dv_M < 0$.

Let's assume from now on that K satisfies conditions i) and ii) above. We further assume that the boundary of M, $\partial M = \Sigma_1 \cup \Sigma_2$. The geometric situation where the following Theorem applies is to a metric with zero scalar curvature on M and with zero mean curvature on Σ_2. The functions K and H below represent the prescribed scalar curvature on M and precribed mean curvature on Σ_2, respectively.

Theorem 2. *Let (M^n, g) be a compact Riemannian manifold with boundary and $n \geq 3$. Assume that $\partial M = \Sigma_1 \cup \Sigma_2$ where $\Sigma_1 \neq \emptyset$. Let $\varphi > 0$ and $H \leq 0$ be smooth functions defined on Σ_1 and Σ_2 respectively. There exists a smooth positive function u satisfying*

$$\Delta u + \lambda K u^{\frac{n+2}{n-2}} = 0 \quad \text{in} \quad M,$$
$$u = \varphi \quad \text{on} \quad \partial \Sigma_1,$$
$$\frac{\partial u}{\partial n} = \lambda H u^{\frac{n}{n-2}} \quad \text{on} \quad \partial \Sigma_2,$$

where λ is a positive real number.

Proof. Consider the set

$$E_\Lambda = \{u \in \mathcal{C}_{\alpha,\beta} : D(u) \leq \Lambda\}$$

where

$$\mathcal{C}_{\alpha,\beta,\varphi} = \left\{u \in H^1(M) : \int_M K|u|^{\alpha+1} + \int_{\Sigma_2} H|u|^{\beta+1} = 1, u - f = 0 \text{ a.e. on } \Sigma_1\right\},$$

and f is any positive harmonic extension of φ.

Claim: The set E_Λ is bounded in $H^1(M)$.

Proof of Claim: Here we follow our previous argument given in [E-S]. It suffices to bound $\int_M u^2 dv_M$ for $u \in E_\Lambda$. By the Poincaré inequality this bound follows from a bound on the average value $\overline{u} = \text{vol}(M)^{-1} \int_M u$. To prove such a bound it is necessary to use the condition $\int_M K < 0$. Replacing u by its absolute value $|u|$ we may assume $u \geq 0$. We then have

$$\overline{u}^{(\alpha+1)} \left(\int_M -K dv_M\right) = -\int_M K u^{\alpha+1} dv_M + \int_M K(u^{\alpha+1} - \overline{u}^{\alpha+1}) dv_M.$$

Since $u \in \mathcal{C}_{\alpha,\beta,\varphi}$ and $H \leq 0$, $-\int_M K u^{\alpha+1} < 0$. Thus condition ii) on K implies

$$(\overline{u})^{\alpha+1} \leq C \int_M |u^{\alpha+1} - \overline{u}^{\alpha+1}| dv_M.$$

From elementary calculus, we see that for $a, b \geq 0$
$$|a^{\alpha+1} - b^{\alpha+1}| \leq (\alpha+1)(a+b)^\alpha |a-b| \leq (\alpha+1)(|a-b|+2b)^\alpha |a-b|$$
$$\leq 2^{\alpha-1}(\alpha+1)|a-b|^{\alpha+1} + 2^{2\alpha-1}|a-b|.$$

Applying this with $a = u$ and $b = \overline{u}$ we get
$$\overline{u}^{\alpha+1} \leq C \int_M |u - \overline{u}|^{\alpha+1} dv_M + C\overline{u}^\alpha \int_M |u - \overline{u}| dv_M.$$

The Sobolev inequality implies then $\overline{u}^{\alpha+1} \leq C(1 + \overline{u}^\alpha)$, which in turn implies that $\overline{u}$ is uniformly bounded for $u \in E_\Lambda$.

The direct method gives us a minimizing function $u_{\alpha,\beta}$ for the functional D in $C_{\alpha,\beta,\varphi}$ for any $\alpha \in (1, \frac{n+2}{n-2})$ and $\beta \in (1, \frac{n}{n-2})$. Standard regularity theory shows that $u_{\alpha,\beta}$ is a smooth function and satisfies the equation
$$\Delta u_{\alpha,\beta} + \lambda_{\alpha,\beta} K u^\alpha \; 0 \quad \text{in} \quad M,$$
$$u_{\alpha,\beta} = \varphi \quad \text{on} \quad \Sigma_1,$$
$$\frac{\partial u_{\alpha,\beta}}{\partial \eta} = \lambda_{\alpha,\beta} H u^\beta \quad \text{on} \quad \Sigma_2,$$

where $\lambda_{\alpha,\beta}$ denote the infimum of $D(u)$ taken over $u \in C_{\alpha,\beta,\varphi}$. Note that $u_{\alpha,\beta} > 0$ in $\overline{M}$ by the strong maximum principle. Observe that $\lambda_{\alpha,\beta} > 0$. In order to see that, we first note that $\lambda_{\alpha,\beta} = D(u_{\alpha,\beta}) \geq 0$. If $D(u_{\alpha,\beta}) = 0$ then $u_{\alpha,\beta}$ is a positive constant $C_{\alpha,\beta}$ and since $\int_M K u_{\alpha,\beta}^{\alpha+1} + \int_{\Sigma_2} H u_{\alpha,\beta}^{\beta+1} = 1$ we have $C_{\alpha,\beta}^{\alpha+1} \int_M K + C_{\alpha,\beta}^{\beta+1} \int_{\Sigma_2} H = 1$ which is a contradiction, because $\int_M K < 0$ and $\int_{\Sigma_2} H \leq 0$. The numbers $\lambda_{\alpha,\beta}$ are bounded and $\lambda_{\alpha,\beta} \to \lambda_{\frac{n+2}{n-2}, \frac{n}{n-2}}$ when $\alpha \to \frac{n+2}{n-2}$ with $\alpha < \frac{n+2}{n-2}$ and $\beta \to \frac{n}{n-2}$ with $\beta < \frac{n}{n-2}$. Thus the sequence $\{u_{\alpha,\beta}\}$, $\alpha \in (1, \frac{n+2}{n-2})$, $\beta \in (1, \frac{n}{n-2})$, is bounded in $H^1(M)$. There exists a subsequence of $\{u_{\alpha,\beta}\}$, say $\{u_{\alpha_i,\beta_i}\}$, such that $u_{\alpha_i,\beta_i} \rightharpoonup u_0 \in H^1(M)$ when $(\alpha_i, \beta_i) \to (\frac{n+2}{n-2}, \frac{n}{n-2})$. The function u_0 is an H^1–weak solution of the equation
$$\Delta u_0 + \lambda_{\frac{n+2}{n-2}, \frac{n}{n-2}} K u_0^{\frac{n+2}{n-2}} = 0 \quad \text{in} \quad M,$$
$$u_0 = \varphi \quad \text{on} \quad \Sigma_1,$$
$$\frac{\partial u_0}{\partial \eta} = \lambda_{\frac{n+2}{n-2}, \frac{n}{n-2}} H u^{\frac{n}{n-2}} \quad \text{on} \quad \Sigma_2.$$

A theorem of Cherrier [Ch] guarantees that the function u_0 is a smooth function. From the Sobolev trace inequality we have
$$\int_{\Sigma_1} \varphi^2 \leq \int_{\partial M} u_{\alpha,\beta}^2 \leq C\|u_{\alpha,\beta}\|_{H^1(M)}^2.$$

Thus $\|u_{\alpha,\beta}\|_{H^1}^2 \geq C$ for all $\alpha \in (1, \frac{n+2}{n-2})$ and $\beta \in (1, \frac{n}{n-2})$. By compactness of the embedding of $L^2(\partial M) \hookrightarrow H^1(M)$ we have that the same bound holds for the limiting function u_0, $\|u_0\|_{H^1}^2 \geq C > 0$. Hence $u_0 \geq 0$ and $u_0 \neq 0$. It is well known that u_0 is a smooth solution and by the strong maximum principle $u_0 > 0$ on $\overline{M}$.

References

[C-S] Cafarelli, L. and Spruk, J, *Variational problems with critical Sobolev growth and positive Dirichlet data*, Indiana Math. Jour. **39** (1990), 1–18.

[Ch] Cherrier, P., *Problèmes de Neumann non linéaires sur les varietés riemanniennes*, Journal of Functional Analysis **57** (1984), 154–206.

[E1] Escobar, J. F., *Conformal Deformation of a Riemannian Metric to a Scalar Flat Metric with Constant Mean Curvature on the Boundary*, Annals of Math. **136** (1992), 1–50.

[E2] Escobar, J. F., *Addendum Conformal deformation of a Riemannian metric to a scalar flat metric with constant mean curvature on the boundary*, Annals of Math. **13** (1994), 749–750.

[E-S] Escobar, J. F., Schoen, R. M., *Conformal metrics with prescribed scalar curvature*, Inventiones Mathematicae **86** (1986), 243–254.

Department of Mathematics
Cornell University
Ithaca, NY 14853

Chord Uniqueness and Controllability: the View from the Boundary, I

Robert Gulliver and Walter Littman

July 11, 2000

SCHOOL OF MATHEMATICS
UNIVERSITY OF MINNESOTA
MINNEAPOLIS, MINNESOTA 55455, USA
gulliver@math.umn.edu, littman@math.umn.edu

Abstract

Consider the problem of a compact, n-dimensional Riemannian manifold-with-boundary $\overline{\Omega}$ and the natural hyperbolic P.D.E. (Riemannian wave equation):

$$(0.1) \qquad \frac{\partial^2 u}{\partial t^2} = \Delta_g u,$$

plus possible lower-order terms, where Δ_g is the Riemannian Laplace operator, or Laplace-Beltrami operator, of Ω. We consider the problem of the control in time T of the wave equation from the boundary $\partial\Omega$ of Ω, by specifying Dirichlet boundary controls on $\partial\Omega \times [0,T]$. The question we address is whether, for any Cauchy data on Ω at the initial time $t = 0$, there is a choice of boundary control which will achieve any prescribed Cauchy data at the terminal time $t = T$.

In order to reduce this controllability question to a computable problem about geodesics on Ω, we pose the question: *are chords unique?* Here, a *chord* is a length-minimizing geodesic of $\overline{\Omega}$ joining two given points of $\partial\Omega$. We assume that any two points of $\partial\Omega$ are connected by at most one (and hence exactly one) chord.

If, in addition, the chords are nondegenerate and $\partial\Omega$ has positive second fundamental form, then the wave equation is controllable from $\partial\Omega$ in any time T greater than the maximum distance in $\overline{\Omega}$ between points of $\partial\Omega$.

This result provides a counterpoint to controllability theorems such as those in [14], [7] and [8], in which the existence of a convex function, and hence— roughly speaking— an upper bound on sectional curvature, is assumed. We require no direct hypothesis on the Riemannian metric in the interior of Ω.

2000 *Mathematics Subject Classification.* Primary 35Lxx, 51–XX, 49–XX, 93–XX.

© 2000 American Mathematical Society

1 Introduction

> Whenever one approaches a subject from two different directions, there is bound to be an interesting theorem explaining their relation.
> *Robert Hartshorne, Notices Amer. Math. Soc. April, 2000, p. 464.*

Although the subject of boundary control of partial differential equations is about a quarter of a century old, and that of Riemannian geometry much older still, there has been relatively little interaction between the two. This is especially surprising in view of the rôle bicharacteristics play in boundary control, which naturally bring to mind geodesics— a basic concept in Riemannian geometry.

One of the *raisons d'être* of this conference is the belief that both subjects have much to gain by closer interaction with one other. We hope that this paper makes a — perhaps modest — contribution in that direction. To be more specific, we believe there is a reservoir of as yet untapped Riemannian-geometric tools which could be applied successfully in boundary control. As one example, we cite Theorem 2 of [6], which is crucial to the proof of Proposition 5.1 below. On the other side, we expect that some of the compelling problems of the theory and applications of boundary control will stimulate the disciplines of Riemannian and Lorentzian geometry to undertake new areas of research.

Since our focus in this paper is on the relationship between Riemannian geometry and boundary control, we shall not attempt here to express controllability in terms of the optimal choice of Sobolev spaces, leaving such questions to other papers such as [9]; nor shall we attempt to find the optimal smoothness of the Riemannian metric and of other coefficients of the hyperbolic equation (see Michael Taylor's forthcoming book [16].)

Consider a compact, n-dimensional Riemannian manifold-with-boundary $\overline{\Omega}$. We assume that $\partial \Omega$ is smooth and nonempty, and that the metric of $\overline{\Omega}$ is smooth, *i.e.*, C^∞. We are interested in the boundary control of the following natural hyperbolic partial differential equation (Riemannian wave equation) on $\Omega \times [0, T]$:

$$(1.1) \qquad \frac{\partial^2 u}{\partial t^2} = \Delta_g u := \sum_{i,j=1}^n \frac{1}{\gamma} \frac{\partial}{\partial x_i} \left(\gamma \, g^{ij}(x) \frac{\partial u}{\partial x_j} \right)$$

for all $(x,t) \in \Omega \times [0,T]$, where $(x_1, \ldots, x_n)$ are arbitrary local coordinates, $g^{ij}(x)$ are the entries of the inverse matrix to the coefficients $g_{ij}(x)$ of the Riemannian metric, and γ is the Riemannian volume integrand: $\gamma(x) = \sqrt{\det(g_{ij}(x))}$. We consider the problem of the control in time T of equation (1.1) from boundary $\partial \Omega$. More precisely, we consider the boundary conditions

(1.2) $$u(x,t) = U(x,t) \text{ for all } (x,t) \text{ on } \partial\Omega \times [0,T],$$

where $U \in H^{1/2}(\partial\Omega \times [0,T])$ is the control, *i.e.* a function which may be chosen as needed. The controllability question is whether, given any initial conditions

(1.3) $$u(x,0) = u_0(x), \qquad \frac{\partial u}{\partial t}(x,0) = u_1(x),$$

with finite energy, there is a choice of controls $U \in H^{1/2}(\partial\Omega \times [0,T])$ such that the solution of (1.1) with initial conditions (1.3) and boundary conditions (1.2) vanishes identically on $\Omega \times [T,\infty)$. Equivalently, we ask whether for some choice of controls U the terminal Cauchy values vanish: $u(x,T) = 0$, $\frac{\partial u}{\partial t}(x,T) = 0$ for all x in Ω.

More generally, we shall consider the hyperbolic equation with additional lower-order terms:

(1.4) $$\frac{\partial^2 u}{\partial t^2} = \Delta_g u + \sum_{i=1}^{n} V^i(x,t) \frac{\partial u}{\partial x_i} + a(x,t)u,$$

where $(V^1, \ldots, V^n)$ are the components, in any local system of coordinates $x = (x_1, \ldots, x_n)$ for $\overline{\Omega}$, of a vector field V on $\overline{\Omega}$, and $a : \overline{\Omega} \to \mathbb{R}$. The first-order term $\sum_{i=1}^{n} V^i(x,t) \frac{\partial u}{\partial x_i}$ is invariant under change of coordinates. We shall assume throughout that $V(x,t)$ and $a(x,t)$ describe real-analytic mappings from t to the space of smooth vector fields and smooth functions, resp., on $\overline{\Omega}$.

In section 3 below, we shall show how this control problem is related to a problem about geodesics of Ω, namely, their maximum length before leaving Ω. In section 4 below, we shall see how this geometric problem is related to the question whether chords are unique. Here, a *chord* is defined to be a length-minimizing curve in $\overline{\Omega}$ joining two given points of $\partial\Omega$. The existence of a chord joining any two points of $\partial\Omega$ is well known, although in general it may have nongeodesic segments lying in $\partial\Omega$ (see *e.g.* pp. 147-148 of [4].) If, however, we assume that $\partial\Omega$ is strictly convex, that is, has positive second fundamental form, then for any length-minimizing curve $\sigma : [s_0, s_1] \to \overline{\Omega}$ between two points of $\overline{\Omega}$, we have $\sigma((s_0, s_1)) \subset \Omega$ (see the proof of Corollary 3.3 or Lemma 4.1 below.) In particular, every chord of $\overline{\Omega}$ is a geodesic in this case. We say that a chord $\sigma : [s_0, s_1] \to \overline{\Omega}$ is *nondegenerate* if there are no conjugate points to $\sigma(s_0)$ along $\sigma((s_0, s_1])$ (see Subsection 4.2 below.)

Another way of viewing these hypotheses of chord uniqueness and nondegeneracy is as follows. Consider any two boundary points p and q. Suppose that, among all the light rays leaving q simultaneously and propagating in Ω (without reflection), only one ray reaches p first. This is equivalent to the uniqueness of chords. Nondegeneracy of chords is equivalent to the statement that, in the situation just described, when a ray leaving q turns out to be a chord from q to p, the boundary point p depends in a diffeomorphic way on the initial direction of the ray at q (see also the proof of Proposition 4.6.)

Theorem 1.1 *Suppose that any two boundary points of the manifold $\overline{\Omega}$ are connected by a unique chord, which is nondegenerate. Assume that $\partial\Omega$ has positive second fundamental form. Then the hyperbolic equation (1.4) is controllable from $\partial\Omega$ in the sense of boundary conditions (1.2) in any time*

$$T > T_0 := \operatorname{diam}_{\overline{\Omega}}(\partial\Omega).$$

Here, the *diameter* of the boundary of Ω is the maximum distance between any two of its points, with respect to the distance measured in $\overline{\Omega}$, that is: the length of the longest chord of Ω. Our convention for the sign of the second fundamental form is such that if Ω is a ball of radius r in $I\!R^n$, with the Euclidean metric, then $\partial\Omega$ has positive second fundamental form $B = r^{-1}ds^2$.

The "uniqueness" of a chord $\gamma : [0, a] \to \overline{\Omega}$ is understood *modulo* reparameterizations $s \mapsto \gamma(As + B)$ $(A, B \in I\!R)$ of the independent variable s.

We shall refer to the infimal value T_0 as the "optimal time of control," even though T_0 itself may not be a control time.

As the reader will verify immediately, Theorem 1.1 follows from Proposition 3.1, Proposition 3.2 and Theorem 4.5 below. For the main geometric result of this paper, some readers will prefer the statement of Theorem 4.5 over Theorem 1.1.

Remark 1 A sufficient condition for the uniqueness of chords, and for their nondegeneracy, is that for any point $q \in \partial\Omega$, the gradient of the distance function $d_{\overline{\Omega}}(\cdot, q)$ is a continuous function of $p \in \partial\Omega$, $p \neq q$. In this case, the optimal time of control T_0 is the maximum of $d_{\overline{\Omega}}(p, q)$ over $p, q \in \partial\Omega$. No further information about the coefficients in the interior of Ω is required. See Proposition 4.6 below.

If, on the other hand, information is available giving upper bounds on sectional curvatures in various regions of Ω, then Corollary 5.2 below provides a useful criterion for the time of boundary controllability.

Another sufficient condition for chord uniqueness is that every geodesic of $\overline{\Omega}$ starting from $\partial\Omega$ will leave $\overline{\Omega}$ before any conjugate point appears. See Proposition 4.7 below.

In Section 2 below, we shall indicate how geometric methods may be introduced for the analysis of a partial differential equation in $I\!R^n$ with coefficients depending on the space variables.

In Section 4, we shall first outline some of the known concepts and results of Riemannian geometry which will be relevant to the proof of our theorems. We shall then introduce the main tools, including especially the chord map, to prove a new geometric result (Theorem 4.5), which will be a key step towards the proof of Theorem 1.1.

In Section 5, we shall present five explicit and, we hope, interesting examples to illuminate the variety of Riemannian manifolds in terms of the existence of a convex function; the uniqueness of chords; the nondegeneracy of certain geodesics; and continuity, as opposed to smoothness, of the gradient of the distance function. We shall also give criteria for the absence of conjugate points (Proposition 5.1), necessary conditions for the existence of a convex function (Proposition 5.3), and necessary conditions for chord uniqueness (Proposition 5.4.)

In part II of this paper, we plan to treat the case where control occurs only on a prescribed open subset of $\partial\Omega$.

The first author would like to acknowledge partial research support by the Max Planck Institute for Mathematics in the Sciences, Leipzig, Germany, by the University of Melbourne, Australia and by Monash University, Australia.

2 Geometry and Partial Differential Equations

This section and Section 4.1 below are of a tutorial character, and will outline some well-known concepts and results of differential geometry. Specifically, we shall illustrate here how Riemannian geometry can be introduced to aid in understanding properties of a hyperbolic P.D.E. In the reverse direction, this framework may be applied to any Riemannian manifold $\overline{\Omega}$ which is diffeomorphic to the closure of a smooth domain in $\mathbb{R}^n$.

Consider a hyperbolic equation with time-independent coefficients in a smooth domain $\Omega \subset \mathbb{R}^n$:

$$(2.1) \qquad \frac{\partial^2 u}{\partial t^2} = \frac{\partial}{\partial x_i}\left(g^{ij}(x)\frac{\partial u}{\partial x_j}\right) + \text{lower order terms}$$

(the summation over $1 \leq i,j,k \leq n$ is implicit where the index appears twice in one term.) For each $x \in \overline{\Omega}$, let $g_{ij}(x)$ be the inverse matrix of $g^{ij}(x)$. Then the metric $ds^2 = g_{ij}(x)\, dx_i\, dx_j$ makes $\overline{\Omega}$ into a Riemannian manifold-with-boundary (see [4].) After adding appropriate lower-order terms, (2.1) becomes the Riemannian wave equation (1.1) above. Observe that, since equations (1.1), (1.4) and (2.1) have the same bicharacteristics, they will also have the same controllability properties (cf. Proposition 3.2 below.)

The *length* of a curve in $\overline{\Omega}$ is the integral of ds, and a *geodesic* is a curve $\sigma = (\sigma_1, \ldots, \sigma_n) : (a, b) \to \overline{\Omega}$ which satisfies the equation

$$(2.2) \qquad \frac{d^2\sigma_k}{ds^2} + \Gamma^k_{ij} \frac{d\sigma_i}{ds} \frac{d\sigma_j}{ds} = 0.$$

Here $\Gamma^k_{ij} = \Gamma^k_{ij}(x)$ are the Christoffel symbols, evaluated at $x = \sigma(s)$. The Christoffel symbols are the coefficients of the Levi-Civita connection of $\overline{\Omega}$, determined by

$$2\, g_{kl}\, \Gamma^l_{ij} = \frac{\partial g_{ik}}{\partial x_j} - \frac{\partial g_{ij}}{\partial x_k} + \frac{\partial g_{kj}}{\partial x_i}.$$

Given a vector field $J = J^i(s)\frac{\partial}{\partial x_i}(\sigma(s))$ along σ, the *covariant derivative* $J'(s) = \frac{DJ}{ds}$ of J with respect to s may be written in terms of the Christoffel symbols:

$$J'(s) = \left[\frac{dJ^k}{ds} + \Gamma^k_{ij}(\sigma(s))J^i(s)\frac{d\sigma_j}{ds}(s)\right] \frac{\partial}{\partial x_k}(\sigma(s)),$$

which is independent of the choice of coordinates. With this notation, the geodesic equation becomes $\sigma_0''(s) = 0$.

The hypothesis of Theorem 1.1 that $\partial\Omega$ have positive second fundamental form at each point $p \in \partial\Omega$ may be computed most easily by making a linear change of coordinates so that the coordinate hyperplane $x_n = \text{const.}$ which passes through p is tangent to $\partial\Omega$ there, and so that the nth coordinate vector is the inward unit normal to $\partial\Omega$ at p. Let $\partial\Omega$ be represented locally as the graph $x_n = f(x_1, \ldots, x_{n-1})$. Then one requires that the symmetric matrix with entries

$$B_{ij} = \Gamma^n_{ij} + \frac{\partial f}{\partial x_i}\Gamma^n_{nj} + \frac{\partial f}{\partial x_j}\Gamma^n_{in} + \frac{\partial f}{\partial x_i}\frac{\partial f}{\partial x_j}\Gamma^n_{nn},$$

$1 \le i, j \le n-1$, evaluated at p, be positive definite. The matrix (B_{ij}) represents the *second fundamental form* of $\partial\Omega$ in these coordinates. Equivalently, if $\overline{\Omega}$ is extended to be a smooth subdomain of a Riemannian manifold M, one requires that any geodesic of M which is tangent to $\partial\Omega$ at p remain outside of $\overline{\Omega}$ to second order at p. This property has been called "pseudo-convexity" in the P.D.E. literature; in this paper, since we deal only with quantities which are invariant under smooth changes of coordinates, the convexity of a domain with respect to the affine structure of $\mathbb{R}^n$ will not be relevant.

3 Bicharacteristics and Controllability

In this section, we shall indicate the relationship between the geodesics of the Riemannian manifold Ω and the boundary controllability of equation (1.1), equation (1.4) or of equation (2.1) above.

For the two propositions below, we assume that $\overline{\Omega}$ is a compact n-dimensional Riemannian manifold-with-boundary. Let $\overline{\Omega}$ be extended to become a subset of an open n-dimensional Riemannian manifold M.

Proposition 3.1 *Let M be a Riemannian manifold. Then the bicharacteristics of equation (1.4) are the graphs in $M \times \mathbb{R}$ of geodesics of M, with unit-speed parameter identified with time $\in \mathbb{R}$.*

Proof. See [5], p. 209. Q.E.D.

Proposition 3.2 *If every bicharacteristic in $\overline{\Omega} \times (0,T)$ enters or leaves $\overline{\Omega} \times (0,T)$ across the lateral boundary $\partial \Omega \times (0,T)$, then boundary control is available in any time $\geq T$. Conversely, if there is a single bicharacteristic in $\Omega \times (0,T)$ that enters $\overline{\Omega} \times [0,T]$ through the open bottom $\Omega \times \{0\}$ and leaves through $\Omega \times \{T\}$, without hitting the lateral boundary, then boundary control in time T is not possible.*

Proof. See [11], where the proof, which is given for a bounded domain in $\mathbb{R}^n$, carries over without difficulty to manifolds. The assumption made in [11] that the coefficients be real-analytic is easily removed in the case of time-independent C^∞ coefficients. See for example [1], p. 1050, where a proof is given (in $\mathbb{R}^n$) with optimal Sobolev spaces. The proof of [11] also works if the lower-order terms (*i.e.* not in the principal part) are real-analytic mappings from t to the space of C^∞ vector fields or C^∞ functions on $\overline{\Omega}$. The proof should then be supplemented by uniqueness theorem 2' of [12]. Recent results in propagation of singularities by M. Taylor (see [16]) further indicate that the required smoothness in x of the coefficients can be reduced to $C^{2,\alpha}$.

The converse follows from the propagation of singularities (see [13].) Namely, initial data can be constructed which is not C^∞ near a point inside Ω such that this singularity is propagated along the bicharacteristic, without being affected by boundary values, to form a singularity at time T inside Ω. Thus the solution cannot have terminal Cauchy conditions $u(\cdot, T) \equiv 0$, $\frac{\partial u}{\partial t}(\cdot, T) \equiv 0$. Q.E.D.

Remark 2 The reader might note, in particular, that if Ω contains a closed geodesic, then boundary control is impossible in any finite time.

Remark 3 It will be observed that if the hypothesis of the first part of Proposition 3.2 holds for a Riemannian manifold $\overline{\Omega}$, then it also holds for any compact subdomain $\overline{\Omega}_1 \subset \Omega$.

Corollary 3.3 *If $\partial\Omega$ has positive second fundamental form, and if $\overline{\Omega}$ is not simply connected, then equation (1.1) is not controllable from the boundary in any finite time.*

Proof. Let γ_1 be a non-contractible closed curve in $\overline{\Omega}$. The infimum of lengths of curves in the homotopy class of γ_1 in $\overline{\Omega}$ is assumed by a closed curve γ_0. Since the second fundamental form of $\partial\Omega$ is positive, γ_0 lies entirely in the interior Ω. Otherwise, pushing off locally from $\partial\Omega$, with distance equal to a smooth nonnegative function of small support on $\partial\Omega$, would decrease length strictly. Therefore, γ_0 is a closed geodesic. The conclusion now follows from Remark 2. **Q.E.D.**

4 Chord Uniqueness and Identification of Geodesics

4.1 Geodesics and Jacobi Fields in a Riemannian Manifold

Let $\overline{\Omega}$ be a smooth, compact Riemannian manifold-with-boundary, of dimension n, and write $<X,Y>$ for the Riemannian metric applied to tangent vectors X, Y at a point of $\overline{\Omega}$. A curve $\sigma : [0, a] \to \overline{\Omega}$ is a *geodesic* if its tangent vector $\sigma'(s)$ has vanishing covariant derivative along $[0, a]$ (see equation (2.2) above.) A curve with constant speed and of shortest length among curves joining its endpoints is a geodesic, as long as it remains in the open manifold Ω, although not all geodesics have shortest length, even when they lie entirely in the open manifold Ω. Given $p, q \in \overline{\Omega}$, write $d(p,q)$ for the infimum of lengths of curves in $\overline{\Omega}$ joining p to q. Since $\overline{\Omega}$ is compact, and assuming that $\partial\Omega$ has positive second fundamental form, any two points p, q of $\overline{\Omega}$ may be joined by a geodesic of minimum length $d(p,q)$ (see [4], pp. 147-148.)

Where convenient, and without loss of generality, we shall assume that $\overline{\Omega}$ is a compact subset of an open Riemannian manifold M, from which it inherits its Riemannian metric. In particular, at each point $x \in \overline{\Omega}$, the tangent space $T_x(\overline{\Omega})$ will be isomorphic to the vector space $\mathbb{R}^n$.

Let $\sigma_0 : [0, a_0] \to \overline{\Omega}$ be a geodesic. Consider the linearized geodesic equation, or *Jacobi equation*, for a vector field $J(s)$ along σ_0 :

$$(4.1) \qquad J''(s) + R(\sigma_0'(s), J(s))\sigma_0'(s) = 0.$$

Here, $R(\cdot, \cdot)\cdot$ is the curvature tensor of the Riemannian manifold $\overline{\Omega}$. For example, if $\overline{\Omega}$ is an open subset of the n-dimensional sphere of radius r in $\mathbb{R}^{n+1}$, then $R(\sigma', J)\tau' = r^{-2}[<\sigma', \tau'> J - <J, \tau'> \sigma']$ (see [4], p. 89.) $J'(s)$ and $J''(s)$ are the first and second covariant derivatives of $J(s)$ with respect to s along σ_0. A solution to (4.1) is called a *Jacobi field*.

We begin by recalling three well-known facts about the Jacobi equation (see e.g. [4].)

First, let $\{\sigma_\lambda : -\varepsilon < \lambda < \varepsilon\}$ be a one-parameter family of geodesics $\sigma_\lambda : [0, a_\lambda] \to \overline{\Omega}$, each parameterized with constant speed. Then the transverse vector field $J(s) = \frac{\partial \sigma_\lambda(s)}{\partial \lambda}|_{\lambda=0}$ is a Jacobi field. Conversely, any Jacobi field along σ_0 arises as the transverse vector field to some family $\{\sigma_\lambda : -\varepsilon < \lambda < \varepsilon\}$ of geodesics.

Next, if $\sigma_0 : [0, a_0] \to \overline{\Omega}$ minimizes the length between its endpoints, then any Jacobi field $J(s)$ along σ_0 with the initial condition $J(0) = 0$ will remain nonzero for $0 < s < a_0$. This is Jacobi's theorem; see [4], p. 248. As a partial converse, if every Jacobi field along σ_0 with the initial value zero remains nonzero on $(0, a_0]$, then σ_0 has minimum length among curves connecting $\sigma_0(0)$ with $\sigma_0(a_0)$ and lying in a sufficiently small neighborhood of $\sigma_0([0, a_0])$. This is proved by invoking the Gauss Lemma and constructing a field of extremals; see [4], p. 71.

Last, a Jacobi field with initial inner products $<J(0), \sigma_0'(0)> = <J'(0), \sigma_0'(0)> = 0$ will remain orthogonal to the tangent vector $\sigma_0'(s)$ on $0 \leq s \leq a_0$.

We define the *sphere bundle* $S(\overline{\Omega})$ of $\overline{\Omega}$ as the set of all (x, w) where $x \in \overline{\Omega}$ and w is a unit tangent vector to $\overline{\Omega}$ at x. Since the value of the constant speed of a geodesic σ will be unimportant, it will be convenient to use initial conditions $(\sigma(0), \sigma'(0)) = (x, w) \in S(\overline{\Omega})$ for the geodesic equation. Then convergence of the initial values $\in S(\overline{\Omega})$ is equivalent to convergence in $C^\infty([0, a_0])$ of the unit-speed geodesics. In the context of the sphere bundle, $\pi : S(\overline{\Omega}) \to \overline{\Omega}$ will denote the natural projection $\pi(x, w) := x$.

For $x \in \overline{\Omega}$, the *exponential map* $\exp_x : T_x\overline{\Omega} \to \overline{\Omega}$ is defined on a star-shaped subset of the tangent space $T_x\Omega$ so that for $(x, w) \in S(\overline{\Omega})$ and $t \geq 0$, $t \mapsto \exp_x(tw) = \gamma(t)$ is the geodesic with initial conditions $\gamma(0) = x$, $\gamma'(0) = w$.

The *distance to the cut point* $c : S(\overline{\Omega}) \to (0, \infty]$ is defined by

$$c(x, w) := \sup\{t : \exp_x(tw) \text{ is defined and } d(x, \exp_x(tw)) = t\}.$$

According to the proof of Theorem 3.1 of [2], the function $c : S(\overline{\Omega}) \to (0, \infty)$ is continuous provided that $\overline{\Omega}$ is compact and that any minimizing geodesic from a point of $\overline{\Omega}$ to a boundary point q is transversal to $\partial\Omega$ at q. We may define a star-shaped open set $\mathcal{E}_x \subset T_x\Omega$ as $\{tw \in T_x\Omega : |w| = 1, 0 \leq t < c(x, w)\}$, and write $\mathcal{W}_x \subset \overline{\Omega}$ for its image under $\exp_x$. Then $\partial \mathcal{W}_x = \exp_x(\partial \mathcal{E}_x)$ is called the *cut locus* of x. Note that $\exp_x$ maps $\mathcal{E}_x$ diffeomorphically onto $\mathcal{W}_x$. A cut point $y \in \partial \mathcal{W}_x$ is either a boundary point of Ω; the first conjugate point along a length-minimizing geodesic from x (see section 4.2 below); or the end point of two distinct length-minimizing geodesics from x.

4.2 Conjugate points

Two points $\gamma(s_1)$ and $\gamma(s_2)$ of a geodesic γ are called *conjugate points* if $s_1 \neq s_2$ and there exists a nontrivial Jacobi field J along γ with $J(s_1) = 0$ and $J(s_2) = 0$.

Although a chord, since it has minimum length, may in general have its endpoints conjugate to each other, no two interior points may be conjugate, by Jacobi's theorem. The case when a chord $\gamma_0 : [0, a_0] \to \overline{\Omega}$ has conjugate endpoints is therefore somewhat special, and we have called such a chord *degenerate*.

Lemma 4.1 *Suppose that $\partial\Omega$ has positive second fundamental form. Let $\gamma : [0, a] \to \overline{\Omega}$ be a geodesic with $\gamma(0) \in \partial\Omega$. Then γ meets $\partial\Omega$ transversally at $\gamma(0)$.*

Proof. Let $\overline{\Omega}$ be extended as a Riemannian manifold to an open manifold M, and extend γ as a geodesic to a longer interval $[-b, d]$, for some positive b, d. If $\gamma'(0)$ were tangent to $\partial\Omega$, since the second fundamental form of $\partial\Omega$ is positive, $\gamma(s)$ could only lie outside $\overline{\Omega}$, to second order as $s \to 0$. This would contradict the assumption that $\gamma([0, a]) \in \overline{\Omega}$. Q.E.D.

Our next result is essentially a consequence of the Gauss Lemma.

Lemma 4.2 *Suppose $\partial\Omega$ has positive second fundamental form. Let $\gamma : [0, a_0] \to \overline{\Omega}$ be a nondegenerate chord with unit speed, and write $p_0 = \gamma(0)$, $w_0 = \gamma'(0)$. For $(p, w) \in S(\overline{\Omega})$, write*

$$\gamma_w(s) := \exp_p(sw).$$

Then there is a neighborhood $\mathcal{V}$ of (p_0, w_0) in $S(\overline{\Omega}) \cap \pi^{-1}(\partial\Omega)$ such that for all $(p, w) \in \mathcal{V}$, w points into Ω, and the maximal segment $\gamma_w([0, a_w])$ lying in $\overline{\Omega}$ has minimum length among those curves in $\overline{\Omega}$ connecting $p = \gamma_w(0)$ to $\gamma_w(a_w)$. In particular, for all $(p, w) \in \mathcal{V}$, γ_w is a chord. Moreover, $a_w \to a_{w_0} = a_0$ as $(p, w) \to (p_0, w_0)$.

Proof. Recall that, since γ is a chord, $\gamma : [0, a] \to \overline{\Omega}$ is an embedding. Since γ is a *nondegenerate* chord, there are no conjugate points to $p_0 = \gamma(0)$ along $\gamma((0, a])$. Let $\overline{\Omega}$ be extended as a Riemannian manifold to an open manifold M, and extend γ as a geodesic to a longer interval $[-b_0, d_0]$ on which $\gamma : [-b_0, d_0] \to M$ is an embedding and has no conjugate-point pairs.

Note that w_0 points into $\partial\Omega$, by Lemma 4.1. If $(p, w) \in S(M) \cap \pi^{-1}(\partial\Omega)$ is sufficiently close to (p_0, w_0), the above properties of γ will also be valid for γ_w, by the continuous dependence of solutions of the Jacobi equation (4.1) on initial conditions. That is, w points into $\partial\Omega$, $\gamma_w([0, a_w])$ is a geodesic in $\overline{\Omega}$ with endpoints on $\partial\Omega$, and there is a longer interval $[-b_w, d_w]$ on which γ_w is an embedding into

M and has no conjugate-point pairs. By Lemma 4.1, a_w depends continuously on $(p, w) \in S(M) \cap \pi^{-1}(\partial\Omega)$.

Fix w so that γ_w has the properties just discussed. Write $y := \gamma_w(-b_w) \in M$. Then as v ranges over a sufficiently small neighborhood of $\gamma_w'(-b_w)$ in the sphere $S_y(M)$, we claim that the geodesics $\gamma_v([0, c_v])$ sweep out a neighborhood $\mathcal{U}_w$ of $\gamma_w((-b_w, d_w))$ diffeomorphically. Here $\gamma_v(0) = y$ and we choose $\gamma_v(c_v)$ to be a point in M near $\gamma_w(d_w)$ and after γ_v leaves $\overline{\Omega}$, such that c_v is a continous function of v and such that $\gamma_v : [0, c_v] \to M$ is an embedding without conjugate-point pairs. In fact, the absence of conjugate points implies that the radial segment $[0, c_v]v$ in $T_y(M)$ lies inside $\mathcal{E}_y$. Therefore, $(v, s) \mapsto \gamma_v(s)$ is a local diffeomorphism on a neighborhood of $\{w\} \times (-b_w, d_w)$, and by suitable restriction, using the embedded property of $\gamma_w([-b_w, d_w])$, we may obtain a diffeomorphism, as claimed. Let $\delta > 0$ be a lower bound for the distance from $\gamma_w([0, a_w])$ to the complement of $\mathcal{U}_w$ in M, for all (p, w) in a neighborhood of (p_0, w_0) in $S(M)$.

We may now apply the Gauss Lemma to show that $\gamma_w([0, a_w])$ is the unique curve of shortest length *inside* $\mathcal{U}_w$ joining $\gamma_w(0)$ to $\gamma_w(a_w)$. Specifically, each point $z \in \mathcal{U}_w$ may be represented uniquely as $z = \gamma_v(s)$ for a unit tangent vector v at y near $\gamma_w'(-b_w)$. This defines a smooth real-valued function $s = s(z)$ on $\mathcal{U}_w$. The Gauss Lemma shows that on $\mathcal{U}_w$, the Riemannian gradient of $s(z)$ equals the tangent vector to γ_v, so that any curve σ has length at least as large as the change in s along σ, as long as σ stays inside $\mathcal{U}_w$. Therefore, any minimizing curve inside $\mathcal{U}_w$ must be everywhere transverse or everywhere tangent to the family of geodesics $\{\gamma_v : v \in S_y(M)\}$. It follows that γ_w is the unique curve inside $\mathcal{U}_w$ of minimum length between $\gamma_w(0)$ and $\gamma_w(a_w)$.

It remains to show that, if $(p, w) \in S(M) \cap \pi^{-1}(\partial\Omega)$ is sufficiently close to (p_0, w_0), then $\gamma_w([0, a_w])$ has shortest length *among all curves in* $\overline{\Omega}$ joining $\gamma_w(0)$ to $\gamma_w(a_w)$, that is, that γ_w is a chord of $\overline{\Omega}$.

Otherwise, there are points p_k in $\partial\Omega$ and $(p_k, w_k) \in S_{p_k}(M)$, with $(p_k, w_k) \to (p_0, w_0)$ as $k \to \infty$, such that γ_{w_k} is not a chord. This means that for each k, there is a unit-speed curve $\sigma_k : [0, C_k] \to \overline{\Omega}$ from $p_k = \gamma_{w_k}(0)$ to $\gamma_{w_k}(a_{w_k})$ which is shorter than $\gamma_{w_k}([0, a_{w_k}])$. Since γ_{w_k} has shortest length inside the open set $\mathcal{U}_{w_k}$, the curve σ_k must include at least one point $\sigma_k(s_k)$ in the complement of $\mathcal{U}_{w_k}$ in $\overline{\Omega}$. After passing to a subsequence, we may assume that $s_k \to s_0$, $C_k \to C_0$ and $(\sigma_k(s_k), \sigma_k'(s_k)) \to (z, v) \in S(\overline{\Omega})$. But for all k, $d(\sigma_k(s_k), \gamma_{w_k}([0, a_{w_k}])) \geq \delta$, and hence $d(z, \gamma([0, a_0])) \geq \delta$.

Define $\sigma(s) := \exp_z((s - s_0)v)$, $0 \leq s \leq C_0$. Then $\sigma_k \to \sigma$ in the C^2 norm as $k \to \infty$, so that $\sigma(0) = p_0$ and $\sigma(C_0) = \gamma(a_0)$. By Lemma 4.1, σ meets $\partial\Omega$ transversally at $\sigma(0)$ and at $\sigma(C_0)$. We may compute that the lengths $l(\sigma) = \lim_k l(\sigma_k) \leq$

$\lim_k l(\gamma_k) = l(\gamma)$. That is, σ is also a chord. The uniqueness of γ now implies that $\gamma = \sigma$, which contradicts the fact that $d(\sigma(s_0), \gamma([0, a_0])) \geq \delta$. This contradiction shows that γ_w is a chord of $\overline{\Omega}$.

The continuity of a_w as a function of $(p, w) \in S(\overline{\Omega})$ follows from Lemma 4.1, as noted above.
$\hfill$ Q.E.D.

Remark 4 For another variation on the proof of Lemma 4.2, see Step 3 of Example 5.5 below. In that example, the original chord is degenerate, but nearby geodesics are free of conjugate points and are shown to be chords.

4.3 The Chord Map, and a Geometric Theorem

Assume that $\overline{\Omega}$ has unique chords. Given $p, q \in \partial\Omega$, $p \neq q$, and given $0 \leq s \leq d(p, q)$, we define the *chord map*

$$\Phi(p, q, s) := (\gamma(s), \gamma'(s)) \in S(\overline{\Omega})$$

where $\gamma : [0, d(p, q)] \to \overline{\Omega}$ is the unique unit-speed chord joining p to q. (Φ will also be defined at the diagonal $p = q$, below.)

The domain of definition of Φ, as given above, may be written as $U := \{(p, q, s) \in \partial\Omega \times \partial\Omega \times \mathbb{R} : p \neq q, 0 \leq s \leq d(p, q)\}$.

Lemma 4.3 *Suppose that any two points of $\partial\Omega$ are connected by a unique chord of $\overline{\Omega}$, and that $\partial\Omega$ has positive second fundamental form. Then $\Phi : U \to S(\overline{\Omega})$ is continuous on U.*

Proof. Consider sequences $p_k \to p_0$, $q_k \to q_0$ and $s_k \to s_0$, where $p_k, q_k \in \partial\Omega$, $q_k \neq p_k$, $0 \leq s_k \leq d(p_k, q_k)$. Write $\gamma_0 : [0, d(p_0, q_0)] \to \overline{\Omega}$ for the unique unit-speed chord joining p_0 to q_0, and write γ_k for the unique unit-speed chord joining p_k to q_k. Then, by definition, $\Phi(p_k, q_k, s_k) = (\gamma_k(s_k), \gamma'_k(s_k))$ and $\Phi(p_0, q_0, s_0) = (\gamma_0(s_0), \gamma'_0(s_0))$. After passing to a subsequence, we may assume that $(\gamma_k(s_k), \gamma'_k(s_k)) \to (x, v)$ in $S(\overline{\Omega})$. Let γ be the the geodesic in $\overline{\Omega}$ with initial conditions $\gamma(s_0) = x$, $\gamma'(s_0) = v$ and having the maximal domain of definition (the value s_0 is determined so that γ enters $\overline{\Omega}$ at $s = 0$.) By the uniqueness of solutions to the initial-value problem for the geodesic equation, we have $\gamma = \gamma_0$. Since the limit is the same for all subsequences, this shows that $\Phi(p_k, q_k, s_k) \to \Phi(p_0, q_0, s_0)$.
$\hfill$ Q.E.D.

The domain U is a $(2n-1)$-dimensional manifold-with-boundary, which is not compact because the diagonal $\{p = q, s = 0\}$ of $\partial\Omega \times \partial\Omega \times \mathbb{R}$ has been omitted. We shall define a compactification $\overline{U}$ of U, by adding to U, in place of each point

$(p, p, 0)$, $p \in \partial\Omega$, a copy of the $(n-2)$-sphere $S_p(M) \cap T_p(\partial\Omega)$ of unit tangent vectors to the boundary of Ω at p. We shall also write this $(n-2)$-dimensional sphere as $S_p(\partial\Omega)$. $\overline{U}$ will become a compact topological manifold-with-boundary, possibly with non-smooth boundary. By abuse of notation, we shall write $(p, v, 0)$ for the point of $\overline{U}\backslash U$ corresponding to $(p, v) \in S_p(\partial\Omega)$. A sequence (p_k, q_k, s_k) from U will converge to the point $(p, v, 0)$ in $\overline{U}\backslash U$ if and only if $q_k \to p$, $s_k \to 0$ and $(p_k, v_k) \to (p, v)$, where v_k is the initial tangent vector to the chord from p_k to q_k. Convergence within $\overline{U}\backslash U$ will be equivalent to convergence in $S(\partial\Omega)$. The *chord map* Φ may then be extended to all of $\overline{U}$ by defining $\Phi(p, v, 0) := (p, v) \in S_p(\partial\Omega)$ for each $(p, v, 0) \in \overline{U}\backslash U$.

Lemma 4.4 *Suppose that any two points of $\partial\Omega$ are connected by a unique chord of $\overline{\Omega}$, and that $\partial\Omega$ has positive second fundamental form. Then $\Phi : \overline{U} \to S(\overline{\Omega})$ is continuous on $\overline{U}$.*

Proof. If $(p_k, q_k, s_k) \to (p_0, q_0, s_0)$, $p_0 \neq q_0$, then the conclusion follows from Lemma 4.3. If $(p_k, v_k, 0) \in \overline{U}\backslash U$, and if $(p_k, v_k, 0) \to (p, v, 0) \in \overline{U}\backslash U$, then the conclusion is immediate from the definition $\Phi(p, v, 0) := (p, v)$.

Suppose $(p_k, q_k, s_k) \to (p, v, 0) \in \overline{U}\backslash U$. Then by definition, $\Phi(p_k, q_k, s_k) = (\gamma_k(s_k), \gamma'_k(s_k))$, where γ_k is the unit-speed chord from p_k to q_k, while $\Phi(p, v, 0) = (p, v) \in S_p(\partial\Omega)$. But $(p_k, \gamma'_k(0)) \to (p, v)$ in $S(\overline{\Omega})$, since $(p_k, q_k, s_k) \to (p, v, 0)$, and hence $(\gamma_k(s_k), \gamma'_k(s_k)) \to (p, v)$. This shows that Φ is continuous on the compact topological manifold-with-boundary $\overline{U}$. Q.E.D.

We are ready to prove the following geometric theorem.

Theorem 4.5 *Assume that the boundary $\partial\Omega$ of the compact Riemannian manifold-with-boundary $\overline{\Omega}$ has positive second fundamental form. Suppose that any two points of $\partial\Omega$ are connected by a unique chord, which is nondegenerate. Then any interior geodesic segment $\gamma : (b - \varepsilon, b + \varepsilon) \to \Omega$ may be extended to a geodesic $\gamma : [s_0, s_1] \to \overline{\Omega}$ which is a chord, that is, which realizes the minimum length between two distinct points $\gamma(s_0)$ and $\gamma(s_1)$ in $\partial\Omega$.*

Proof. Without loss of generality, we may assume that Ω is connected; otherwise, we may work in the connected component of Ω containing $\gamma(b)$. The conclusion is immediate if the dimension $n = 1$; we shall assume from now on $n \geq 2$.

We shall show that the image $\Phi(\overline{U})$ of the chord map is both open and closed in $S(\overline{\Omega})$. Since $S(\overline{\Omega})$ is a bundle with fiber S^{n-1}, $n \geq 2$, over the connected space $\overline{\Omega}$, $S(\overline{\Omega})$ is connected; it will then follow that Φ is surjective.

To show that $\Phi(\overline{U})$ is open, first consider any point $(p_0, q_0, s_0) \in U$ and denote $(x_0, v_0) := \Phi(p_0, q_0, s_0)$. Write γ_0 for the chord from $p_0 = \gamma_0(0)$ to $q_0 = \gamma_0(a_0)$, and let $w_0 = \gamma_0'(0)$. Then $(x_0, v_0) = (\gamma_0(s_0), \gamma_0'(s_0))$, by definition of Φ. By Lemma 4.1, $\gamma_0'(0)$ cannot be tangent to $\partial\Omega$, and in fact must point into Ω. Meanwhile, from Lemma 4.2 we know that there is a neighborhood $\mathcal{V}$ of (p_0, w_0) in $S(\overline{\Omega}) \cap \pi^{-1}(\partial\Omega)$ such that every $(p, w) \in \mathcal{V}$ is the initial datum $(\gamma_w(0), \gamma_w'(0))$ for a chord $\gamma_w : [0, a_w] \to \overline{\Omega}$. So in our case, for every $(p, w) \in \mathcal{V}$, $(p, w) = (\gamma_w(0), \gamma_w'(0))$ where γ_w is a chord. Now consider a unit vector $(x, v) \in S(\overline{\Omega})$ close to (x_0, v_0). Let $\gamma : [0, a] \to \overline{\Omega}$ be the unit-speed geodesic with initial conditions $\gamma(s_1) = x$, $\gamma'(s_1) = v$ and having maximal interval of definition $[0, a]$, choosing the value $s_1 > 0$ so that γ enters $\overline{\Omega}$ at $s = 0$. Note that since $\gamma_0'(0)$ is not tangent to $\partial\Omega$, if (x, v) is sufficiently close to (x_0, v_0), then γ must enter $\overline{\Omega}$ nontangentially. Further, $(p, w) := (\gamma(0), \gamma'(0))$ will be close to (p_0, w_0); in particular, we may achieve that $(p, w) \in \mathcal{V}$. But this implies that γ is a chord. Therefore $\Phi(p, \gamma(a), s_1) = (x, v)$. But (x, v) was an arbitrary point of $S(\overline{\Omega})$ near (x_0, v_0), so this shows that the restriction of Φ to U is an open mapping.

Next, consider a point $(p_0, w_0, 0) \in \overline{U}\backslash U$, and recall that $\Phi(p_0, w_0, 0) = (p_0, w_0) \in S_{p_0}(\partial\Omega)$. Consider any point $(x, v) \in S(\overline{\Omega})$ close to (p_0, w_0). If $x \in \partial\Omega$ and the vector v points into Ω, or if x is not a boundary point, we form the geodesic $\gamma : [0, a] \to \overline{\Omega}$ with $\gamma(s_1) = x$, $\gamma'(s_1) = v$, $\gamma(0) \in \partial\Omega$. Since (x, v) is close to (p_0, w_0), we find that also $(p, w) := (\gamma(0), \gamma'(0))$ is close to (p_0, w_0), and we conclude as before that (x, v) is in the image of Φ. If x is a boundary point and v points out of Ω, we choose $p = x$, $w = -v$ and the conclusion follows in the same way.

This shows that the image of $\Phi : \overline{U} \to S(\Omega)$ is an open subset of $S(\Omega)$.

In order to show that $\Phi(\overline{U})$ is closed, we may apply Lemma 4.4 and recall that $\overline{U}$ is compact. Q.E.D.

Remark 5 The reader might observe that under the hypotheses of Theorem 4.5, Φ in fact maps $\overline{U}$ homeomorphically onto $S(\overline{\Omega})$. This observation implies a necessary topological condition for any manifold $\overline{\Omega}$ satisfying the hypotheses of the theorem.

4.4 Criteria for Chord Uniqueness

We shall conclude this section with two criteria which imply certain of the hypotheses of Theorem 1.1. The first is especially appropriate in a situation where observations about $\overline{\Omega}$ can only be made from its boundary:

Proposition 4.6 *Assume that $\partial\Omega$ has positive second fundamental form. For fixed $q \in \partial\Omega$, let $\phi : \overline{\Omega} \to [0, \infty)$ be given by $\phi(x) := d_{\overline{\Omega}}(x, q)^2$. (1) If, for each $q \in \partial\Omega$, $\nabla\phi$ is continuous along $\partial\Omega$, then any two points of $\partial\Omega$ are connected by a unique* chord. (2) If, moreover, for each $q \in \partial\Omega$, $\nabla\phi$ is continuously differentiable along $\partial\Omega$, then the chords are nondegenerate.

Proof. Note that the distance function $\zeta := \sqrt{\phi}$ is a solution of the Riemannian Hamilton-Jacobi equation $<\nabla\zeta,\nabla\zeta> \equiv 1$, $\zeta(q) = 0$. The unit vector $\nabla\zeta(p)$ is the final tangent vector to a chord from q to p.

(1) Of course, in general the first derivatives of the Lipschitz function ζ need not exist everywhere on $\partial\Omega\backslash\{q\}$. Our assumption here implies that $\nabla\zeta(p)$ does exist at each $p \in \partial\Omega$, $p \neq q$, and is continuous there. Then the chord from q to p is unique. Namely, if there were two chords γ and σ from q to p, then they would have distinct terminal tangent vectors, by the uniqueness of solutions to the Cauchy problem for the Jacobi equation (4.1). But then the chord from q to a point $x \in \partial\Omega$ near p would jump from a chord near γ to a chord near σ, as x moves past p in the direction of the difference of final tangent vectors $\gamma'(\zeta(p)) - \sigma'(\zeta(p))$, contradicting continuity of $\nabla\zeta(p)$.

(2) Now suppose that for all $q \in \partial\Omega$, the gradient $\nabla\zeta$ is continuously differentiable along $\partial\Omega\backslash\{q\}$. In terms of the chord map Φ, this means that $(p, \nabla\zeta(p)) = \Phi(q, p, \zeta(p))$ is continuously differentiable as a function of $p \in \partial\Omega$, $p \neq q$, for any fixed $q \in \partial\Omega$. But the chord γ from q to p, since it is a solution of the system (2.2) of ODE's, depends smoothly on its terminal conditions $(\gamma(\zeta(p)), \gamma'(\zeta(p))) = \Phi(q, p, \zeta(p))$. In consequence, its initial values $(\gamma(0), \gamma'(0)) = \Phi(q, p, 0) \in S_q(\overline{\Omega})$ depend on p in a C^1 manner. For similar reasons, $\Phi(q, p, \zeta(p))$ depends smoothly on $\Phi(q, p, 0))$. Thus, the correspondence $p \mapsto \Phi(q, p, 0))$ is a C^1 local diffeomorphism from $\partial\Omega\backslash\{q\}$ to $S_q(\overline{\Omega})$.

Consider a chord γ_0 from q to $p_0 \in \partial\Omega$, and let $\{\gamma_\lambda : -\varepsilon < \lambda < \varepsilon\}$ be a smooth one-parameter family of geodesics starting from $q = \gamma_\lambda(0)$, with $\frac{\partial}{\partial\lambda}\gamma'_\lambda(0) \neq 0$ at $\lambda = 0$. Then for small λ, the geodesic γ_λ is also a chord, from q to a point $p_\lambda \in \partial\Omega$. Namely, the C^1 local diffeomorphism $p \mapsto \Phi(q, p, 0))$ maps a neighborhood of p_0 in $\partial\Omega$ onto a neighborhood of $\gamma'(0)$ in $S_q(\overline{\Omega})$, by the inverse function theorem. Recall that γ_0 meets $\partial\Omega$ transversely at $p_0 = \gamma_0(\zeta(p_0))$. Since $\Phi(q, \cdot, 0)$ is a C^1 local diffeomorphism, and since $\frac{\partial}{\partial\lambda}\Phi(q, p_\lambda, 0) = \frac{\partial}{\partial\lambda}\gamma'_\lambda(0) \neq 0$ at $\lambda = 0$, it follows that $\frac{\partial}{\partial\lambda}p_\lambda \neq 0$ at $\lambda = 0$. Now any normal Jacobi field J along γ_0, with $J(0) = 0$, arises from a one-parameter family $\{\gamma_\lambda : -\varepsilon < \lambda < \varepsilon\}$, as the variation vector field $J(s) = \frac{\partial}{\partial\lambda}\gamma_\lambda(s)$ at $\lambda = 0$. Since $J(0) = 0$, we may assume that for all λ, $\gamma_\lambda(0) = q$. But $J(\zeta(p_0))$ is the component orthogonal to $\gamma'_0(\zeta(p_0))$ of $\frac{\partial}{\partial\lambda}p_\lambda$ at $\lambda = 0$, which is nonzero, using what we have just shown and transversality. Therefore $J(s) \neq 0$ at $s = \zeta(p_0)$, and since γ_0 is a chord, at all $0 < s < \zeta(p_0)$ as well. This shows that γ_0 is nondegenerate. **Q.E.D.**

Remark 6 Somewhat surprisingly, it is not true that the continuity of $\nabla \phi$ implies disconjugacy of the chords of $\overline{\Omega}$. That is, the stronger hypothesis of part (2) of Proposition 4.6 is required to imply the stronger conclusion. See Example 5.5 below.

The second, rather different criterion concludes that chords are unique, assuming a condition which implies their nondegeneracy (compare Proposition 5.1 below):

Proposition 4.7 *Consider a Riemannian manifold-with-boundary $\overline{\Omega}$, whose boundary has positive second fundamental form. Suppose for all $q \in \partial\Omega$, each geodesic starting from q leaves $\overline{\Omega}$ strictly before any conjugate point along the geodesic. Then $\overline{\Omega}$ has unique chords (which are nondegenerate).*

Proof. Let $\overline{\Omega}$ be extended to become the closure of an open subset of an open Riemannian manifold M. Consider $p_0, q_0 \in \partial\Omega$, and write $a_0 = d(p_0, q_0)$. Let $\gamma_0 : [0, a_0] \to \overline{\Omega}$ be a chord from $q_0 = \gamma_0(0)$ to $p_0 = \gamma_0(a_0)$. Write $w_0 = \gamma'_0(0)$. It follows from Lemma 4.1 that w_0 points into Ω. Since γ_0 is nondegenerate, according to Lemma 4.2 there is a neighborhood $\mathcal{V}$ of (q_0, w_0) in $S(\overline{\Omega}) \cap \pi^{-1}(\partial\Omega)$ such that for all $(q,w) \in \mathcal{V}$, $\gamma_w : [0, a_w] \to \overline{\Omega}$, defined by $\gamma_w(s) = \exp_q(sw)$, is a chord. Moreover, using Lemma 4.1, $a_w \to a_0$ as $(q,w) \to (q_0, w_0)$. Another application of Lemma 4.1 shows that $a_w \in (0, \infty)$ is smooth as a function of (q,w). Meanwhile, since $\gamma_0 = \gamma_{w_0}$ is nondegenerate, $\exp_{q_0}$ maps a neighborhood $\mathcal{U}$ of w_0 in $T_{q_0}\overline{\Omega}$ diffeomorphically onto a neighborhood of p_0 in M. We may assume that $\mathcal{U}$ is a subset of $\mathcal{V}$. In particular, for p in a neighborhood of p_0 in $\partial\Omega$, there is a unique $w(p) \in \mathcal{U} \cap S_{q_0}M$ such that $\gamma_{w(p)}(a_{w(p)}) = p$, and $w(p)$ varies smoothly as a function of p. Also, $\gamma_{w(p)}$ is a chord, so that $\zeta(p) = a_{w(p)}$. Therefore $\nabla \zeta(p) = \gamma'_{w(p)}(a_{w(p)})$ is smooth near p_0. Chord uniqueness now follows from Proposition 4.6. **Q.E.D.**

5 Examples

5.1 Manifold with a Big Bulge

Let $\overline{\Omega}$ be a Riemannian manifold-with-boundary, diffeomorphic to the closed ball of $\mathbb{R}^2$, which contains in its interior a smooth subdomain $\overline{D}$ isometric to the closed hemisphere of the unit sphere $S^2 \subset \mathbb{R}^3$. We assume that $\partial\Omega$ has positive second fundamental form. Such a manifold may be constructed as a hypersurface of revolution in $\mathbb{R}^3$, with a generating curve which begins orthogonally to the axis of revolution with a quarter-circle of radius 1, and ends at a moment when it is travelling away from the axis of rotation. Alternatively, such an example may be constructed by "pasting in" an isometric copy of the closed hemisphere in the interior of a reference manifold whose boundary has positive second fundamental form.

We claim that $\overline{\Omega}$ will not have unique chords. Namely, any geodesic σ of $\overline{\Omega}$ which enters D, crossing ∂D transversally at x, will remain inside D for a distance exactly π, until it crosses ∂D transversally at y. The points x and y will be conjugate along σ. These properties follow from the behavior of geodesics (great circles) of S^2, or from the Jacobi equation (4.1) above. Also, if σ is tangent to ∂D at one point, then it remains forever inside ∂D. This implies that no chord of $\overline{\Omega}$ can meet $\overline{D}$.

Observe that the curves in the annulus $\overline{\Omega}\backslash\overline{D}$ joining two points $p, q \in \partial\Omega$ fall into an infinite sequence of homotopy classes. Since ∂D is a geodesic, there is a curve of minimum length in each of these homotopy classes. Now if q remains fixed while p moves along $\partial\Omega$ and a point $\hat{p}$ over p moves along the universal covering space of $\partial\Omega$, these minimum lengths will vary continuously; each will assume the value 0 once, at a moment when $\hat{p}$ is one of the sequence of points over q, while all the others will be positive. It follows that for a specific choice of $p \neq q$, two of these minimum lengths for homotopy classes will coincide and provide the minimum length among all homotopy classes. Meanwhile, we have seen that all chords must lie in $\overline{\Omega}\backslash\overline{D}$. Thus, there will be at least two chords joining q to p.

The nonuniqueness of chords also follows from Theorem 4.5 above, assuming the chords are nondegenerate. For if $\overline{\Omega}$ had unique and nondegenerate chords, then every interior geodesic would be part of a chord; but a geodesic segment inside D cannot be a segment of a chord, as we have seen above. Alternatively, one may form a simpler argument from the existence of closed geodesics in the boundary of D. These closed geodesics also show that the wave equation (1.1) on $\overline{\Omega}$ is not controllable from $\partial\Omega$ (cf. Remark 2.)

The manifold $\overline{\Omega}$ also does not allow any convex functions, as follows from Proposition 5.3 below using a domain slightly larger than D.

In the context of boundary control of hyperbolic equations, a similar example was introduced and discussed in detail by Yao ([17]).

5.2 Manifold with a Bulge of Moderate Size

Let $\overline{\Omega}$ be a spherical cap, of intrinsic radius $R < \frac{\pi}{2}$, in the unit n-sphere S^n. Then $\partial\Omega$ has positive second fundamental form. Also, $\overline{\Omega}$ has unique chords, which are nondegenerate and have lengths $\leq 2R$, so that Theorem 1.1 may be applied to show that the spherical wave equation (1.1) may be controlled from the boundary in any time $T > T_0 = 2R$. Note that the requirement $R < \frac{\pi}{2}$ is sharp, since the normal curvatures of $\partial\Omega$ equal $\cot R$, which becomes negative for $R > \frac{\pi}{2}$.

A second method for proving controllability in the recent literature has been to show that certain Carleman estimates hold, relying on the quantitative properties of a strictly convex function on $\overline{\Omega}$. A real-valued function v on $\overline{\Omega}$ is said to be *convex* if

$\nabla^2 v(X, X) \geq 0$ for all tangent vectors X. It is *strictly convex* if $\nabla^2 v(X, X) > 0$ for all $X \neq 0$. For example, Lasiecka-Triggiani-Yao [8] show that boundary control in any time $> T_1$ is possible, provided there is a uniformly convex function $v : \overline{\Omega} \to \mathbb{R}$, with $T_1 := 2\frac{\max |\nabla v|}{c_0}$, where the uniform convexity of v is quantified by the minimum value $c_0 > 0$ of the Hessian $\nabla^2 v(X, X)$ over unit tangent vectors X to $\overline{\Omega}$. Note that, with certain other hypotheses strengthened, Yao ([17]) requires only a coercive vector field, which need not be a gradient, rather than a convex function. The paper [17] introduced the use of geometric methods of Bochner type to the control of hyperbolic P.D.E. See also Tataru's paper [14]. In order to apply the method of [8], an apparently optimal choice of convex function for the spherical-cap example of this subsection would be $v(x) := -\cos r(x)$, where $r(x)$ is the distance from x to the center x_0. Since $\nabla^2 v(X, X) \equiv -v <X, X>$ (see [3], p. 173), and since $|\nabla r| \equiv 1$, we compute $c_0 = \cos R$ and $T_1 = 2 \tan R$. But this estimate blows up as $R \to \frac{\pi}{2}$, so that the requirement $R < \frac{\pi}{2}$ is again seen to be sharp. We have not found a way to improve on the control time T_1 using Yao's result with an appropriate coercive vector field. By comparison, the time of control which follows from Theorem 1.1 in this case is $T_0 = 2R < 2 \tan R = T_1$.

A far more general class of examples may be given, with analogous properties:

Proposition 5.1 *Let $\overline{\Omega}$ be a smooth, compact subdomain of a Riemannian manifold M, whose sectional curvatures at $x \in M$ are bounded above by $f(r(x))$, where $r(x) = d(x, x_0)$, for some $x_0 \in \Omega$. Assume that $\exp_{x_0}$, the exponential map of M at x_0, is defined and injective on the closed ball $\overline{B}_R(0) \subset T_{x_0} M$, and that $r(x) \leq R$ on $\overline{\Omega}$. We assume that $f : [0, R] \to \mathbb{R}$ satisfies: (1) f is monotone decreasing; (2) the solution u_1 of the O.D.E. $u'' + fu = 0$ with initial conditions $u_1(0) = 1, u_1'(0) = 0$ remains positive on $[0, R]$; and (3) the solution u_2 of the same O.D.E. with initial conditions $u_2(0) = 0, u_2'(0) = 1$ has positive first derivative on $[0, R]$. Suppose also that $\partial \Omega$ has positive second fundamental form. Then $\overline{\Omega}$ has unique chords, which are nondegenerate.*

Proof. According to Theorem 2 of [6], any geodesic of $\overline{B}_R(x_0) = \exp_{x_0}(\overline{B}_R(0))$ has length at most $2R$ and is free of conjugate points, provided that all hypotheses of this proposition are satisfied, except perhaps the hypothesis of positive second fundamental form. Thus, a unit-speed geodesic γ entering $\overline{B}_R(x_0)$ at $\gamma(0)$ will leave $\overline{B}_R(x_0)$ at $\gamma(s_1)$, where $s_1 \leq 2R$, and where $\gamma([0, s_1])$ has no conjugate points. That is, with the Riemannian metric of M, $\overline{B}_R(x_0)$ fulfills the hypotheses of Proposition 4.7, and so must have unique chords. We may now apply Theorem 4.5 to $\overline{B}_R(x_0)$, and conclude that any geodesic arc in $\overline{B}_R(x_0)$ is the restriction of a chord of $\overline{B}_R(x_0)$, and therefore is free of conjugate points.

In particular, $\overline{\Omega}$ satisfies the hypotheses of Proposition 4.7, and so has unique chords. Since each chord of $\overline{\Omega}$ is the restriction of a geodesic $\gamma([0, s_1])$ of M without conjugate points, it must be nondegenerate. **Q.E.D.**

Remark 7 Consider any compact subdomain $\overline{\Omega}_1 \subset \Omega$, and any solution u of (1.1) having finite energy. It follows from Remark 3 above and from Tataru's trace theorem [15] that the trace of the conormal derivative of u will be in $L^2(\partial\Omega_1 \times (0,T))$. This gives us boundary control for either Neumann or Robin controls in optimal Sobolev spaces.

For example, under the hypotheses of Proposition 5.1, we have boundary control for either Neumann or Robin controls in optimal Sobolev spaces for $\Omega \times (0,T)$.

Corollary 5.2 *Suppose that for some radius R, the Riemannian manifold-with-boundary $\overline{\Omega}$ satisfies the hypotheses of Proposition 5.1. Then the wave equation (1.1) on $\overline{\Omega}$ is controllable from $\partial\Omega$ in any time $T > T_0 = 2R$.*

Proof. In order to compute the optimal time T_0 for boundary control, according to Theorem 1.1 above, we only need to know an upper bound for the diameter of $\partial\Omega$ in the Riemannian distance of $\overline{\Omega}$. But each chord of $\overline{\Omega}$ is a segment of a geodesic of $\overline{B}_R(x_0)$ which has length $\leq 2R$, as shown in the proof of Proposition 5.1. Hence $T_0 = \mathrm{diam}_{\overline{\Omega}}(\partial\Omega) \leq 2R$. Optimality of T_0 follows from Proposition 3.2. **Q.E.D.**

The reader will note that Corollary 5.2 may be proved more directly, without referring to Proposition 4.7, since Theorem 2 of [6] implies that any geodesic has length at most $2R$, and the conclusion follows from Proposition 3.2.

The spherical-cap example just considered is the special case of Proposition 5.1 with sectional curvatures $\equiv f(r) \equiv 1$. In this special case, the precise interval allowed for either of the conditions (2) or (3) of Proposition 5.1 is $R < \frac{\pi}{2}$. When $f(r)$ is not constant, however, these maximum intervals may differ: compare Example 5.4 below.

The reader will notice that if $\overline{\Omega}$ has curvature satisfying the hypotheses of Proposition 5.1, then the function $r(x)$ has convex level sets. This follows from condition (3), along with the Rauch Comparison Theorem in its differentiated form (see [4], pp. 215–217.) In fact, for a tangent vector V to the distance sphere $\partial B_{r_1}(x_0)$, the Rauch Comparison Theorem implies that the second fundamental form of $\partial B_{r_1}(x_0)$ satisfies the inequality

$$B(V, V) \geq \frac{u_2'(r_1)}{u_2(r_1)} |V|^2.$$

The presumably optimal choice of convex function is then $v(x) := \psi(r(x))$, where $\psi'(r) \equiv u_2(r)$. Thus, by the methods of [8], one may prove boundary control in any

time greater than
$$T_1 = \frac{2u_2(R)}{\min\{u_2'(r) : 0 \leq r \leq R\}}.$$

Again in this general case, it may be seen that $T_1 > T_0 = 2R$, unless $f(r) \equiv 0$.

5.3 Disconjugate Cross Sections, but No Convex Function: the Frisbee

For a rotationally symmetric Riemannian manifold, we may use the term *cross section* for a geodesic through the center of symmetry of the manifold. The language is suggested by the special case $n = 2$, where this geodesic cuts the manifold into two congruent pieces. This curve would be called a "diameter" in elementary geometry, a term we shall avoid, since $\text{diam}_{\overline{\Omega}}(\partial\Omega)$ may be substantially less than the length of a cross section in a general rotationally symmetric manifold, and in the example we are about to present in particular.

Example 5.1, and the first part of Example 5.2, above, deal with rotationally symmetric manifolds of constant sectional curvature one. We saw in that situation that three properties are lost simultaneously as the radius R increases beyond the critical radius $\frac{\pi}{2}$!: chord uniqueness, the existence of a convex function and disconjugacy (the absence of any pair of conjugate points) of the cross sections. The first two properties are major hypotheses in theorems about boundary controllability of (1.1) (see Theorem 1.1 and [8].) The third property is intuitively related to disconjugacy of all geodesics which start at the boundary, which implies the first property, assuming that the boundary has positive second fundamental form (Proposition 4.7 above.)

In order to compare the various hypotheses of Theorem 1.1 above with the hypothesis of the existence of a convex function whose gradient is outward along $\partial\Omega$, as required by [8], for example, we shall consider an example with nonconstant sectional curvatures. We shall construct a rotationally symmetric manifold whose cross sections are locally length-minimizing, but which does not support a convex function. Chords will not be unique. This example is a large flat n-disk surrounded by a moderate-sized region of positive curvatures and a thin region of negative curvature, resembling the inside surface of the flying toy known as the Frisbee (apologies to the Wham-O Corporation.)

Let $\overline{\Omega}$ be diffeomorphic to the ball B^n and radially symmetric. Then the Riemannian metric may be given in spherical coordinates by $ds^2 = dr^2 + u(r)^2 d\theta^2$, where $d\theta^2$ denotes the Riemannian metric of the unit $(n-1)$–sphere. The parameter $r = r(x)$ is then the distance from $x \in \overline{\Omega}$ to the center x_0. The sectional curvatures of $\overline{\Omega}$ at x will lie between $K_1(r)$ and $K_2(r)$, where u satisfies the Jacobi equation $\frac{d^2u}{dr^2} + K_1(r)u = 0$; and where $K_2(r) = u^{-2}\left(1 - \left(\frac{du}{dr}\right)^2\right)$, as may be verified using the Gauss equations for the hypersurfaces $r = $ const. In the special case $n = 2$, of course, there is only one sectional curvature at each point, the Gauss curvature, which equals $K_1(r)$.

For our example, we choose $\overline{\Omega}$ to be the closed ball $r \leq R_2$ in $\mathbb{R}^n$. Let $u(r) = r, 0 \leq r \leq R_0$; $u(r) = R_0 \frac{\cos(r-r_0)}{\cos(R_0-r_0)}, R_0 < r \leq R_1$; and $u(r) = u(R_1) \frac{\cosh\beta(r-r_1)}{\cosh\beta(R_1-r_1)}$, $R_1 > r \leq R_2$. Here r_0, r_1, $R_0 < R_1 < R_2$ and β are positive constants to be determined. Then u will have a Lipschitz-continuous first derivative provided that r_0 and r_1 are chosen so that $-\tan(R_0 - r_0) = \frac{1}{R_0}$, $\beta\tanh\beta(R_1 - r_1) = -\tan(R_1 - r_0)$, $-\pi/2 < R_0 - r_0 < 0$ and $0 < R_1 - r_0 < \arctan\beta$. The radial-normal sectional curvatures of Ω, in a distributional sense, are then $K_1(r) = 0$, $0 \leq r \leq R_0$; $K_1(r) = 1$, $R_0 < r \leq R_1$; and $K_1(r) = -\beta^2$, $R_1 < r \leq R_2$. The boundary of Ω will have positive second fundamental form provided that $\frac{du}{dr}(R_2) > 0$, that is, $R_2 > r_1$. We require also that the boundary of the ball $D \subset \Omega$ described by the inequality $r < R_1$ have negative second fundamental form, that is, $0 < R_1 - r_0 < \pi/2$.

To be specific, based on casual examination of a Frisbee, we choose $\beta = 6$. For mathematical convenience, we choose $R_0 \gg 1$, which entails that r_0 be slightly greater than R_0. We shall further choose R_1 so that ∂D has small negative normal curvatures, thus $0 < R_1 - r_0 \ll 1$, and choose R_2 to give $\partial\Omega$ small positive normal curvatures: $0 < R_2 - r_1 \ll 1$.

Then any cross section σ of Ω, given by $0 \leq r \leq R_2$ with two antipodal points of S^{n-1} as spherical coordinate, is a geodesic of $\overline{\Omega}$ without conjugate points. Namely, a Jacobi field J which starts out at the center $x_0 = \sigma(0)$ with $|J(0)| = 1$ and $J'(0) = 0$ will have length $|J(r)| = 1$ for all $0 \leq r \leq R_0$, and $|J(r)| = \cos(r - R_0)$ for all $R_0 \leq r \leq R_1$. But $0 < R_1 - R_0 \ll 1$, which implies that $|J(r)|$ remains close to 1 on the interval $[0, R_1]$. We assume that $\tan(R_1 - R_0) < \beta$; then $|J(r)| > 0$ on the final interval $[R_1, R_2]$ as well. Since, for all r, $J'(r)$ is a scalar multiple of $J(r)$, the length of any normal Jacobi field $J(r)$ satisfies the scalar Jacobi equation $\frac{d^2|J|}{dr^2} + K_1(r)|J| = 0$, and a Sturm theorem shows that J cannot have two zeroes in the interval $[-R_2, R_2]$. This implies that a cross section cannot have conjugate points. This also implies that the cross sections of $\overline{\Omega}$ have shortest length in a C^0 neighborhood.

We claim that there can be no strictly convex function $v : \overline{\Omega} \to \mathbb{R}$ which has positive normal derivative on $\partial\Omega$. In support of our claim, we shall first show

Proposition 5.3 *Suppose a Riemannian manifold Ω contains a subdomain $\overline{D}$ whose boundary has negative first fundamental form. If $w : \overline{\Omega} \to \mathbb{R}$ is subharmonic on D, the Hessian of w is nonnegative at points of ∂D, and w is constant on ∂D, then w is constant on $\overline{D}$.*

Proof. We integrate Δw over D, and find that

$$\int_{\partial D} \frac{\partial w}{\partial \nu} = \int_D \Delta w \geq 0,$$

where ν is the outward unit normal vector to ∂D. In particular, either there is a point p_0 of ∂D at which $\frac{\partial w}{\partial \nu}$ is positive; or else w is harmonic everywhere in D, and hence constant on $\overline{D}$ (since it has constant boundary values.)

Let X be any nonzero tangent vector to ∂D at p_0. Since w is constant on ∂D, we compute that the Hessian $\nabla^2 w(X, X) = -\frac{\partial w}{\partial \nu} B(X, X) < 0$, where B is the second fundamental form of ∂D. This contradicts local convexity of w at p_0. **Q.E.D.**

Since a convex function is *a fortiori* subharmonic, our claim follows from Proposition 5.3 by symmetrizing a given strictly convex function v in the rotationally-symmetric manifold $\overline{\Omega}$ constructed above, to form a rotationally invariant, strictly convex function $w : \overline{\Omega} \to \mathbb{R}$. Recall that Ω contains the subdomain D described by the inequality $r < R_1$, whose boundary has negative second fundamental form. Rotational symmetry implies that w is constant on ∂D. Proposition 5.3 now implies that w is constant on D, which contradicts the strict convexity of w.

In order to investigate chord uniqueness for this Frisbee example, it will be useful to have the following proposition.

Proposition 5.4 *Assume that $\partial \Omega$ has positive second fundamental form. Suppose there is a subdomain $\overline{D} \subset \Omega$ such that ∂D has* negative *second fundamental form. If the dimension $n > 2$, assume further that $\overline{\Omega}$ and $\overline{D}$ are rotationally symmetric. Then $\overline{\Omega}$ does not have unique chords.*

Proof. We first consider the case $n = 2$. Choose a point $p \in \partial \Omega$. Suppose on the contrary that any two points of $\partial \Omega$ are connected by a unique chord. Then, according to Lemma 4.4, the chord map $\Phi : \overline{U} \to S(\overline{\Omega})$ is continuous. This means in particular that as a second point $q \in \partial \Omega$ moves around the connected component of p in $\partial \Omega$, the chord joining p to q varies continuously, and the initial tangent vector to the chord sweeps out the half-circle of $S_p(\overline{\Omega})$ from one tangent vector v to $\partial \Omega$ at p to the other, $-v$. The chords themselves sweep out all of $\overline{\Omega}$ continuously. It follows that there is a first point $q_1 \neq p$ so that the chord from p to q_1 meets $\overline{D}$, at a point $x \in \partial D$, where it is tangent to ∂D. But ∂D has negative second fundamental form at x, implying that this chord would lie inside $D \cup \{x\}$ near x. However, q_1 was the *first* point such that the chord from p to q_1 meets $\overline{D}$, which implies that the chord from p to q_1 lies in $\overline{\Omega} \backslash D$, a contradiction.

For the case $n > 2$, we have assumed that $\overline{\Omega}$ and $\overline{D}$ are rotationally symmetric. Suppose, contrary to what we want to show, that $\overline{\Omega}$ has unique chords. Choose

$p \in \partial\Omega$, and choose a totally geodesic two-dimensional submanifold $\overline{N}$ of $\overline{\Omega}$ which contains p ($\overline{N}$ is the image of an appropriate two-dimensional subspace ot $T_{x_0}(\Omega)$ under the exponential mapping at the center x_0 of symmetry.) Then $\overline{N}$ is the fixed-point set of an isometry f of $\overline{\Omega}$ with itself, and is totally geodesic. Since $\overline{D}$ is rotationally symmetric, N meets ∂D orthogonally, implying that $N \cap \overline{D}$ is a subdomain of N whose boundary has negative second fundamental form. For the same reasons, $\partial N = \partial\Omega \cap \overline{N}$ has positive second fundamental form. By the $n = 2$ case of this Lemma, which we have already proved, $\overline{N}$, considered as a Riemannian 2-manifold with the metric induced from $\overline{\Omega}$, does not have unique chords.

Now any curve in $\overline{N}$ from p to another point $q \in \partial N$ is also a curve of $\overline{\Omega}$, and thus $d_{\overline{N}}(p, q) \geq d_{\overline{\Omega}}(p, q)$. Meanwhile, any chord of $\overline{\Omega}$ from p to any point $q \in \partial N$ must lie in $\overline{N}$, since otherwise its image under f would be another chord, violating uniqueness. Thus the chord of $\overline{\Omega}$ joining p to q is a chord of $\overline{N}$, and is therefore the unique chord in $\overline{N}$ from p to q. But this is a contradiction of the $n = 2$ case. Q.E.D.

We may now conclude from Proposition 5.4 that the Frisbee example constructed above does not have unique chords. For D in the hypothesis of Proposition 5.4, we may choose the subdomain described by $r < R_1$.

Finally, observe that, for the Frisbee example, the wave equation (1.1) is not controllable from the boundary. In fact, the sphere $\{r = r_0\}$ is totally geodesic as a submanifold of Ω, and therefore any of the sphere's own great circles will be closed geodesics of Ω. Any one of these closed geodesics suffices to make boundary controllability impossible (see Remark 2.)

Remark 8 Although the example constructed in this section is only $C^{1,1}$, *i.e.* the g_{ij} are only Lipschitz continuous, we may smooth the function $u(r)$ in small neighborhoods of R_0 and of R_1, so that $u''(r)$ remains monotone in each of the small neighborhoods. This C^∞ example will enjoy the same properties we have demonstrated for the original $C^{1,1}$ example. The metric of the C^∞ example will be arbitrarily close in the C^1 norm or in the $C^{1,\alpha}$ norm ($0 < \alpha < 1$) to the metric of the original example.

5.4 Convex Function but Nonunique Chords: the Salt Shaker

In this rotationally symmetric example, positive sectional curvature $\equiv 1$ is concentrated in a ball $B_{R_0}(x_0)$ near the center of $\overline{\Omega}$, while the sectional curvature $K_1(r)$ which affects Jacobi fields along a cross section becomes identically zero outside that ball. This allocation of curvatures is opposite to Example 5.3. Moreover, we shall construct the metric so that there is a convex function $v: \overline{\Omega} \to \mathbb{R}$, although there will be conjugate points along the cross sections, and chords will *not* be unique. Control is possible in a finite time, but the optimal time of control may be much less than $\text{diam}_{\overline{\Omega}}(\partial\Omega)$. The example is a truncated cone, topped off at the smaller end with a spherical cap. This resembles a design for salt shakers which are commonly found in American roadside diners, for example.

Let Ω be diffeomorphic to the ball $B_{R_1}(x_0)$, with the Riemannian metric $ds^2 = dr^2 + u(r)^2 d\theta^2$. We choose specifically $u(r) = \sin r$, $0 \leq r \leq R_0$; and $u(r) = \cos R_0(r - R_0) + \sin R_0$, $R_0 < r \leq R_1$. Any choice of $0 < R_0 < \frac{\pi}{2}$ will suffice. We require $R_1 > R_0 + \cot R_0$: we claim that this will imply that there are conjugate points along any cross section. The extreme sectional curvatures at $x \in \overline{\Omega}$ are $K_1(r) = K_2(r) = 1$, $0 \leq r \leq R_0$; and $K_1(r) = 0$, $K_2(r) = \tan^2 R_0 (r - R_0 + \tan R_0)^{-2}$, $R_0 < r \leq R_1$. Here $r = r(x)$. There is a Jacobi field $J(s)$ along any cross section $\sigma : [-R_1, R_1] \to \overline{\Omega}$, with length $|J(s)| = u_1(|s|)$, where u_1 is the solution of the scalar Jacobi equation $\frac{d^2 u_1}{dr^2} + K_1(r)u_1 = 0$ with initial conditions $u_1(0) = 1$, $u_1'(0) = 0$. We may compute that $u_1(r_1) = 0$, where $r_1 = R_0 + \cot R_0 < R_1$, so that the points $\sigma(-r_1)$ and $\sigma(r_1)$ are conjugate, as claimed.

It follows that chords of $\overline{\Omega}$ are not unique. Namely, let p and q be two opposite points of $\partial\Omega$, that is, $(R_1, \pm\theta_0) \in [0, R_1] \times S^{n-1}$ in spherical coordinates. If the chord joining p and q were unique, then it would necessarily be the cross section $\sigma(s) = (|s|, \frac{s}{|s|}\theta_0)$ which joins them, since this is the only curve which is symmetric under all reflections of $\overline{\Omega}$ that fix q and p. But σ has a pair of conjugate points $\sigma(\pm r_1)$ in the open interval $(-R_1, R_1)$, and therefore cannot have shortest length.

In particular, Theorem 1.1 above does not apply to this salt-shaker example.

On the other hand, $\overline{\Omega}$ *does* support a convex function v with $\frac{\partial v}{\partial \nu} > 0$ on $\partial\Omega$. For example, the function v may be constructed in the rotationally-symmetric form $v = \psi(r)$, with $\psi(0) = 0$ and $\frac{d\psi}{dr} = u(r)$. We may apply the results of Lasiecka-Triggiani-Yao [8] to obtain boundary control in any time greater than $T_1 := 2\frac{\max |\nabla v|}{c_0}$, where c_0 is a positive lower bound on convexity of v, as in the second paragraph of subsection 5.2 above. With the convex function v just constructed, we find $c_0 = \cos R_0$ and $\max |\nabla v| = u(R_1)$, so that $T_1 = 2(L + \tan R_0)$. Here we have written $L := R_1 - R_0$ for convenience.

In order to find an estimate for the *optimal* time of control T_0 for this example, we may apply Proposition 3.2: T_0 is the maximum length of any geodesic γ in $\overline{\Omega}$. This requires also some explicit geometrical computations. Note that $\overline{\Omega}$ has no closed geodesics, since $v(\gamma(s))$ is a convex function of s. Any unit-speed geodesic γ has a unique point where $r(\gamma(s))$ assumes a minimum value, which we write as $\lambda := r(\gamma(0))$. Since $\overline{\Omega}$ is rotationally symmetric, γ lies in a totally geodesic two-dimensional submanifold. Therefore, to compute the length of γ, we may assume with no loss of generality that $\overline{\Omega}$ has dimension $n = 2$. Write $\gamma(\pm s_1)$ for the points where γ meets $\partial \Omega$. Then the length of γ is $2s_1 = 2s_1(\lambda)$.

If $\lambda \geq R_0$, then γ lies entirely in the flat subset $A := \{x : R_0 \leq r(x) \leq R_1\}$ of $\overline{\Omega}$, and we may compute its length by working in the euclidean annulus $\widetilde{A} := \{y \in \mathbb{R}^2 : \tan R_0 \leq |y| \leq L + \tan R_0\}$, since $\widetilde{A}$ has the same Riemannian universal cover as A. We find that the length of γ is maximized if $\lambda = R_0$. Therefore in this case, γ has length at most $2\sqrt{L(L + 2 \tan R_0)}$, by the Pythagorean theorem. Similarly, the length of any geodesic segment crossing A from one boundary component to the other is at most $\sqrt{L(L + 2 \tan R_0)}$.

If $\lambda \leq R_0$, then a central segment $\gamma([-s_0, s_0])$ of the geodesic is a small circle in the spherical cap $\{x : 0 \leq r(x) \leq R_0\}$. Its length $2s_0$ is less than or equal to $2R_0$. This leads to the rough estimate $T_0 \leq 2R_0 + 2\sqrt{L(L + 2 \tan R_0)}$. This estimate has the same asymptotic behavior as T_1 in the limit as $L \to \infty$, for fixed $R_0 < \frac{\pi}{2}$. However, in the approach $R_0 \to \frac{\pi}{2}$ to the non-controllable geometry, this estimate is substantially better than T_1.

Note that for fixed L, if R_0 is close to $\frac{\pi}{2}$, then the conclusion $T_0 = \mathrm{diam}_{\overline{\Omega}}(\partial \Omega)$ of Theorem 1.1 may easily fail. Namely, if the inner radius $\tan R_0$ of the Euclidean annulus $\widetilde{A}$ is large enough, then $\mathrm{diam}_{\overline{\Omega}}(\partial \Omega)$ will be the length of a chord of the outer circle, which has radius $L + \tan R_0$, subtending an arc of length $\pi[\sin R_0 + L \cos R_0]$. As $R_0 \to \frac{\pi}{2}$, this chord length approaches π, which is much less than the length $2\sqrt{L(L + 2 \tan R_0)}$ of the geodesic tangent to the circle $\{r = R_0\}$, which is a lower bound for the optimal time of control T_0.

This example also serves as an illustration for Proposition 5.1 above, with $f(r) \equiv 1$, $0 \leq r \leq R_0$; and $f(r) \equiv 0$, $R_0 < r \leq R_1$. Condition (3) of that proposition holds on any interval $[0, R]$, $R < \infty$, and in fact $u_2 = u$ has positive derivative on $[0, \infty)$. However, condition (2) of Proposition 5.1 fails when $R \geq r_1$ (recall that we required $R_1 > r_1 := R_0 + \cot R_0$.)

Remark 8 above applies to this example, as well as to Example 5.3, showing that there are C^∞ manifolds, in every C^1 neighborhood of the salt-shaker example, enjoying the same properties.

5.5 Nonsmooth $\nabla \zeta$: the Last Bite of the Bagel

In this last example, we shall contrast the hypotheses of the two parts of Proposition 4.6, by constructing an example where the gradient $\nabla \zeta$ of the distance function ζ from a boundary point q is continuous but not C^1, and where one of the chords is degenerate. In this two-dimensional example, chords are not unique (since the manifold is not simply connected, there are minimizing curves in each homotopy class of curves from q to p, whose lengths become equal when p is moved around $\partial\Omega$); and, indeed, the wave equation (1.1) is not controllable from the boundary (since there are closed geodesics, cf. Remark 2.) The reader may find it interesting to make the further effort to find a similar example, such as a subdomain of $\overline{\Omega}$, with these additional properties. Most of our effort in this example is devoted to identifying the distance function ζ from a point $q \in \partial\Omega$.

Consider a torus of revolution M in (x, y, z)-space, obtained by rotating a circle of radius 1 in the open half-space $\{y > 0\}$ of the (y, z)-plane about the z-axis. At the maximum distance $R > 2$ from the z-axis, M contains a plane circle γ of radius R, which is a geodesic in M. The closed subdomain $\overline{\Omega} \subset M$ will be the "last bite of the bagel" bounded by two circles (of radius 1) in planes containing the z-axis, chosen so that the distance $\ell = L(0)$ between them, measured along γ, equals $\pi\sqrt{R}$. Then, in light of the Jacobi equation (4.1), and since the Gauss curvature equals $\frac{1}{R}$ at points of γ, there is a Jacobi field along γ which has zeroes $\{q, p_0\}$ at the point q where it enters $\overline{\Omega}$ and at the point p_0 where it leaves $\overline{\Omega}$. That is, p_0 and q are conjugate points along γ. Write σ_0 for the segment of γ from q to p_0.

Our next aim is to prove, in three steps, that σ_0 is one of a foliating family of chords σ_μ of $\overline{\Omega}$ starting at q. Let $\overline{\Omega}$ be parameterized by

$$X(\phi, \theta) = \begin{pmatrix} w(\phi)\cos\theta \\ w(\phi)\sin\theta \\ \sin\phi \end{pmatrix},$$

where $w(\phi) = R - 1 + \cos\phi$, $-\pi \leq \phi \leq \pi$, $0 \leq \theta \leq \theta_0 := \frac{\pi}{\sqrt{R}}$. Then the unit-speed geodesic arc $\sigma_0(s) = X(0, \frac{s}{R})$. Recall Clairaut's relation for a geodesic σ on a surface of revolution:

$$w(\phi(s))\sin\beta(s) \equiv \text{const.},$$

where $w(\phi(s))$ is, as above, the distance in $\mathbb{R}^3$ from $\sigma(s)$ to the z-axis, and where $\beta(s)$ is the signed angle from the tangent vector $\sigma'(s)$ to the generating curve $\{\theta = \text{const.}\}$ through $\sigma(s)$ (see [4].) In particular, since $w(\phi) > 0$, if $\beta(0) \in (0, \pi)$ then for all s, we have $\beta(s) \in (0, \pi)$.

Step 1: *We shall show that σ_0 is one leaf of a foliation $\{\gamma_\lambda\}$ of $\overline{\Omega}\setminus\{q, p_0\}$.*

For each $\lambda \in [-\pi, \pi]$, let $\gamma_\lambda(s) = X(\phi(s), \theta(s))$ be the unit-speed geodesic of M with the initial conditions $\theta(0) = \frac{\pi}{2\sqrt{R}}$, $\theta'(0) = 1$, $\phi(0) = \lambda$, $\phi'(0) = 0$. Then σ_0 is the restriction of γ_0 to the interval $-\frac{\pi}{2\sqrt{R}} \leq s \leq \frac{\pi}{2\sqrt{R}}$, up to reparameterization. Note that each of the circles $\{\phi = \text{const.} \in (0, \pi)\}$ has constant positive geodesic curvature in the direction of increasing ϕ. It follows that for $0 < \lambda < \pi$, along γ_λ, $\phi(s)$ reaches a maximum at $s = 0$, with maximum value $\phi(0) = \lambda$. Clairaut's relation implies that $0 < \frac{w(\lambda)}{R} \le \sin\beta(s)$ everywhere along γ_λ. In particular, $\frac{d\theta}{ds} = \frac{\sin\beta}{w(\phi)}$ is uniformly positive, implying that γ_λ will reach $\partial\Omega$ in either direction from $\gamma_\lambda(0)$ in a bounded distance.

Let ν be the unit normal vector to γ_λ, pointing in the direction of increasing ϕ. Then ν is parallel along γ_λ. Since the dimension $n = 2$, the Jacobi field $J(s) := \frac{\partial\gamma_\lambda}{\partial\lambda}(s)$ along γ_λ may be written as $J(s) = \mathcal{J}(s)\nu(s)$, where $\mathcal{J}(s)$ is a real-valued function. The system of O.D.E.s which form the Jacobi equation (4.1) becomes a single O.D.E.:

(5.1) $$\mathcal{J}''(s) + K(\gamma_\lambda(s))\mathcal{J}(s) = 0.$$

We claim that the family $\{\gamma_\lambda : 0 \le \lambda \le \pi\}$ foliates the upper half $\{0 \le \phi \le \pi\}$ of $\overline{\Omega}$, except at q and at p_0. Our claim is easily verified if $\ell = \pi\sqrt{R}$ is replaced by a small positive value ℓ, so that $\overline{\Omega}$ is replaced by its subdomain $\overline{\Omega}^{(\ell)} = \left\{X(\phi,\theta) : -\pi \le \phi \le \pi, \left|\theta - \frac{\pi}{2\sqrt{R}}\right| \le \frac{\ell}{2R}\right\}$. As ℓ increases, the first moment when our claim fails would imply that one of the geodesics γ_λ, $0 \le \lambda \le \pi$ has a pair of conjugate points, one on each component of $\partial\Omega^{(\ell)} = \{\theta = \frac{\pi}{2\sqrt{R}} \pm \frac{\ell}{2R}\}$. However, since the family is a foliation for all smaller values of ℓ, each of the curves γ_λ minimizes the length between its endpoints, by a well-known argument about fields of extremals. But the curve $\{\phi = \text{const.}\}$ between the endpoints of γ_λ has length less than the length ℓ of γ_0, so that γ_λ must have length $< \ell \le \pi\sqrt{R}$. At the same time, the Gauss curvature $K = \frac{\cos\phi}{w(\phi)}$ along γ_λ is at most $\frac{1}{R}$, with equality only along γ_0. It follows from the Jacobi equation (5.1) and a Sturm comparison theorem that no conjugate points are possible unless $\lambda = 0$ and $\ell = \pi\sqrt{R}$. This proves our claim.

By symmetry, the family $\{\gamma_\lambda : -\pi \le \lambda \le 0\}$ foliates the lower half $\{-\pi \le \phi \le 0\}$ of $\overline{\Omega}\setminus\{q,p_0\}$. As a result, the entire family $\{\gamma_\lambda : \lambda \in (-\pi,\pi]\}$ foliates all of $\overline{\Omega}\setminus\{q,p_0\}$. This smooth foliation of $\overline{\Omega}\setminus\{q,p_0\}$ extends to a continuous foliation of all of $\overline{\Omega}$. Namely, for λ near 0, the point where γ_λ meets one component of $\partial\Omega$ is a homeomorphism as a function of $\lambda \in \mathbb{R}/2\pi$. This follows from the absence of conjugate points along $\gamma_\lambda \cap \overline{\Omega}$ for $\lambda \ne 0$. It follows from the field-of-extremals argument that each of the geodesics γ_λ, for $-\pi < \lambda \le \pi$, and $\gamma_0 = \sigma_0$ in particular, is a chord of $\overline{\Omega}$.

Step 2: *We shall show that σ_0 is one of a family of geodesics σ_μ, starting from q and ending at distinct points $p_\mu \in \partial\Omega$, and each without conjugate points.*

Write $v_0 := \sigma_0'(0) \in T_q M$. The family of geodesics σ_μ will be defined so that σ_μ starts at q with initial direction making an angle μ with v_0. Given $\mu \in \left(-\frac{\pi}{2},\frac{\pi}{2}\right)$, let $\sigma_\mu(s) := \exp_q(sv_\mu)$, where $v_\mu := (\cos\mu)v_0 + (\sin\mu)\frac{\partial X}{\partial\phi}(0,0)$. The unit-speed geodesic σ_μ enters $\overline{\Omega}$ at q and leaves $\overline{\Omega}$ after a distance $L(\mu)$ at a point near p_0, which we shall write as $p_\mu := \sigma_\mu(L(\mu)) =: X(f(\mu), \theta_0)$. By the smooth dependence of solutions of (2.2) on initial data, p_μ and $f(\mu)$ will be smooth functions of μ.

We need to show that $f(\mu)$ is an increasing function on a neighborhood of 0. Note that $\frac{df}{d\mu} = 0$ at $\mu = 0$, since the Jacobi field $J(s) = \mathcal{J}(s)\nu(s) = \frac{\partial \sigma_\mu}{\partial \mu}(s)$ vanishes at $\mu = 0$, $s = \pi\sqrt{R}$. Here, the real-valued function $\mathcal{J}(s)$ satisfies equation (5.1) along σ_μ. Since $\mathcal{J}(s) = \sqrt{R}\sin\frac{s}{\sqrt{R}}$ when $\mu = 0$, we have $\mathcal{J}' = \frac{d\mathcal{J}}{ds} = -1$ at $\mu = 0$, $s = \pi\sqrt{R}$. By continuity, $\mathcal{J}' < 0$ for μ close to 0 and for s close to $\pi\sqrt{R}$. As a consequence, $\cos\beta_\mu(s) < 0$ for all such μ and s. Namely, $\mathcal{J}'\nu = J'$, which is the covariant derivative $\frac{D}{\partial\mu}\sigma'_\mu$, and thus

$$\frac{\partial}{\partial\mu}\cos\beta_\mu(s) = \frac{\partial}{\partial\mu}<\sigma'_\mu(s), \frac{\partial X}{\partial\phi}> = <\frac{D}{\partial\mu}\sigma'_\mu(s), \frac{\partial X}{\partial\phi}> = \mathcal{J}'(s)<\nu, \frac{\partial X}{\partial\phi}>,$$

which is negative. Integrating from $\mu = 0$, where $\cos\beta_0(s) \equiv 0$, we find that $\cos\beta_\mu(s) < 0$ for μ small and positive and for s close to $\pi\sqrt{R}$.

In order to treat quantitative properties of the family $\{\sigma_\mu\}$, we shall compare M to the 2-sphere $\widetilde{M}$ of radius $\sqrt{R}$, that is, with constant Gauss curvature $\widetilde{K} \equiv \frac{1}{R}$. Choose $\tilde{q} \in \widetilde{M}$, and identify the tangent space $T_{\tilde{q}}\widetilde{M}$ with T_qM in an isometric way. Write the corresponding family of geodesics in $\widetilde{M}$ as $\tilde{\sigma}_\mu$. Since the family of geodesics (great circles) $\tilde{\sigma}_\mu$ have no conjugate points out to distance $\pi\sqrt{R}$, the same is true of the family $\{\sigma_\mu\}$, by the Rauch Comparison Theorem (see [4], pp. 215–217.)

Now choose $\varepsilon > 0$ and write $\theta_\varepsilon = \theta_0 - \varepsilon$: for μ in some interval around 0, by continuity of the family $\{\sigma_\mu\}$, the open geodesic arcs $\sigma_\mu((0, \pi\sqrt{R}))$ cross the circle $\theta = \theta_\varepsilon$. For fixed ε, write $s_\mu = s_\mu^{(\varepsilon)} < \pi\sqrt{R}$ for the unique parameter such that $\sigma_\mu(s_\mu)$ lies on the circle $\theta = \theta_\varepsilon$. We may estimate the length s_μ of $\sigma_\mu([0, s_\mu])$ using the first variation formula for the length of a curve:

$$\frac{ds_\mu}{d\mu} = [<V, \sigma'_\mu>]_0^{s_\mu} + \int_0^{s_\mu} <V, \sigma''_\mu> ds,$$

where V is the variation vector field (see [4].) For this variation, V has normal component $<V, \nu> = \mathcal{J}$, while its tangential component is chosen so that V remains tangent to the ϕ-coordinate curves, and in particular, $V(s_\mu)$ is tangent to the curve $\sigma_\mu(s_\mu)$ of endpoints. This implies that $<V(s), \sigma'_\mu(s)> = \mathcal{J}(s)\cot\beta_\mu(s)$. For μ small and positive, and ε close to 0, we have $\mathcal{J}(s_\mu) > 0$ and $\cos\beta_\mu(s_\mu) < 0$, so that $<V, \sigma'_\mu> < 0$ at $s = s_\mu$. Meanwhile, all other terms in the first variation formula vanish, implying that $\frac{ds_\mu}{d\mu} < 0$, and hence that $s_\mu = s_\mu^{(\varepsilon)} < s_0^{(\varepsilon)} = R\theta_\varepsilon$.

Letting $\varepsilon \to 0$, we see that $L(\mu) = s_\mu^{(0)} < R\theta_0 = \pi\sqrt{R}$. But the curves $\sigma_\mu([0, \pi\sqrt{R}])$ have no conjugate points. This implies in particular that $\frac{df}{d\mu}(\mu) > 0$ for small $\mu > 0$, while as we have seen, $\frac{df}{d\mu}(0) = 0$.

Observe that the symmetry of $\mathbb{R}^3$ under the reflection $z \mapsto -z$, corresponding to the symmetry $\phi \mapsto -\phi$ of $\overline{\Omega}$, implies that $f(\mu)$ is an odd function. Thus $f(\mu)$ is strictly increasing on a neighborhood of $\mu = 0$.

Step 3: *We shall show that each σ_μ is a chord of $\overline{\Omega}$, for sufficiently small μ.*

By Step 1 above, we know that σ_0 is the unique chord of $\overline{\Omega}$ from q to p_0. We shall argue, analogously to the proof of Lemma 4.2, to show that for sufficiently small $|\mu|$, σ_μ is the unique chord from q to p_μ. Suppose otherwise: then for some sequence $\mu_k \to 0$, there is a unit-speed chord $\tau_{\mu_k} : [0, b_k] \to \overline{\Omega}$ from q to p_{μ_k}, distinct from σ_{μ_k}. As $k \to \infty$, by compactness of $S(\overline{\Omega})$, a subsequence, still denoted τ_{μ_k}, converges to a geodesic $\tau_0 : [0, b_0] \to \overline{\Omega}$ from q to $p_0 = \tau_0(b_0)$. Since, by Clairaut's relation, τ_0 is transverse to the circle $\{\theta = \theta_0\}$, we have $b_0 = \lim_{k \to \infty} b_k$. By continuity of distance, $d(q, p_0) = \lim_{k \to \infty} d(q, p_{\mu_k}) = \lim_{k \to \infty} b_k = b_0$, which shows that τ_0 is a chord from q to p_0. The uniqueness of σ_0 now implies that $\tau_0 = \sigma_0$. As a consequence, the initial tangent vectors $\tau'_{\mu_k}(0) \to \tau'_0(0) = v_0$, so for large k there is $\widehat{\mu}_k$ close to 0 with $\tau'_{\mu_k}(0) = v_{\widehat{\mu}_k}$, and thus $\tau_{\mu_k} = \sigma_{\widehat{\mu}_k}$. But τ_{μ_k} leaves $\overline{\Omega}$ at $p_{\mu_k} = X(f(\mu_k), \theta_0)$, while $\sigma_{\widehat{\mu}_k}$ leaves $\overline{\Omega}$ at $p_{\widehat{\mu}_k} = X(f(\widehat{\mu}_k), \theta_0)$. Since the function $f(\mu)$ is strictly increasing, we see that $\widehat{\mu}_k = \mu_k$. This shows that $\tau_\mu = \sigma_\mu$ after all, so that σ_μ is the unique chord from q to p_μ. In particular, $L(\mu) = d(q, p_\mu)$ for sufficiently small $|\mu|$.

Conclusion:

As we have seen in Step 2, on a neighborhood of $\mu = 0$, f is a homeomorphism whose inverse $\mu = f^{-1}(\phi)$ has an infinite derivative at $\phi = f(0) = 0$. In particular, $f(\mu) = O(\mu^3)$, with formal first differentiation (*i.e.* $\frac{df}{d\mu}(\mu) = O(\mu^2)$.) Since for all μ near 0, $\sigma_\mu([0, L(\mu)])$ is a chord of $\overline{\Omega}$, we see that the distance function ζ from q has gradient $\nabla \zeta(p_\mu) = \sigma'_\mu(L(\mu))$. By the definition of β_μ, we have

$$\nabla \zeta(p_\mu) = \cos \beta_\mu(L(\mu)) \frac{\partial X}{\partial \phi}(f(\mu), \theta_0) + \frac{\sin \beta_\mu}{R}(L(\mu)) \frac{\partial X}{\partial \theta}(f(\mu), \theta_0).$$

Next, we apply Clairaut's relation: $w(f(\mu)) \sin \beta_\mu(L(\mu)) = w(0) \sin \beta_\mu(0) = R \cos \mu$. But $w(f(\mu)) = R + O(f(\mu)^2) = R + O(\mu^6)$, which leads to $\beta_\mu(L(\mu)) = \frac{\pi}{2} + \mu + O(\mu^3)$, and thereby $\nabla \zeta(p_\mu) = -\sin \mu \frac{\partial X}{\partial \phi} + \frac{\cos \mu}{R} \frac{\partial X}{\partial \theta} + O(\mu^3)$. To summarize, for μ near 0, the coefficients of $\nabla \zeta(p_\mu)$ depend locally diffeomorphically on μ, while $p_\mu = X(f(\mu), \theta_0)$ is characterized by the coordinate $\phi = f(\mu)$. Since f^{-1} has an infinite derivative at $\phi = 0$, the restriction of $\nabla \zeta$ to $\partial \Omega$ is not differentiable at p_0.

References

[1] Bardos, Claude, Gilles Lebeau and Jeffrey Rauch: Sharp sufficient conditions for the observation, control, and stabilization of waves from the boundary. *SIAM J. Control Optim.* **30** (1992), 1024–1065.

[2] Chavel, Isaac: *Riemannian Geometry, a Modern Introduction.* Cambridge University Press, Cambridge 1993.

[3] Choe, Jaigyoung and Robert Gulliver: Isoperimetric inequalities on minimal submanifolds of space forms. *Manuscripta Math.* **77** (1992), 169-189.

[4] Do Carmo, Manfredo P.: *Riemannian Geometry.* Birkhäuser, Boston 1992.

[5] Duff, G. F. D.: *Partial Differential Equations.* University of Toronto Press, 1956.

[6] Gulliver, Robert: On the variety of manifolds without conjugate points. *Trans. Amer. Math. Soc.* **210,** 185-201 (1975).

[7] Lasiecka, Irene, Roberto Triggiani and Peng-Fei Yao: Exact controllability for second-order hyperbolic equations with variable coefficient-principal part and first-order terms. *Nonlinear Anal.* **30** (1997), 111–122.

[8] Lasiecka, Irene, Roberto Triggiani and Peng-Fei Yao: Inverse/observability estimates for second-order hyperbolic equations with variable coefficients. *J. Math. Anal. Applications* **235** (1999), 13-57.

[9] Lasiecka, Irene, Roberto Triggiani and Peng-Fei Yao: An observability estimate in $L_2 \times H^{-1}$ for second-order hyperbolic equations with variable coefficients. S. Chen, X. Li, J. Yong, X. Zhou, eds.: *Control of Distributed Parameter and Stochastic Systems,* Kluwer Academic Publishers, Dordrecht 1999.

[10] Lebeau, Gilles: Contrôle analytique. I. Estimations à priori. *Duke Math. J.* **68** (1992), 1–30.

[11] Littman, Walter: Near-optimal-time boundary controllability for a class of hyperbolic equations. *Control Problems for Systems Described by Partial Differential Equations and Applications* (Gainesville, Fla., 1986), 307–312, Lecture Notes in Control and Inform. Sci. **97,** Springer, Berlin-New York, 1987.

[12] Littman, Walter: Remarks on global uniqueness theorems for partial differential equations. These *Proceedings*.

[13] Ralston, James: Gaussian beams and the propagation of singularities. *Studies in Partial Differential Equations* (W. Littman, ed.), *MAA Studies in Mathematics* **23** (1982), 206–248.

[14] Tataru, Daniel: Boundary controllability of conservative PDEs. *Appl. Math. Optim.* **31** (1995), 257–295.

[15] Tataru, Daniel: On the regularity of boundary traces for the wave equation. *Ann. Scuola Norm. Sup. Pisa Cl. Sci. (4)* **26** (1998), 185–206.

[16] Taylor, Michael: *Tools for P.D.E.: Pseudodifferential Operators, Paradifferential Operators, and Layer Potentials*. In press.

[17] Yao, Peng-Fei: On the observability inequalities for exact controllability of wave equations with variable coefficients. *SIAM J. Control Optim.* **37** (1999), 1568–1599.

Nonlinear Boundary Stabilization of a System of Anisotropic Elasticity with Light Internal Damping

Mary Ann Horn

ABSTRACT. Controllability and stabilization of more complex, connected systems becomes more and more of an issue as results on single components become more numerous. An anisotropic system of elasticity is considered containing light internal damping which is physically motivated. To uniformly stabilize the system, nonlinear boundary damping is introduced, acting via traction forces. With this additional damping, exponential decay of the energy with respect to time is established. Techniques used to prove the required estimates include multiplier methods, sharp trace regularity results and a uniqueness argument relying on the Korn's inequality for the system.

1. Introduction

Much attention has been focused on the questions of controllability and stabilizability of elastic systems in recent years. Numerous affirmative answers using boundary controllers have been achieved for wave and plate equations, particularly in the 1980's and 1990's. However, one of the basic questions, whether stabilization was possible for systems of three-dimensional elasticity, had little in the way of positive answers until this decade. One of the first results achieved by Lagnese [7] considered a linear elastodynamic system with boundary control acting as traction forces. In this case, with velocity feedback control on the entire boundary, the domain must satisfy strict geometric conditions.

Lagnese later considered the system of linear elasticity for a homogeneous, isotropic body [8]. He obtained uniform stability through the use of nonlinear boundary control, however, to minimize difficulties arising from mathematical technicalities in the proof, the feedback control was modified. As opposed to using only velocity feedback control, he included the tangential derivative of the displacement as well. In addition, at the start of the analysis, the assumption that the body remains in a state of plane strain was imposed, thus reducing the problem to two dimensions. As in his early work, strong geometric conditions which force the domain to be star-shaped must be assumed.

1991 *Mathematics Subject Classification.* Primary 93D15, 93B52; Secondary 35B37, 74B05.
The author was supported in part by NSF Grant DMS-9803547.

More recently, the question of stabilizability for the system of isotropic elasticity was addressed in [4]. In this work, the goals were twofold. First, to eliminate the strong geometric constraints on the domain, particularly in the case when control was active on the entire boundary. Second, to use only velocity feedback control acting through the boundary conditions as a traction force, thus omitting the tangential component of the displacement required in [8]. In order to obtain this result, the proof is based on a pseudodifferential analysis which permits certain boundary traces of the solution to be expressed in terms of other traces modulo lower order interior terms. Such sharp trace regularity results had previously proven themselves to be of critical importance in the study of numerous elastic systems, including the wave equation [11] and both linear and nonlinear plate equations [3, 10]. Additionally, the proof required use of a uniqueness result for the system of linear elasticity with overdetermined homogeneous boundary conditions. While Dehman and Robbiano's result for the static system of isotropic elasticity was utilized [2], an analogous result for the dynamic system of isotropic elasticity by Isakov [6] could also have been used to facilitate the proof.

It is precisely the tangential derivative of the solution which is the primary foil of the multiplier approach. Including this derivative in the boundary conditions eliminates the mathematical difficulties within the proof, but gives rise to boundary conditions which are not physically motivated. However, if the tangential derivative is not included, its norm arises in the proof and must be bounded in terms of the norm of the control. This cannot be done through traditional methods without imposing severe geometric restrictions, as can be seen in the following two cases.

In the context of thermoelasticity, Rivera and Olivera considered an anisotropic system of elasticity coupled with the heat equation [14]. To focus on their result, consider the case when control is acting over the entire boundary. In this case, the assumptions made in Rivera-Olivera require that the domain be star-shaped (see (1.9), (1.10) in [14]). Under this assumption, the claim made in the paper is that linear velocity feedback uniformly stabilizes the anisotropic system with natural boundary conditions *without including the tangential derivative of the solution in the boundary feedback*. However, within their proof, an additional assumption must be imposed. The domain cannot simply be star-shaped, it must be spherical. This additional constraint is very subtle and occurs at the level of application of the divergence theorem (see (2.31) in [14]). Without restricting the domain to be a sphere, it is not possible to obtain the conclusion of Lemma 2.1. However, even with the restrictive assumption that the domain be a sphere, the results of Rivera and Olivera were novel and interesting and they were among the first to establish positive results for anisotropic systems.

Alabau and Komornik considered an anisotropic system of elasticity and have established uniform energy decay rates when feedback control is acting via natural and physically implementable boundary conditions [1]. While valid for the non-isotropic system, their results require even more stringent geometric assumptions than those in the work of Lagnese. As in [14], Alabau and Komornik must assume that the domain is a sphere, once again leaving the question of uniform stability open in the case of more general domains. The assumption that the domain is a sphere arises due to the combination of vector fields used within the computations, first $\vec{h} = x - x_0$ with $\vec{h} \cdot \nu \geq 0$ on the controlled portion of the boundary, Γ_1 which is then combined with the assumption $\vec{h} = R\nu$ on Γ_1.

While the results of Alabau and Komornik, as well as those contained within this paper, make use of the technique of multipliers, the proofs diverge at the level of eliminating norms of the traces of the solution on the boundary. Alabau and Komornik take the same route seen in many earlier papers, adding geometric restrictions and eliminating problematic traces at an early stage. In this work, the sharp trace regularity results used do not require such additional geometric constraints. Derivation of such trace regularity results relies critically on microlocal analysis. It should be noted, however, that this technique does require the use of a compactness/uniqueness argument which is a proof by contradiction. Thus, the precise decay estimates derived by Alabau and Komornik cannot be obtained through the arguments contained within this paper.

Thus, the goal of this paper is to consider the stabilization of an anisotropic system on general domains. To this end, nonlinear boundary feedback control defined in terms of the velocity is applied along a portion of the boundary of the domain. (Results for linear boundary feedback can be viewed as a subcase of this problem.) While geometric conditions are obviously necessary on uncontrolled portions of the boundary, they may be eliminated on the controlled portions. To prove this, it is necessary to apply sharp trace regularity results initially obtained in [4]. A second issue is the fact that unique continuation results, while available for the linear system of isotropic elasticity, are not easily found for anisotropic systems. Within the system under consideration within this paper, a light interior damping term is introduced. Physically, all elastic bodies have some inherent damping, thus it is reasonable to include such a term in the model.

2. Statement of the Problem

To formulate the system of elasticity, the following definitions are needed. Let n be a positive integer and let (a_{ijkl}) be a tensor such that

$$(2.1) \qquad a_{ijkl} = a_{klij} = a_{jikl} \quad 1 \leq i, j, k, l \leq n.$$

For some $\alpha > 0$, the above tensor satisfies the ellipticity condition,

$$(2.2) \qquad a_{ijkl}\varepsilon_{ij}\varepsilon_{kl} \geq \alpha \varepsilon_{ij}\varepsilon_{ij}$$

holds for every symmetric tensor ε_{ij}.

Let Ω be a nonempty bounded domain in $\mathbb{R}^n$ with sufficiently smooth boundary $\partial\Omega$. If the function $u = (u_1, ..., u_n) : \Omega \to \mathbb{R}^n$, denotes the displacement vector, the strain tensor, ε_{ij}, and the stress-strain relation can be expressed as

$$(2.3) \qquad \varepsilon_{ij}(u) = \frac{1}{2}\left(\frac{\partial u_i}{\partial x_j} + \frac{\partial u_j}{\partial x_i}\right) \text{ and } \sigma_{ij}(u) = a_{ijkl}\varepsilon_{kl}(u).$$

With the above definitions in mind, consider the system of anisotropic elasticity with light internal damping defined in the domain Ω with sufficiently smooth boundary $\partial\Omega = \partial\Omega_0 \cup \partial\Omega_1$:

$$(2.4.a) \qquad u_{tt} - \nabla \cdot \sigma(u) + b(x)u_t = 0 \quad \text{in } \Omega \times (0, T),$$
$$(2.4.b) \qquad u = 0 \quad \text{on } \partial\Omega_0 \times (0, T),$$
$$(2.4.c) \qquad \sigma(u)\nu = -g(u_t) \quad \text{on } \partial\Omega_1 \times (0, T),$$
$$(2.4.d) \qquad u(x, 0) = \phi_0(x); \ u_t(x, 0) = \phi_1(x) \quad \text{in } \Omega,$$

where ν represents the unit outward normal vector to $\partial\Omega$ and $b(x) > 0$ for all $x \in \Omega$. To facilitate the proofs, the assumption $\partial\bar{\Omega}_0 \cap \partial\bar{\Omega}_1 = \emptyset$ is imposed to avoid the necessity of compatibility conditions at the junction of $\partial\Omega_0$ and $\partial\Omega_1$.

Throughout this discussion, the coefficients a_{ijkl} are assumed to be independent of both time and space. For this system, the question addressed is that of uniform stabilization when a control defined in terms of velocity feedback is acting through a portion of the boundary. Within this problem, the control function, g, is a continuous vector-valued function, with each component assumed to be monotone increasing and zero at the origin. Additionally, $g \in C^1(\mathbb{R})$ is subject to the following constraints:

$$
\begin{aligned}
g(s) \cdot s &> 0, \quad \text{for } s \neq 0, \\
m|s| \leq |g(s)| &\leq M|s|, \quad \text{for } |s| > 1,
\end{aligned}
\tag{2.5}
$$

for two positive constants, m and M. Notice that for this system, growth conditions are imposed on g away for large values of $|s|$, but are not required near the origin.

2.1. Wellposedness of the System. For this problem, wellposedness results have been shown to hold using standard semigroup theory [13]. Alternatively, wellposedness can also be established via elliptic theory [15]. Before stating these results, the function space $\mathcal{H}^1_{\partial\Omega_0}(\Omega)$ is defined to be

$$
\mathcal{H}^1_{\partial\Omega_0}(\Omega) \equiv \left\{ v \in [H^1(\Omega)]^n : v = 0 \text{ on } \partial\Omega_0 \right\}.
\tag{2.6}
$$

THEOREM 2.1. *(Existence of strong solutions.)* Let $(\phi_0(x), \phi_1(x)) \in [H^2(\Omega)]^n \cap \mathcal{H}^1_{\partial\Omega_0}(\Omega) \times [H^1(\Omega)]^n$ satisfy the following compatibility condition:

$$\sigma(\phi_0)\nu = -g(\phi_1) \text{ on } \partial\Omega_1.$$

Then there exists a unique solution,

$$(u(x,t), u_t(x,t)) \in C\left([0,T]; [H^2(\Omega)]^n \cap \mathcal{H}^1_{\partial\Omega_0}(\Omega) \times [H^1(\Omega)]^n\right)$$

satisfying system (2.4).

THEOREM 2.2. *(Existence of weak solutions.)* Let $(\phi_0(x), \phi_1(x)) \in \mathcal{H}^1_{\partial\Omega_0}(\Omega) \times [L^2(\Omega)]^n$. *Then there exists a unique solution (in the sense of distributions),*

$$(u(x,t), u_t(x,t)) \in C\left([0,T]; \mathcal{H}^1_{\partial\Omega_0}(\Omega) \times [L^2(\Omega)]^n\right)$$

satisfying system (2.4).

2.2. Uniform Decay of the Energy. For system (2.4), the natural energy is defined by

$$
E(t) \equiv \frac{1}{2}\left\{\|u_t\|^2_{[L^2(\Omega)]^n} + (\sigma(u), \varepsilon(u))_\Omega\right\}
\tag{2.7}
$$

which is topologically equivalent to the usual norm on $[H^1(\Omega)]^n \times [L^2(\Omega)]^n$. The goal of this paper is to show that the energy of the solution to the system of elasticity decays uniformly to zero with respect to the initial energy. Because of the nonlinear control, to state the stability result, the following definitions are required. Let the function $\mathcal{G}(x)$ be a concave, strictly increasing function which is zero at the origin and satisfies

$$
\mathcal{G}(s \cdot g(s)) \geq |s|^2 + |g(s)|^2 \text{ for } |s| \leq 1.
\tag{2.8}
$$

(Such a function may be easily constructed. For further details, see [9].) Define

$$\tilde{\mathcal{G}}(x) \equiv \mathcal{G}\left(\frac{x}{\text{meas}(\partial\Omega \times (0,T))}\right). \tag{2.9}$$

Since $\tilde{\mathcal{G}}$ is monotone increasing, for every constant $c \geq 0$, the operator $cI + \tilde{G}$ is invertible. Defining

$$p(x) \equiv (cI + \tilde{\mathcal{G}})^{-1}(kx), \tag{2.10}$$

where k is a positive constant, the function $p(x)$ is a positive, continuous, strictly increasing function with $p(0) = 0$. With these definitions, the main result of this paper is stated in the following theorem.

THEOREM 2.3. *(Uniform stability.) Let u be a solution of system (2.4). If $\vec{h}(x) = x - x_0$, where $x_0 \in \mathbb{R}^n$, satisfies the assumption,*

$$\vec{h}(x) \cdot \nu \leq 0 \text{ on } \partial\Omega_0, \tag{2.11}$$

then for some $T_0 > 0$,

$$E(t) \leq \mathcal{S}\left(\frac{t}{T_0} - 1\right) \quad \forall t > T_0, \tag{2.12}$$

where $\mathcal{S}(t) \to 0$ as $t \to 0$ and is the contraction semigroup which solves the differential equation,

$$\begin{aligned} \mathcal{S}_t(t) + q(\mathcal{S}(t)) &= 0, \\ \mathcal{S}(0) &= E(0), \end{aligned} \tag{2.13}$$

and $q(x)$ is defined by

$$q(x) \equiv x - (I - p)^{-1}(x) \text{ for } x > 0. \tag{2.14}$$

REMARK 2.4. If the control function is linear, e.g., $g(u_t) \equiv u_t$, then, under the assumptions of the above theorem, the energy of the system decays exponentially with respect to the initial energy, i.e., there exist constants $C > 0$ and $\omega > 0$ such that

$$E(t) \leq Ce^{-\omega t}E(0), \; \forall t > 0. \tag{2.15}$$

3. Proof of Theorem 2.3

Associated with the system of elasticity in (2.4) is the following variational formulation. Let $\phi \in [H^1(\Omega)]^n$. Then any solution u of system (2.4) satisfies the variational equality,

$$(u_{tt}, \phi)_\Omega + (\sigma(u), \varepsilon(u))_\Omega + (bu_t, \phi)_\Omega + (g(u_t), \phi)_{\partial\Omega_1} - (\sigma(u)\nu, \phi)_{\partial\Omega_0} = 0, \tag{3.1}$$

which takes into account the boundary conditions in (2.4.b) and (2.4.c).

3.1. Dissipativity Inequality. The first step in the proof of Theorem 2.3 is to establish that the energy of the system is dissipative. Note that if $b(x) \equiv 0$ and $g \equiv 0$, the resulting system of elasticity is a conservative system. The dissipativity inequality establishes an important relationship between the energy at any two different times, however, alone it is not sufficient to guarantee exponential decay of the energy.

LEMMA 3.1. *Let u be a weak solution of system (2.4). Then for any $s < t$, the following inequality is satisfied:*

$$E(t) + \int_s^t \int_\Omega b(x)|u_t|^2 dxdt + \int_s^t (g(u_t), u_t)_{\partial \Omega_1} dt = E(s). \tag{3.2}$$

PROOF. Let $\phi \equiv u_t$ in the variational formulation (3.1) and integrate with respect to time. □

3.2. Stability Estimates. To establish the uniform stability of system (2.4) stated in Theorem 2.3, the critical step lies in proving the following stability estimate. Upon establishing the following lemma, the proof of uniform stabilization follows through the use of the techniques of Lasiecka and Tataru [9] for nonlinear control problems.

LEMMA 3.2. *(Stability Estimate.) Let u be a strong solution to the system of elasticity defined in (2.4). Assume*

$$\vec{h}(x) \cdot \nu \leq 0 \text{ on } \partial\Omega_0.$$

Then there exists a sufficiently large time T and a constant $C_T(E(0))$, dependent upon T and possibly dependent upon the initial energy $E(0)$ such that the following estimate is satisfied:

$$E(T) \leq C_T(E(0)) \left\{ \int_0^T \int_\Omega b(x)|u_t|^2 dxdt + \int_0^T \int_{\partial\Omega_1} \left(|u_t|^2 + |g(u_t)|^2\right) dxdt \right\}. \tag{3.3}$$

Proof of Theorem 2.3. Assuming the validity of the above lemma, the proof of Theorem 2.3 is straightforward. Define $\Sigma_A \equiv \{(x,t) \in \partial\Omega \times (0,T) : |u_t| \leq 1\}$ and $\Sigma_B \equiv \{(x,t) \in \partial\Omega \times (0,T) : |u_t| > 1\}$. Then, if $\mathcal{G}$ is a concave, strictly increasing function which is zero at the origin and is such that

$$\mathcal{G}(s \cdot g(s)) \geq |s|^2 + |g(s)|^2, \text{ for all } |s| \leq 1, \tag{3.4}$$

the following inequality holds:

$$\begin{aligned}\int_0^T \int_{\partial\Omega_1} (|u_t|^2 &+ |g(u_t)|^2) dxdt \\ &\leq \int_{\Sigma_A} \mathcal{G}(u_t \cdot g(u_t)) dxdt + C \int_{\Sigma_B} u_t \cdot g(u_t) dxdt \\ &\leq \int_0^T \int_{\partial\Omega} \left(\mathcal{G}(u_t \cdot g(u_t)) + C u_t \cdot g(u_t)\right) dxdt.\end{aligned} \tag{3.5}$$

Using Jensen's inequality and the fact that $\mathcal{G}$ is strictly increasing,

$$\begin{aligned}\int_0^T \int_{\partial\Omega_1} (|u_t|^2 &+ |g(u_t)|^2) dxdt \\ &\leq \mathcal{G}\left(\int_0^T \int_{\partial\Omega} u_t \cdot g(u_t) dxdt\right) + C \int_0^T \int_{\partial\Omega} u_t \cdot g(u_t) dxdt \\ &\leq \mathcal{G}\left(\int_0^T \int_\Omega b(x)|u_t|^2 dxdt + \int_0^T \int_{\partial\Omega} u_t \cdot g(u_t) dxdt\right) \\ &\quad + C \int_0^T \int_{\partial\Omega} u_t \cdot g(u_t) dxdt.\end{aligned} \tag{3.6}$$

Defining $\mathcal{F} \equiv \int_\Omega b(x)|u_t|^2 dxdt + \int_0^T \int_{\partial\Omega} u_t \cdot g(u_t) dxdt$, the stability inequality (3.3) becomes

$$E(T) \leq C_T(E(0))\left(C\mathcal{F} + \mathcal{G}(\mathcal{F})\right). \tag{3.7}$$

Therefore, since $\mathcal{G}$ is monotone,

$$\frac{1}{C_T(E(0))}(CI + \mathcal{G})^{-1} E(T) \leq \mathcal{F} \equiv E(0) - E(T), \tag{3.8}$$

implying that for an appropriately defined monotone function p,

(3.9) $$E(T) + p(E(T)) \leq E(0).$$

With the monotonicity of p, the results of Lasiecka and Tataru [9] can now be applied to obtain uniform stability of the elasticity system defined in (2.4). To do so, the following lemma is required.

LEMMA 3.3. ([9], Lemma 3.3.) *Let p be a positive, increasing function such that $p(0) = 0$. Since p is increasing, we can define an increasing function q, $q(x) \equiv x - (I+p)^{-1}(x)$. Consider a sequence s_m of positive numbers which satisfies*

(3.10) $$s_{m+1} + p(s_{m+1}) \leq s_m.$$

Then $s_m \leq \mathcal{S}(m)$ where $\mathcal{S}(t)$ is a solution of the differential equation

(3.11) $$\frac{d}{dt}\mathcal{S}(t) + q(\mathcal{S}(t)) = 0, \ \mathcal{S}(0) = s_0.$$

Moreover, if $p(x) > 0$ for $x > 0$, then $\lim_{t\to\infty} \mathcal{S}(t) = 0$.

Applying the result in (3.9) on the time interval $[mT, m(T+1)]$ yields

(3.12) $$E(m(T+1)) + p(E(m(T+1))) \leq E(mT)$$

for $m = 0, 1, \ldots$ Thus, the result of Lemma 3.3 can be applied with

$$s_m \equiv E(mT), \ s_0 \equiv E(0).$$

This yields

$$E(mT) \leq \mathcal{S}(m), \ m = 0, 1, 2, \ldots$$

Setting $t = mT + \tau$ and recalling the evolution property gives the inequality

$$E(t) \leq E(mT) \leq \mathcal{S}(m) \leq \mathcal{S}\left(\frac{t-\tau}{T}\right) \leq \mathcal{S}\left(\frac{t}{T} - 1\right) \text{ for all } t > T,$$

which completes the proof of Theorem 2.3.

4. Proof of Lemma 3.2: Stability Estimate

4.1. Preliminary Estimate. Proving the required stability estimate stated in (3.3) can be broken into three primary steps. The first involves the technique commonly known as the multiplier method, which results in a preliminary bound on the energy.

LEMMA 4.1. *Let u be a solution to the system of elasticity defined in (2.4). Assume*

$$\vec{h}(x) \cdot \nu \leq 0 \text{ on } \partial\Omega_0.$$

Then there is a constant C such that for any $0 < \epsilon < 1/4$, the following estimate is satisfied:

(4.1) $$\begin{aligned}\int_0^T E(t)dt \leq C \ \Big\{ &E(0) + E(T) + \int_0^T \int_\Omega b(x)|u_t|^2 dxdt \\ &+ \int_0^T \int_{\partial\Omega_1} \left(|u_t|^2 + |g(u_t)|^2\right) dxdt \\ &+ \int_0^T \int_{\partial\Omega_1} |\nabla u|^2 dxdt + l.o.t.(u) \Big\},\end{aligned}$$

where

$$l.o.t.(u) \equiv \int_0^T \|u\|^2_{[H^{1-\epsilon}(\Omega)]^n} dt$$

and refers to terms involving norms of u which are strictly lower order with respect the the energy norm.

PROOF. *First multiplier.* Let $\phi = \nabla u \vec{h}$ be the test function in the variational formulation (3.1). Then

(4.2)
$$\begin{aligned}(u_{tt}, \nabla u \vec{h})_\Omega &+ (\sigma(u), \varepsilon(\nabla u \vec{h}))_\Omega + (bu_t, \nabla u \vec{h})_\Omega \\ &+ (g(u_t), \nabla u \vec{h})_{\partial \Omega_1} - (\sigma(u)\nu, \nabla u \vec{h})_{\partial \Omega_0} = 0.\end{aligned}$$

Note that
$$\varepsilon(\nabla u \vec{h}) = \varepsilon(u) + \mathcal{R}(u),$$

where the components of the first order tensor $\mathcal{R}(u)$ are given by

$$\mathcal{R}_{ij}(u) \equiv \frac{1}{2}(h_k u_{i,kj} + h_k u_{j,ki}),$$

and in the above, the summation convention is used to indicate summation with respect to k. Therefore,

(4.3)
$$\begin{aligned}(\sigma(u), \varepsilon(\nabla u \vec{h}))_\Omega &= (\sigma(u), \varepsilon(u))_\Omega + (\sigma(u), \mathcal{R}(u))_\Omega \\ &= (\sigma(u), \varepsilon(u))_\Omega + \tfrac{1}{2}(\sigma_{ij}(u), u_{j,ki} h_k + u_{i,kj} h_k)_\Omega \\ &= (\sigma(u), \varepsilon(u))_\Omega + (\sigma_{ij}(u), u_{i,kj} h_k)_\Omega,\end{aligned}$$

since $\sigma(u)$ is a symmetric tensor. Following the initial estimates for the isotropic system described in [4], the only need is to include the light damping term and to incorporate the nonlinear control. Thus, integrating (4.2) by parts and applying the divergence theorem results in the inequality,

(4.4)
$$\begin{aligned}\tfrac{n}{2} \int_0^T \int_\Omega |u_t|^2 dx dt &+ \int_0^T (\sigma(u), \varepsilon(u))_\Omega dt + \int_0^T (\sigma_{ij}, u_{i,kj} h_k)_\Omega dt + \int_0^T (bu_t, \nabla u \vec{h})_\Omega dt \\ &\leq -(u_t, \nabla u \vec{h})_\Omega|_0^T + \tfrac{1}{2}\int_0^T \int_{\partial \Omega_1} |u_t|^2 \vec{h} \cdot \nu dx dt - \int_0^T (g(u_t), \nabla u \vec{h})_{\partial \Omega_1} dt,\end{aligned}$$

once the boundary condition $u = 0$ on $\partial \Omega_0$ and the fact that $\vec{h}$ is a radial vector with $\vec{h} \cdot \nu \leq 0$ on $\partial \Omega_0$ are taken into account. Noting that

$$2\sigma_{ij} u_{i,kj} h_k + n\sigma(u) \cdot \varepsilon(u) = \nabla \cdot (\sigma(u) \cdot \varepsilon(u)\vec{h}),$$

the above inequality can be further simplified by applying the divergence theorem,

(4.5)
$$\begin{aligned}\tfrac{n}{2}\int_0^T \int_\Omega |u_t|^2 dx dt &+ \int_0^T (\sigma(u), \varepsilon(u))_\Omega dt - \tfrac{n}{2}\int_0^T (\sigma(u), \varepsilon(u))_\Omega dt \\ &\leq C\left(E(0) + E(T)\right) + \tfrac{1}{2}\int_0^T \int_{\partial \Omega_1} |u_t|^2 \vec{h} \cdot \nu dx dt \\ &\quad - \int_0^T (bu_t, \nabla u \vec{h})_\Omega dt - \int_0^T (g(u_t), \nabla u \vec{h})_{\partial \Omega_1} dt \\ &\quad - \int_0^T (\sigma(u), \varepsilon(u)\vec{h} \cdot \nu)_{\partial \Omega} dt.\end{aligned}$$

Second Multiplier. Let $\phi = u$ be the test function in the variational formulation (3.1). Then integrating the result with respect to time gives

(4.6)
$$\begin{aligned}\int_0^T (\sigma(u), \varepsilon(u))_\Omega dt &= \int_0^T \|u_t\|^2_{[L^2(\Omega)]^n} dt - (u_t, u)_\Omega|_0^T - \int_0^T (bu_t, u)_\Omega dt \\ &\quad + \int_0^T (\sigma(u)\nu, u)_{\partial \Omega_0} dt - \int_0^T (g(u_t), u)_{\partial \Omega_1}.\end{aligned}$$

Final Identity. Multiplying the above equality by $(n-1)/2$ and adding the result to (4.5) yields the desired inequality,

$$\frac{1}{2}\int_0^T \|u_t\|^2_{[L^2(\Omega)]^n} dt + \frac{1}{2}\int_0^T (\sigma(u), \varepsilon(u))_\Omega dt$$

$$\leq C\left\{E(0) + E(T) + \int_0^T \int_{\partial\Omega_1} \left(|u_t|^2 + |g(u_t)|^2\right) dxdt + \int_0^T \int_\Omega b|u_t|^2 dxdt\right.$$

$$(4.7) \qquad\qquad\qquad\qquad\qquad \left. + \int_0^T \int_{\partial\Omega_1} |\nabla u|^2 dxdt + l.o.t.(u)\right\},$$

once terms have been bounded by absorbing them into the energy and trace theory has been applied to estimate boundary terms. $\square$

4.2. Boundary Traces. To eliminate the integral involving boundary traces of the displacement, the following theorem, initially proven in [5] is applied. Note that the proof in [5] is a general proof which does not require the assumption of isotropy stated in the main result of [5]. Due to the use of a partition of unity and a flattening of the boundary procedure, the primary requirement in the analysis of the symbol corresponding to the system of elasticity is the ellipticity of the spatial operator. Under this assumption, which is satisfied by the anisotropic system under consideration, the analysis previously done in [5] is also valid.

THEOREM 4.2. *(Trace regularity [5].) Let u be the solution to the system,*

$$\begin{aligned} u_{tt} - \nabla \cdot \sigma(u) &= f && \text{in } \Omega \times (0, T), \\ u &= 0 && \text{on } \partial\Omega_0 \times (0, T), \\ \sigma(u)\nu &= g && \text{on } \partial\Omega_1 \times (0, T), \\ u(x,0) = \phi_0(x); \; u_t(x,0) &= \phi_1(x) && \text{in } \Omega, \end{aligned}$$

and let $0 < \alpha < T/2$. Then u satisfies the following inequality,

$$(4.8) \quad \|\nabla u\tau\|_{[L^2(\alpha, T-\alpha;\partial\Omega)]^n} \leq C\{\|u_t\|_{[L^2(0,T;\partial\Omega_1)]^n} + \|u\|_{[L^2(0,T;H^{1/2+\epsilon}(\Omega))]^n}$$
$$+\|f\|_{[H^{-1/2}(0,T;\Omega)]^n} + \|g\|_{[L^2(0,T;\partial\Omega_1)]^n}\}.$$

Applying the result of Lemma 4.1 on the interval $(\alpha, T - \alpha)$ yields

$$\int_\alpha^{T-\alpha} E(t)dt \leq C \left\{E(\alpha) + E(T-\alpha) + \int_\alpha^{T-\alpha} \int_\Omega b(x)|u_t|^2 dxdt \right.$$
$$+ \int_\alpha^{T-\alpha} \int_{\partial\Omega_1} \left(|u_t|^2 + |g(u_t)|^2\right) dxdt$$
$$\left. + \int_\alpha^{T-\alpha} \int_{\partial\Omega_1} |\nabla u|^2 dxdt + l.o.t.(u)\right\},$$

Noting that $\nabla u \equiv \nabla u \nu + \sum_{k=1}^{n-1} \nabla u \tau_k$, where $\{\nu, \tau_k\}_{k=1}^{n-1}$ form an orthonormal set of normal and tangent vectors, the following holds:

$$\int_\alpha^{T-\alpha} \int_{\partial\Omega_1} |\nabla u|^2 dxdt \leq \int_\alpha^{T-\alpha} \int_{\partial\Omega_1} |\nabla u\nu|^2 dxdt + \sum_{k=1}^{n-1} \int_\alpha^{T-\alpha} \int_{\partial\Omega_1} |\nabla u\tau_k|^2 dxdt.$$

To obtain a bound for the normal derivative in terms of the boundary condition in (2.4.c), denote

$$\vec{d}_k \equiv \nabla u \tau_k$$

and write

$$\sigma(u)\nu = -g(u_t),$$
$$\nabla u\tau_k = \vec{d}_k,$$

resulting in the algebraic linear system
$$A\vec{u} = (-g(u_t), \vec{d}_1, ..., \vec{d}_{n-1})^T,$$
where $\vec{u} \equiv (u_{1,1}, ..., u_{1,n}, u_{2,1}, ..., u_{2,n}, ..., u_{n,1}, ..., u_{n,n})$ and the determinant of the matrix A is nonzero. Solving the above system and integrating the result over $\partial\Omega_1 \times (\alpha, T-\alpha)$ yields the inequality,

$$\sum_{k=1}^{n} \int_\alpha^{T-\alpha} \int_{\partial\Omega_1} |D_k u|^2 dx dt \leq C \int_\alpha^{T-\alpha} \int_{\partial\Omega_1} \left\{ |g(u_t)|^2 + \sum_{k=1}^{n-1} |\vec{d}|^2 \right\} dx dt.$$

Application of this estimate, together with Theorem 4.2 and the dissipativity of the energy to the estimate in (4.1) gives

$$(T-2\alpha)E(T) \leq C \left\{ E(0) + \int_0^T \int_\Omega b|u_t|^2 dx dt + \int_0^T \int_{\partial\Omega_1} \left(|u_t|^2 + |g(u_t)|^2 \right) dx dt \right.$$

(4.9)
$$\left. + \int_0^T \int_{\partial\Omega_1} |\sigma(u)\nu|^2 dx dt + l.o.t.(u) \right\}$$

Recalling the dissipativity identity (3.2) and the boundary condition in (2.4.c) yields
$$(T-2\alpha)E(T) + C_1 E(T)$$
$$\leq \left\{ \int_0^T \int_\Omega b|u_t|^2 dx dt + \int_0^T \int_{\partial\Omega_1} \left(|u_t|^2 + |g(u_t)|^2 \right) dx dt + l.o.t.(u) \right\}.$$

Taking T to be sufficiently large results in the completion of the proof of the following lemma.

LEMMA 4.3. *Let u be a solution of the system of elasticity in (2.4). Assume*
$$\vec{h}(x) \cdot \nu \leq 0 \text{ on } \partial\Omega_0.$$
Then for some sufficiently large time T, there exists a constant $C(T)$ such that for any $0 < \epsilon < 1/4$,

(4.10)
$$E(T) \leq C(T) \left\{ \int_0^T \int_\Omega b|u_t|^2 dx dt + \int_0^T \int_{\partial\Omega_1} \left(|u_t|^2 + |g(u_t)|^2 \right) dx dt + l.o.t.(u) \right\}.$$

4.3. Estimation of Lower Order Terms. To complete the proof of Lemma 3.2, the lower order terms remaining on the right hand side of the inequality in (4.10) must be bounded in terms of norms of the control and the light internal damping. Thus, the following lemma must be proven.

LEMMA 4.4. *Let u be a solution to the system of elasticity in (2.4). Then there exists a constant, $C_T(E(0))$, dependent upon T and possibly upon the initial energy, $E(0)$, such that the following estimate is satisfied:*

(4.11)
$$\begin{aligned} l.o.t.(u) &\equiv \int_0^T \|u\|^2_{[H^{1-\epsilon}(\Omega)]^n} dt \\ &\leq C_T(E(0)) \left\{ \int_0^T \int_\Omega b|u_t|^2 dx dt + \int_0^T \int_{\partial\Omega_1} \left(|u_t|^2 + |g(u_t)|^2 \right) dx dt \right\}. \end{aligned}$$

PROOF. By using a compactness-uniqueness argument, the proof proceeds by contradiction. Assume that (4.11) is not valid. Then there exists a sequence of solutions $\{u_m(t)\}_{m=1}^{\infty}$ satisfying system (2.4) with initial data $(\phi_{m,0}, \phi_{m,1}) \in \mathcal{H}^1_{\partial\Omega_0}(\Omega) \times [L^2(\Omega)]^n$ such that

(4.12) $$\frac{l.o.t.(u_m)}{\int_0^T \int_\Omega b|u_{m,t}|^2 dxdt + \int_0^T \int_{\partial\Omega_1} (|u_{m,t}|^2 + |g(u_{m,t})|^2) \, dxdt} \to \infty \text{ as } m \to \infty.$$

Let $E_m(t)$ denote the energy of the system satisfied by u_m at time t. Since $u_m(t)$ satisfies system (2.4), Lemma 4.3 can be applied. Therefore,

$$E_m(t) \leq C(T) \text{ uniformly in } m.$$

Thus, the sequence has the following convergence properties:

$$u_m \to u \quad \text{in } [L^\infty(0,T; H^1(\Omega))]^n \text{ weakly star},$$
$$u_{m,t} \to u_t \quad \text{in } [L^\infty(0,T; L^2(\Omega))]^n \text{ weakly star}.$$

Therefore, since the lower order terms are compact with respect to the energy norm, it can be inferred that

$$l.o.t.(u_m) \to l.o.t.(u) \text{ as } m \to \infty.$$

Case 1: Assume $l.o.t.(u) \neq 0$. Under the assumptions in (4.12), this implies that

$$u_{m,t} \to 0 \quad \text{in } [L^2(0,T; \Omega)]^n,$$
$$u_{m,t} \to 0 \quad \text{in } [L^2(0,T; \partial\Omega_1)]^n,$$
$$g(u_{m,t}) \to 0 \quad \text{in } [L^2(0,T; \partial\Omega_1)]^n.$$

Hence, by passing the limit on the original system in (2.4), the limit function u must satisfy

(4.13) $$\begin{aligned} \nabla \cdot \sigma(u) &= 0 &&\text{in } \Omega \times (0,T), \\ u &= 0 &&\text{on } \partial\Omega_0 \times (0,T), \\ \sigma(u)\nu &= 0 &&\text{on } \partial\Omega_1 \times (0,T), \end{aligned}$$

and, therefore, by Korn's inequality, $u \equiv 0$ [12]. However, this contradicts the assumption that $l.o.t.(u) \neq 0$.

Case 2: Assume $l.o.t.(u) = 0$. Define $c_m \equiv l.o.t.(u_m)$ and $\tilde{u}_m \equiv u_m/c_m$. Then

$$l.o.t.(\tilde{u}_m) = 1,$$

(4.14) $$\frac{1}{c_m^2} \int_0^T \int_\Omega b|u_{m,t}|^2 dxdt + \frac{1}{c_m^2} \int_0^T \int_{\partial\Omega_1} (|u_{m,t}|^2 + |g(u_{m,t})|^2) \, dxdt \to 0,$$

which implies

(4.15) $$\begin{aligned} \tilde{u}_{m,t} &\to 0 &&\text{in } [L^2(0,T; \Omega)]^n, \\ \tilde{u}_{m,t} &\to 0 &&\text{in } [L^2(0,T; \partial\Omega)]^n, \\ \tfrac{1}{c_m} g(u_{m,t}) &\to 0 &&\text{in } [L^2(0,T; \partial\Omega_1)]^n. \end{aligned}$$

Noting that $\tilde{u}_m$ satisfies the system,

(4.16) $$\begin{aligned} \tilde{u}_{m,tt} - \nabla \cdot \sigma(\tilde{u}_m) + b(x)\tilde{u}_{m,t} &= 0 &&\text{in } \Omega \times (0,T), \\ \tilde{u}_m &= 0 &&\text{on } \partial\Omega_0 \times (0,T), \\ \sigma(\tilde{u}_m)\nu &= -\tfrac{1}{c_m} g(u_{m,t}) &&\text{on } \partial\Omega_1 \times (0,T), \end{aligned}$$

the estimate in (4.10) and the convergence in (4.15) yield

(4.17) $$\tfrac{1}{c_m^2} E_m(T) \leq C_T \left\{ \tfrac{1}{c_m^2} \int_0^T \int_\Omega b|u_{m,t}|^2 dxdt \right. \\ \left. + \tfrac{1}{c_m^2} \int_0^T \int_{\partial\Omega_1} (|u_{m,t}|^2 + |g(u_{m,t})|^2) \, dxdt + l.o.t.(\tilde{u}_m) \right\},$$

where the right-hand side is uniformly bounded for all m. Therefore,

(4.18) $\qquad \|\tilde{u}_m\|_{[H^1(\Omega)]^n} \leq C$ and $\|\tilde{u}_{m,t}\|_{[L^2(\Omega)]^n} \leq C$,

resulting in the convergence,

(4.19) $\qquad \begin{aligned} \tilde{u}_m &\to \tilde{u} &&\text{in } [L^\infty(0,T;H^1(\Omega))]^n \text{ weakly star,} \\ \tilde{u}_{m,t} &\to \tilde{u}_t &&\text{in } [L^\infty(0,T;L^2(\Omega))]^n \text{ weakly star,} \end{aligned}$

which implies

(4.20) $\qquad l.o.t.(\tilde{u}_m) \to l.o.t.(\tilde{u}) = 1.$

Additionally, the above convergence properties imply that $\tilde{u}$ satisfies (4.13). Therefore, as in the first case, $\tilde{u}$ must be identically zero, contradicting $l.o.t.(\tilde{u}) = 1$. □

4.4. Completion of the Proof of Lemma 3.2. Substituting the results of Lemma 4.4 into the inequality stated in Lemma 4.3 yields the desired stability estimate,

$$E(T) \leq C(T) \left\{ \int_0^T \int_\Omega b|u_t|^2 dxdt + \int_0^T \int_{\partial\Omega_1} \left(|u_t|^2 + |g(u_t)|^2\right) dxdt \right\}.$$

References

[1] Alabau, F. and V. Komornik. Boundary observability, controllability and stabilization of linear elastodynamic systems. *SIAM J. Control Optim.* **37** (1999), no. 2, 521–542.
[2] Dehman, B. and L. Robbiano. La propriété du prolongement unique pour un système elliptique. *J. Math. Pures Appl.* **72** (1993), no. 5, 475–492.
[3] Horn, M. A. and I. Lasiecka. Asymptotic behavior with respect to thickness of boundary stabilizing feedback for the Kirchhoff plate. *J. Diff. Eq.* **114** (1994), no. 2, 396–433.
[4] Horn, M. A. Implications of sharp trace regularity results on boundary stabilization of the system of linear elasticity. *J. Math. Anal. Appl.* **223** (1998), 126–150.
[5] Horn, M. A. Sharp trace regularity for the solutions of the equations of dynamic elasticity. *J. Math. Sys. Est. Cont.* **8** (1998), no. 2, 217–219.
[6] Isakov, V. On uniqueness and stability in the Cauchy problem. In *Optimization methods in partial differential equations (South Hadley, MA, 1996)*, Contemp. Math. **209** (1997), 131–146.
[7] Lagnese, J. E. Boundary stabilization of linear elastodynamic systems. *SIAM J. Control Optim.* **21** (1983), no. 6, 968–984.
[8] Lagnese, J. E. Uniform asymptotic energy estimates for solutions of the equations of dynamic plane elasticity with nonlinear dissipation at the boundary. *Nonlinear Analysis* **16** (1991), no. 1, 35-54.
[9] Lasiecka, I. and D. Tataru. Uniform boundary stabilization of semilinear wave equations with nonlinear boundary damping. *Differential and Integral Equations* **6** (1993), 507–533.
[10] Lasiecka, I. and R. Triggiani. Sharp trace estimates of solutions to Kirchhoff and Euler-Bernoulli equations. *Appl. Math. Optim.* **28** (1993), 277–306.
[11] Lasiecka, I. and R. Triggiani. Uniform stabilization of the wave equation with Dirichlet or Neumann feedback control without geometric conditions. *Appl. Math. Optim.* **25** (1992), 189–224.
[12] Oleinik, O. A., A. S. Shamaev and G. A. Yosifian. *Mathematical Problems in Elasticity and Homogenization*, North-Holland, Amsterdam, 1992.
[13] Pazy, A. *Semigroups of Linear Operators and Applications to Partial Differential Equations*, Springer-Verlag, New York, 1993.
[14] Rivera, J. E. M. and M. L. Olivera. Stability in inhomogeneous and anisotropic thermoelasticity. *Bollettina U. M. I.* **11-A** (1997), no. 7, 115–127.
[15] Taylor, M. *Pseudodifferential Operators*, Princeton University Press, Princeton, New Jersey, 1981.

Department of Mathematics, Vanderbilt University, Nashville, Tennessee 37240
E-mail address: horn@math.vanderbilt.edu

CARLEMAN ESTIMATE WITH THE NEUMANN BOUNDARY CONDITION AND ITS APPLICATIONS TO THE OBSERVABILITY INEQUALITY AND INVERSE HYPERBOLIC PROBLEMS

[1] Victor Isakov and [2] Masahiro Yamamoto

[1] Department of Mathematics and Statistics, Wichita State University
Wichita Kansas 67260-0033 USA
e-mail: isakov@twsuvm.uc.twsu.edu
[2] Department of Mathematical Sciences
The University of Tokyo
3-8-1 Komaba, Meguro, Tokyo 153 Japan
e-mail : myama@ms.u-tokyo.ac.jp

ABSTRACT. We consider an initial value problem with the homogeneous Neumann boundary condition:

$$(Pu)(x,t) \equiv \frac{\partial^2 u}{\partial t^2}(x,t) - \Delta u(x,t) - \sum_{|\alpha| \leq 1} a_\alpha(x,t)(D^\alpha u)(x,t) = 0 \quad \text{in } \Omega \times (0,T)$$

$$\frac{\partial u}{\partial \nu}(x,t) = 0 \quad \text{on } \partial\Omega \times (0,T).$$

Here $\Omega \subset \mathbb{R}^n$, $n \geq 2$, is a bounded domain with C^2-boundary $\partial\Omega$ and $a_\alpha \in L^\infty(\Omega \times (0,T))$ for $|\alpha| = 1$ and a_α is in some L^γ-space. We establish a Carleman estimate for P under Neumann boundary condition on the whole boundary and Dirichlet condition on a suitable subboundary. For the existence of a weight function in the Carleman estimate, the subboundary where the Dirichlet data are prescribed, is subject to some geometric restrictions. Next we apply the Carleman estimate to prove the observability inequality

$$\|u(\cdot,0)\|^2_{H^1(\Omega)} + \left\|\frac{\partial u}{\partial t}(\cdot,0)\right\|^2_{L^2(\Omega)} \leq C\|u\|^2_{H^1(\Gamma \times (0,T))}$$

under some conditions on $\Gamma \subsetneq \partial\Omega$ and $T > 0$. Finally we apply the Carleman estimate to the uniqueness and the stability in determining source terms or coefficients from Dirichlet data on the subboundary.

2000 *Mathematics Subject Classification.* Primary 35Lxx, 49–XX, 93–XX.

© 2000 American Mathematical Society

§1. Introduction.

In this paper, we first establish a Carleman estimate for the hyperbolic operator of the second order with the homogeneous Neumann boundary condition:

$$(Pu)(x,t) \equiv u''(x,t) - \Delta u(x,t) - \sum_{|\alpha| \leq 1} a_\alpha(x,t)(D^\alpha u)(x,t),$$

(1.1) $\qquad x \in \Omega, 0 < t < T$

and

$$\frac{\partial u}{\partial \nu}(x,t) = 0, \qquad x \in \partial\Omega, 0 < t < T.$$

Here $x = (x_1, ..., x_n)$, $u' = \partial_t u = \frac{\partial u}{\partial t}$, $u'' = \partial_t^2 u = \frac{\partial^2 u}{\partial t^2}$, $\Omega \subset \mathbb{R}^n$, $n \geq 2$, is a bounded domain whose boundary is of class C^2, $\nu(x) = (\nu_1(x), ..., \nu_n(x))$ is the outward unit normal vector to $\partial\Omega$ at x, $\nabla = \left(\frac{\partial}{\partial x_1}, ..., \frac{\partial}{\partial x_n}\right)$, $\frac{\partial u}{\partial \nu} = \sum_{i=1}^n \frac{\partial u}{\partial x_i} \nu_i = \nabla u \cdot \nu$, $\alpha = (\alpha_1, ..., \alpha_n, \alpha_{n+1})$ is a multi-index with nonnegative integer components and $|\alpha| = \alpha_1 + \cdots + \alpha_n + \alpha_{n+1}$,

$$D^\alpha = \left(\frac{\partial}{\partial x_1}\right)^{\alpha_1} \cdots \left(\frac{\partial}{\partial x_n}\right)^{\alpha_n} \left(\frac{\partial}{\partial t}\right)^{\alpha_{n+1}}.$$

In the case of $n = 1$, the observability inequality is easily obtained by the D'Alembert formula and here we mainly consider the case of $n \geq 2$.

Furthermore we will use the following notation: $\partial_j = \frac{\partial}{\partial x_j}$, $\partial_i \partial_j = \frac{\partial^2}{\partial x_i \partial x_j}$, $\partial_i^2 = \frac{\partial^2}{\partial x_i^2}$, and $\Box = \partial_t^2 - \Delta$.

Throughout this paper, we assume

(1.2) $\qquad a_\alpha \in L^\infty(\Omega \times (0,T)), \qquad |\alpha| = 1,$

(1.3) $\qquad a_0 \in L^{n+1}(\Omega \times (0,T)),$

and

$$\partial\Omega = \overline{\Gamma_0 \cup \Gamma}, \quad \Gamma_0 \cap \Gamma = \emptyset,$$

(1.4) $\qquad \Gamma_0$ and Γ are relatively open subsets of $\partial\Omega$.

We set

$$h_{n+1}(t) = -\beta\left(t - \frac{T}{2}\right)^2, \quad \varphi(x,t) = \sum_{j=1}^n h_j(x_j) + h_{n+1}(t),$$

$$Q(\delta) = \{(x,t) \in \Omega \times (0,T); \varphi(x,t) > \delta\},$$

for given $h_1, ..., h_n \in C^2(\mathbb{R})$ and $\delta \geq 0$.

Let $h_1 = h_1(x_1)$, ..., $h_n = h_n(x_n)$, $h_{n+1} = h_{n+1}(t)$ and $c \in \mathbb{R}$, $\mu > 0$, $\delta > 0$ satisfy

(1.5) $\qquad 2(\partial_i^2 h_i)(x_i) + (\Box\varphi)(x,t) - c > \mu, \quad 1 \leq i \leq n, (x,t) \in \overline{Q(\delta)}$

(1.6) $$(\Delta + \partial_t^2)\varphi(x,t) + c > \mu, \quad (x,t) \in \overline{Q(\delta)}$$

(1.7) $$(-\Box\varphi(x,t) + c)(|\nabla\varphi|^2 - |\partial_t\varphi|^2) > \mu, \quad (x,t) \in \overline{Q(\delta)}$$

(1.8) $$\nabla\varphi(x) \cdot \nu(x) = 0, \quad x \in \Gamma_0$$

and

(1.9) $$\sum_{j=1}^{n}(\partial_j h_j)(x_j)(\partial_j^2 h_j)(x_j)\nu_j(x) \geq 0, \quad x \in \Gamma_0.$$

We can state a Carleman estimate as our first main result:

Theorem 1. *Let $\delta \geq 0$ and D be a subdomain of $\Omega \times (0,T)$ such that $D \supset Q(\delta)$ and let D have the cone property. Let φ, h_i, $1 \leq i \leq n+1$ satisfy (1.5) - (1.9). Then there exists a constant $M = M(h_1, ..., h_n, \beta, T, \delta, a_\alpha) > 0$ such that*

$$\lambda \int_{Q(\delta)} e^{2\lambda\varphi}(|\nabla u|^2 + |u'|^2)dxdt + \lambda^3 \int_{Q(\delta)} e^{2\lambda\varphi}|u|^2 dxdt$$

$$\leq M\left(\int_D e^{2\lambda\varphi}|(Pu)(x,t)|^2 dxdt\right.$$

$$+ \lambda \int_{D\setminus Q(\delta)} e^{2\lambda\varphi}(|\nabla u|^2 + |u'|^2)dxdt + \lambda^3 \int_{D\setminus Q(\delta)} e^{2\lambda\varphi}|u|^2 dxdt$$

$$+ \int_{\partial D \cap (\Gamma \times (0,T))} \{\lambda e^{2\lambda\varphi}(\nabla\varphi \cdot \nu)(-|\nabla u|^2 + |u'|^2) + \lambda^3 e^{2\lambda\varphi}|u|^2\}d\sigma$$

$$+ \int_{\partial D \cap (\partial\Omega \times (0,T))} \lambda e^{2\lambda\varphi}\{2((\nabla\varphi \cdot \nabla u) - \varphi' u')$$

(1.10)
$$\left.+ (2\lambda(|\nabla\varphi|^2 - |\varphi'|^2) - c)u\}\frac{\partial u}{\partial \nu}d\sigma\right)$$

for $\lambda > M$, provided that u satisfies

(1.11) $$\begin{cases} Pu \in L^2(D), \quad u \in H^1(D), \quad u \in H^1(\partial D), \\ u = |\nabla u| = u' = 0 \quad \text{on } \partial D \setminus (\partial\Omega \times (0,T)). \end{cases}$$

This is a Carleman estimate for the hyperbolic operator P with the Neumann boundary condition where Cauchy data are prescribed only on the subboundary $\Gamma \subsetneq \partial\Omega$. Carleman estimates for $u \in C_0^\infty(Q(\delta))$ or $u \in H_0^2(Q(\delta))$ for hyperbolic operators, have been proved by Hörmander [9], Isakov [11], [12], [13], [14]. For functions with non-zero boundary data, we refer to Kubo [24], Lavrent'ev, Romanov and Shishat·skiĭ[31], Tataru [35]. Tataru [35] established a Carleman estimate with non-zero Dirichlet boundary condition within general framework. Our proof is based on an inequality in Lavrent'ev, Romanov and Shishat·skiĭ[31], and was inspired by Kazemi and Klibanov [17].

Carleman estimates are useful in several important problems for a hyperbolic equation. In this paper, we first apply our Carleman estimate to Lipschitz stability in determining initial values by means of Dirichlet data on a subboundary Γ under the homogeneous Neumann boundary condition. More precisely,

Theorem 2. *In addition to* (1.5) - (1.9), *we assume that* $T > 0$ *and* $h_1, ..., h_n$ *satisfy*

(1.12) $\qquad h_1(x_1) + \cdots + h_n(x_n) > 0, \qquad x = (x_1, ..., x_n) \in \overline{\Omega}$

and

(1.13) $\qquad T > \dfrac{2}{\sqrt{\beta}} \max_{x \in \overline{\Omega}} (h_1(x_1) + \cdots h_n(x_n))^{\frac{1}{2}},$

while the coefficients satisfy:

(1.14) $\qquad a_0 \leq c_1 \quad$ *in* $\Omega \times (0,T)$ *with some* $c_1 \in \mathbb{R}$

and

(1.15) $\qquad a_0' \in L^\infty(0,T; L^r(\Omega)), \qquad r = \dfrac{n}{2}.$

Then there exists a constant $C = C(\Omega, T, \Gamma, h_1, ..., h_n, \beta, a_\alpha) > 0$, *independent of* u, *such that*

(1.16) $\qquad \|u(\cdot, 0)\|^2_{H^1(\Omega)} + \|u'(\cdot, 0)\|^2_{L^2(\Omega)} \leq C \|u\|^2_{H^1(\Gamma \times (0,T))},$

if $u \in H^2(\Omega \times (0,T))$ *satisfies* $(Pu)(x,t) = 0$ *with*

(1.17) $\qquad \dfrac{\partial u}{\partial \nu}(x,t) \equiv \nabla u(x,t) \cdot \nu(x) = 0, \quad x \in \partial\Omega, 0 < t < T.$

Remark 1. Recently we have learned that Lasiecka, Triggiani and Zhang [30] obtained a similar result to Theorem 2, under different conditions. In some cases, their results are more complete. In particular, in [30], the estimate (1.16) is improved as

(1.16') $\quad \|u(\cdot, 0)\|^2_{H^1(\Omega)} + \|u'(\cdot, 0)\|^2_{L^2(\Omega)} \leq C(\|u\|^2_{L^2(\Gamma \times (0,T))} + \|u'\|^2_{L^2(\Gamma \times (0,T))}).$

The improvement such as (1.16)' is very helpful for researches in the corresponding inverse problems as are discussed in Sections 5 and 6. Still we think that the observability inequality with the homogeneous Neumann condition has no complete solution.

An inequality reverse to (1.16) is very difficult, and we can refer to Avalos [1], Lasiecka and Triggiani [26] and Tataru [36]. By [26] and [36], we can prove

$$\|u\|^2_{H^\gamma(\partial\Omega \times (0,T))} \leq C(\|u(\cdot, 0)\|^2_{H^1(\Omega)} + \|u'(\cdot, 0)\|^2_{L^2(\Omega)})$$

where $0 < \gamma < 1$ depends on the geometry of Ω and the form of P. Since $\gamma < 1$, this inequality cannot be reverse to (1.16).

Now we give two cases where the conditions (1.5) - (1.9) are satisfied. See also [30] for more detailed results.

Corollary 1. *(Isakov and Yamamoto [16]) We further assume*

(1.18) $$\Gamma_0 \subset \{x = (x_1, ..., x_{n-1}, 0); x_1, ..., x_{n-1} \in \mathbb{R}\}.$$

Let Γ_0 and Γ be of class C^2 and we assume (1.2), (1.3), (1.14), and (1.15). We arbitrarily take

(1.19) $$x_0 = (x_0^{(1)}, ..., x_0^{(n-1)}, 0) \notin \overline{\Omega}.$$

Let $u \in H^2(\Omega \times (0,T))$ satisfy $Pu = 0$ and (1.17). If

(1.20) $$T > 2 \sup_{x \in \Omega} |x - x_0|,$$

then there exists a constant $C = C(\Omega, T, \Gamma, a_\alpha, x_0) > 0$ such that

(1.21) $$\|u(\cdot, 0)\|_{H^1(\Omega)}^2 + \|u'(\cdot, 0)\|_{L^2(\Omega)}^2 \le C \|u\|_{H^1(\Gamma \times (0,T))}^2.$$

Indeed, we can take $\beta \in (0,1)$ such that

$$T > \frac{2 \sup_{x \in \Omega} |x - x_0|}{\sqrt{\beta}},$$

which is possible by (1.20). We set $h_i(x_i) = (x_i - x_i^{(0)})^2$, $1 \le i \le n$, $h_{n+1}(t) = -\beta \left(t - \frac{T}{2}\right)^2$, namely, $\varphi(x,t) = |x - x_0|^2 - \beta \left(t - \frac{T}{2}\right)^2$. Then, by $0 < \beta < 1$, we can directly verify that φ satisfies (1.5) - (1.9) and (1.12), (1.13) with $c = 4 - 2\beta - 2n - \mu$ and $0 < \mu < 4 - 2\beta$, noting that $\nu_1 = \cdots = \nu_{n-1} = 0$ and $\nu_n = -1$ on Γ_0 by (1.18).

As another example of φ satisfying (1.5) - (1.9) and (1.12), (1.13), we can prove

Corollary 2. *Let a bounded domain $\Omega \subset \mathbb{R}^2$ have C^2-boundary $\partial \Omega$ and let $\partial \Omega = \overline{\Gamma_0 \cup \Gamma}$ and $\Gamma_0 \cap \Gamma = \emptyset$. Moreover let $\overline{\Gamma_0}$ be parametrized by $s \in I$: an interval in $\mathbb{R} \longrightarrow (x_1(s), x_2(s))$ such that the both maps $s \longrightarrow x_1(s)$ and $s \longrightarrow x_2(s)$ are one to one on I. Moreover let Γ_0 be a concave or convex curve: the segment $\overrightarrow{xy} \subset \overline{\Omega}$ for any $x, y \in \Gamma_0$ or $\overrightarrow{xy} \subset \mathbb{R}^2 \setminus \overline{\Omega}$ for any $x, y \in \Gamma_0$. Then there exists a constant $T_0 = T_0(\Omega, \Gamma) > 0$ such that if $T > T_0$, then for some constant $C = C(\Omega, T, \Gamma, a_\alpha) > 0$ the estimate (1.16) holds for a solution u to $Pu = 0$ with (1.17).*

Remark 2. We can prove a similar result to Corollary 2 in general dimensions. Our choice of T_0 may not be optimal in general. One can conjecture that T_0 should be optimal in the case where the pseudo-convex funtion is given by the squared Euclidean distance on a radial vector field, like in Corollary 1, but the optimality of T_0 is an open problem.

In the case of a disk $\Omega = \{x \in \mathbb{R}^2; |x| < \rho\}$, if Γ_0 is strictly smaller than the quarter of $\partial \Omega$, then Γ_0 is parametrized in one-to-one manner, that is, Corollary 2 is applicable. The characterization of Γ in Corollary 2 may be not the best possible. In fact, in the case of a disk, for example, we can expect that (1.16) should be true if Γ is strictly larger than the half circle. In the Dirichlet case: $u_{|\partial \Omega \times (0,T)} = 0$ in place of (1.17), we refer to the formula (3) (p.35) in Komornik [23], p.55 in Lions

[32], and the geometric optics condition by Bardos, Lebeau and Rauch [2], Burq [6].

The Lipschitz stability such as (1.16) is called an observability inequality, and is well studied in the Dirichlet case. We can refer to three methodologies.

Multiplier method.
Ho [8], Komornik [23], Lions [32], Triggiani [38].

Carleman estimate.
Cheng, Isakov, Yamamoto and Zhou [7], Kazemi and Klibanov [17], Klibanov and Malinsky [22], Lasiecka and Triggiani [27], Tataru [34], Zhang [40]. In particular, the papers [17] and [22] are motivating to our work, and solve the observability inequality in the Dirichlet case using an inequality in Lavrent'ev, Romanov and Shishat·skiĭ[31, p.124].

Microlocal analysis.
Bardos, Lebeau and Rauch [2] and Burq [6].

We further refer to Lasiecka, Triggiani and Yao [28], [29] about combination of a Carleman estimate with differential geometry.

On the other hand, the observability inequality is difficult for $\Gamma \subsetneq \partial\Omega$ in the Neumann case (1.17). For the present case, we refer to Isakov and Yamamoto [15], [16], Lasiecka and Triggiani [25]. In [15] and [25], different methods from our papar yield observability inequalities.

The proof of Theorem 2 is essentially based on Theorem 1, a Carleman estimate which is adjusted to the Neumann case (1.17). From the technical point of view, we can mention the advantage and the disadvantage of our present methodology:

Disadvantage. We have to assume restrictive conditions on geometry of $\Gamma \subsetneq \partial\Omega$ and the length of $T > 0$ for the observability inequality (1.16). Such restrictions on Γ and T are required for the existence of φ meeting (1.5) - (1.9) and (1.12) - (1.13). We do not know whether Theorem 2 can, in general, yield the best possible conditions for the observability in the Neumann case.

Advantage.
(1) It is sufficient to assume less regularity for the boundary $\partial\Omega$ and the coefficients a_α. The method by the microlocal analysis is, in principle, applicable to the Neumann case, but more smoothness of the boundary and the coefficients in (1.1) is necessary. In particular, L^γ-coefficient a_0 may make the application of microlocal analysis critical. The cited paper [2] gives a geometrical optics condition for the observability, and provides valuable heuristic arguments and suggests that probably exact controllability holds for any convex non-controllable part of the boundary. Unfortunately, the conditions of the main Theorems 3.3 and 3.4 in [2] regarding non-diffractive points of controllable part of the boundary, are too restrictive, most likely not necessary, and hard to be checked in many situation.
(2) We can discuss (1.1) with lower order terms. The classical multiplier method is very difficult for including the first order terms $a_\alpha D^\alpha u$ for $|\alpha| = 1$.

Moreover, thanks to the large parameter $\lambda > 0$ in (1.10), a Carleman estimate is useful for obtaining the stability and the uniqueness in inverse problems of determining a coefficient or a source term in a hyperbolic operator (1.1). In the case where we assume the Dirichlet data and the Neumann data on the whole boundary $\partial\Omega \times (0, T)$, we refer to Bukhgeim [3], [4], Bukhgeim and Klibanov [5], Imanuvilov and Yamamoto [10], Isakov [11] - [14], Khaĭdarov [18], [19], Klibanov [20], [21], Yamamoto [39]. When in the Neumann case (1.17) we discuss inverse problems of determining a_α in (1.1) from $u|_{\Gamma \times (0,T)}$, Theorem 1 is indispensable. This paper is motivated not only by an application to the observability inequality, but also by the application to inverse hyperbolic problems. The application to the inverse problems is discussed in Sections 5 and 6.

In this paper we set
$$\|u\|_2(Q) = \left(\int_Q u^2 \, dx dt\right)^{\frac{1}{2}}$$
for $Q \subset \mathbb{R}^n \times \mathbb{R}$.

§2. Proof of Theorem 1.

First Step. We will prove Theorem 1 for $Pu = \Box u \equiv u'' - \Delta u$. First we show

Lemma 1. *Let* $u = u(x,t) \in C^2(\overline{D})$ *and* $\varphi(x_1, ..., x_n, t) = \sum_{j=1}^n h_j(x_j) + h_{n+1}(t)$ *where* $h_j \in C^2(\mathbb{R})$, $1 \le j \le n+1$ *and let* $c \in \mathbb{R}$. *Then for any* $\lambda > 0$, *we have*

$$\begin{aligned}
& e^{2\lambda\varphi}|(\Box u)(x,t)|^2 \\
& \ge 2\lambda e^{2\lambda\varphi} \sum_{i=1}^n \{2(\partial_i^2 h_i)(x_i) + \Box\varphi - c\}|\partial_i u(x,t)|^2 \\
& + 2\lambda e^{2\lambda\varphi}(\Delta\varphi + \partial_t^2\varphi + c)|\partial_t u(x,t)|^2 \\
& + 4\lambda^3 e^{2\lambda\varphi}\left((-\Box\varphi + c)(|\nabla\varphi|^2 - |\partial_t\varphi|^2) + O\left(\frac{1}{\lambda}\right)\right)|u(x,t)|^2
\end{aligned}$$
(2.1)
$$+ \sum_{i=1}^n (\partial_i U_i)(x,t) + (\partial_t V)(x,t), \quad (x,t) \in D.$$

Here
$$\begin{aligned}
U_i(x,t) = & \; 2\lambda e^{2\lambda\varphi}\Big\{(\partial_i h_i)(x_i)(|\nabla u|^2 - |\partial_t u|^2) + R_1(x,t,u,\lambda)(\partial_i u)(x,t) \\
& + \{-2\lambda^2(\partial_i h_i)(x_i)(|\nabla\varphi|^2 - |\partial_t\varphi|^2) + 2\lambda(\partial_i h_i)(x_i)(\partial_i^2 h_i)(x_i)\}u^2 \Big\},
\end{aligned}$$
(2.2)
$$1 \le i \le n$$

and
$$\begin{aligned}
V(x,t) = & \; -2\lambda e^{2\lambda\varphi}\Big\{(\partial_t h_{n+1})(t)(|\nabla u|^2 - |\partial_t u|^2) + R_1(x,t,u,\lambda)(\partial_t u)(x,t) \\
& + \{-2\lambda^2(\partial_t h_{n+1})(t)(|\nabla\varphi|^2 - |\partial_t\varphi|^2) - 2\lambda(\partial_t h_{n+1})(t)(\partial_t^2 h_{n+1})(t)\}u^2 \Big\},
\end{aligned}$$
(2.3)

where

(2.4) $\quad R_1(x,t,u,\lambda) = -2(\nabla\varphi \cdot \nabla u) + 2(\partial_t\varphi)(\partial_t u) + (-2\lambda(|\nabla\varphi|^2 - |\partial_t\varphi|^2) + c)u.$

For the proof, we refer to Lemma 1 (p.124) in Lavrent'ev, Romanov and Shishat·skiĭ[31].

Let

(2.5) $\quad\quad\quad\quad\quad\quad\quad\quad\quad\quad u \in C^2(\overline{D}).$

Integrating (2.1) over D and applying (1.5) - (1.7) for large $\lambda > 0$, we have

$$\int_D e^{2\lambda\varphi}|\Box u(x,t)|^2 dx dt$$

$$\geq 2\lambda \left(\int_{Q(\delta)} + \int_{D\setminus Q(\delta)}\right) e^{2\lambda\varphi}\left\{\sum_{i=1}^n \{2(\partial_i^2 h_i)(x_i) + \Box\varphi - c\}|\partial_i u(x,t)|^2\right.$$

$$\left. + (\Delta\varphi + \partial_t^2\varphi + c)|\partial_t u(x,t)|^2 \right\} dx dt$$

$$+ 4\lambda^3 \left(\int_{Q(\delta)} + \int_{D\setminus Q(\delta)}\right)$$

$$e^{2\lambda\varphi}\left((-\Box\varphi + c)(|\nabla\varphi|^2 - |\partial_t\varphi|^2) + O\left(\frac{1}{\lambda}\right)\right)|u(x,t)|^2 dx dt$$

$$+ \sum_{i=1}^n \int_D (\partial_i U_i)(x,t) dx dt + \int_D (\partial_t V)(x,t) dx dt$$

$$\geq 2\lambda\mu \int_{Q(\delta)} e^{2\lambda\varphi}(|\nabla u(x,t)|^2 + |u'(x,t)|^2) dx dt$$

$$+ 2\lambda^3\mu \int_{Q(\delta)} e^{2\lambda\varphi}|u(x,t)|^2 dx dt$$

$$- M_1\lambda \int_{D\setminus Q(\delta)} e^{2\lambda\varphi}(|\nabla u|^2 + |u'|^2) dx dt - M_1\lambda^3 \int_{D\setminus Q(\delta)} e^{2\lambda\varphi}|u|^2 dx dt$$

(2.6)

$$+ \sum_{i=1}^n \int_{\partial D} U_i\nu_i d\sigma + \int_{\partial D\setminus(\partial\Omega\times(0,T))} V\nu_{n+1} d\sigma.$$

Here we have integrated by parts in the integrals over D. Here and henceforth $\nu = (\nu_1, ..., \nu_n, \nu_{n+1})$ is the unit outward normal vector to ∂D at (x,t). By (2.2),

(2.3) and (2.5), we see

$$\int_{\partial D} \sum_{i=1}^{n} U_i \nu_i d\sigma$$

$$= \int_{\partial D \cap (\partial \Omega \times (0,T))} \left\{ 2\lambda (\nabla \varphi \cdot \nu)(|\nabla u|^2 - |u'|^2) + 2\lambda R_1 \sum_{i=1}^{n} (\partial_i u_i) \nu_i \right.$$

$$\left. - \left(4\lambda^3 e^{2\lambda \varphi} (\nabla \varphi \cdot \nu)(|\nabla \varphi|^2 - |\partial_t \varphi|^2) - 4\lambda^2 \sum_{i=1}^{n} (\partial_i h_i)(\partial_i^2 h_i) \nu_i \right) \right\} e^{2\lambda \varphi} u^2 d\sigma$$

(2.7)

$$+ \int_{\partial D \setminus (\partial \Omega \times (0,T))} \sum_{i=1}^{n} U_i \nu_i d\sigma.$$

Hence, by (1.8) and (1.9), we have

$$\int_D \sum_{i=1}^{n} \partial_i U_i dx dt$$

$$\geq \int_{\partial D \cap (\Gamma \times (0,T))} \left\{ 2\lambda (\nabla \varphi \cdot \nu)(|\nabla u|^2 - |u'|^2) + 2\lambda R_1 \frac{\partial u}{\partial \nu} \right.$$

$$\left. -4\lambda^3 (\nabla \varphi \cdot \nu)(|\nabla \varphi|^2 - |\partial_t \varphi|^2) u^2 + 4\lambda^2 \sum_{i=1}^{n} (\partial_i h_i)(\partial_i^2 h_i) \nu_i u^2 \right\} e^{2\lambda \varphi} d\sigma$$

(2.8)

$$+ \int_{\partial D \setminus (\partial \Omega \times (0,T))} \sum_{i=1}^{n} U_i \nu_i d\sigma.$$

On the other hand, by (2.5), we have

(2.9) $$\int_D \partial_t V dx dt = \int_{\partial D \setminus (\partial \Omega \times (0,T))} V \nu_{n+1} d\sigma.$$

Using (2.8) and (2.9) in (2.6), we obtain

$$\int_D e^{2\lambda\varphi}|\Box u(x,t)|^2\,dx\,dt$$

$$+M_1\lambda\int_{D\setminus Q(\delta)} e^{2\lambda\varphi}(|\nabla u|^2+|u'|^2)\,dx\,dt$$

$$+M_1\lambda^3\int_{D\setminus Q(\delta)} e^{2\lambda\varphi}|u|^2\,dx\,dt$$

$$\geq 2\lambda\mu\int_{Q(\delta)} e^{2\lambda\varphi}(|\nabla u(x,t)|^2+|u'(x,t)|^2)\,dx\,dt$$

$$+2\lambda^3\mu\int_{Q(\delta)} e^{2\lambda\varphi}|u(x,t)|^2\,dx\,dt$$

$$+\int_{\partial D\cap(\Gamma\times(0,T))} 2\lambda e^{2\lambda\varphi}(\nabla\varphi\cdot\nu)(|\nabla u|^2-|u'|^2)\,d\sigma$$

$$+2\int_{\partial D\cap(\partial\Omega\times(0,T))} \lambda e^{2\lambda\varphi}R_1\frac{\partial u}{\partial\nu}\,d\sigma$$

$$-\int_{\partial D\cap(\Gamma\times(0,T))} 4\lambda^3 e^{2\lambda\varphi}(\nabla\varphi\cdot\nu)(|\nabla\varphi|^2-|\partial_t\varphi|^2)u^2\,d\sigma$$

$$+\int_{\partial D\cap(\Gamma\times(0,T))} 4\lambda^2 e^{2\lambda\varphi}\sum_{i=1}^n(\partial_i h_i)(\partial_i^2 h_i)\nu_i u^2\,d\sigma$$

(2.10) $$+\int_{\partial D\setminus(\partial\Omega\times(0,T))}\sum_{i=1}^n U_i\nu_i\,d\sigma+\int_{\partial D\setminus(\partial\Omega\times(0,T))} V\nu_{n+1}\,d\sigma$$

for $u\in C^2(\overline{D})$. Since $C^2(\overline{D})$ is dense in $H^2(D)$, we can see that (2.10) holds for all $u\in H^2(D)$.

Second Step. We can take a limit passage, and we prove (2.10) for u satisfying

(2.11) $$\begin{cases} \Box u\in L^2(D),\quad u\in H^1(D),\quad u\in H^1(\partial D) \\ u=|\nabla u|=u'=0 \quad\text{on}\ \partial D\setminus(\partial\Omega\times(0,T)). \end{cases}$$

In fact, let $u\in H^1(D)$ satisfy (2.11). Then, by a known result (e.g. Lemma 4.4 in Tataru [35]), we can choose $u_n\in H^2(D)$ such that

(2.12) $$\begin{cases} u_n\longrightarrow u\quad\text{in}\ H^1(D), \\ \Box u_n\longrightarrow \Box u\quad\text{in}\ L^2(D), \\ u_n\longrightarrow u\quad\text{in}\ H^1(\partial D). \end{cases}$$

Moreover by (2.11) and the third condition in (2.12), we have

$$\nabla u_n, u_n, u_n'\longrightarrow 0\quad\text{in}\ L^2(\partial D\setminus(\partial\Omega\times(0,T))).$$

Since (2.10) holds for u_n, $n\in\mathbb{N}$, noting (2.11), (2.2) and (2.3), we let n tend to ∞ in (2.10) with $u=u_n$ to obtain (1.10) for u satisfying (1.11) with $P=\Box$.

Third Step. Next we have to prove (1.10) for P given by (1.1) whose coefficients satisfy (1.2) and (1.3).

By (1.10) for $P = \Box$ and the triangle inequality, we obtain

$$\int_D e^{2\lambda\varphi}|\Box u|^2 dx dt$$
$$\leq \int_D e^{2\lambda\varphi}|Pu|^2 dx dt + C \sum_{|\alpha|=1} \int_D e^{2\lambda\varphi}|a_\alpha D^\alpha u|^2 dx dt$$
(2.13) $$+C \int_D e^{2\lambda\varphi}|a_0 u|^2 dx dt.$$

For $|\alpha| = 1$, we have

$$\int_D e^{2\lambda\varphi}|a_\alpha D^\alpha u|^2 dx dt \leq \|a_\alpha\|_{L^\infty(\Omega\times(0,T))}^2 \left(\int_{Q(\delta)} + \int_{D\setminus Q(\delta)}\right) e^{2\lambda\varphi}|D^\alpha u|^2 dx dt.$$

Next, for $|\alpha| = 0$ and $n \geq 2$, by Hölder's inequality, Sobolev embedding and (1.3), we see

$$\left(\int_D e^{2\lambda\varphi}|a_0 u|^2 dx dt\right)^{\frac{1}{2}} \leq \|a_0\|_{L^{n+1}(\Omega\times(0,T))} \|e^{\lambda\varphi}u\|_{L^{\frac{2n+2}{n-1}}(D)}$$
(2.14)
$$\leq C\|e^{\lambda\varphi}u\|_{H^1(D)},$$

so,

$$\int_D e^{2\lambda\varphi}|a_0 u|^2 dx dt \leq C^2 \Bigg\{ \lambda^2 \left(\int_{Q(\delta)} + \int_{D\setminus Q(\delta)}\right) e^{2\lambda\varphi}|u|^2 dx dt$$
$$+ \sum_{|\alpha|\leq 1} \left(\int_{Q(\delta)} + \int_{D\setminus Q(\delta)}\right) e^{2\lambda\varphi}|D^\alpha u|^2 dx dt \Bigg\}.$$

We substitute these inequalities in (2.13) and taking λ large, we absorb the second and the third integrals of the right side of (2.13) over $Q(\delta)$ into the left side. Thus the proof of Theorem 1 is complete.

§3. Proof of Theorem 2.
Let

(3.1) $$E(t) = \int_\Omega \{|u'(x,t)|^2 + |\nabla u(x,t)|^2 + (1 + c_1 - a_0(x,t))|u(x,t)|^2\} dx.$$

First we prove the energy estimate.

Lemma 2. We assume (1.2), (1.3) and (1.14), (1.15). Let u satisfy $Pu = 0$ in $\Omega \times (0,T)$ and (1.17). Then there exist constants $C_1 > 0$ and $C_2 > 0$ depending on a_α, $|\alpha| \leq 1$, Ω and T, such that

$$C_1 E(0) \leq E(t) \leq C_2 E(0), \qquad 0 \leq t \leq T.$$

Proof of Lemma 2. Multiplying $Pu = 0$ with u', integrating over Ω, in view of Green's formula and (1.17), we have

$$\frac{d}{dt}\int_\Omega \{|u'(x,t)|^2 + |\nabla u(x,t)|^2\}dx = 2\sum_{|\alpha|=1}\int_\Omega a_\alpha(D^\alpha u)u'\,dx + \int_\Omega a_0(u^2)'\,dx.$$

On the other hand,

$$\frac{d}{dt}\int_\Omega (1+c_1-a_0)u^2\,dx = -\int_\Omega a_0'u^2\,dx + \int_\Omega (1+c_1-a_0)(u^2)'\,dx.$$

Hence

(3.2) $$\frac{d}{dt}E(t) = 2\sum_{|\alpha|=1}\int_\Omega a_\alpha(D^\alpha u)u'\,dx - \int_\Omega a_0'u^2\,dx + \int_\Omega (1+c_1)(u^2)'\,dx.$$

Moreover, by (1.2) and Cauchy-Bunyakovskii's inequality, we have

(3.3) $$\left|\sum_{|\alpha|=1}\int_\Omega a_\alpha(D^\alpha u)u'\,dx\right| \leq C_3\int_\Omega \{|u'(x,t)|^2 + |\nabla u(x,t)|^2\}dx.$$

Furthermore, for $n \geq 3$, by (1.15), Sobolev's embedding, Cauchy-Bunyakovskii's inequality and Hölder's inequality, we see

(3.4) $$\left|-\int_\Omega a_0'u^2\,dx\right| \leq \|a_0'(\cdot,t)\|_{L^{\frac{n}{2}}(\Omega)}\|u^2(\cdot,t)\|_{L^{\frac{n}{n-2}}(\Omega)} \leq C_4\|u(\cdot,t)\|_{H^1(\Omega)}^2.$$

For $n = 2$, we can estimate similarly. Moreover we see

(3.5) $$\left|\int_\Omega (1+c_1)(u^2)'\,dx\right| \leq C_5\int_\Omega \{|u|^2 + |u'|^2\}dx \leq C_6 E(t),$$

because $1 + c_1 - a_0 > 0$. The combination of (3.1) - (3.3) yields

(3.6) $$\left|\frac{dE(t)}{dt}\right| \leq C_7 E(t), \quad 0 \leq t \leq T.$$

Therefore by Gronwall's inequality, we see that $E(t) \leq CE(0)$.
Exchanging t and 0, we complete the proof of Lemma 2.
Now we return to the proof of Theorem 2. In view of (1.12) and (1.13), we have

$$\varphi > 0 \quad \text{on } \overline{\Omega} \times \left\{\frac{T}{2}\right\}, \quad \varphi < 0 \quad \text{on } \overline{\Omega} \times \{0, T\}.$$

Therefore there exist small $\varepsilon > 0$ and $\varepsilon_1 > 0$ such that

(3.7) $$\varphi < -\varepsilon \quad \text{on } \overline{\Omega} \times ((0, 2\varepsilon_1) \cup (T - 2\varepsilon_1, T))$$

and

(3.8) $$\varphi > 2\varepsilon \quad \text{on } \overline{\Omega} \times \left(\frac{T}{2} - \varepsilon_1, \frac{T}{2} + \varepsilon_1\right).$$

To apply Theorem 1, we make use of a cut-off function χ: $0 \leq \chi \leq 1$. Let $\chi \in C^\infty(\mathbb{R})$ and

(3.9) $\chi = 0$ on $(0, \varepsilon_1) \cup (T - \varepsilon_1, T)$, $\chi = 1$ on $(2\varepsilon_1, T - 2\varepsilon_1)$.

We set

(3.10) $$v = \chi u.$$

In Theorem 1, we set $D = \Omega \times (0, T)$ and $\delta = \varepsilon$. By (3.9), we see that v satisfies (1.11). In view of (1.17), we note that

$$\frac{\partial v}{\partial \nu} = \chi \frac{\partial u}{\partial \nu} = 0 \quad \text{on } \partial\Omega \times (0, T).$$

Hence Theorem 1 yields

$$\sum_{|\alpha| \leq 1} \lambda^{3-2|\alpha|} \|e^{\lambda\varphi} D^\alpha v\|_2^2(Q(\varepsilon))$$

$$\leq M_1 \|e^{\lambda\varphi} Pv\|_2^2(\Omega \times (0,T)) + M_1 \sum_{|\alpha| \leq 1} \lambda^{3-2|\alpha|} \|e^{\lambda\varphi} D^\alpha v\|_2^2((\Omega \times (0,T)) \setminus Q(\varepsilon))$$

$$+ M_1 \int_{\Gamma \times (0,T)} \{\lambda e^{2\lambda\varphi}(\nabla\varphi \cdot \nu)(-|\nabla v|^2 + |v'|^2) + \lambda^3 e^{2\lambda\varphi}|v|^2\} d\sigma$$

$$\leq M_1 \|e^{\lambda\varphi} Pv\|_2^2(\Omega \times (0,T)) + M_1 \sum_{|\alpha| \leq 1} \lambda^{3-2|\alpha|} \|e^{\lambda\varphi} D^\alpha u\|_2^2((\Omega \times (0,T)) \setminus Q(\varepsilon))$$

(3.11)

$$+ M_1 \int_{\Gamma \times (0,T)} \{\lambda e^{2\lambda\varphi}(\nabla\varphi \cdot \nu)(-|\nabla u|^2 + |u'|^2) + e^{2\lambda\varphi}(\lambda|(\nabla\varphi \cdot \nu)| + \lambda^3)|u|^2\} d\sigma.$$

Moreover by $Pu = 0$, we obtain

(3.12) $$Pv = \chi Pu + A_1 u = A_1 u,$$

where

(3.13) $$A_1 u = 2\chi' u' + (\chi'' - a_1 \chi')u$$

and a_1 is the coefficient of the term $\partial_t u$ in P. Hence in terms of (3.9), (3.7) and Lemma 2, we have

$$\|e^{\lambda\varphi} Pv\|_2^2(\Omega \times (0,T)) = \int_{\Omega \times ((\varepsilon_1, 2\varepsilon_1) \cup (T - 2\varepsilon_1, T - \varepsilon_1))} |A_1 u|^2 e^{2\lambda\varphi} dx dt$$

$$\leq e^{-2\lambda\varepsilon} \int_{\Omega \times ((\varepsilon_1, 2\varepsilon_1) \cup (T - 2\varepsilon_1, T - \varepsilon_1))} |A_1 u|^2 dx dt$$

(3.14)

$$\leq C_8 e^{-2\lambda\varepsilon} \int_{\Omega \times (0,T)} (|u'|^2 + |u|^2) dx dt \leq C_9 e^{-2\lambda\varepsilon} E(0).$$

On the other hand, since $\varphi \leq \varepsilon$ in $(\Omega \times (0,T)) \setminus Q(\varepsilon)$, we have

$$\sum_{|\alpha|\leq 1} \lambda^{3-2|\alpha|} \|e^{\lambda\varphi} D^\alpha u\|_2^2 ((\Omega \times (0,T)) \setminus Q(\varepsilon))$$

$$\leq e^{2\lambda\varepsilon} \sum_{|\alpha|\leq 1} \lambda^{3-2|\alpha|} \|D^\alpha u\|_2^2 ((\Omega \times (0,T)) \setminus Q(\varepsilon))$$

(3.15)
$$\leq C_{10} e^{2\lambda\varepsilon} \lambda^3 E(0)$$

by Lemma 2. Furthermore (3.8) implies

$$Q(2\varepsilon) \supset \Omega \times \left(\frac{T}{2} - \varepsilon_1, \frac{T}{2} + \varepsilon_1\right),$$

and so

$$[\text{the left side of } (3.11)] \geq \sum_{|\alpha|\leq 1} \lambda^{3-2|\alpha|} \|e^{\lambda\varphi} D^\alpha u\|_2^2 (Q(2\varepsilon))$$

$$\geq M \sum_{|\alpha|\leq 1} \lambda^{3-2|\alpha|} \|e^{\lambda\varphi} D^\alpha u\|_2^2 \left(\Omega \times \left(\frac{T}{2} - \varepsilon_1, \frac{T}{2} + \varepsilon_1\right)\right)$$

$$\geq M\lambda e^{4\lambda\varepsilon} \sum_{|\alpha|\leq 1} \|D^\alpha u\|_2^2 \left(\Omega \times \left(\frac{T}{2} - \varepsilon_1, \frac{T}{2} + \varepsilon_1\right)\right)$$

(3.16)
$$\geq C_{11} \lambda \varepsilon_1 e^{4\lambda\varepsilon} E(0)$$

by $Q(2\varepsilon) \subset Q(\varepsilon)$, (3.8) and Lemma 2. Applying (3.14) - (3.16) in (3.11), we obtain

$$C_{11} \lambda \varepsilon_1 e^{4\lambda\varepsilon} E(0)$$
$$\leq C_9 e^{-2\lambda\varepsilon} E(0) + C_{10} e^{2\lambda\varepsilon} \lambda^3 E(0)$$
(3.17)
$$+ M_1 \int_{\Gamma \times (0,T)} \{\lambda e^{2\lambda\varphi} (\nabla\varphi \cdot \nu)(-|\nabla u|^2 + |u'|^2) + e^{2\lambda\varphi}(\lambda |(\nabla\varphi \cdot \nu)| + \lambda^3)|u|^2\} d\sigma.$$

Taking $\lambda > 0$ sufficiently large, we complete the proof of Theorem 2.

§4. Proof of Corollary 2.

We have to prove that there exist $h_1 = h_1(x_1)$, $h_2 = h_2(x_2)$ in $C^2(\mathbb{R})$ and $\beta > 0$, $T > 0$ satisfying (1.5) - (1.9) and (1.12) - (1.13).

We can assume that

(4.1)
$$\Omega \subset \{(x_1, x_2); |x_1| < \rho_1, 0 < x_2 < \rho_2\}$$

for some $\rho_1 > 0$, $\rho_2 > 0$. By a congruent transform, we may further assume that

(4.2)
$$\Gamma_0 = \{(x_1, f(x_1)); 0 < x_1 < a\}$$

where $0 < a < \rho_1$, and

(4.3) $\quad f \in C^2[0, a], \quad f(0) \equiv b_0 > 0, \quad f'(x_1) > 0, f''(x_1) \geq 0 \quad 0 \leq x_1 \leq a$

by the assumption for Γ_0. In this section, f' denotes the derivative with respect to the independent variable of f.

We set $b_1 = f(a)$. Since $f'(x_1) > 0$ and $f(x_1) > 0$, $0 \leq x_1 \leq a$, there exists an inverse function f^{-1} defined on $[b_0, b_1]$. For sufficiently large $\rho_3 > 0$ such that $\rho_3 > \rho_1$, we can take $\widetilde{f}$ defined on $[-\rho_3, \rho_3]$ such that

(4.4) $\quad \begin{cases} \widetilde{f}(x_1) = f(x_1), & 0 \leq x_1 \leq a \\ \widetilde{f}'(x_1) > 0, \widetilde{f}''(x_1) \geq 0, & -\rho_3 \leq x_1 \leq \rho_3 \\ \widetilde{f} \in C^2[-\rho_3, \rho_3], & \widetilde{f}(-\rho_3) \leq 0, \widetilde{f}(\rho_3) \geq \rho_2. \end{cases}$

In fact, we can extend f outside $[0, a]$ by a polynomial function, keeping it in the class C^2. Then

$$\widetilde{f} : x_1 \in [-\rho_3, \rho_3] \longrightarrow x_2 \in [\widetilde{f}(-\rho_3), \widetilde{f}(\rho_3)]$$

is one to one. We note that $[\widetilde{f}(-\rho_3), \widetilde{f}(\rho_3)] \supset [0, \rho_2]$. We choose $h_1 \in C^2[-\rho_3, \rho_3]$ such that

(4.5) $\quad h_1(x_1), h_1'(x_1), h_1''(x_1) > 0, \quad x \in [-\rho_3, \rho_3].$

We set

(4.6) $\quad h_2(x_2) = \int_{\widetilde{f}(-\rho_3)}^{x_2} \widetilde{f}'(\widetilde{f}^{-1}(s)) h_1'(\widetilde{f}^{-1}(s)) ds, \quad \widetilde{f}(-\rho_3) < x_2 < \widetilde{f}(\rho_3).$

Then, by (4.6), we have

$$h_2'(x_2) = \widetilde{f}'(\widetilde{f}^{-1}(x_2)) h_1'(\widetilde{f}^{-1}(x_2))$$

$$h_2''(x_2) = \widetilde{f}''(\widetilde{f}^{-1}(x_2))(\widetilde{f}^{-1})'(x_2) h_1'(\widetilde{f}^{-1}(x_2))$$
$$+ \widetilde{f}'(\widetilde{f}^{-1}(x_2)) h_1''(\widetilde{f}^{-1}(x_2))(\widetilde{f}^{-1})'(x_2), \quad \widetilde{f}(-\rho_3) \leq x_2 \leq \widetilde{f}(\rho_3),$$

and so

(4.7) $\quad h_2(x_2) \geq 0, \quad h_2'(x_2) > 0, \quad h_2''(x_2) > 0, \quad \widetilde{f}(-\rho_3) \leq x_2 \leq \widetilde{f}(\rho_3)$

and

(4.8) $\quad \begin{cases} h_2'(f(x_1)) = f'(x_1) h_1'(x_1), \\ h_2''(f(x_1)) = \dfrac{f''(x_1)}{f'(x_1)} h_1'(x_1) + h_1''(x_1), & 0 \leq x_2 \leq a. \end{cases}$

Moreover we choose $\beta > 0$ and $T > 0$ such that

(4.9)
$$0 < \beta < \min\left\{\frac{(h_1'(x_1))^2 + (h_2'(x_2))^2}{4(h_1(x_1) + h_2(x_2))}, \frac{1}{2}h_1''(x_1), \frac{1}{2}h_2''(x_2)\right\},$$
$$|x_1| \leq \rho_1, 0 \leq x_2 \leq \rho_2$$

and

(4.10) $$T > \frac{2}{\sqrt{\beta}}\sqrt{h_1(x_1) + h_2(x_2)}, \quad |x_1| \leq \rho_1, 0 \leq x_2 \leq \rho_2.$$

Then we will prove that these h_1, h_2, β and $T > 0$ satisfy (1.5) - (1.9) and (1.12) - (1.13). Noting that

(4.11) $$\begin{pmatrix} \nu_1 \\ \nu_2 \end{pmatrix} \text{ and } \begin{pmatrix} -f'(x_1) \\ 1 \end{pmatrix} \text{ are parallel on } \Gamma_0,$$

we see that (1.5) - (1.9), (1.12) and (1.13) are satisfied if

(4.12) $$h_1''(x_1) - h_2''(x_2) - 2\beta - c > 0, \quad |x_1| \leq \rho_1, 0 \leq x_2 \leq \rho_2,$$

(4.13) $$h_2''(x_2) - h_1''(x_1) - 2\beta - c > 0, \quad |x_1| \leq \rho_1, 0 \leq x_2 \leq \rho_2,$$

(4.14) $$h_1''(x_1) + h_2''(x_2) - 2\beta + c > 0, \quad |x_1| \leq \rho_1, 0 \leq x_2 \leq \rho_2,$$

$$(h_1''(x_1) + h_2''(x_2) + 2\beta + c)\left((h_1'(x_1))^2 + (h_2'(x_2))^2 - 4\beta^2\left(t - \frac{T}{2}\right)^2\right) > 0,$$

(4.15)
for $|x_1| \leq \rho_1, 0 \leq x_2 \leq \rho_2, 0 < t < T$, $h_1(x_1) + h_2(x_2) - \beta\left(t - \frac{T}{2}\right)^2 > 0$,

(4.16) $$h_2'(f_1(x_1)) = f'(x_1)h_1'(x_1), \quad 0 \leq x_1 \leq a,$$

(4.17) $$-h_1'(x_1)h_1''(x_1)f'(x_1) + h_2'(f(x_1))h_2''(f(x_1)) \geq 0, \quad 0 \leq x_1 \leq a,$$

(4.18) $$h_1(x_1) + h_2(x_2) - \frac{\beta}{4}T^2 < 0, \quad |x_1| \leq \rho_1, 0 \leq x_2 \leq \rho_2$$

and

(4.19) $$h_1(x_1) + h_2(x_2) > 0, \quad |x_1| \leq \rho_1, 0 \leq x_2 \leq \rho_2.$$

First by (4.5) - (4.8) and (4.10), we verify that (4.18) and (4.19) are satisfied. Second we can take $c \in \mathbb{R}$ satisfying (4.12) - (4.14). In fact, in view of (4.9), we have
$$h_1''(x_1) + h_2''(x_2) - 2\beta > |h_1''(x_1) - h_2''(x_2)| + 2\beta,$$
for $|x_1| \leq \rho_1$ and $0 \leq x_2 \leq \rho_2$. Therefore there exists a constant $c \in \mathbb{R}$ such that
$$h_1''(x_1) + h_2''(x_2) - 2\beta > -c > |h_1''(x_1) - h_2''(x_2)| + 2\beta$$
for $|x_1| \leq \rho_1$ and $0 \leq x_2 \leq \rho_2$. The condition (4.14) is now straightforward. Moreover $-c > |h_1''(x_1) - h_2''(x_2)| + 2\beta$ implies
$$h_1''(x_1) - h_2''(x_2) - c - 2\beta \geq -|h_1''(x_1) - h_2''(x_2)| - c - 2\beta > 0$$
for $|x_1| \leq \rho_1$ and $0 \leq x_2 \leq \rho_2$, which is (4.12). Similarly (4.13) follows. The condition (4.15) is seen from $\beta > 0$, (4.9) and (4.14).

On the other hand, (4.16) is nothing but (4.8). Finally we have to verify (4.17). Since $f''(x_1) \geq 0$, $0 \leq x_1 \leq a$ by (4.3), we have
$$-h_1'(x_1)h_1''(x_1)f'(x_1) + h_2'(f(x_1))h_2''(f(x_1)) = f''(x_1)(h_1'(x_1))^2 \geq 0,$$
in view of (4.8). Thus setting $T_0 = T$ given by (4.10), we apply Theorem 2 to complete the proof of Corollary 2.

§5. Application of the Carleman estimate to inverse hyperbolic problems.

Let $\Omega \subset \mathbb{R}^n$, $n \geq 2$, be a bounded domain and its boundary $\partial\Omega$ be of class C^2. We consider two systems:

(5.1)
$$\begin{cases} u''(x,t) = \Delta u(x,t) - p(x)u(x,t), & x \in \Omega, \ 0 < t < T \\ u(x, T/2) = a(x), \quad u'(x, T/2) = g(x), & x \in \Omega \\ \dfrac{\partial u}{\partial \nu}(x,t) = \xi(x,t), & x \in \partial\Omega, \ 0 < t < T. \end{cases}$$

(5.2)
$$\begin{cases} y''(x,t) = \Delta y(x,t) - p(x)y(x,t) + f(x)R(x,t), & x \in \Omega, \ 0 < t < T \\ y(x, T/2) = 0, \quad y'(x, T/2) = g(x), & x \in \Omega \\ \dfrac{\partial y}{\partial \nu}(x,t) = 0, & x \in \partial\Omega, \ 0 < t < T. \end{cases}$$

Here $a \in H^1(\Omega)$, $\xi \in L^2(\partial\Omega \times (0,T))$, R are fixed suitably, and in (5.2) also $p \in L^\infty(\Omega)$ is given. In (5.1) p and g are unknown while f and g are unknown in (5.2).

Remark 3. In (5.1) and (5.2), we regard $t = T/2$ as an initial moment. This is not essential, because the change of independent variables $t \longrightarrow t - \frac{T}{2}$ transforms $t = \frac{T}{2}$ to $t = 0$.

Remark 4. In (5.1) and (5.2) we can include more general lower order terms like in (1.1) under suitable regularity on coefficients. For brevity, we here consider only a term of the zeroth order with t-independent L^∞-coefficient p.

In this section, we discuss the uniqueness in multidimensional inverse hyperbolic problems. More precisely,

Uniqueness in determining coefficients.
Let $\Gamma \subset \partial\Omega$. Let $u = u(p,g)$ satisfy (5.1) in the sense below. Does $u_{|\Gamma \times (0,T)}$ determine p uniquely? In other words, does

$$u(q,h)_{|\Gamma \times (0,T)} = u(p,g)_{|\Gamma \times (0,T)}$$

imply $p(x) = q(x)$, $g(x) = h(x)$, $x \in \Omega$?

Uniqueness in determining source terms.
Let $\Gamma \subset \partial\Omega$. Let $y = y(f,g)$ satisfy (5.2) in the sense below. Does $y_{|\Gamma \times (0,T)}$ determine f uniquely? In other words, does

$$y(f,g)_{|\Gamma \times (0,T)} = 0$$

imply $f(x) = 0$ and $g(x) = 0$, $x \in \Omega$?

For the uniqueness in multidimensional inverse problems with a single observation, Bukhgeim [3], [4], Bukhgeim and Klibanov [5] proposed a useful methodology on the basis of Carleman estimates. As papers discussing inverse problems by Carleman estimates, we can refer to Isakov [11] - [14], Khaĭdarov [18], [19], Klibanov [20], [21], Kubo [24]. Except for Kubo [24], Dirichlet data and Neumann data on the whole $\partial\Omega \times (0,T)$ are required for the uniqueness. Kubo [24] is most related with this section among the above mentioned papers, but as $\partial\Omega \setminus \Gamma$, the flat boundary is exclusively considered like in Corollary 1. Although it is remarked in [24] that the uniqueness can be proved for observations on a more general subboundary $\Gamma \subsetneq \partial\Omega$, the characterization for such subboundary Γ is not clear so far from the viewpoint of the method in [24].

For the stability in our inverse problems, we refer to Yamamoto [39] in the case of the Dirichlet boundary condition $u_{|\partial\Omega \times (0,T)} = 0$ and Imanuvilov and Yamamoto [10] in the case of the Neumann boundary condition $\frac{\partial u}{\partial \nu}|_{\partial\Omega \times (0,T)} = 0$ by means of the observations $u_{|\partial\Omega \times (0,T)}$ on the whole lateral boundary $\partial\Omega \times (0,T)$.

Now we are ready to state the uniqueness results for the inverse problems.

Theorem 3. *(Uniqueness in determining source terms in (5.2))* Let $h_1, ..., h_n \in C^2(\mathbb{R})$ and

(5.3) $$h_{n+1}(t) = -\beta \left(t - \frac{T}{2}\right)^2$$

satisfy (1.5) - (1.9) and (1.12). Moreover let

$$R \in W^{3,\infty}(\Omega \times (0,T)), \quad \frac{\partial R}{\partial \nu}(x,t) = 0, \quad x \in \partial\Omega, \ 0 < t < T,$$

(5.4) $$|R(x,T/2)| \geq r_0 > 0, \quad x \in \overline{\Omega}$$

with some constant $r_0 > 0$,

(5.5) $$p \in L^\infty(\Omega)$$

and

(5.6) $$T > \frac{2}{\sqrt{\beta}} \max_{x \in \overline{\Omega}} (h_1(x_1) + \cdots + h_n(x_n))^{\frac{1}{2}}.$$

If the weak solution $y = y(f,g)$ to (5.2) satisfies

(5.7) $$y(f,g), y(f,g)' \in H^2(\Omega \times (0,T))$$

and

(5.8) $$y(f,g)(x,t) = 0, \qquad x \in \Gamma, \ 0 < t < T,$$

then $f(x) = 0$ and $g(x) = 0$, $x \in \Omega$.

Theorem 4. *(Uniqueness in determining coefficients in (5.1)) We assume (1.5) - (1.9), (1.12), (5.6) and*
$$q, p \in L^\infty(\Omega).$$

Let

(5.9) $$u(q,h) - u(p,g), u(q,h)' - u(p,g)' \in H^2(\Omega \times (0,T))$$

and either of $u(q,h)$ and $u(p,g)$ satisfy

(5.10) $$u \in W^{3,\infty}(\Omega \times (0,T)).$$

Moreover let

(5.11) $$|a(x)| \geq a_0 > 0, \quad x \in \overline{\Omega}$$

with some constant $a_0 > 0$. If
$$u(q,h)(x,t) = u(p,g)(x,t), \qquad x \in \Gamma, \ 0 < t < T,$$

then $p(x) = q(x)$, $g(x) = h(x)$, $x \in \Omega$.

Remark 5. In view of (5.4) and (5.10), in order that the regularity conditions (5.7) and (5.9) are satisfied, it is necessary that $f \in H^1(\Omega)$ and $p - q \in H^1(\Omega)$ respectively (e.g. Lions and Magenes [33]).

Theorem 4 follows from Theorem 3. In fact, without loss of generality, we may assume that $u(q,h) \in W^{3,\infty}(\Omega \times (0,T))$. Setting $y = u(q,h) - u(p,g)$, $R = u(q,h)$ and $f = p - q$ and denoting $h - g$ again by g, we obtain (5.2). Since $R(x,T/2) = u(p,g)(x,T/2) = a(x)$, $x \in \Omega$, the conditions (5.10) and (5.11) imply (5.4). By $u(q,h)(x,t) = u(p,g)(x,t)$, $x \in \Gamma$, $0 < t < T$, the conclusion $q = p$ and $g = h$ in Ω follows from Theorem 3. Thus it is sufficient to prove Theorem 3.

Proof of Theorem 3.

First Step. For fixed $0 < \eta < \delta$, we define a function $\kappa = \kappa_{\delta,\eta}(t)$ such that

(5.12) $$\kappa \in C_0^\infty(\mathbb{R}), \quad 0 \leq \kappa(t) \leq 1, \, t \in \mathbb{R}$$

and

(5.13) $$\kappa(t) = \begin{cases} 1, & |t - T/2| \leq \sqrt{\delta^2 - \eta^2} \\ 0, & |t - T/2| > \delta. \end{cases}$$

Moreover we define $\widetilde{R}(x,t)$, $x \in \Omega$, $0 < t < T$ by

(5.14) $$\widetilde{R}(x,t) = R(x, T/2) + \kappa(t)(R(x,t) - R(x, T/2)), \quad x \in \Omega, \, 0 < t < T.$$

Then by (5.14) we see
$$|\widetilde{R}(x,t) - R(x, T/2)| = |\kappa(t)(R(x,t) - R(x, T/2))|$$
$$\leq C\delta \|R\|_{C^1(\overline{\Omega} \times [0,T])}, \quad x \in \overline{\Omega}, \, |t - T/2| \leq \delta.$$

Therefore by the assumption (5.4), we can choose sufficiently small $\delta > 0$ such that

(5.15) $$\delta = \frac{T}{2\sqrt{m}} \quad \text{for some } m \in \mathbb{N}$$

and

(5.16) $$R(x,t) \neq 0, \quad \widetilde{R}(x,t) \neq 0, \quad x \in \overline{\Omega}, \, |t - T/2| \leq \delta.$$

Moreover by (5.12) - (5.14), we have

(5.17) $$\widetilde{R}(x,t) = \begin{cases} R(x,t), & x \in \overline{\Omega}, \, |t - T/2| \leq \sqrt{\delta^2 - \eta^2} \\ R(x, T/2), & x \in \overline{\Omega}, \, |t - T/2| > \delta. \end{cases}$$

Second Step. We set

(5.18) $$z(x,t) = \frac{y(x,t)}{\widetilde{R}(x,t)}, \quad x \in \Omega, \, |t - T/2| \leq \delta.$$

The function z is well-defined due to (5.6), (5.7) and (5.16), and

(5.19) $$z', z \in H^2(\Omega \times (0,T))$$

by (5.7) and (5.4). Let us define a differential operator A_1 of the first order in (x,t) by

(5.20) $$(A_1 v)(x,t) = 2 \frac{\nabla R(x,t)}{R(x,t)} \cdot \nabla v(x,t) - 2 \frac{R'(x,t)}{R(x,t)} v'(x,t)$$
$$+ \left(\frac{\Delta R(x,t)}{R(x,t)} - q(x) - \frac{R''(x,t)}{R(x,t)} \right) v(x,t), \quad x \in \Omega, \, |t - T/2| \leq \delta.$$

Its coefficients are all in $W^{1,\infty}(\Omega \times (-\delta+T/2, \delta+T/2))$ by (5.4). Since $\frac{\partial R}{\partial \nu}|_{\partial\Omega \times (0,T)} = 0$ from (5.4), direct calculations show

(5.21) $$z''(x,t) = \Delta z(x,t) + (A_1 z)(x,t) + f(x), \quad x \in \Omega, |t - T/2| < \delta$$

(5.22) $$z(x, T/2) = 0, \quad x \in \Omega$$

(5.23) $$\frac{\partial z}{\partial \nu}(x,t) = 0, \quad x \in \partial\Omega, |t - T/2| < \delta$$

and

(5.24) $$z(x,t) = 0, \quad x \in \Gamma, |t - T/2| < \delta.$$

Furthermore set

(5.25) $$w(x,t) = z'(x,t), \quad x \in \Omega, |t - T/2| < \delta.$$

Then

(5.26) $$w \in H^2\left(\Omega \times \left(-\delta + \frac{T}{2}, \delta + \frac{T}{2}\right)\right)$$

by (5.19). Since $z(x,t) = \int_{T/2}^{t} w(x,s)dx$, $x \in \Omega$, $|t - T/2| \leq \delta$ by (5.22), we have

(5.27) $$(\Box w)(x,t) = (A_1 w)(x,t) + \frac{\partial A_1}{\partial t}\left(\int_{T/2}^{t} w(x,s)ds\right),$$

(5.28) $$\frac{\partial w}{\partial \nu}(x,t) = 0, \quad x \in \partial\Omega, |t - T/2| < \delta$$

and

(5.29) $$w(x,t) = 0, \quad x \in \Gamma, |t - T/2| < \delta.$$

Here we note that

(5.30) $$\left(\frac{\partial A_1}{\partial t}v\right)(x,t) = 2\left(\frac{\nabla R(x,t)}{R(x,t)}\right)' \cdot \nabla v(x,t) - 2\left(\frac{R'(x,t)}{R(x,t)}\right)' v'(x,t)$$
$$+ \left(\frac{\Delta R(x,t)}{R(x,t)} - \frac{R''(x,t)}{R(x,t)}\right)' v(x,t), \quad x \in \Omega, |t - T/2| \leq \delta.$$

Third Step. Let us set

(5.31) $$\rho = \max_{x \in \overline{\Omega}}(h_1(x_1) + \cdots + h_n(x_n))^{\frac{1}{2}}.$$

By (1.12) we note that $h_1(x_1) + \cdots + h_n(x_n) > 0$ for $x \in \overline{\Omega}$. Moreover we set

(5.32) $$c(\varepsilon) = \rho^2 - \beta\delta^2 + \varepsilon^2$$

for sufficiently small $\varepsilon > 0$. We define a cut-off function $\chi \in C^\infty(\overline{\Omega} \times [0,T])$ such that

(5.33) $$0 \leq \chi(x,t) \leq 1, \quad x \in \overline{\Omega}, 0 \leq t \leq T$$

and

(5.34) $$\chi(x,t) = \begin{cases} 1, & (x,t) \in Q(c(3\varepsilon)) \\ 0, & (x,t) \in Q(c(\varepsilon)) \setminus Q(c(2\varepsilon)). \end{cases}$$

We note by the definition of $Q(c(\varepsilon))$ and (5.32) that

(5.35) $$Q(c(\varepsilon)) \subset \Omega \times \left(\frac{T}{2} - \delta, \frac{T}{2} + \delta\right).$$

In this step, we will apply Theorem 1 with $D = Q(c(\varepsilon)) = Q(\delta)$ and $a_\alpha = 0$ for $|\alpha| \leq 1$ to

(5.36) $$v = \chi w \quad \text{in } Q(c(\varepsilon)).$$

First by (5.26) we have

(5.37) $$v \in H^2\left(\Omega \times \left(\frac{T}{2} - \delta, \frac{T}{2} + \delta\right)\right).$$

Next, noting (5.34) and $\frac{\partial v}{\partial \nu} = \frac{\partial \chi}{\partial \nu}w$, we have

(5.38) $$v = 0 \quad \text{on } \Gamma \times \left(-\delta + \frac{T}{2}, \delta + \frac{T}{2}\right)$$

(5.39) $$\frac{\partial v}{\partial \nu} = 0 \quad \text{on } \left(\partial\Omega \times \left(-\delta + \frac{T}{2}, \delta + \frac{T}{2}\right)\right)$$
$$\cap \left(Q(c(3\varepsilon)) \cup \{Q(c(\varepsilon)) \setminus Q(c(2\varepsilon))\}\right).$$

By the Leibniz formula,

$$\Box v = \chi(\Box w) + A_2 w \quad \text{in } Q(c(\varepsilon))$$

where A_2 is a differential operator of the first order in (x,t) whose coefficients are in $L^\infty(Q(c(\varepsilon)))$. Therefore by (5.27) we obtain

$$(\Box v)(x,t) = \chi(x,t)(A_1 w)(x,t) + \chi(x,t)A_1'\left(\int_{T/2}^t w(x,s)ds\right) + (A_2 w)(x,t),$$

in $Q(c(\varepsilon))$.

Therefore $\Box w \in L^2(Q(c(\varepsilon)))$ and we can take a constant $M > 0$ such that

(5.40)
$$|(\Box v)(x,t)|^2 \leq M(|\nabla w(x,t)|^2 + |w'(x,t)|^2 + |w(x,t)|^2)$$
$$+ M\left|\int_{T/2}^t (|\nabla w(x,s)|^2 + |w(x,s)|^2)ds\right|, \quad (x,t) \in Q(c(\varepsilon)).$$

Here and henceforth $M > 0$ denotes a generic constant which is independent of $\lambda > 0$.

We apply Theorem 1 to v, and from (5.40), in view of (5.35), (5.38) and (5.39), we obtain

(5.41)
$$\int_{Q(c(\varepsilon))} e^{2\lambda\varphi(x,t)}(|\nabla w(x,t)|^2 + |w'(x,t)|^2 + |w(x,t)|^2)dxdt$$
$$+ \int_{Q(c(\varepsilon))} e^{2\lambda\varphi(x,t)}\left|\int_{T/2}^t (|\nabla w(x,s)|^2 + |w(x,s)|^2)ds\right|dxdt$$
$$+ \int_{(\partial\Omega\times(-\delta+T/2,\delta+T/2))\cap(\overline{Q(c(2\varepsilon))\setminus Q(c(3\varepsilon))})} \lambda e^{2\lambda\varphi}(|\nabla v| + |v'| + \lambda|v|)\left|\frac{\partial v}{\partial \nu}\right|d\sigma$$
$$\geq M^{-1}\lambda \int_{Q(c(\varepsilon))} e^{2\lambda\varphi(x,t)}(|\nabla v(x,t)|^2 + |v'(x,t)|^2 + |v(x,t)|^2)dxdt.$$

Splitting $Q(c(\varepsilon))$ into $Q(c(3\varepsilon))$ and its complement, we conclude

[the left hand side of (5.41)]

(5.42)
$$\leq \int_{Q(c(3\varepsilon))} e^{2\lambda\varphi(x,t)}(|\nabla w(x,t)|^2 + |w'(x,t)|^2 + |w(x,t)|^2)dxdt$$
$$+ \int_{Q(c(3\varepsilon))} e^{2\lambda\varphi(x,t)}\left|\int_{T/2}^t (|\nabla w(x,s)|^2 + |w(x,s)|^2)ds\right|dxdt$$
$$+ M_1 e^{2\lambda c(3\varepsilon)} + M_1 \lambda^2 e^{2\lambda c(3\varepsilon)},$$

because, by (5.26) and (5.37), we see that

$$\int_{Q(c(\varepsilon))\setminus Q(c(3\varepsilon))} (|\nabla w(x,t)|^2 + |w'(x,t)|^2 + |w(x,t)|^2)dxdt$$
$$+ \int_{Q(c(\varepsilon))\setminus Q(c(3\varepsilon))} \left|\int_{T/2}^t (|\nabla w(x,s)|^2 + |w(x,s)|^2)ds\right|dxdt \leq M_1$$

and

$$\int_{(\partial\Omega\times(-\delta+T/2,\delta+T/2))\cap(Q(c(2\varepsilon))\setminus Q(c(3\varepsilon)))} (|\nabla v|+|v'|+\lambda|v|)\left|\frac{\partial v}{\partial \nu}\right|d\sigma \le M_1$$

(are bounded uniformly in $\lambda > 0$). On the other hand, we have a lemma which can be easily proved but is essential for the application of the Carleman estimate to an inverse problem:

Lemma 3.

$$\int_{Q(c)} e^{2\lambda\varphi(x,t)}\left|\int_{T/2}^{t}|g(x,s)|^2 ds\right| dx\, dt \le M_2 \int_{Q(c)} |g(x,t)|^2 e^{2\lambda\varphi(x,t)} dx\, dt$$

for $g \in L^2(Q(c))$.

Noting that

$$\left(t-\frac{T}{2}\right)\frac{\partial\varphi}{\partial t}(x,t) = -2\beta\left(t-\frac{T}{2}\right)^2 \le 0, \quad x \in \overline{\Omega}, \, 0 \le t \le T,$$

we can prove the lemma by the same way as in Isakov [13], Klibanov [21], for instance.

Application of Lemma 3 to (5.42) yields

[the left hand side of (5.41)]

$$\le M \int_{Q(c(3\varepsilon))} e^{2\lambda\varphi(x,t)}(|\nabla w(x,t)|^2 + |w'(x,t)|^2 + |w(x,t)|^2) dx\, dt$$

(5.43)

$$+ M_1(1+\lambda^2)e^{2\lambda c(3\varepsilon)}.$$

On the other hand, by (5.34) we have

[the right hand side of (5.41)]

(5.44)

$$\ge M^{-1}\lambda \int_{Q(c(3\varepsilon))} e^{2\lambda\varphi(x,t)}(|\nabla w(x,t)|^2 + |w'(x,t)|^2 + |w(x,t)|^2) dx\, dt.$$

Therefore (5.43) and (5.44) imply

$$\lambda \int_{Q(c(3\varepsilon))} e^{2\lambda\varphi(x,t)}(|\nabla w(x,t)|^2 + |w'(x,t)|^2 + |w(x,t)|^2) dx\, dt \le M_3(1+\lambda^2)e^{2\lambda c(3\varepsilon)},$$

for large $\lambda > 0$. Noting that $Q(c(4\varepsilon)) \subset Q(c(3\varepsilon))$, we have

$$\lambda \int_{Q(c(4\varepsilon))} e^{2\lambda\varphi(x,t)}(|\nabla w(x,t)|^2 + |w'(x,t)|^2 + |w(x,t)|^2) dx\, dt$$

$$\le \lambda \int_{Q(c(3\varepsilon))} e^{2\lambda\varphi(x,t)}(|\nabla w(x,t)|^2 + |w'(x,t)|^2 + |w(x,t)|^2) dx\, dt.$$

Therefore

$$\lambda \int_{Q(c(4\varepsilon))} e^{2\lambda\varphi(x,t)}(|\nabla w(x,t)|^2 + |w'(x,t)|^2 + |w(x,t)|^2)dxdt \leq M(1+\lambda^2)e^{2\lambda c(3\varepsilon)}.$$

Hence since $\varphi > c(4\varepsilon)$ in $Q(c(4\varepsilon))$, we obtain

$$\lambda e^{2\lambda c(4\varepsilon)} \int_{Q(c(4\varepsilon))} (|\nabla w(x,t)|^2 + |w'(x,t)|^2 + |w(x,t)|^2)dxdt \leq M(1+\lambda^2)e^{2\lambda c(3\varepsilon)}$$

for large $\lambda > 0$. After dividing the both hand sides by $\lambda e^{2\lambda c(4\varepsilon)}$, noting that $c(3\varepsilon) - c(4\varepsilon) < 0$, we let $\lambda > 0$ tend to ∞ to obtain $w(x,t) = 0$, $(x,t) \in Q(c(4\varepsilon))$. By (5.41), (5.36), (5.25) and (5.18) we obtain $y(x,t) = 0$, $(x,t) \in Q(c(4\varepsilon))$. Therefore (5.2) implies $f(x)R(x,t) = 0$ and $g(x) = 0$, $(x,t) \in Q(c(4\varepsilon))$. By (5.35) and (5.16), we see that $R(x,t) \neq 0$, $(x,t) \in Q(c(4\varepsilon))$, and so we obtain $f(x) = g(x) = 0$ in $(x,t) \in Q(c(4\varepsilon))$, namely,

(5.45) $\quad f(x) = g(x) = 0 \quad$ in $\{x \in \Omega; h_1(x_1) + \cdots + h_n(x_n) \geq \rho^2 - \beta\delta^2\}$,

because $\varepsilon > 0$ is arbitrary.

Fourth Step. Henceforth we set

$$\Omega(k) = \{x \in \Omega; h_1(x_1) + \cdots + h_n(x_n) \geq \rho^2 - k\beta\delta^2\}$$

for $k \in \mathbb{N} \cup \{0\}$. Since $h_1(x_1) + \cdots + h_n(x_n) > 0$ for $x \in \overline{\Omega}$, we see that

(5.46) $\quad\quad\quad\quad\quad \Omega(k) = \Omega \quad$ if $\rho^2 - k\beta\delta^2 \leq 0$.

By (5.45) we have

(5.47) $\quad\quad y''(x,t) = \Delta y(x,t) - p(x)y(x,t), \quad x \in \Omega(1), 0 < t < T$

(5.48) $\quad\quad\quad\quad\quad y(x,0) = 0, \quad\quad x \in \Omega$

(5.49) $\quad\quad\quad\quad\quad \dfrac{\partial y}{\partial \nu}(x,t) = 0, \quad\quad x \in \partial\Omega, 0 < t < T$

and

(5.50) $\quad\quad\quad\quad\quad y(x,t) = 0, \quad\quad x \in \Gamma, 0 < t < T.$

In this step, we will prove

(5.51) $\quad\quad\quad\quad\quad f(x) = g(x) = 0, \quad x \in \Omega(2).$

We choose $\eta > 0$ such that $0 < \eta < \delta$. Here δ satisfies (5.15) and (5.16). Henceforth we set

(5.52) $$c_1(\eta) = \rho^2 - 2\beta\delta^2 + \beta\eta^2.$$

We recall that $\widetilde{R}$ is defined by (5.14). Then we can see

(5.53) $$\widetilde{R}(x,t)f(x) = \begin{cases} 0, & (x,t) \in Q(c_1(\eta)) \cap (\Omega(1) \times \mathbb{R}) \\ R(x,t)f(x), & (x,t) \in Q(c_1(\eta)) \cap ((\Omega \setminus \Omega(1)) \times \mathbb{R}). \end{cases}$$

In fact, as direct calculations show, if $(x,t) \in Q(c_1(\eta)) \cap ((\Omega \setminus \Omega(1)) \times \mathbb{R})$, then $|t - T/2| \le \sqrt{\delta^2 - \eta^2}$, and so (5.17) implies that $\widetilde{R}(x,t)f(x) = R(x,t)f(x)$ for $(x,t) \in Q(c_1(\eta)) \cap ((\Omega \setminus \Omega(1)) \times \mathbb{R})$. Moreover from (5.45), we can directly see that $\widetilde{R}(x,t)f(x) = 0$ if $(x,t) \in Q(c_1(\eta)) \cap (\Omega(1) \times \mathbb{R})$.

Furthermore

(5.54) $$\widetilde{R}(x,t) \ne 0, \quad (x,t) \in \overline{Q(c_1(\eta))}$$

by (5.16) and (5.17).

Therefore (5.47) and (5.53) yield

(5.55) $$y''(x,t) - \Delta y(x,t) + p(x)y(x,t) = \widetilde{R}(x,t)f(x), \quad (x,t) \in Q(c_1(\eta)).$$

Here since $Q(c_1(\eta)) \subset \Omega \times (\frac{T}{2} - \sqrt{2}\delta, \frac{T}{2} + \sqrt{2}\delta)$, we have

$$\partial Q(c_1(\eta)) \cap (\partial\Omega \times (0,T)) \subset \partial\Omega \times \left(\frac{T}{2} - \sqrt{2}\delta, \frac{T}{2} + \sqrt{2}\delta\right),$$

so that

$$\frac{\partial y}{\partial \nu}(x,t) = 0, \quad (x,t) \in \partial Q(c_1(\eta)) \cap (\partial\Omega \times (0,T)),$$

and

$$y(x,t) = 0, \quad (x,t) \in \partial Q(c_1(\eta)) \cap (\Gamma \times (0,T))$$

because $\delta > 0$ is sufficiently small. Therefore in view of (5.54), we can repeat the argument in Third Step to the system (5.55) with (5.48) - (5.50). Hence, since we can take an arbitrary small $\eta > 0$ provided that η satisfies $0 < \eta < \delta$, we obtain (5.51).

Fifth Step. We will complete the proof of Theorem 3. Repeating k-times the argument in Fourth Step, we see that

(5.56) $$f(x) = g(x) = 0, \quad x \in \Omega(k).$$

We have

$$Q(\rho^2 - k\beta\delta^2) \subset \Omega \times \left(\frac{T}{2} - \sqrt{k}\delta, \frac{T}{2} + \sqrt{k}\delta\right)$$

and so we can actually repeat the argument until $k \in \mathbb{N}$ satisfies $\delta\sqrt{k} \le \frac{T}{2} < \delta\sqrt{k+1}$, namely, $k = m$ by (5.15). Then $\rho^2 - m\beta\delta^2 = \rho^2 - \frac{\beta T^2}{4} < 0$ by (5.6). Therefore we see that $f(x) = g(x) = 0$, $x \in \Omega$. Thus the proof of Theorem 3 is complete.

§6. Stability in a linear inverse hyperbolic problem.

In this section, we continue to discuss an inverse hyperbolic problem, especially, a stability estimate in determining $f = f(x)$ and $g = g(x)$ in (5.2). The key is an observability inequality (Theorem 2). A similar argument for determining coefficients in (5.1) requires more delicate estimates and we omit it here in order to keep the present paper concise.

We can state stability in a linear inverse hyperbolic problem of determining f in (5.2) as follows:

Theorem 5. *Let $h_1, ..., h_n \in C^2(\mathbb{R})$ and $h_{n+1}(t) = -\beta\left(t - \frac{T}{2}\right)^2$ satisfy (1.5) - (1.9) and (1.12). Moreover let R satisfy (5.4) and*

$$(6.1) \qquad \partial_t R, \partial_t^2 R, \partial_t^3 R \in L^1(0, T; H^{\gamma-1}(\Omega)), \qquad \gamma > \max\left\{1, \frac{n-1}{2}\right\}$$

and let $p \in L^\infty(\Omega)$. Furthermore let

$$(6.2) \qquad T > \frac{2}{\sqrt{\beta}} \max_{x \in \overline{\Omega}}(h_1(x_1) + \cdots + h_n(x_n))^{1/2}.$$

Then there exists a constant $C = C(\Omega, T, \Gamma, h_1, ..., h_n, \beta, p, R) > 0$ such that

$$(6.3) \qquad \|f\|_{H^1(\Omega)} + \|g\|_{H^2(\Omega)} \leq C(\|y(f,g)'\|_{H^1(\Gamma \times (0,T))} + \|y(f,g)''\|_{H^1(\Gamma \times (0,T))}),$$

whenever the right hand side is finite.

For the case of the homogeneous Dirichlet boundary condition, we refer to Yamamoto [39] where the two-sided estimate was proved between $\|f\|_{L^2(\Omega)}$ and $\left\|\left(\frac{\partial y}{\partial \nu}\right)'\right\|_{L^2(\partial\Omega \times (0,T))}$ under suitable conditions on Γ and T. Unlike the Dirichlet case, in the Neumann case under consideration, the regularity of Dirichlet traces of the solutions is delicate (e.g. Avalos [1], Lasiecka and Triggiani [26], Tataru [36]) and the reverse inequality to (6.3) is extremely difficult.

Proof. Henceforth $C > 0$ denotes generic constants depending only on Ω, T, Γ, $h_1, ..., h_n$, β, p and R.

First Step. we show

Lemma 4. *Let*

$$(6.4) \qquad \gamma > \max\left\{1, \frac{n-1}{2}\right\}.$$

Then there exists $\delta > 0$ such that

$$(6.5) \qquad \|fg\|_{H^{-\frac{1}{2}+\delta}(\Omega)} \leq C_\delta \|f\|_{H^1(\Omega)} \|g\|_{H^{\gamma-1}(\Omega)}.$$

Indeed we see

$$(6.6) \qquad \delta \begin{cases} = \frac{2\gamma-1}{2}, & n = 1 \\ < \frac{2\gamma-1}{2}, & n = 2 \\ = \frac{2\gamma-n+1}{2}, & n \geq 3. \end{cases}$$

Proof. We will prove the lemma in the case of $n \geq 3$. The proof in the case of $n = 1, 2$ is similar. Henceforth we may assume that $\gamma < \frac{n}{2}$ without loss of generality. Then we have

(6.7) $$0 < \delta < \frac{1}{2}$$

by (6.6). By the Sobolev embedding, we see that

$$H^{\gamma-1}(\Omega) \subset L^{\frac{2n}{n+2-2\gamma}}(\Omega), \qquad H^1(\Omega) \subset L^{\frac{2n}{n-2}}(\Omega).$$

Hence, by the Hölder inequality, we have

(6.8) $$\|fg\|_{L^{\frac{n}{n-\gamma}}(\Omega)} \leq \|f\|_{L^{\frac{2n}{n-2}}(\Omega)} \|g\|_{L^{\frac{2n}{n+2-2\gamma}}(\Omega)} \leq C\|f\|_{H^1(\Omega)} \|g\|_{H^{\gamma-1}(\Omega)}.$$

By the Sobolev embedding, we see

$$H^{\frac{1}{2}-\delta}(\Omega) \subset L^{\frac{2n}{n-1+2\delta}}(\Omega),$$

which implies

$$H^{-\frac{1}{2}+\delta}(\Omega) \supset L^{\frac{2n}{n+1-2\delta}}(\Omega),$$

by the duality. Direct calculation shows us that $\frac{2n}{n+1-2\delta} = \frac{n}{n-\gamma}$, that is, $H^{-\frac{1}{2}+\delta}(\Omega) \supset L^{\frac{n}{n-\gamma}}(\Omega)$. Therefore (6.8) yields the conclusion of the lemma.

Second Step. We show

Lemma 5. Let $p \in L^\infty(\Omega)$ and let $\gamma > \max\left\{1, \frac{n-1}{2}\right\}$. We consider

(6.9) $$\begin{cases} z''(x,t) = \Delta z(x,t) - p(x)z(x,t) + f(x)R''(x,t), & x \in \Omega, 0 < t < T \\ z(x, T/2) = z'(x, T/2) = 0, & x \in \Omega \\ \dfrac{\partial z}{\partial \nu}(x,t) = 0, & x \in \partial\Omega, 0 < t < T. \end{cases}$$

Then there exists a constant $C = C(\Omega, T, \gamma, p, R) > 0$ such that

(6.10) $$\|z\|_{C^1([0,T];H^{\frac{1}{2}+\delta}(\Omega))} + \|z\|_{C^2([0,T];H^{-\frac{1}{2}+\delta}(\Omega))} + \|z\|_{C([0,T];H^{\frac{3}{2}+\delta}(\Omega))}$$
$$\leq C\|f\|_{H^1(\Omega)}(\|R''\|_{L^1(0,T;H^{\gamma-1}(\Omega))} + \|R'''\|_{L^1(0,T;H^{\gamma-1}(\Omega))}).$$

Here $\delta = \delta(\gamma) > 0$ is defined by (6.6).

Proof. Setting $z_1 = z'$, we have
(6.11)
$$\begin{cases} z_1''(x,t) = \Delta z_1(x,t) - p(x)z_1(x,t) + f(x)R'''(x,t), & x \in \Omega, 0 < t < T \\ z_1(x, T/2) = 0, \quad z_1'(x, T/2) = f(x)R''(x, T/2), & x \in \Omega \\ \dfrac{\partial z_1}{\partial \nu}(x,t) = 0, & x \in \partial\Omega, 0 < t < T. \end{cases}$$

By Lemma 4, we have

$$\|fR'''\|_{L^1(0,T;H^{-\frac{1}{2}+\delta}(\Omega))} \leq C\|f\|_{H^1(\Omega)}\|R'''\|_{L^1(0,T;H^{\gamma-1}(\Omega))}$$

and

$$\|fR''(\cdot,T/2)\|_{H^{-\frac{1}{2}+\delta}(\Omega)} \leq C\|f\|_{H^1(\Omega)}\|R''(\cdot,T/2)\|_{H^{\gamma-1}(\Omega)}$$
$$\leq C\|f\|_{H^1(\Omega)}\|R''\|_{W^{1,1}(0,T;H^{\gamma-1}(\Omega))}$$
$$\leq C\|f\|_{H^1(\Omega)}(\|R''\|_{L^1(0,T;H^{\gamma-1}(\Omega))} + \|R'''\|_{L^1(0,T;H^{\gamma-1}(\Omega))}).$$

Therefore

(6.12)
$$\|z_1\|_{C([0,T];H^{\frac{1}{2}+\delta}(\Omega))} + \|z_1\|_{C^1([0,T];H^{-\frac{1}{2}+\delta}(\Omega))}$$
$$\leq C\|f\|_{H^1(\Omega)}(\|R''\|_{L^1(0,T;H^{\gamma-1}(\Omega))} + \|R'''\|_{L^1(0,T;H^{\gamma-1}(\Omega))})$$

(e.g. Theorem 9.5 (p.292) in Lions and Magenes [33]). Finally we have to verify

$$\|z\|_{C([0,T];H^{\frac{3}{2}+\delta}(\Omega))} \leq C\|f\|_{H^1(\Omega)}(\|R''\|_{L^1(0,T;H^{\gamma-1}(\Omega))} + \|R'''\|_{L^1(0,T;H^{\gamma-1}(\Omega))}).$$

This follows directly from (6.12), $H^{-\frac{1}{2}+\delta}(\Omega) \supset L^2(\Omega)$ and $(\Delta-1)z = z'' + (p-1)z - fR''$ by noting that $\Delta - 1$ with $\mathcal{D}(\Delta-1) = \{u \in H^2(\Omega); \frac{\partial u}{\partial \nu} = 0\}$ is an isomorphism between $H^{\frac{3}{2}+\delta}(\Omega)$ and $H^{-\frac{1}{2}+\delta}(\Omega)$. Thus the proof of Lemma 5 is complete.

Third Step. Setting $v = v(f,g) = y(f,g)''$, we have
(6.13)
$$\begin{cases} v''(x,t) = \Delta v(x,t) - p(x)v(x,t) + f(x)R''(x,t), & x \in \Omega, 0 < t < T \\ v(x,T/2) = f(x)R(x,T/2), v'(x,T/2) = \Delta g(x) - p(x)g(x) + f(x)R'(x,T/2), & x \in \Omega \\ \frac{\partial v}{\partial \nu}(x,t) = 0, & x \in \partial\Omega, 0 < t < T. \end{cases}$$

In relation with (6.13), we introduce
(6.14)
$$\begin{cases} \psi''(x,t) = \Delta\psi(x,t) - p(x)\psi(x,t), & x \in \Omega, 0 < t < T \\ \psi(x,T/2) = f(x)R(x,T/2), \psi'(x,T/2) = \Delta g(x) - p(x)g(x) + f(x)R'(x,T/2), & x \in \Omega \\ \frac{\partial \psi}{\partial \nu}(x,t) = 0, & x \in \partial\Omega, 0 < t < T \end{cases}$$

and

(6.15)
$$\begin{cases} z''(x,t) = \Delta z(x,t) - p(x)z(x,t) + f(x)R''(x,t), & x \in \Omega, 0 < t < T \\ z(x,T/2) = z'(x,T/2) = 0, & x \in \Omega \\ \frac{\partial z}{\partial \nu}(x,t) = 0, & x \in \partial\Omega, 0 < t < T. \end{cases}$$

Then

(6.16)
$$v = \psi + z.$$

By the trace theorem and Lemma 5, we have

(6.17)
$$\|z''\|_{C([0,T];H^{-1+\delta}(\partial\Omega))} + \|z'\|_{C([0,T];H^{\delta}(\partial\Omega))}$$
$$\leq C\|f\|_{H^1(\Omega)}(\|R''\|_{L^1(0,T;H^{\gamma-1}(\Omega))} + \|R'''\|_{L^1(0,T;H^{\gamma-1}(\Omega))})$$

and

(6.18)
$$\|\nabla z\|_{C([0,T];H^{\delta}(\partial\Omega))} + \|(\nabla z)'\|_{C([0,T];H^{-1+\delta}(\partial\Omega))}$$
$$\leq C\|f\|_{H^1(\Omega)}(\|R''\|_{L^1(0,T;H^{\gamma-1}(\Omega))} + \|R'''\|_{L^1(0,T;H^{\gamma-1}(\Omega))}).$$

We define an operator $K : H^1(\Omega) \longrightarrow H^1(\Gamma \times (0,T))$ by

(6.19)
$$(Kf)(x,t) = z(x,t), \qquad x \in \Gamma, \, 0 < t < T.$$

Since the embedding $H^\delta(\Gamma) \longrightarrow L^2(\Gamma)$ is compact, we see by (6.17) and (6.18) (e.g. Theorem III.2.1 in Temam [37]) that

(6.20) $\qquad\qquad\qquad K$ is a compact operator.

On the other hand, in view of (6.2), we apply Theorem 2 to (6.14), so that

(6.21)
$$\|fR(\cdot, T/2)\|_{H^1(\Omega)} \leq C\|\psi\|_{H^1(\Gamma \times (0,T))}$$

and

(6.22)
$$\|(\Delta - p)g + fR'(\cdot, T/2)\|_{L^2(\Omega)} \leq C\|\psi\|_{H^1(\Gamma \times (0,T))}.$$

Similarly set $v_1 = v_1(f,g) = y(f,g)'$, and we have
(6.23)
$$\begin{cases} v_1''(x,t) = \Delta v_1(x,t) - p(x)v_1(x,t) + f(x)R'(x,t), & x \in \Omega, \, 0 < t < T \\ v_1(x, T/2) = g(x), \, v_1'(x, T/2) = f(x)R(x, T/2), & x \in \Omega \\ \dfrac{\partial v_1}{\partial \nu}(x,t) = 0, & x \in \partial\Omega, \, 0 < t < T. \end{cases}$$

We introduce

(6.24)
$$\begin{cases} \psi_1''(x,t) = \Delta \psi_1(x,t) - p(x)\psi(x,t), & x \in \Omega, \, 0 < t < T \\ \psi_1(x, T/2) = g(x), \, \psi_1'(x, T/2) = f(x)R(x, T/2), & x \in \Omega \\ \dfrac{\partial \psi_1}{\partial \nu}(x,t) = 0, & x \in \partial\Omega, \, 0 < t < T \end{cases}$$

and
(6.25)
$$\begin{cases} z_1''(x,t) = \Delta z_1(x,t) - p(x)z_1(x,t) + f(x)R'(x,t), & x \in \Omega, \, 0 < t < T \\ z_1(x, T/2) = z_1'(x, T/2) = 0, & x \in \Omega \\ \dfrac{\partial z_1}{\partial \nu}(x,t) = 0, & x \in \partial\Omega, \, 0 < t < T. \end{cases}$$

Then

(6.26) $$v_1 = \psi_1 + z_1.$$

We define an operator $K_1 : H^1(\Omega) \longrightarrow H^1(\Gamma \times (0,T))$ by

(6.27) $$(K_1 f)(x,t) = z_1(x,t), \quad x \in \Gamma, 0 < t < T.$$

Then, similarly to (6.20), we can prove

(6.28) $$K_1 \text{ is a compact operator.}$$

In terms of Theorem 2, we have

(6.29) $$\|g\|_{H^1(\Omega)} \leq C\|\psi_1\|_{H^1(\Gamma \times (0,T))}.$$

Therefore since $|R(x, T/2)| \geq r_0 > 0$ for almost all $x \in \overline{\Omega}$ by (5.4), we obtain

$$\|f\|_{H^1(\Omega)} \leq C\|\psi\|_{H^1(\Gamma \times (0,T))} = C\|v - z\|_{H^1(\Gamma \times (0,T))}$$

by means of (6.16) and (6.21). Consequently by the triangle inequality and $v = y(f)''$, we have

(6.30) $$\|f\|_{H^1(\Omega)} \leq C\|y(f,g)''\|_{H^1(\Gamma \times (0,T))} + C\|Kf\|_{H^1(\Gamma \times (0,T))}.$$

Next by (6.22) and (6.29), we have

(6.31) $$\|g\|_{H^2(\Omega)} \leq C\|\psi_1\|_{H^1(\Gamma \times (0,T))} + C\|f\|_{H^1(\Omega)} + C\|\psi\|_{H^1(\Gamma \times (0,T))}.$$

Hence (6.26), (6.30) and (6.31) imply

(6.32) $$\begin{aligned}&\|f\|_{H^1(\Omega)} + \|g\|_{H^2(\Omega)} \\ &\leq C(\|y(f,g)'\|_{H^1(\Gamma \times (0,T))} + \|y(f,g)''\|_{H^1(\Gamma \times (0,T))}) \\ &+ C(\|Kf\|_{H^1(\Gamma \times (0,T))} + \|K_1 f\|_{H^1(\Gamma \times (0,T))}).\end{aligned}$$

When we take away the second term in (6.32), we can complete the proof. By (5.4) and a usual a-priori estimate (e.g. [33]), we see that

(6.33) $$\begin{aligned}&\|y(f,g)\|_{C([0,T];H^1(\Omega))} + \|y(f,g)\|_{C^1([0,T];L^2(\Omega))} \\ &\leq C(\|Rf\|_{L^1(0,T;L^2(\Omega))} + \|g\|_{L^2(\Omega)}) \\ &\leq C\|R\|_{L^1(0,T;L^\infty(\Omega))}\|f\|_{L^2(\Omega)} + C\|g\|_{L^2(\Omega)}.\end{aligned}$$

Hence the trace theorem implies that

(6.34) $$\|y(f,g)\|_{L^2(\Gamma \times (0,T))} \leq C(\|f\|_{L^2(\Omega)} + \|g\|_{L^2(\Omega)}),$$

where $C > 0$ depends on $\|R\|_{L^1(0,T;L^\infty(\Omega))}$.

Let us apply the compactness-uniqueness argument. Contrarily assume that the inequality (6.3) does not hold. Then there exist $f_n \in H^1(\Omega)$, $g_n \in H^2(\Omega)$, $n \geq 1$ such that

(6.35) $$\|f_n\|_{H^1(\Omega)} + \|g_n\|_{H^2(\Omega)} = 1, \quad n \geq 1$$

and

(6.36) $$\lim_{n\to\infty}(\|y(f_n,g_n)'\|_{H^1(\Gamma\times(0,T))} + \|y(f_n,g_n)''\|_{H^1(\Gamma\times(0,T))}) = 0.$$

By (6.35), we can extract subsequences, denoted again by $\{f_n\}_{n\geq 1}$ and $\{g_n\}_{n\geq 1}$, such that f_n and g_n, $n \geq 1$ converge to some elements $f_0 \in H^1(\Omega)$ and $g_0 \in H^2(\Omega)$ weakly in $H^1(\Omega)$ and $H^2(\Omega)$, respectively. Then (6.20) and (6.28) yield that

(6.37) $$\lim_{m,n\to\infty}\|Kf_n - Kf_m\|_{H^1(\Gamma\times(0,T))} = 0, \quad \lim_{m,n\to\infty}\|K_1 f_n - K_1 f_m\|_{H^1(\Gamma\times(0,T))} = 0.$$

On the other hand, it follows from (6.32) that

$$\|f_n - f_m\|_{H^1(\Omega)} + \|g_n - g_m\|_{H^2(\Omega)}$$
$$\leq C\|(y(f_n,g_n) - y(f_m,g_m))'\|_{H^1(\Gamma\times(0,T))} + C\|(y(f_n,g_n) - y(f_m,g_m))''\|_{H^1(\Gamma\times(0,T))}$$
$$+ C\|Kf_n - Kf_m\|_{H^1(\Gamma\times(0,T))} + C\|K_1 f_n - K_1 f_m\|_{H^1(\Gamma\times(0,T))}$$
$$\leq C\|y(f_n,g_n)'\|_{H^1(\Gamma\times(0,T))} + C\|y(f_m,g_m)'\|_{H^1(\Gamma\times(0,T))}$$
$$+ C\|y(f_n,g_n)''\|_{H^1(\Gamma\times(0,T))} + C\|y(f_m,g_m)''\|_{H^1(\Gamma\times(0,T))}$$
$$+ C\|Kf_n - Kf_m\|_{H^1(\Gamma\times(0,T))} + C\|K_1 f_n - K_1 f_m\|_{H^1(\Gamma\times(0,T))}, \quad m,n \geq 1.$$

Therefore by (6.36) and (6.37), we see that

$$\lim_{m,n\to\infty}(\|f_n - f_m\|_{H^1(\Omega)} + \|g_n - g_m\|_{H^2(\Omega)}) = 0,$$

namely, $\lim_{n\to\infty}\|f_n - f_0\|_{H^1(\Omega)} = 0$ and $\lim_{n\to\infty}\|g_n - g_0\|_{H^2(\Omega)} = 0$. By (6.35) we obtain

(6.38) $$\|f_0\|_{H^1(\Omega)} + \|g_0\|_{H^2(\Omega)} = 1.$$

On the other hand, in view of (6.34), we have

$$\lim_{n\to\infty}\|y(f_n,g_n) - y(f_0,g_0)\|_{L^2(\Gamma\times(0,T))} \leq C \lim_{n\to\infty}(\|f_n - f_0\|_{L^2(\Omega)} + \|g_n - g_0\|_{L^2(\Omega)}) = 0,$$

and by $y(f_n, g_n)(\cdot, 0) = 0$, the limit (6.36) implies $\lim_{n\to\infty}\|y(f_n,g_n)\|_{L^2(\Gamma\times(0,T))} = 0$. Hence

$$y(f_0, g_0)(x, t) = 0, \quad x \in \Gamma, 0 < t < T.$$

On the other hand, by $f_0 \in H^1(\Omega)$ and $g_0 \in H^2(\Omega)$, we use a known result (e.g. [33]) that $y(f_0, g_0), y(f_0, g_0)' \in H^2(\Omega \times (0,T))$. Therefore in Theorem 3, the conditions (5.7) and (5.8) are satisfied, so that $f_0 = g_0 = 0$ follows. This contradicts (6.38). Thus the proof of Theorem 5 is complete.

Acknowldgement. Victor Isakov is in part supported by the NSF Grant DMS 98-03397. Masahiro Yamamoto is partially supported by Sanwa Systems Development Co., Ltd. The both authors thank Professor Irena Lasiecka and Professor Roberto Triggiani (University of Virginia, Charlottesville) for valuable comments and discussions. Furthermore they are very grateful to the reviewers for useful comments.

References

1. G. Avalos, *Sharp regularity estimates for solutions of the wave equation and their traces with prescribed Neumann data*, Appl. Math. Optim. **35** (1997), 203–219.
2. C. Bardos, G. Lebeau and J. Rauch, *Sharp sufficient conditions for the observation, control, and stabilization of waves from the boundary*, SIAM J. Control and Optimization **30** (1992), 1024–1065.
3. A.L. Bukhgeim, *Volterra Equations and Inverse Problems*, VSP, Utrecht, 1999.
4. A.L. Bukhgeim, *Introduction to the Theory of Inverse Problems*, VSP, Utrecht, 2000.
5. A.L. Bukhgeim and M.V. Klibanov, *Global uniqueness of a class of multidimensional inverse problems*, English translation, Soviet Math. Dokl. **24** (1981), 244–247.
6. N. Burq, *Contrôlabilité exacte des ondes dans des ouverts peu réguliers*, Asymptotic Analysis **14** (1997), 157–191.
7. J. Cheng, V. Isakov, M. Yamamoto and Q. Zhou, *Lipschitz stability in the lateral Cauchy problem for elasticity system*, preprint: UTMS 99-33, Graduate School of Mathematical Sciences, The University of Tokyo (1999).
8. L.F. Ho, *Observabilité frontière de l'équation des ondes*, C. R. Acad. Sci. Paris Sér.I Math. **302** (1986), 443–446.
9. L. Hörmander, *Linear Partial Differential Operators*, Springer-Verlag, Berlin, 1963.
10. O. Yu. Imanuvilov and M. Yamamoto, *Global uniqueness and stability in determining coefficients of wave equations*, preprint: UTMS 2000-13, Graduate School of Mathematical Sciences, The University of Tokyo (2000).
11. V. Isakov, *A nonhyperbolic Cauchy problem for $\Box_b\Box_c$ and its applications to elasticity theory*, Comm. Pure Appl. Math. **39** (1986), 747–767.
12. V. Isakov, *Uniqueness of the continuation across a time-like hyperplane and related inverse problems for hyperbolic equations*, Commun. in Partial Differential Equations **14** (1989), 465–478.
13. V. Isakov, *Inverse Source Problems*, American Mathematical Society, Providence, Rhode Island, 1990.
14. V. Isakov, *Inverse Problems for Partial Differential Equations*, Springer-Verlag, Berlin, 1998.
15. V. Isakov and M. Yamamoto, *Exact observability in a wave equation by Dirichlet data on subboundary*, preprint: UTMS 99-35, Graduate School of Mathematical Sciences, The University of Tokyo (1999).
16. V. Isakov and M. Yamamoto, *Lipschitz stability in a lateral Cauchy problem of a hyperbolic equation with the homogenuous Neumann boundary condition*, preprint: UTMS 99-36, Graduate School of Mathematical Sciences, The University of Tokyo (1999).
17. M.A. Kazemi and M.V. Klibanov, *Stability estimates for ill-posed Cauchy problems involving hyperbolic equations and inequalities*, Appl. Anal. **50** (1993), 93–102.
18. A. Khaĭdarov, *Carleman estimates and inverse problems for second order hyperbolic equations*, English translation, Math. USSR Sbornik **58** (1987), 267–

277.
19. A. Khaĭdarov, *On stability estimates in multidimensional inverse problems for differential equations*, English translation, Soviet. Math. Dokl. **38** (1989), 614–617.
20. M.V. Klibanov, *Inverse problems in the "large" and Carleman bounds*, English translation, Differential Equations **20** (1984), 755–760.
21. M.V. Klibanov, *Inverse problems and Carleman estimates*, Inverse Problems **8** (1992), 575–596.
22. M.V. Klibanov and J. Malinsky, *Newton-Kantorowich method for three-dimensional potential inverse scattering problem and stability of the hyperbolic Cauchy problem with time-dependent data*, Inverse problems **7** (1991), 577–596.
23. V. Komornik, *Exact Controllability and Stabilization, the Multiplier Method*, Masson, Paris, 1994.
24. M. Kubo, *Uniqueness in inverse hyperbolic problems - Carleman estimate for boundary value problems -*, Doctoral dissertation, Department of Mathematics, Kyoto University, 1998.
25. I. Lasiecka and R. Triggiani, *Exact controllability of the wave equation with Neumann boundary control*, Appl. Math. Optim. **19** (1989), 243–290.
26. I. Lasiecka and R. Triggiani, *Sharp regularity theory for second order hyperbolic equations of Neumann type, Part I: L_2 nonhomogeneous data*, Annali Mat. Pura Appl. **197** (1990), 285–367.
27. I. Lasiecka and R. Triggiani, *Carleman estimates and exact boundary controllability for a system of coupled, nonconservative second-order hyperbolic equations*, in "Partial Differential Equation Methods in Control and Shape Analysis" (Lecture Notes in Pure and Applied Mathematics Vol.188) (1997), Marcel Dekker, Inc., New York, 215–243.
28. I. Lasiecka, R. Triggiani and P.F. Yao, *Exact controllability for second-order hyperbolic equations with variable coefficient-principal part and first-order term*, Nonliner Analysis Theory, Methods and Applications **30** (1997), 111-122.
29. I. Lasiecka, R. Triggiani and P.F. Yao, *Inverse/observability estimates for second-order hyperbolic equations with variable coefficients*, J. Math. Anal. Appl **235** (1999), 13–57.
30. I. Lasiecka, R. Triggiani and X. Zhang, *Nonconservative wave equations with unobserved Neumann B.C.: global uniqueness and observability in one shot*, this volume of *Contemporary Mathematics*.
31. M.M. Lavrent'ev, V.G. Romanov and S.P. Shishat·skiĭ, *Ill-posed Problems of Mathematical Physics and Analysis*, English translation (the original: Nauka Moscow, 1980), American Mathematical Society, Providence Rhode Island, 1986.
32. J.-L. Lions, *Contrôlabilité Exacte Perturbations et Stabilisation de Systèmes Distribués*, Vol.1, Masson, Paris, 1988.
33. J.-L. Lions and E. Magenes, *Non-homogeneous Boundary Value Problems and Applications*, Springer-Verlag, Berlin, 1972.
34. D. Tataru, *Boundary controllability for conservative PDEs*, Appl. Math. Optim. **31** (1995), 257–295.
35. D. Tataru, *Carleman estimates and unique continuation for solutions to boundary value problems*, J. Math. Pures Appl. **75** (1996), 367–408.

36. D. Tataru, *On regularity of boundary traces for the wave equation*, Ann. Scuola Norm. Sup. Pisa Cl. Sci. (4) **36** (1998), 185–206.
37. R. Temam, *Navier-Stokes Equations*, North-Holland, Amsterdam, 1979.
38. R. Triggiani, *Exact boundary controllability on $L^2(\Omega) \times H^{-1}(\Omega)$ of the wave equation with Dirichlet boundary control acting on a portion of the boundary $\partial\Omega$, and related problems*, Appl. Math. Optim. **18** (1988), 241–277.
39. M. Yamamoto, *Uniqueness and stability in multidimensional hyperbolic inverse problems*, J. Math. Pures Appl. **78** (1999), 65–98.
40. X. Zhang, *Exact controllability of the semilinear distributed parameter system and some related problems*, Doctoral dissertation, Mathematics Institute, Fudan University, 1998.

Nonconservative Wave Equations with Unobserved Neumann B.C.: Global Uniqueness and Observability in One Shot

I. Lasiecka, R. Triggiani and X. Zhang

ABSTRACT. We consider a second-order hyperbolic equation on an open bounded domain Ω in $\mathbb{R}^n$, with C^1-boundary $\Gamma = \partial\Omega = \overline{\Gamma_0 \cup \Gamma_1}$, $\Gamma_0 \cap \Gamma_1 = \emptyset$, subject to Neumann boundary conditions on the *entire* boundary Γ. Here, Γ_0 (unobserved/uncontrolled part) and Γ_1 (observed/controlled part) are relatively open subsets of Γ. The principal part is of constant coefficients, while the energy level $(H^1(\Omega)-)$ terms may be variable in both space and time, and of low regularity $L_\infty(Q)$. Verifiable geometric conditions are imposed on the unobserved portion Γ_0. Then: we first establish a Carleman-type inequality for $H^{1,1}(Q)$-solutions of the hyperbolic equation *with no interior lower-order* terms. From here, we deduce *global uniqueness* results for $H^{1,1}(Q)$ solutions of the hyperbolic equation satisfying Neumann B.C. on all of Γ, and Dirichlet B.C. on Γ_0, over a time T greater than an explicit time T_0. T_0 is optimal if, e.g., Γ_0 is flat. Finally we obtain continuous observability (or stabilization) inequalities with an *explicit* constant. A three-part appendix, of purely geometric nature, provides several independent approaches leading to various general classes of triples $\{\Omega, \Gamma_0, \Gamma_1\}$ which satisfy the geometric conditions of Section 1, and, more relevantly, the geometric conditions of the far more general Section 10: see Theorem C.1 in Appendix C. In particular: Γ_0 may be flat; Γ_0 may be either convex or concave; Γ_0 may be logarithm convex or concave; etc. In the case of a disk, we can take the unobserved part Γ_0 to be as close to a half-circumference as we please: an indication that our results in Section 10 are sharp. Finally, in line with the AMS Conference at the University of Colorado, we point out throughout some open geometric questions, as well as some potential extensions which would require geometric methods. Extension of the fundamental Lemma 3.1 to the case of variable (in space) coefficients of the principal part has already been accomplished [L-T-Y-Z.1] by means of Bochner's techniques in Riemann geometry, in the style of [L-T-Y.1–3], [Y.1].

1. Introduction. Problem statement

1.1. Problem statement. Assumptions. Consequences. Dynamical problem. Let Ω be an open bounded domain in $\mathbb{R}^n$ with boundary $\partial\Omega = \Gamma$ of class C^1, consisting of the closure of two disjoint parts: Γ_0 (uncontrolled or unobserved part) and Γ_1 (controlled or observed part), both relatively open in

---

1991 *Mathematics Subject Classification.* 35,93,49.

First and second authors' research partially supported by the National Science Foundation under grant DMS-9804056, and by the Army Research Office under grant DAAH04-1-0059. Third author's research partially supported by the NSF of China under grant #19901024.

© 2000 American Mathematical Society

$\Gamma : \partial\Omega = \overline{\Gamma_0 \cup \Gamma_1}$, $\Gamma_0 \cap \Gamma_1 = \emptyset$. Let $\nu = [\nu_1, \ldots, \nu_n]$ be the unit outward normal vector on Γ, and let $\frac{\partial}{\partial \nu} = \nabla \cdot \nu$ denote the corresponding normal derivative. In this note we consider the following purely Neumann problem for a second-order hyperbolic equation in the unknown $w(t,x)$:

(1.1.1a) $$\begin{cases} w_{tt} - \Delta w = F(w) + f & \text{in } Q = (0,T] \times \Omega; \\ w(0, \cdot) = w_0, \ w_t(0, \cdot) = w_1 & \text{in } \Omega; \\ \left.\dfrac{\partial w}{\partial \nu}\right|_{\Sigma} \equiv 0 & \text{in } \Sigma = (0,T] \times \Gamma. \end{cases}$$
(1.1.1b)
(1.1.1c)

In (1.1.1a) we have set

(1.1.1d) $$F(w) = q_1(t,x)w + q_2(t,x)w_t + q_3(t,x) \cdot \nabla w,$$

subject to the following standing *assumption* on the coefficients: q_1, q_2, $|q_3| \in L_\infty(Q)$, so that the following pointwise estimate holds true:

(1.1.2a) $$|F(w)| \leq C_T[w_t^2 + |\nabla w|^2 + w^2], \quad (t,x) \in Q.$$

Remark 1.1.1. In effect we could relax the standing assumption on the lower-order coefficient q_1, and just require $q_1 \in L_p(Q)$ for $p = n+1$, $n \geq 2$, by using a Sobolev embedding theorem to estimate

(1.1.2b) $$\begin{aligned} \left\{\int_Q e^{2\tau\phi}|q_1 w|^2 dQ\right\}^{\frac{1}{2}} &\leq \|q_1\|_{L^p(Q)} \|e^{\tau\phi}w\|_{L^{\frac{2n+2}{n-1}}(Q)} \\ &\leq C\|e^{\tau\phi}w\|_{H^1(Q)}. \end{aligned}$$

We shall set

(1.1.2c) $$r \equiv \|q_1\|_{L^{n+1}(Q)} + \|q_2\|_{L^\infty(Q)} + \|q_3\|_{L^\infty(Q)}.$$

Theorem 10.1.1a will give the observability/stabilization inequality with an *explicit* constant of the order of Ce^{Cr^2}, where C is a generic constant, and r is defined by (1.1.2c). □

Moreover, we assume throughout that the non-homogeneous term f satisfies

(1.1.3) $$f \in L_2(Q).$$

Main assumptions. In addition to the standing assumptions (1.1.2), (1.1.3), on the first-order operator F, and on the forcing term f, the following assumption is postulated throughout Section 9 of this paper:

(A.1) Given the triple $\{\Omega, \Gamma_0, \Gamma_1\}$, $\partial\Omega = \overline{\Gamma_0 \cup \Gamma_1}$, there exists a strictly convex function $d : \bar{\Omega} \to \mathbb{R}$, of class $C^3(\bar{\Omega})$, such that if we introduce the (conservative) vector field $h(x) = [h_1(x), \ldots, h_n(x)] \equiv \nabla d(x)$, $x \in \Omega$, then the following two properties hold true:

(i)

(1.1.4) $$\left.\frac{\partial d}{\partial \nu}\right|_{\Gamma_0} = \nabla d \cdot \nu = h \cdot \nu = 0 \text{ on } \Gamma_0; \qquad h = \nabla d;$$

(ii) the (symmetric) Hessian matrix $\mathcal{H}_d$ of $d(x)$ [i.e., the Jacobian matrix J_h of $h(x)$] is strictly positive definite on $\bar{\Omega}$: there exists a constant $\rho_0 > 0$ such that for all $x \in \bar{\Omega}$

(1.1.5)
$$\mathcal{H}_d(x) = J_h(x) = \begin{bmatrix} d_{x_1 x_1}, & \cdots, & d_{x_1 x_n} \\ \vdots & & \vdots \\ d_{x_n x_1}, & \cdots, & d_{x_n x_n} \end{bmatrix} = \begin{bmatrix} \dfrac{\partial h_1}{\partial x_1}, & \cdots, & \dfrac{\partial h_1}{\partial x_n} \\ \vdots & & \vdots \\ \dfrac{\partial h_n}{\partial x_1}, & \cdots, & \dfrac{\partial h_n}{\partial x_n} \end{bmatrix} \geq \rho_0 I.$$

A working assumption throughout Section 9, to be later relaxed in our final results of Section 10, is that $d(x)$ has no critical point on $\bar{\Omega}$:

(A.2)

(1.1.6)
$$\inf_{x \in \Omega} |h(x)| = \inf_{x \in \Omega} |\nabla d(x)| = p > 0.$$

Without loss of generality as far as assumptions (A.1) and (A.2) are concerned, and for purposes of Eqn. (1.1.8b) below, we may always translate $d(x)$ as to make it positive on $\bar{\Omega}$: $\min d(x) = m > 0$ on $\bar{\Omega}$.

Remark 1.1.2. Assumption (A.1) is due to the Neumann B.C. Assumption (A.2) is needed for the validity of the key estimate (1.1.15b) which, in turn, is responsible for the elimination of the interior lower-order term in the final Carleman estimate of Theorem 5.1(ii), Eqn. (5.2). Assumption (A.1) was introduced in [Tr.1, Section 5] in dealing with a corresponding second-order hyperbolic problem with Neumann homogeneous B.C. on Γ_0 (uncontrolled/unobserved part of the boundary) and Dirichlet homogeneous B.C. on Γ_1 (controlled/observed part of the boundary); but it was not investigated in detail. Assumption (A.1) is much less restrictive than one would expect at first. Various classes of triples $\{\Omega, \Gamma_0, \Gamma_1\}$ satisfying assumption (A.1) and, moreover, also (A.2), are given in Appendix A through Appendix C, which constitute an intrinsic part of this paper. See, in particular, Theorem A.2.1, Lemma A.2.2, Appendix B.2, via a constructively geometric approach; Theorem A.3.2 resting on a conformal mapping approach in the 2-dimensional case; Theorem A.4.1, based on a perturbation approach valid in any dimension; finally, the conclusive Theorem C.1 regarding the validity of the far more relaxed geometrical setting of Section 10. □

Remark 1.1.3. In effect, assumption (1.1.6) is needed to hold true only for $x \in \Gamma_0$ (uncontrolled or unobserved part of the boundary Γ):

(1.1.6′)
$$\inf_{x \in \Gamma_0} |\nabla d(x)| = p > 0,$$

for then a critical point of $d(x)$ at a point (necessarily interior) of Ω, or at a point $x \in \Gamma_1$ (controlled or observed part of Γ) can always be eliminated, by smoothly redefining locally $d(x)$ while preserving the positivity condition (1.1.5). □

Preliminary scaling. Assumptions (A.1) and (A.2) above yield, in effect, a full family of strictly convex functions $\{ad(x) + b\}$ for all constants $a > 0$ and b (scaling and translation of $d(x)$), each member of which satisfies (1.1.4), (1.1.5),

(1.1.6). By translating and rescaling $d(x)$, with some $a > 1$, if necessary, we shall then operate a preliminary choice of 'normalization', in order to achieve the following outcomes:

(1.1.7) $$\min_{\overline{\Omega}} d(x) = m > 0, \qquad \rho_0 \geq 2.$$

The first condition is only for convenience: we think, however, of m as arbitrarily small, in order not to deteriorate the threshold time T_0 in (1.1.8b) below. The second condition $\rho_0 \geq 2$ will allow us to automatically verify properties (p_1) and (p_2) below in (1.1.12) and (1.1.14), which are a consequence of (A.1). For instance, in the case where Γ_0 is *flat*, we can take $d(x) = |x-z|^2$, for some $z \in \mathbb{R}^n$ just *outside* Ω on the hyperplane containing Γ_0 (Fig. A.1, Appendix A.1), and then (A.1), (A.2) are satisfied, along with (1.1.7) where, in fact, $\rho_0 = 2$. See more details on this case at the end of this Section 1.1 below.

Pseudo-convex function. Having chosen, on the strength of assumption (A.1), a strictly convex potential function $d(x)$ satisfying the preliminary scaling condition (1.1.7), we next introduce the pseudo-convex function $\phi : \Omega \times \mathbb{R} \to \mathbb{R}$ of class C^3 by setting:

(1.1.8a) $$\phi(x,t) = d(x) - c\left(t - \frac{T}{2}\right)^2; \qquad 0 \leq t \leq T,\ x \in \Omega,$$

where $T > 0$ and $0 < c < 1$ are selected as follows. We define first T_0 by setting

(1.1.8b) $$T_0^2 \equiv 4 \max_{x \in \Omega} d(x).$$

Let $T > T_0$ be given. By (1.1.8b), there exists $\delta > 0$ such that

(1.1.8c) $$T^2 > 4 \max_{x \in \overline{\Omega}} d(x) + 4\delta.$$

For this $\delta > 0$, there exists a constant c, $0 < c < 1$ such that

(1.1.8d) $$cT^2 > 4 \max_{x \in \overline{\Omega}} d(x) + 4\delta.$$

Henceforth, let $\phi(x,t)$ be defined by (1.1.8a) with T and c chosen as described above, unless otherwise explicitly noted. Such function $\phi(x,t)$ has the following properties:

(a) for the constant $\delta > 0$, fixed in (1.1.8c), we have via (1.1.8d),

(1.1.9)
$$\phi(x,0) \equiv \phi(x,T) = d(x) - c\frac{T^2}{4} \leq \max_{\overline{\Omega}} d(x) - c\frac{T^2}{4} \leq -\delta,\ \text{uniformly in}\ x \in \Omega;$$

(b) there are t_0 and t_1, with $0 < t_0 < \frac{T}{2} < t_1 < T$, say chosen symmetrically about $\frac{T}{2}$, such that

(1.1.10) $$\min_{x \in \overline{\Omega}, t \in [t_0, t_1]} \phi(x,t) \geq \sigma,\qquad 0 < \sigma < m,$$

since $\phi\left(x, \frac{T}{2}\right) = d(x) \geq m > 0$, under present choice: indeed, we take

$$m - c(t_1 - \frac{T}{2})^2 \equiv \sigma > 0,\ \text{or}\ t_1 - \frac{T}{2} = \sqrt{\frac{m-\sigma}{c}}.$$

We remark that the property $\sigma > -\delta$, $\delta > 0$, here achieved from (1.1.10), having imposed $\sigma > 0$, will be critically invoked in going from Eqn. (6.5) to Eqn. (6.6) below.

Consequences of assumption (A.1), (A.2) and the scaling condition (1.1.7). Let $d(x)$ be a strictly convex (potential) function provided by assumptions (A.1), (A.2) and satisfying the scaling choice (1.1.7): $\rho_0 \geq 2$. It then follows readily that any such $d(x)$ fulfills automatically the following two properties (p_1) and (p_2): there exists a function $\alpha(x) \in C^1(\bar{\Omega})$, in fact, take

(1.1.11) $$\alpha(x) \equiv \Delta d(x) - 2c - 1 + k \in C^1(\bar{\Omega}),$$

for a constant $0 < k < 1$ such that

(p_1)

(1.1.12) $$\Delta d(x) - 2c - \alpha(x) \equiv 1 - k > 0, \quad x \in \bar{\Omega};$$

(p_2) if we define γ by

(1.1.13) $$\gamma \equiv \alpha(x) - 2c - \Delta d(x) \equiv -4c - 1 + k < 0, \quad x \in \bar{\Omega},$$

the Hessian matrix $\mathcal{H}_d$ of any such $d(x)$ satisfies the following inequality for all $x \in \bar{\Omega}$:

(1.1.14a) $\quad 2\mathcal{H}_d(x) + [\alpha(x) - 2c - \Delta d(x)]I = 2\mathcal{H}_d(x) + \gamma I$

$$= \begin{bmatrix} 2d_{x_1 x_1} + \gamma & 2d_{x_1 x_2} & \cdots & 2d_{x_1 x_n} \\ 2d_{x_2 x_1} & 2d_{x_2 x_2} + \gamma & \cdots & 2d_{x_2 x_n} \\ \vdots & & \ddots & \vdots \\ 2d_{x_n x_1} & 2d_{x_n x_2} & \cdots & 2d_{x_n x_n} + \gamma \end{bmatrix} \geq \rho I, \ \forall\, x \in \bar{\Omega},$$

for some constant $\rho > 0$, by virtue of assumption (1.1.5) and (1.1.13):

(1.1.14b) $\quad \rho \equiv 2\rho_0 + \gamma = 2\rho_0 + (-4c - 1 + k) > 0 \text{ for } 1 + 2(2c - \rho_0) < k < 1.$

We note that positivity of ρ in (1.1.14b) is obtained, since the constant c was selected below (1.1.8a) as $0 < c < 1$, and, moreover, by the scaling choice (1.1.7) we have $\rho_0 \geq 2$, so that $2c - \rho_0 < 0$ and then (1.1.14b) is achieved.

In addition, by additional rescaling, if necessary, we shall show below that any strictly convex function $d(x)$ provided by assumptions (A.1), (A.2), and satisfying the preliminary scaling condition (1.1.7), and, possibly, additional scaling, fulfills also the following property (p_3):

(p_3) noting via (1.1.12) that

(1.1.15a) $\quad 6c + \Delta d(x) - \alpha(x) \equiv 8c + 1 - k, \quad x \in \bar{\Omega},$

and, moreover, recalling (1.1.10) and $\nabla d = h$ from (1.1.4), we have that the following inequality holds true, by virtue of assumption (1.1.6), and of (1.1.13), (1.1.15a):

(1.1.15b) $$\begin{cases} (2c + \Delta d - \alpha)|\nabla d|^2 + 2\mathcal{H}_d \nabla d \cdot \nabla d - (6c + \Delta d - \alpha)4c^2\left(t - \dfrac{T}{2}\right)^2 \\ \equiv (4c + 1 - k)|\nabla d|^2 + 2\mathcal{H}_d \nabla d \cdot \nabla d - (8c + 1 - k)4c^2\left(t - \dfrac{T}{2}\right)^2 \\ \geq \beta_1 > 0, \qquad\qquad \forall (x,t) \in \text{ set } Q^*(\sigma^*), \end{cases}$$

for a constant $\beta_1 > 0$, where the set $Q^*(\sigma^*)$ is defined by

(1.1.16) $\qquad Q^*(\sigma^*) \equiv \{(x,t) : x \in \Omega, \; 0 \le t \le T, \; \phi^*(x,t) \ge \sigma^* > 0\},$

for a constant σ^* chosen to satisfy $0 < \sigma^* < \sigma$, see (1.1.10), where in turn the function $\phi^*(x,t)$ is defined by

(1.1.17) $\qquad \phi^*(x,t) \equiv d(x) - c^2\left(t - \dfrac{T}{2}\right)^2, \quad x \in \Omega, \; 0 \le t \le T.$

Since $0 < c < 1$, we note, via (1.1.8a), (1.1.17), that

(1.1.18) $\qquad \phi^*(x,t) \ge \phi(x,t), \qquad x \in \Omega, \; 0 \le t \le T.$

Thus, if we define, in agreement with (1.1.17), the set $Q(\sigma)$ by

(1.1.19) $\qquad Q(\sigma) \equiv \{(x,t) : x \in \Omega, \; 0 \le t \le T, \; \phi(x,t) \ge \sigma > 0\},$

we see, since $0 < \sigma^* < \sigma$, and by virtue of (1.1.10), that

(1.1.20) $\qquad [t_0, t_1] \times \Omega \subset Q(\sigma) \subset Q^*(\sigma^*) \subset [0, T] \times \Omega,$

see Figure 1 and Figure 2. The point of the set $Q^*(\sigma^*)$ is twofold: (1) it is a convenient subset of $[0,T] \times \Omega$ where to require the validity of inequality (1.1.15b); moreover, *it is comparable*, in the sense described by (1.1.20), *with the set $Q(\sigma)$ in (1.1.19), which instead is defined in terms of the level surface $\phi(x,t) = \sigma$, related to the original pseudo-convex function*. All this will be seen in the proof of Theorem 5.1.

A class where properties $(p_1) =$ (1.1.12), $(p_2) =$ (1.1.14), and $(p_3) =$ (1.1.15b) are fulfilled without rescaling: the radial field case. Consider the special but important case where the unobserved boundary Γ_0 is *flat*. Then, take a point $z \in \mathbb{R}^n$ just *outside Ω on the hyperplane containing Γ_0* (Fig. A.1, Appendix A.1). Define $d(x) = |x - z|^2$, so $d(x) \ge m > 0$ on $\bar{\Omega}$, as desired. Assumptions (A.1) and (A.2) hold true, and moreover the constant ρ_0 in (1.1.5) is: $\rho_0 = 2$; moreover, $T > 2$ (diameter of Ω):

(1.1.21a)
(1.1.21b)
(1.1.21c)
$$\begin{cases} h(x) \equiv \nabla d(x) = 2(x - z) = \text{a radial field centered at } z; \\ |h(x)|^2 = |\nabla d(x)|^2 = 4|x-z|^2 = 4d(x), \; T_0^2 = 4\max|x-z|^2; \\ d_{x_i x_i} \equiv 2, \; i = 1, \dots, n; \; \Delta d(x) = 2n = 2\dim\Omega; \\ d_{x_i x_j} \equiv 0, \; i \ne j; \; \rho_0 = 2. \end{cases}$$

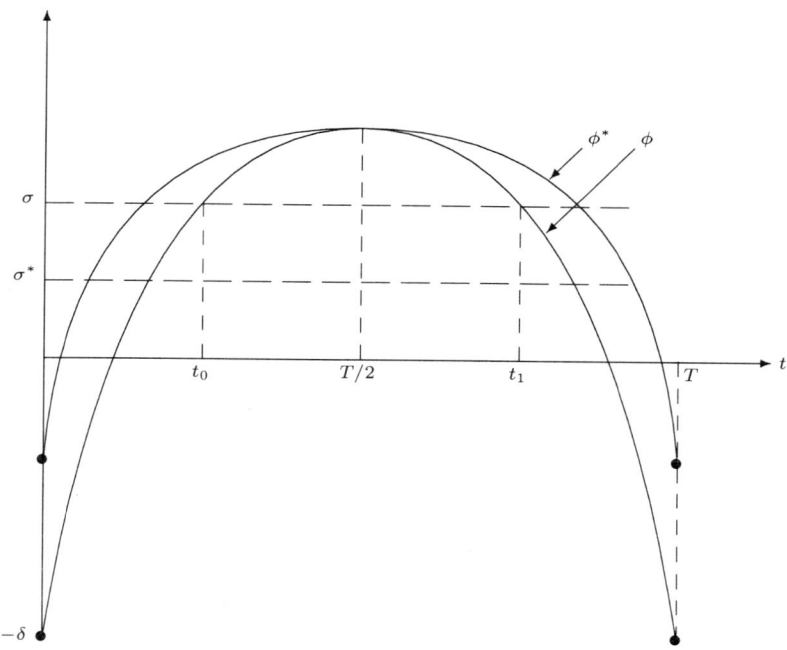

FIGURE 1: ϕ AND ϕ^* AT FIXED x, SAY WHERE $|x - z| = $ min OVER Ω

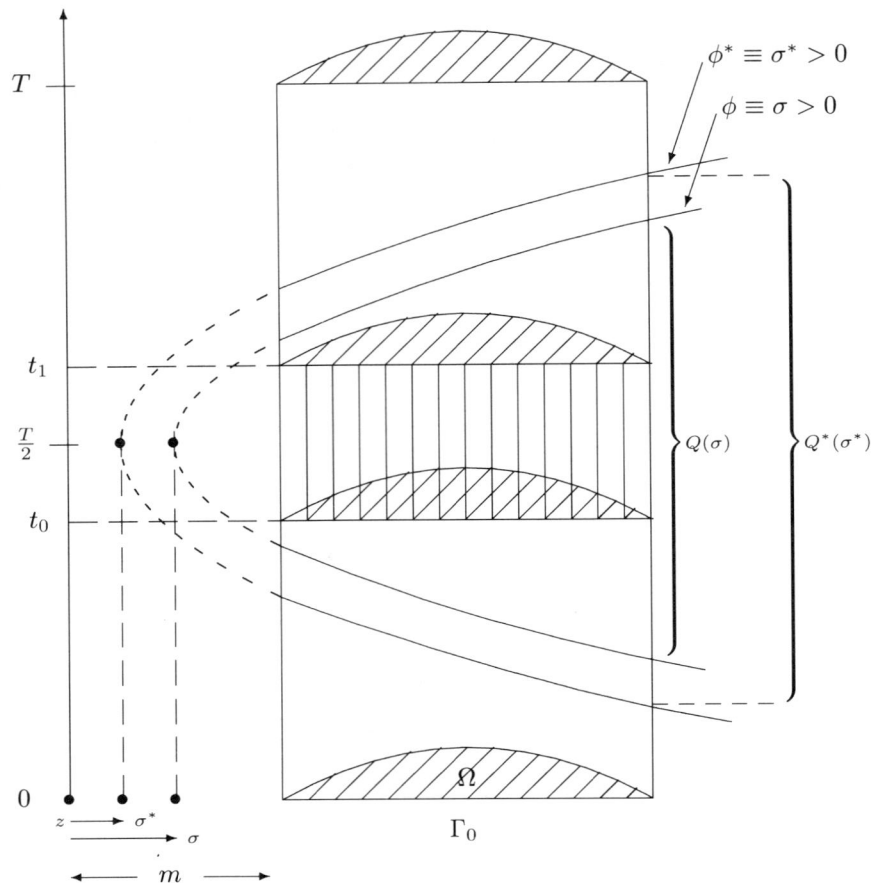

FIGURE 2: Use $d(x) = |x - z|^2$. At $t = 0$ and $t = T$, no point of $x \in \Omega$ belongs to $Q^*(\sigma^*)$, by virtue of (1.1.8D): $\phi(x, 0) = \phi\left(x, \frac{T}{2}\right) \leq \max d(x) - c\frac{T^2}{4} \leq -\delta$, where $0 < \sigma^* < \sigma < m$, and so (1.1.16) is violated.

We now verify the validity of (p_1), (p_2), (p_3) with no rescaling of $d(x)$:

$(p_1) = (1.1.12)$; $(p_2) = (1.1.14)$. Since now $\rho_0 = 2$, then (1.1.7) is satisfied and, as seen below (1.1.14), (p_1) and (p_2) are fulfilled.

$(p_3) = (1.1.15b)$. We premise verification of (p_3) with the following

Claim: *When $\rho_0 > 2c$, in particular, in the present radial vector field, we always have the following estimate*, which refers to the key expression in (1.1.15b); by (1.1.5) with $\rho_0 > 2c$:

$$(1.1.22) \quad (4c+1-k)|\nabla d|^2 + 2\mathcal{H}_d \nabla d \cdot \nabla d - (8c+1-k)4c^2\left(t-\frac{T}{2}\right)^2$$

$$\geq (4c+1-k+4c)|\nabla d|^2 - (8c+1-k)4c^2\left(t-\frac{T}{2}\right)^2$$

$$(1.1.23) \quad = (8c+1-k)\left[|\nabla d|^2 - 4c^2\left(t-\frac{T}{2}\right)^2\right].$$

We now verify (p_3). In addition, in the present case of a radial vector field, we make use of the (lucky) identity $|\nabla d(x)|^2 = 4d(x)$, see (1.1.21b), and obtain

$$(1.1.24) \quad |\nabla d|^2 - 4c^2\left(t-\frac{T}{2}\right)^2 = 4\left[d(x) - c^2\left(t-\frac{T}{2}\right)^2\right] = 4\phi^*(x,t),$$

recalling the function $\phi^*(x,t)$ defined in (1.1.17). Thus, for the present radial vector field, we obtain by (1.1.23) and (1.1.24):

$$(1.1.25) \ (4c+1-k)|\nabla d|^2 + 2\mathcal{H}_d \nabla d \cdot \nabla d - (8c+1-k)4c^2\left(t-\frac{T}{2}\right)^2$$

$$\geq 4(8c+1-k)\phi^*(x,t).$$

We conclude: *the requirement* $(p_3) = (1.1.15b)$ *always holds true with no rescaling of* $d(x) = |x-z|^2$ *in the present radial vector field case*, with controllability time $T > 2$ (diameter of Ω), which is optimal.

1.2. Rescaling of $d(x)$. Fulfillment of condition $(p_3) = (1.1.15b)$ in general, under assumptions (A.1), (A.2). Let $d(x)$ be a function satisfying (A.1), (A.2). Then property $(p_1) = (1.1.12)$ is always satisfied with the choice of $\alpha(x)$ as in (1.1.11).

Rescaling $d(x)$. Here the point is the following: On the positive side, rescaling $d(x)$ will always allow us to achieve also properties $(p_2) = (1.1.14)$ (by enforcing the choice $\rho_0 \geq 2$ in (1.1.7)) and, moreover, $(p_3) = (1.1.15\text{a-b})$, in fact, even (1.2.1) below. In fact, by rescaling $d(x)$ more, one can always achieve, by virtue of assumption (1.1.6), the validity of inequality (1.1.15b) even on *the entire cylinder* $[0,T] \times \Omega \equiv Q$:

$$(1.2.1) \quad (4c+1-k)|\nabla d|^2 + 2\mathcal{H}_d \nabla d \cdot \nabla d - (8c+1-k)4c^2\left(t-\frac{T}{2}\right)^2$$

$$\geq \beta_1 > 0. \ \forall \ (x,t) \in \Omega \times [0,T],$$

a stronger, and less desirable, requirement than (1.1.15b), as noted below. This is established below. Yet, on the negative side, rescaling $d(x)$ deteriorates the minimal observability time T_0 in (1.1.8b) while keeping $0 < c < 1$ fixed and close to 1. Thus, rescaling $d(x)$ calls for care. We have seen above that: *there is no need to rescale $d(x)$ in the special but important case where $d(x) = |x-z|^2$, z a fixed point just outside Ω*, which is relevant when the unobserved boundary Γ_0 is flat. See Appendix A.1.

Generally, if we rescale the original function $d(x)$, i.e., if we replace $d(x)$ by $d_{\text{new}}(x) = ad(x)$ for a constant $a > 1$, then:

(1.2.2)
$$\begin{cases} \nabla d_{\text{new}} = a\nabla d, \ |\nabla d_{\text{new}}|^2 = a^2|\nabla d|^2; \ \mathcal{H}_{d_{\text{new}}}(x) = a\mathcal{H}_d(x), \ \rho_{0,\text{new}} = a\rho_0; \\ T^2_{0,\text{new}} = aT_0^2; \ T^2_{\text{new}} = aT^2; \ \mathcal{H}_{d_{\text{new}}}\nabla d_{\text{new}} \cdot \nabla d_{\text{new}} = a^3 \mathcal{H}_d \nabla d \cdot \nabla d, \end{cases}$$

and we *can then keep the same constant* $0 < c < 1$, and close to 1, of the original $d(x)$. We note that the time T was arbitrary $> T_0$. Thus, $\sqrt{a}\,T$ is arbitrary $> T_{0,\text{new}}$. Thus, in the critical expression on the left of, say (1.2.1), we have that the first two space terms grow, after rescaling, by a^2 and a^3 respectively, while the worst case time term (on $[0,T]$), i.e., $T^2 > T_0^2$, deteriorates only by a factor a. Thus, (1.2.1) can always be achieved with $a > 1$ sufficiently large, at the price of deteriorating the original T_0 in (1.1.8b) into $\sqrt{a}\,T_0$. More precisely,

Achievement, in general, of estimate $(p_3) = (1.1.15b)$ **on** $Q^*(\sigma^*)$ **under rescaling of** $d(x)$; **indeed, of the stronger estimate (1.2.1).** We now establish the following

Claim. Given a function $d(x) \geq 0$ satisfying assumptions (A.1), (A.2). Rescale it, if necessary, with a rescaling factor $a > 1$ sufficiently large as to achieve the condition

(1.2.3) $$|\nabla d(x)|^2 - 4d(x) \geq 0, \quad \forall\, x \in \bar{\Omega},$$

in addition to $\rho_0 \geq 2$ in (1.1.7). [This can always be done by (1.2.2)]. Then, in fact, inequality (1.2.1) on the entire cylinder $[0,T] \times \Omega \equiv Q$ holds true. Thus, *a-fortiori*, property $(p_3) = (1.1.15b)$ on the set $Q^*(\sigma^*)$ is fulfilled.

Proof of Claim. We can first achieve $\rho_0 \geq 2$ in (1.1.7). Thus, by the Claim at the end of Section 1.1, we obtain that inequality (1.1.23) holds true for $x \in \Omega$, $0 \leq t \leq T$:

(1.2.4) $(4c + 1 - k)|\nabla d|^2 + 2\mathcal{H}_d \nabla d \cdot \nabla d - (8c + 1 - k)4c^2 \left(t - \dfrac{T}{2}\right)^2$

$$\geq (8c + 1 - k)\left[|\nabla d|^2 - 4c^2\left(t - \dfrac{T}{2}\right)^2\right].$$

Next, recalling (1.1.17) for ϕ^* and for $x \in \Omega$, $0 \leq t \leq T$, we obtain:

(1.2.5) $$|\nabla d(x)|^2 - 4c^2\left(t - \dfrac{T}{2}\right)^2$$

$$= |\nabla d(x)|^2 - 4d(x) + 4\left[d(x) - c^2\left(t - \dfrac{T}{2}\right)^2\right]$$

$$= |\nabla d(x)|^2 - 4d(x) + 4\phi^*(x,t).$$

Then, assumption (1.2.3) used in (1.2.5) yields, by use of (1.2.4),

$$(1.2.6) \quad (4c + 1 - k)|\nabla d|^2 + 2\mathcal{H}_d \nabla d \cdot \nabla d - (8c + 1 - k)4c^2 \left(t - \frac{T}{2}\right)^2$$

$$\geq 4(8c + 1 - k)\phi^*(x, t) \quad \text{on } Q = (0, T] \times \Omega,$$

and (1.2.6) establishes (1.2.1), as claimed. □

Remark 1.2.1. The proof in Sections 5 and 6 will markedly simplify if we assumed the validity of (1.2.1) on the whole cylinder $[0, T] \times \Omega$, rather than the validity of (1.1.15b) on its smaller subset $Q^*(\sigma^*)$. See Remark 5.1. □

2. Main results under assumptions (A.1) and (A.2)

In this section we state our main results under assumptions (A.1) [(1.1.4), (1.1.5)] and (A.2) = (1.1.6). In Section 10, assumption (A.2) will be suitably relaxed, in fact in many cases it will be dispensed with altogether. While it is possible to construct interesting examples of triples $\{\Omega, \Gamma_0, \Gamma_1\}$ which satisfy assumptions (A.1) and (A.2)—see Appendix A through Appendix C—nevertheless, assumption (A.2) introduces undesirable limitations in some key examples, such as Ω = disk. These can be eliminated by increasing "by ϵ" the observed boundary Γ_0, making it non-connected, see Example B.1 in Appendix B. Thus, omitting assumption (A.2) is a worthwhile endeavor. This will be done in Section 10.

Essentially, the setting of Section 10 will consist of splitting the original domain Ω into two subdomains Ω_1 and Ω_2: $\Omega = \Omega_1 \cup \Omega_2$, $\Omega_1 \cap \Omega_2 \neq \emptyset$, where the present framework of Section 1, based on assumptions (A.1) and (A.2), applies to each subdomain Ω_i separately, $i = 1, 2$. This avoids imposing assumption (A.2) on the entire domain Ω. The corresponding proofs in Section 10 become much more complicated, even though to each subproblem on Ω_i, we apply the key results of the preceding sections such as Corollary 4.3, obtained under assumptions (A.1) and (A.2). Thus, a separate treatment of the more general setting (without assumption (A.2)), is to be postponed until Section 10. This is justified in name of clarity. In conclusion, the most general results of this paper may essentially be obtained from those given below in this section, by replacing assumption (A.2) with a similar assumption on Ω_1 and Ω_2, however, often a more relaxed condition. See Section 10. The validity of the setting of Section 10 for large classes of triples $\{\Omega, \Gamma_0, \Gamma_1\}$ is provided by Theorem C.1 in Appendix C.

2.1. Continuous observability. Global uniqueness.

We first list our main continuous observability inequality, and related global uniqueness result of our treatment in Sections 1 through 9. In the next subsection, we shall present a corresponding uniform stabilization result. All these results are obtained under assumptions (A.1) and (A.2). In effect, they are derived from a corresponding Carleman-type estimate to be established in Section 6 (Theorem 6.1), which actually holds true in a more general setting (that of hypotheses (H.1) = (4.17), (H.2) = (4.18), and (H.3) = (4.19), which is a fortiori satisfied by the checkable assumptions (A.1) and (A.2)). Throughout this paper we introduce

$$(2.1.1) \quad E(t) = \int_\Omega [w_t^2(t) + |\nabla w(t)|^2 + w^2(t)]d\Omega = \|\{w(t), w_t(t)\}\|^2_{H^1(\Omega) \times L_2(\Omega)}.$$

Theorem 2.1.1. With reference to problem (1.1.1), let F satisfy (1.1.2). Let assumptions (A.1) and (A.2) hold true, so that there is a (coercive) conservative vector field $h(x) = \nabla d(x) \in [C^2(\bar{\Omega})]^n$ such that $h \cdot \nu = 0$ on Γ_0. Let $\Gamma_1 = \Gamma \setminus \Gamma_0$ and let $T_0 > 0$ be the constant defined in (1.1.8b). Then

(a) for all $T > T_0$, the following continuous observability inequality holds true for $H^{1,1}(Q)$-solutions: there exists a constant $C_T > 0$ such that, with $\Sigma_1 \equiv (0,T] \times \Gamma_1$, we have:

$$(2.1.2) \qquad C_T E(0) \leq \int_0^T \int_{\Gamma_1} [w_t^2 + w^2] d\Sigma_1 + \int_0^T \int_\Omega f^2 dQ.$$

(b) *A-fortiori*, the following global uniqueness result holds true: Let $T > T_0$ and let w be an $H^{1,1}(Q)$-solution of problem (1.1a) with $f \equiv 0$ along with the B.C.

$$(2.1.3) \quad \left.\frac{\partial w}{\partial \nu}\right|_\Sigma = 0 \quad \text{and} \quad w|_{\Sigma_1} = 0 \quad \text{where } h \cdot \nu = 0 \text{ on } \Gamma_0, \ \Sigma_1 \equiv (0,T] \times \Gamma_1.$$

Then, in fact, $w \equiv 0$ in Q (in fact, $w \equiv 0$ in $\mathbb{R}_t \times \Omega$). □

Indeed, we shall first prove the uniqueness statement of part (b) in Theorem 7.1 [as a direct consequence of the Carleman estimates of Theorem 6.1 for $H^{2,2}(Q)$-solutions], as supplemented by Section 8, which provides the extension to $H^{1,1}(Q)$-solutions. Next, part (b) will be used to establish part (a), in Theorem 9.2, by virtue also of the trace Lemma 9.1.

Duality between continuous observability and exact controllability. We now explain the control-theoretic terminology so far used. Consider the mixed hyperbolic problem:

$$(2.1.4a)$$
$$(2.1.4b)$$
$$(2.1.4c)$$
$$\begin{cases} v_{tt} - \Delta v = F_1(v) & \text{in } Q; \\ v(0,\cdot) = 0, \ v_t(0,\cdot) = 0 & \text{in } \Omega; \\ \left.\dfrac{\partial v}{\partial \nu}\right|_{\Sigma_0} \equiv 0; \ \left.\dfrac{\partial v}{\partial \nu}\right|_{\Sigma_1} \equiv g & \text{in } \Sigma, \end{cases}$$

where F_1 is the first-order differential operator such as F in (1.1.1d), satisfying therefore hypothesis (1.1.2) [and Remark 1.1.1]. Problem (2.1.4) is called exactly controllable in the space $H^1(\Omega) \times L_2(\Omega)$ within the class of $L_2(0,T;L_2(\Gamma_1))$-controls, for $0 < T < \infty$, in case: given such $T > 0$ and given any pair $\{v_{0,T}, v_{1,T}\} \in H^1(\Omega) \times L_2(\Omega)$, there exists a control function $g \in L_2(0,T;L_2(\Gamma_1))$ on the universal time interval $[0,T]$, such that the corresponding solution to problem (2.1.4) satisfies the terminal condition: $v(T,\cdot) = v_{0,T}; \ v_t(T,\cdot) = v_{1,T}$.

In other words, the input-solution map $g \to \mathcal{L}_T g \equiv \{v(T,\cdot), v_t(T,\cdot)\}$ of problem (2.1.4) is surjective:

$$(2.1.5) \qquad \mathcal{L}_T : L_2(0,T;L_2(\Gamma_1)) \supset \mathcal{D}(\mathcal{L}_T) \text{ onto } H^1(\Omega) \times L_2(\Omega).$$

By a standard Functional Analysis result [T-L.1, p. 235], the surjectivity condition (2.1.5) is *equivalent* to the condition that the adjoint $\mathcal{L}_T^*$ be bounded below: there exists $C_T > 0$ such that the following continuous observability inequality holds true:

$$(2.1.6) \qquad \|\mathcal{L}_T^* z\|_{L_2(0,T;L_2(\Gamma_1))} \geq C_T \|z\|_{H^1(\Omega) \times L_2(\Omega)},$$

for all $z \in \mathcal{D}(\mathcal{L}_T^*) \subset L_2(0,T;L_2(\Gamma_1))$, so that the left-hand side is finite. Conditions (2.1.6) turns out to be the counterpart, modulo *l.o.t.*, to inequality (2.1.2) for $f \equiv 0$

(e.g., [L-T.2–4], [L-T-Y.1], [Tr.1]). Thus, inequality (2.1.2) of Theorem 2.1.1 is the crux in establishing the exact controllability property for problem (2.1.4), as defined before, on a universal time $T > T_0$.

2.2. Uniform stabilization. In this section we consider the damped problem

(2.2.1a) $\quad\begin{cases} w_{tt} = \Delta w + Fw & \text{in } (0,T] \times \Omega = Q; \end{cases}$

(2.2.1b) $\quad w(0, \cdot) = w_0, w_t(0, \cdot) = w_1 \quad \text{in } \Omega;$

(2.2.1c) $\quad \left.\dfrac{\partial w}{\partial \nu}\right|_{\Sigma_0} \equiv 0 \quad \text{in } (0,T] \times \Gamma_0 = \Sigma_0;$

(2.2.1d) $\quad \left.\dfrac{\partial w}{\partial \nu}\right|_{\Sigma_1} = -w_t \quad \text{in } (0,T] \times \Gamma_1 = \Sigma_1.$

Theorem 2.2.1. With reference to problem (2.2.1), let F satisfy hypothesis (1.1.2).

(i) Then, problem (2.2.1) is well posed on $Y = H^1(\Omega) \times L_2(\Omega)$ in the semigroup sense: the map

(2.2.2) $\quad \{w_0, w_1\} \to e^{A_N t}\{w_0, w_1\} \equiv \{w(t), w_t(t)\} : Y \to C([0,T]; Y),$

defines a strongly continuous semigroup $e^{A_N t}$ on Y.

(ii) Let assumptions (A.1) and (A.2) hold true, so that there exists a (coercive) conservative vector field $h(x) = \nabla d(x) \in [C^2(\Omega)]^n$ such that $h \cdot \nu = 0$ on Γ_0. Let $\Gamma_1 = \Gamma \backslash \Gamma_0$. Then: the following estimates holds: for $T > T_0$, T_0 defined in (1.1.8b), there is a constant $c_T > 0$ such that

(2.2.3) $\quad c_T E(T) \leq \int_0^T \int_{\Gamma_1} \left[\left(\dfrac{\partial w}{\partial \nu}\right)^2 + w^2\right] d\Sigma_1.$

We now explain the terminology of the present subsection's title. Let $F \equiv 0$ in (2.2.1a). It then follows readily from inequality (2.2.3) via the dissipation identity (obtained by multiplying problem (2.2.1) by w_t and integrating by parts) that $E(T) \leq c_T E(0)$ for some $c_T < 1$. Or, in semigroup terms, $\|e^{A_N T}\| \leq C_T < 1$ in the uniform norm of $\mathcal{L}(Y)$. A standard semigroup result then yields that: there exist constants $a > 0$ and $M \geq 1$, such that $\|e^{A_N t}\| \leq Me^{-at}$, $t \geq 0$. Thus, the feedback $-w_t$ in (2.2.1d) uniformly stabilizes, when $F \equiv 0$, the corresponding problem where $-w_t$ is replaced by zero, which is a conservative, energy preserving problem.

2.3. Literature and overview. Since the inception of this topic, continuous observability/stabilization inequalities for second-order hyperbolic equations have been established, almost exclusively, in the case where the unobserved/uncontrolled part Γ_0 of the boundary Γ is subject to homogeneous *Dirichlet*, rather than Neumann, B.C. To begin with, this is the case for the original uniform stabilization (hence [R.1], exact controllability) results of the wave equation: in $H^1(\Omega) \times L_2(\Omega)$ with Neumann $L_2(\Sigma)$-boundary feedback [C.1], [Lag.1], [Tr.2], [L-T.3] and in $L_2(\Omega) \times H^{-1}(\Omega)$ with Dirichlet $L_2(\Sigma)$-boundary feedback [L-T.1], [L-T.3], [L-T-Y.3]. Moreover, this is the case also for most of the subsequent works aimed at a *direct* establishment of continuous observability inequalities (which then, by duality, yield exact controllability results, without passing through the generally more, or even much more, demanding stabilization problem): [H.1], [L-T.2], [Li.1–2]. See also an account in [K.1]. An excellent summary of the earlier literature is given in [R.2]. A first exception to the above statement is paper [Tr.1], which deals also with the observability/controllability issue of the *wave* equation subject to homogeneous *Neumann* B.C. on Γ_0, though under Dirichlet control on Γ_1. It was this reference [Tr.1] that introduced the vector field condition, $h \cdot \nu \equiv 0$ on Γ_0 in (1.1.4), of parallelism along Γ_0, for a coercive vector field h on Ω, see (1.1.5), to obtain continuous observability estimates: in the last step, appeal to compactness/uniqueness (Holmgren) is made, to absorb an interior lower-order term. Though several classes of examples were given in [Tr.1], this geometrical condition (A.1) (i.e., (1.1.4) and (1.1.5)) due to the homogeneous Neumann B.C. on Γ_0 was not analyzed there.

In the present paper, the inclusion of an energy level term F with coefficients both space and time dependent produces serious additional difficulties when coupled with the geometric condition (A.1) in particular on Γ_0.

First, the original Holmgren unique continuation result across the non-characteristic surface is not available with only $L_\infty(Q)$-regularity of the coefficients of the energy level terms. However, a local unique continuation result of $H_m^{(loc)}$ solutions across a strongly pseudo-convex surface is available for (linear) differential operators of order m with, say, real C^1-coefficients of the principal part and L_∞-coefficient of lower-order terms [Ho.1, Theorem 8.9.1, p. 224]. (A non-characteristic surface is strongly pseudo-convex, but not conversely.) Furthermore, we refer to [Ho.1-2], [Ta.2-3] for some very recent references where at least partial analyticity is needed, and [I.1] for a recent research monograph on this subject. (If these coefficients were *only space dependent*, appeal to the unique continuation results for the corresponding *elliptic* problem [Ho.1, Theorem 17.2.6, p. 14] would do the job.) A contemporaneous work [Lit.3] provides a much welcome *global* uniqueness theorem, say for hyperbolic second-order equations, starting, however, from a *local* uniqueness result. Thus, this is a result, proved in [Lit.3] by purely geometric means, that is a counterpart of the 'triangle lemma' in [I.1, Lemma 3.4.6, p. 67], which by contrast is established by analytic means. On the other hand, a main goal of the present paper is precisely to *eliminate* from the continuous observability/stabilization estimates *interior lower-order terms*, thus the need of appealing to external (until now apparently non-existent) *global* unique continuation results, to absorb them. This is achieved by virtue of the additional requirement (A.2) = (1.1.6) on the vector field h, which is responsible for obtaining the key estimate (1.1.15b), in the treatment of Sections 1 through 9 (or the additional requirement (A.2i) = (10.1.4), which is responsible for the key estimate (10.1.14) in the treatment of Section 10). In fact, a sub-goal of the present paper is to obtain, *a-fortiori*, directly, new global unique continuation results for $T > T_0$, from the preliminary Carleman estimates established this time without lower-order terms. The consequent global unique continuation results in Theorem 7.1 are obtained in precisely the form that is then needed to eliminate the interior lower-order terms arising, however, from a different source: the goal, this time, of eliminating the traditional star-shaped restrictions of the literature on the observed/controlled part Γ_1 of the boundary Γ. This step requires Lemma 9.1, Eqn. (9.1), to control the tangential trace of the solutions to Eqn. (1.1.11) in terms of their normal and velocity traces, and hence the global uniqueness Theorem 7.1 (and Remark 8.1) to eliminate the corresponding interior lower-order term in (9.1): see Step 2, in the proof of Theorem 9.2. Alternatively, if one wishes to retain the star-shaped geometrical conditions

$h \cdot \nu \geq 0$ on the observed/controlled portion Γ_1 of the boundary Γ, this method has then the virtue over the literature of yielding an *explicit* constant in the continuous observability/stabilization estimates of the order Ce^{Cr^2}, where C is a constant and r is the Hilbert norm of the involved coefficients, see (1.1.2c). This is the case, e.g., in the more general setting of Section 10, see Theorem 10.1.1.

Second, the further effort in Section 10 to weaken in many cases (including the case of flat Γ_0) the geometrical conditions (A.1), (A.2), relies on a domain decomposition $\Omega = \Omega_1 \cup \Omega_2$, $\Omega_1 \cap \Omega_2 \neq \emptyset$, with consequent cut-off functions χ_i on each Ω_i. It is because of the Neumann (rather than Dirichlet) B.C., that these cut-off functions have to be non-trivially selected as to be, among other features, only time-dependent on a small interior layer of the boundary Γ, see (10.2.10), (10.2.11) and Remark 10.2.1. A relevant reference for general second-order hyperbolic equations giving sharp (geometric optics) sufficient conditions for observability/stabilization at the energy $H^1 \times L_2$-level (as in the present paper) is the [B-L-R.1] and its precursor [Lit.1] for first-order hyperbolic systems. A comparison of techniques in the literature is attempted in the introduction of [L-T-Y.1], as well as of [I-Y.1], to which we refer. Generally, the geometric optics conditions are not readily checkable.

Conceptually, this paper is motivated by the desire to give a self-contained treatment of continuous observability/stabilization inequalities without passing through the preliminary traditional step of *first establishing these inequalities polluted by lower-order terms*, as done essentially in almost all of the literature, including [Lag.1], [B-L-R], [L-T.2–4], [Lio.1–2], [Ta.1], etc. To this end, we would then seek to introduce an additional degree of flexibility to the explicit computational treatment of [L-T.4] in obtaining Carleman estimates. The inspiration for this comes from the Russian literature, in particular [L-R-S.1, Lemma 1, p. 124], which is apparently not well known outside its original circle. This gives a complicated, yet very useful, *pointwise* Carleman-type inequality with, apparently, one further degree of freedom over [L-T.4], and other literature, where the benefits for the estimates are obtained not pointwise, but after integration on Q. In [L-R-S.1] Lemma 1, p. 124 of this reference is used precisely to obtain unique continuation results (pp. 133–142). This aforementioned result [L-R-S.1, Lemma 1, p. 124] forms also the basic starting point in other investigations on inverse problems and on stability estimates for ill-posed Cauchy problems involving hyperbolic equations and inequalities. See [K-K.1] for the latter, and references therein for the former.

Comparison with [K-K.1]. The main result of [K-K.1] is an *a-priori* stability estimate for $H^{2,2}(Q)$-solutions of a hyperbolic problem such as (1.1.1) and (1.1.2), which in particular yields uniqueness $w \equiv 0$ in Q of such solutions, if *zero Dirichlet* B.C. $w|_\Sigma \equiv 0$ is imposed on the *whole boundary* Γ, while zero Neumann B.C. $\frac{\partial w}{\partial \nu}|_{\Sigma_1} \equiv 0$ is imposed on a subportion Γ_1 where $h \cdot \nu \geq 0$, $h(x) = x - x_0$. Yet the stability estimate in [K-K.1, Theorem 2.2 or Theorem 3.1] is not quite the continuous observability estimate, as it needs to be extended to $H^{1,1}(Q)$-solutions. Apart from the higher *a-priori* regularity ($H^{2,2}(Q)$ rather than $H^{1,1}(Q)$) that [K-K.1] requires of the wave equation solution, its main estimate [K-K.1, Theorem 3.1, p. 101] would yield continuous observability inequalities in the following two 'classical' cases of the mid-eighties: (i) either for the case where $w|_\Sigma \equiv 0$ on the *entire* boundary Γ, while the *observed* portion of the boundary Γ_1 satisfies the (star-shaped) geometrical condition $h \cdot \nu \geq 0$, $h(x) = x - x_0$; (ii) or else for the case where $w|_{\Sigma_0} \equiv 0$ and $\frac{\partial w}{\partial \nu}|_{\Sigma_1} \equiv 0$, with $h \cdot \nu \geq 0$, $h(x) = x - x_0$, on the *observed* portion of the boundary Γ_1 (in our present notation) with, in addition, a *full* $H^{1,1}(\Sigma_1)$-norm on Σ_1. (In this case, the passage from $H^{2,2}(Q)$ to $H^{1,1}(Q)$ solutions is technical, see our Section 8.)

By contrast, our present paper assumes the definitively more challenging and non-classical B.C. $\frac{\partial w}{\partial \nu}|_{\Sigma_0} \equiv 0$ as well on the *unobserved* portion Γ_0 of the boundary Γ, and, moreover, it manages to dispense altogether—in the final estimates (2.1.2), or (10.1.20)—of the tangential H^1-norm on Γ_1 with no star-shaped geometrical condition imposed on the observed portion Γ_1; or else, if a star-shaped condition is imposed on the observed portion Γ_1, then the resulting observability constant is *explicit* (and of the order of Ce^{Cr^2}, where C is a generic constant and r is the norm in (1.1.2c) on the involved coefficients).

In terms of control rather than observation, the preceding considerations can be expressed as follows:

Case (i) above in [K-K.1] refers to Dirichlet control on Γ_1, while $w|_{\Sigma_0} \equiv 0$ on Γ_0.

Case (ii) above in [K-K.1] refers to Neumann control on Γ_1, while $w|_{\Sigma_0} \equiv 0$ on Γ_0.

The present paper refers to Neumann control on Γ_1, while $\frac{\partial w}{\partial \nu}|_{\Sigma_0} \equiv 0$ on Γ_0.

The cut-off functions in [K-K.1] are not really needed (as shown in an unpublished report by the present authors, Fall 1998). By contrast, our present paper requires the choice of sophisticated cut-off functions, see Remark 10.2.1, to achieve its maximally claimed generality in Theorem 10.1.1, in addition to a new version of the fundamental pointwise Lemma 3.1, over [L-R-S.1, Lemma 1, p. 124].

The method of [K-K.1] suitably refined permits to obtain an *explicit* constant in the corresponding observability/stabilization inequalities [Z.1–3], which has beneficial consequences in semilinear problems, in the style of the present paper.

A key feature of the pointwise Carleman-type estimate given by [L-R-S.1, Lemma 1, p. 124] is that the interior lower-order term comes with a controlled 'right' sign $c\tau^3$ (as in our Eqns. (4.21) and (4.26)) in a suitable set in time and space (akin to our $Q^*(\sigma^*)$).

For the aforementioned reasons stated at the outset of this subsection, the Neumann (rather than Dirichlet) problem is technically much more demanding. In our first effort to the Neumann problem of the present paper (Fall 1998), we also took an approach that relied, in its starting point, on the pointwise Carleman-type inequality for C^2-solutions of problem (1.1.1), (1.1.2) given by [L-R-S.1, Lemma 1, p. 124]. This approach led to a sharp result, precisely the one of our present paper, in the case of a *flat unobserved/uncontrolled boundary* Γ_0. Here, the special feature $|\nabla d(x)|^2 = 4|x - z|^2$ noted in (1.1.21a) was useful. An apparently somewhat similar treatment was carried out in the almost contemporaneous, and surely independent, work of [I-Y.1], presented at the Colorado's Conference. However, in the general case of a *curved unobserved/uncontrolled boundary* Γ_0, the approach based on [L-R-S.1, Lemma 1, p. 124] of both our first Fall 1998 effort and [I-Y.1] inherits unfortunately additional geometrical conditions. For instance, in the case where Ω is the $\mathbb{R}^2$-unit disk centered at the origin, this method required the uncontrolled/unobserved portion Γ_0 of the boundary Γ to be arbitrarily close to $\frac{1}{4}$ of the circumference Γ (if connected, see Example B.1.1 in Appendix B).

By contrast, we expect by known control theory results [Lio.1–2], [Tr.1], that Γ_0 should be arbitrarily close to $\frac{1}{2}$ of the circumference Γ. The improvement from $\frac{1}{4}$ of Γ to $\frac{1}{2}$ of Γ is highly non-trivial and is achieved in our Section 10.

In order to relax the additional geometrical conditions, our present second effort (Spring 1999) obtains another more suitable pointwise Carleman-type estimate for C^2-solutions of problems (1.1.1), (1.1.2) in a form (our present Lemma 3.1), which is a sufficiently noteworthy variation of [L-R-S.1, Lemma 1, p. 124] to warrant an explicit, complete proof. This we provide in Section 3. Building upon this new pointwise estimate of our Lemma 3.1, we are thus able to relax the geometric conditions on Γ_0 when Γ_0 is curved, even in the framework of Sections 1 through 9, and more so in the much less restrictive setting of Section 10, at the price of serious additional technical difficulties (see cut-off functions χ_i of Section 10.2). It is through the treatment of Section 10 (of which there is no counterpart either in our first Fall 1998 effort, or in [I-Y.1]) that we are finally able to reach our goal to increase, in the case of an $\mathbb{R}^2$-disk, the unobserved/uncontrolled portion Γ_0 of the boundary from almost $\frac{1}{4}$ to almost $\frac{1}{2}$ of the whole circumference Γ. A similar result holds for the $\mathbb{R}^n$-sphere for any n. See Appendix C, Theorem C.1, Fig. C.2. In common with the original Russian approach, our present methodology shares the key observation and the benefit of penalizing the interior lower-order term with a controlled 'right' sign (our Eqns. (4.21) and (4.24) with $\tau^3\tilde{\beta}$ or $\tau^3\beta$, $\tilde{\beta}, \beta > 0$) on a suitable time and space—set $Q^*(\sigma^*)$, which may, at worst, be the entire cylinder $Q = (0,T] \times \Omega$. This then allows one *to drop* such interior lower-order term, see the one-paragraph argument below (5.18), in the proof of the final Carleman estimate, Theorem 5.1, part (ii), Eqn. (5.2), or in the proof of Theorem 10.4.1, Step 2. This step is the main virtue of all these [L-R-S]-based, or [L-R-S]-inspired approaches: the elimination of the interior lower-order term in the final Carleman estimate. As the method accomodates, with no extra difficulty, energy level terms which may be space as well as *time* dependent and of low regularity $L_\infty(Q)$, the resulting global uniqueness result Theorem 7.1 for $T > T_0$ and $H^{1,1}(Q)$-solutions is, apparently, new. Here, time-dependence prevents appealing to elliptic theory. In the case of *flat* Γ_0, the uniqueness time T_0 is optimal. In general, for Γ_0 curved, the time T_0 of observability/controllability/global uniqueness is subject to scaling, as explained in Section 1.2, and is not necessarily guaranteed (nor do we expect it) to be optimal. We finally remark that, in the Neumann case, an additional technical difficulty not present in the Dirichlet case is the passage from $H^{2,2}(Q)$-solutions to $H^{1,1}(Q)$-solutions for the final Carleman estimates. This step is carried out in Section 8, via an approximation argument. This step is non-trivial, since finite energy solutions subject to Neumann B.C. do not produce (in dimension dim $\Omega \geq 2$) H^1-traces on the boundary, see [L-T.5].

This paper leads naturally to some geometrical open questions. To keep in line with the spirit of the AMS-IMS-SIAM Summer Research Conference held at the University of Colorado, Boulder, June 27–July 2, 1998, these are duly noted. One is the fulfillment of assumptions (A.1), (A.2) of Section 1, or (A.1i), (A.2i) of Section 10: see the appendices. Another is the possibility of extending the present paper to the general case where $(-\Delta)$ is replaced by a strongly elliptic operator with C^1-space dependent coefficients. Such an extension will have to rest on the counterpart of the basic Lemma 3.1 of this paper, given, however, in terms of a corresponding Riemann metric, as in [L-T-Y.1–3], [Y.1].

The first key step of such an extension—the proof of the basic Lemma 3.1 to this variable coefficient situation—has been (December 1999) recently carried out successfully [L-T-Y-Z.1]. It uses Bochner's techniques in Riemann geometry, in the style of [L-T-Y.1–3], [Y.1], which closely patterns the present proof in the Euclidean environment, thus confirming the validity of the aim of the Colorado's Conference.

3. A fundamental lemma

The starting point of our proofs is the following pointwise estimate. This is a sufficiently noteworthy variation of a result in the literature [L-R-S.1, Lemma 1, p. 124] to warrant an explicit proof. Such estimate will then be applied to (smooth) solutions of the wave equation (1.1.1a). With $t \in \mathbb{R}_t$ and $x = [x_1, \ldots, x_n] \in \mathbb{R}_x^n$, we shall indicate the partial derivatives as follows: $\frac{\partial}{\partial x_j}\frac{\partial}{\partial t} f = f_{tx_j}$, etc. [K-K.1], by contast, uses [L-R-S.1, Lemma 1, p. 124].

Lemma 3.1. Let

(3.1) $\quad w(t,x) \in C^2(\mathbb{R}_t \times \mathbb{R}_x^n); \; \ell(t,x) \in C^3((\mathbb{R}_t \times \mathbb{R}_x^n); \; \psi(t,x) \in C^2$

in t and C^1 in x

be three given functions and set $\theta(t,x) = e^{\ell(t,x)}$. Let $\epsilon > 0$ be arbitrary.

Then, the following pointwise inequality holds true

$$\theta^2(w_{tt} - \Delta w)^2 - \frac{\partial M}{\partial t} + \operatorname{div} V \geq -8v_t \nabla \ell_t \cdot \nabla v + 2(\Delta \ell + \ell_{tt} - \psi)v_t^2$$

(3.2) $\quad + 2\left(\psi - \frac{\epsilon}{2} - \Delta\ell + \ell_{tt}\right)|\nabla v|^2 + 4\left(\sum_{i,j=1}^n \ell_{x_i x_j} v_{x_i} v_{x_j}\right) + \theta^2 \tilde{B} w^2,$

where we have set

$M = M(w) \equiv \theta^2\{-2\ell_t(w_t^2 + |\nabla w|^2) + 4\nabla\ell \cdot \nabla w\, w_t$

(3.3) $\quad + 2(-2\ell_t^2 + 2|\nabla\ell|^2 + \psi)w_t w + (-2A\ell_t - 2\ell_t^3 + 2\ell_t|\nabla\ell|^2 - \psi_t)w^2\};$

(3.4a) $\quad V = [V_1, \ldots, V_j, \ldots, V_n];$

(3.4b) $\quad V_j = V_j(w) \equiv 2\theta^2\Big\{\ell_{x_j}(w_t^2 - |\nabla w|^2) - 2w_{x_j}(\ell_t w_t - \nabla\ell \cdot \nabla w)$

$\qquad + 2\left(|\nabla\ell|^2 - \ell_t^2 + \frac{\psi}{2}\right)w_{x_j}w + \ell_{x_j}(|\nabla\ell|^2 - \ell_t^2 - A)w^2\Big\},$

and, moreover,

(3.5) $\quad A \equiv (\ell_t^2 - \ell_{tt}) - |\nabla\ell|^2 + \Delta\ell - \psi;$

(3.6)

$$\tilde{B} = 2A\psi - 2\left[\sum_{j=1}^n \frac{\partial}{\partial x_j}((A+\psi)\ell_{x_j}) - \frac{\partial}{\partial t}((A+\psi)\ell_t) - \frac{1}{\epsilon}|\nabla\psi|^2 + \psi_{tt}\right].$$

[(3.6) shows the need for ψ to be C^2 in t, and C^1 in x.]

Proof. Step 1. We let $v(t,x) = \theta(t,x)w(t,x) = e^{\ell(t,x)}w(t,x)$, $(t,x) \in Q = (0,T] \times \Omega$. By direct differentiation, we get

$$
\begin{cases}
(3.7) \quad \theta w_{tt} = v_{tt} - 2\ell_t v_t + (\ell_t^2 - \ell_{tt})v \\
(3.8) \quad \theta w_{x_j x_j} = v_{x_j x_j} - 2\ell_{x_j} v_{x_j} + (\ell_{x_j}^2 - \ell_{x_j x_j})v, \; j = 1,\ldots,n \\
(3.9) \quad \theta \Delta w = \theta \sum_{j=1}^n w_{x_j x_j} = \Delta v - 2\nabla\ell \cdot \nabla v + (|\nabla\ell|^2 - \Delta\ell)v.
\end{cases}
$$

Thus, from (3.7) and (3.9) we obtain

$$(3.9a) \quad \theta^2[w_{tt} - \Delta w]^2 = \{[v_{tt} - 2\ell_t v_t + (\ell_t^2 - \ell_{tt})v]$$
$$- [\Delta v - 2\nabla\ell \cdot \nabla v + (|\nabla\ell|^2 - \Delta\ell)v]\}^2$$
$$(3.9b) \quad = |I_1 + I_2 + I_3|^2.$$

Here we have set (after adding and subtracting ψv)

$$(3.10) \quad I_1 = v_{tt} - \Delta v + Av, \; I_2 = -2\ell_t v_t + 2\nabla\ell \cdot \nabla v; \; I_3 = \psi v,$$

where A is defined by (3.5). From (3.9b), we obtain

$$(3.11) \quad \theta^2[w_{tt} - \Delta w]^2 \geq 2(I_1 I_2 + I_2 I_3 + I_1 I_3).$$

Step 2. With reference to (3.10), we shall prove in this Step 2 that

(3.12)
$$2I_1 I_2 = \frac{\partial}{\partial t}\{-2\ell_t[v_t^2 + |\nabla v|^2 + Av^2] + 4v_t \nabla\ell \cdot \nabla v\}$$
$$- 2\sum_{j=1}^n \frac{\partial}{\partial x_j}\{2v_{x_j}\nabla\ell \cdot \nabla v - \ell_{x_j}|\nabla v|^2 - 2\ell_t v_t v_{x_j} + \ell_{x_j} v_t^2 - A\ell_{x_j} v^2\}$$
$$- 8v_t \nabla\ell_t \cdot \nabla v + 2(\Delta\ell + \ell_{tt})v_t^2 + 4\sum_{i,j}^n \ell_{x_i x_j} v_{x_i} v_{x_j}$$
$$- 2(\Delta\ell - \ell_{tt})|\nabla v|^2 - 2\left[\sum_{j=1}^n \frac{\partial}{\partial x_j}(A\ell_{x_j}) - \frac{\partial}{\partial t}(A\ell_t)\right]v^2.$$

Proof of (3.12). From the definitions of I_1, I_2 in (3.10), we obtain after using $2v_t v_{tt} = \frac{\partial}{\partial t}(v_t^2)$, $2vv_t = \frac{\partial}{\partial t}(v^2)$, and $2v\nabla v = \nabla(v^2)$:

$$(3.13) \quad 2I_1 I_2 = 2(v_{tt} - \Delta v + Av)(-2\ell_t v_t + 2\nabla\ell \cdot \nabla v)$$
$$= -2\ell_t \frac{\partial}{\partial t}(v_t^2) - 2A\ell_t \frac{\partial}{\partial t}(v^2) + 2A\nabla\ell \cdot \nabla(v^2)$$
$$+ 4v_{tt}\nabla\ell \cdot \nabla v + 4\Delta v v_t \ell_t - 4\Delta v \nabla\ell \cdot \nabla v.$$

But, with reference to the last three terms in (3.13), we have by direct computations:

(3.14a) $$4v_{tt}\nabla\ell\cdot\nabla v = 4\frac{\partial}{\partial t}(v_t\nabla\ell\cdot\nabla v) - 4v_t\nabla\ell_t\cdot\nabla v - 4v_t\nabla\ell\cdot\nabla v_t$$

(3.14b) $$= 4\sum_{j=1}^{n}\frac{\partial}{\partial t}(\ell_{x_j}v_{x_j}v_t) - 4v_t\nabla\ell_t\cdot\nabla v - 2\nabla\ell\cdot\nabla(v_t^2);$$

(3.15) $$4\Delta v v_t \ell_t = 4\sum_{j=1}^{n}\frac{\partial}{\partial x_j}(v_t\ell_t v_{x_j}) - 2\ell_t\frac{\partial}{\partial t}(|\nabla v|^2) - 4v_t\nabla\ell_t\cdot\nabla v;$$

$$-4\Delta v\nabla\ell\cdot\nabla v = -4\sum_{i,j=1}^{n}v_{x_jx_j}\ell_{x_i}v_{x_i} = -4\sum_{i,j=1}^{n}\frac{\partial}{\partial x_j}(v_{x_j}\ell_{x_i}v_{x_i})$$

(3.16) $$+ 4\sum_{i,j=1}^{n}\ell_{x_ix_j}v_{x_i}v_{x_j} + 2\sum_{i,j=1}^{n}\ell_{x_i}\frac{\partial}{\partial x_i}(v_{x_j}^2).$$

Next, we substitute (3.14), (3.15) and (3.16) into (3.13), thus obtaining, after a rearrangement of terms

(3.17)
$$2I_1I_2 = -2\ell_t\frac{\partial}{\partial t}(v_t^2) - 2A\ell_t\frac{\partial}{\partial t}(v^2) + 4\frac{\partial}{\partial t}(v_t\nabla\ell\cdot\nabla v) - 2\ell_t\frac{\partial}{\partial t}(|\nabla v|^2)$$
$$+ 2A\sum_{j=1}^{n}\ell_{x_j}\frac{\partial}{\partial x_j}(v^2) - 2\sum_{j=1}^{n}\ell_{x_j}\frac{\partial}{\partial x_j}(v_t^2)$$
$$+ 4\sum_{j=1}^{n}\frac{\partial}{\partial x_j}(v_t\ell_t v_{x_j}) - 4\sum_{i,j=1}^{n}\frac{\partial}{\partial x_j}(v_{x_j}\ell_{x_i}v_{x_i}) + 2\sum_{i,j=1}^{n}\ell_{x_j}\frac{\partial}{\partial x_j}(v_{x_i}^2)$$
$$- 4v_t\nabla\ell_t\cdot\nabla v - 4v_t\nabla\ell_t\cdot\nabla v + 4\sum_{i,j=1}^{n}\ell_{x_ix_j}v_{x_i}v_{x_j}.$$

Next, regarding the first four-term group in (3.17) (with $\frac{\partial}{\partial t}$), we can rewrite as follows:

(3.18) $$-2\ell_t\frac{\partial}{\partial t}(v_t^2) - 2A\ell_t\frac{\partial}{\partial t}(v^2) + 4\frac{\partial}{\partial t}(v_t\nabla\ell\cdot\nabla v) - 2\ell_t\frac{\partial}{\partial t}(|\nabla v|^2)$$
$$= \frac{\partial}{\partial t}\{-2\ell_t(v_t^2 + Av^2 + |\nabla v|^2) + 4v_t\nabla\ell\cdot\nabla v\}$$
$$+ 2\ell_{tt}v_t^2 + 2v^2\frac{\partial}{\partial t}(A\ell_t) + 2\ell_{tt}|\nabla v|^2.$$

Similarly, regarding the second five-term group in (3.17) (with $\frac{\partial}{\partial x_j}$), we can rewrite as follows:

(3.19)
$$2A\sum_{j=1}^{n}\ell_{x_j}\frac{\partial}{\partial x_j}(v^2) - 2\sum_{j=1}^{n}\ell_{x_j}\frac{\partial}{\partial x_j}(v_t^2)$$
$$+ 4\sum_{j=1}^{n}\frac{\partial}{\partial x_j}(v_t\ell_t v_{x_j}) - 4\sum_{i,j=1}^{n}\frac{\partial}{\partial x_j}(v_{x_j}\ell_{x_i}v_{x_i}) + 2\sum_{i,j=1}^{n}\ell_{x_j}\frac{\partial}{\partial x_j}(v_{x_i}^2)$$
$$= \sum_{j=1}^{n}\frac{\partial}{\partial x_j}\left\{2A\ell_{x_j}v^2 - 2\ell_{x_j}v_t^2 + 4v_t\ell_t v_{x_j} - 4v_{x_j}\sum_{i=1}^{n}\ell_{x_i}v_{x_i} + 2\ell_{x_j}\sum_{i=1}^{n}v_{x_i}^2\right\}$$
$$- 2\sum_{j=1}^{n}\left[\frac{\partial}{\partial x_j}(A\ell_{x_j})\right]v^2 + 2\left(\sum_{j=1}^{n}\ell_{x_j x_j}\right)v_t^2 - 2\left(\sum_{j=1}^{n}\ell_{x_j x_j}\right)|\nabla v|^2.$$

Substituting (3.18) and (3.19) into (3.17) yields (3.12), as desired.

Step 3. With reference to (3.10) we shall prove in this Step 3 that

(3.20) $\quad 2I_1 I_3 = \frac{\partial}{\partial t}[2\psi v v_t - \psi_t v^2] + [\psi_{tt} + 2A\psi]v^2 - 2\psi v_t^2$
$$- 2\sum_{j=1}^{n}\frac{\partial}{\partial x_j}(\psi v_{x_j}v) + 2v\nabla\psi\cdot\nabla v + 2\psi|\nabla v|^2$$

(3.21) $\quad \geq \frac{\partial}{\partial t}[2\psi v v_t - \psi_t v^2] + \left[\psi_{tt} + 2A\psi - \frac{1}{\epsilon}|\nabla\psi|^2\right]v^2$
$$- 2\psi v_t^2 + [2\psi - \epsilon]|\nabla v|^2 - 2\sum_{j=1}^{n}\frac{\partial}{\partial x_j}(\psi v_{x_j}v),$$

where, in (3.21), $\epsilon > 0$ is arbitrary.

Proof of (3.20), (3.21). From the definition of I_1 and I_3 in (3.10), we obtain

(3.22) $\quad 2I_1 I_3 = 2(v_{tt} - \Delta v + Av)\psi v = 2\psi v v_{tt} - 2\psi\Delta v v + 2A\psi v^2,$

where

(3.23) $$2\psi v v_{tt} = 2\frac{\partial}{\partial t}(\psi v v_t) - 2\psi_t v v_t - 2\psi v_t^2;$$

(3.24) $\quad -2\psi v \Delta v = -2\sum_{i=1}^{n}\psi v v_{x_i x_i}$
$$= -2\sum_{j=1}^{n}\frac{\partial}{\partial x_j}(\psi v v_{x_j}) + 2v\nabla\psi\cdot\nabla v + 2\psi|\nabla v|^2.$$

Finally, to obtain (3.20), we insert (3.23) and (3.24) into (3.22) and use $-2\psi_t v v_t = -\frac{\partial}{\partial t}(\psi_t v^2) + \psi_{tt} v^2$. Then, the estimate $2v\nabla\psi\cdot\nabla v \geq -\epsilon|\nabla v|^2 - \frac{1}{\epsilon}|\nabla\psi|^2 v^2$, for $\epsilon > 0$, used in the penultimate term in (3.20) yields (3.21).

Step 4. With reference to (3.10), we shall prove in this Step 4 that

$$(3.25) \quad 2I_2 I_3 = \frac{\partial}{\partial t}[-2\ell_t \psi v^2] + \sum_{j=1}^n \frac{\partial}{\partial x_j}(2\psi \ell_{x_j} v^2)$$
$$+ 2\left[\frac{\partial}{\partial t}(\ell_t \psi) - \sum_{j=1}^n \frac{\partial}{\partial x_j}(\psi \ell_{x_j})\right] v^2.$$

Proof of (3.25). From the definition of I_2 and I_3 in (3.10), we obtain

$$(3.26) \quad 2I_2 I_3 = 2[-2\ell_t v_t + 2\nabla \ell \cdot \nabla v]\psi v = -2\ell_t \psi \frac{\partial}{\partial t}(v^2) + 2\psi \nabla \ell \cdot \nabla(v^2),$$

where

$$(3.27) \quad -2\ell_t \psi \frac{\partial}{\partial t}(v^2) = \frac{\partial}{\partial t}(-2\ell_t \psi v^2) + \left[\frac{\partial}{\partial t}(2\ell_t \psi)\right] v^2$$

$$(3.28) \quad 2\psi \nabla \ell \cdot \nabla(v^2) = 2\sum_{j=1}^n \psi \ell_{x_j} \frac{\partial(v^2)}{\partial x_j}$$
$$= 2\sum_{j=1}^n \frac{\partial}{\partial x_j}(\psi \ell_{x_j} v^2) - 2\sum_{j=1}^n \left[\frac{\partial}{\partial x_j}(\psi \ell_{x_j})\right] v^2.$$

Substituting (3.27) and (3.28) into (3.26) yields (3.25), as desired.

Step 5. With reference to (3.11), in this Step 5 we prove that

$$(3.29)$$
$$\theta^2 [w_{tt} - \Delta w]^2 \geq \frac{\partial}{\partial t}\left\{-2\ell_t(v_t^2 + |\nabla v|^2) + 4v_t \nabla \ell \cdot \nabla v\right.$$
$$\left. + 2\psi v_t v - 2\ell_t(A + \psi)v^2 - \psi_t v^2\right\}$$
$$- 2\sum_{j=1}^n \frac{\partial}{\partial x_j}\bigg[2v_{x_j} \nabla \ell \cdot \nabla v - \ell_{x_j}|\nabla v|^2$$
$$- 2\ell_t v_{x_j} v_t + \ell_{x_j} v_t^2 + \psi v_{x_j} v - (A + \psi)\ell_{x_j} v^2\bigg]$$
$$- 8v_t \nabla \ell_t \cdot \nabla v + 2(\Delta \ell + \ell_{tt} - \psi)v_t^2$$
$$+ 2\left(\psi - \frac{\epsilon}{2} - \Delta \ell + \ell_{tt}\right)|\nabla v|^2 + 4\sum_{i,j=1}^n \ell_{x_i x_j} v_{x_i} v_{x_j}$$
$$+ \left\{2A\psi - 2\bigg[\sum_{j=1}^n \frac{\partial}{\partial x_j}((A + \psi)\ell_{x_j})\right.$$
$$\left. - \frac{\partial}{\partial t}((A + \psi)\ell_t)\bigg] - \frac{1}{\epsilon}|\nabla \psi|^2 + \psi_{tt}\right\}v^2.$$

Proof of (3.29). We return to (3.11), where we use (3.12) for $2I_1I_2$, (3.21) for $2I_1I_3$, and (3.25) for $2I_2I_3$. Combining all '$\frac{\partial}{\partial t}$-terms,' all '$\frac{\partial}{\partial x_j}$-terms,' all $|\nabla v|^2$-terms and all v^2-terms, we arrive at (3.29), as desired.

Step 6. Henceforth, we specialize (3.29) with

(3.30) $\begin{cases} v = \theta w, \ \theta = e^\ell; \text{ hence } v_t = \theta[w_t + \ell_t w]; \\ v_{x_j} = \theta[w_{x_j} + \ell_{x_j} w]; \ |\nabla v|^2 = \theta^2 \sum_{j=1}^n [w_{x_j} + \ell_{x_j} w]^2. \end{cases}$

Then, the terms under $\frac{\partial}{\partial t}$ in (3.29) become via (3.30):

$$(3.31) \quad \frac{\partial}{\partial t}\left\{-2\ell_t(v_t^2 + |\nabla v|^2) + 4v_t\sum_{j=1}^n \ell_{x_j} v_{x_j} + 2\psi v_t v - 2\ell_t(A+\psi)v^2 - \psi_t v^2\right\}$$

$$= \frac{\partial}{\partial t}\left\{\theta^2\left[-2\ell_t(w_t^2 + |\nabla w|^2) + 4w_t \nabla\ell \cdot \nabla w \right.\right.$$

$$\left.\left. + 2(2|\nabla\ell|^2 - 2\ell_t^2 + \psi)w_t w + (2\ell_t|\nabla\ell|^2 - 2\ell_t^3 - 2A\ell_t - \psi_t)w^2\right]\right\}.$$

Proof of (3.31). Using (3.30) in the terms under $\frac{\partial}{\partial t}$ in (3.29) we obtain

$$(3.32) \quad \frac{\partial}{\partial t}\left\{-2\ell_t(v_t^2 + |\nabla v|^2) + 4\sum_{j=1}^n \ell_{x_j} v_{x_j} v_t + 2\psi v_t v - 2\ell_t(A+\psi)v^2 + \psi_t v^2\right\}$$

$$= \frac{\partial}{\partial t}\left\{\theta^2\left[-2\ell_t\left((w_t + \ell_t w)^2 + \sum_{j=1}^n(w_{x_j} + \ell_{x_j} w)^2\right)\right.\right.$$

$$+ 4\sum_{j=1}^n \ell_{x_j}(w_{x_j} + \ell_{x_j} w)(w_t + \ell_t w)$$

$$\left.\left. + 2\psi(\ell_t w + w_t)w - 2\ell_t(A+\psi)w^2 - \psi_t w^2\right]\right\}$$

$$(3.33) \quad = \frac{\partial}{\partial t}\left\{\theta^2\left[-2\ell_t(w_t^2 + |\nabla w|^2) + 4\sum_{j=1}^n \ell_{x_j} w_{x_j} w_t\right.\right.$$

$$\left.\left. + 2(-2\ell_t^2 + 2|\nabla\ell|^2 + \psi)w_t w + (-2A\ell_t - 2\ell_t^3 + 2\ell_t|\nabla\ell|^2 - \psi_t)w^2\right]\right\},$$

after two cancellations: of $2\psi\ell_t w^2$ and of $4\ell_t \sum_j \ell_{x_j} w_{x_j} w$, and (3.31) is proved.

Step 7. Under the specialization $v = \theta w$ as in (3.30), we have that the block of 'divergence terms' in (3.29) (modulo the coefficient -2) becomes

(3.34) $\displaystyle\sum_{j=1}^{n}\frac{\partial}{\partial x_j}\Big[2v_{x_j}\nabla\ell\cdot\nabla v - \ell_{x_j}|\nabla v|^2$

$\qquad\qquad - 2\ell_t v_{x_j} v_t + \ell_{x_j} v_t^2 + \psi v_{x_j} v - (A+\psi)\ell_{x_j} v^2\Big]$

$\quad = \displaystyle\sum_{j=1}^{n}\frac{\partial}{\partial x_j}\Big\{\theta^2\Big[2w_{x_j}\nabla\ell\cdot\nabla w - \ell_{x_j}|\nabla w|^2 - 2\ell_t w_{x_j} w_t + \ell_{x_j} w_t^2$

$\qquad\qquad + 2\Big(|\nabla\ell|^2 - \ell_t^2 + \dfrac{\psi}{2}\Big)w_{x_j} w + \ell_{x_j}(|\nabla\ell|^2 - \ell_t^2 - A)w^2\Big]\Big\}.$

Proof of (3.34). Using (3.30) for v_{x_j} and v_{x_i}, we preliminarily compute

(3.35) $\quad 2v_{x_j}\nabla\ell\cdot\nabla v = 2v_{x_j}\displaystyle\sum_{i=1}^{n}\ell_{x_i}v_{x_i} = 2\theta[\ell_{x_j}w + w_{x_j}]\displaystyle\sum_{i=1}^{n}\ell_{x_i}\theta[\ell_{x_i}w + w_{x_i}]$

$\qquad\qquad = 2\theta^2(\ell_{x_j}w + w_{x_j})(|\nabla\ell|^2 w + \nabla\ell\cdot\nabla w).$

Next, we use (3.35), as well as the expressions in (3.30) for $|\nabla v|$, v_{x_j}, v_t, and v, into the left side of (3.34) to rewrite it as

(3.36) $\displaystyle\sum_{j=1}^{n}\frac{\partial}{\partial x_j}[2v_{x_j}\nabla\ell\cdot\nabla v - \ell_{x_j}|\nabla v|^2$

$\qquad\qquad - 2\ell_t v_{x_j} v_t + \ell_{x_j} v_t^2 + \psi v_{x_j} v - (A+\psi)\ell_{x_j} v^2]$

$\quad = \displaystyle\sum_{j=1}^{n}\frac{\partial}{\partial x_j}\Big[2\theta^2(\ell_{x_j}w + w_{x_j})(|\nabla\ell|^2 w + \nabla\ell\cdot\nabla w)$

$\qquad\qquad - \ell_{x_j}\theta^2\displaystyle\sum_{i=1}^{n}(\ell_{x_i}w + w_{x_i})^2 - 2\ell_t\theta(\ell_{x_j}w + w_{x_j})\theta(\ell_t w + w_t)$

$\qquad\qquad + \ell_{x_j}\theta^2(\ell_t w + w_t)^2 + \psi\theta(\ell_{x_j}w + w_{x_j})\theta w - (A+\psi)\ell_{x_j}\theta^2 w^2\Big]$

(3.37) $\quad = \displaystyle\sum_{j=1}^{n}\frac{\partial}{\partial x_j}\Big\{\theta^2\Big[2|\nabla\ell|^2\ell_{x_j}w^2 + 2\ell_{x_j}\nabla\ell\cdot\nabla ww + 2|\nabla\ell|^2 w_{x_j}w$

$\qquad\qquad + 2\nabla\ell\cdot\nabla w w_{x_j} - \ell_{x_j}(|\nabla\ell|^2 w^2 + |\nabla w|^2 + 2w\nabla\ell\cdot\nabla w)$

$\qquad\qquad - 2\ell_t^2\ell_{x_j}w^2 - 2\ell_t\ell_{x_j}ww_t - 2\ell_t^2 ww_{x_j} - 2\ell_t w_i w_{x_j} + \ell_{x_j}\ell_t^2 w^2$

$\qquad\qquad + \ell_{x_j}w_t^2 + 2\ell_t\ell_{x_j}ww_t + \psi\ell_{x_j}w^2 + \psi w_{x_j}w - (A+\psi)\ell_{x_j}w^2\Big]\Big\}.$

After a cancellation of $2\ell_{x_j}\nabla\ell\cdot\nabla ww$, $2\ell_t\ell_{x_j}ww_t$, and $\psi\ell_{x_j}w^2$, then (3.37) becomes (3.34) as desired.

Step 8. We finally insert (3.31) and (3.34) in the right side of (3.29), recall the definition of M, V, and $\tilde{B}$ in (3.3), (3.4), and (3.6), respectively, and finally arrive at

$$(3.38) \quad \theta^2(w_{tt} - \Delta w)^2 - \frac{\partial M}{\partial t} + \operatorname{div} V \geq -8v_t \nabla \ell_t \cdot \nabla v + 2(\Delta \ell + \ell_{tt} - \psi)v_t^2$$

$$+ 2\left(\psi - \frac{\epsilon}{2} - \Delta \ell + \ell_{tt}\right)|\nabla v|^2 + 4\sum_{i,j=1}^{n} \ell_{x_i x_j} v_{x_i} v_{x_j} + \tilde{B}v^2,$$

which is precisely the sought-after Eqn. (3.2). The proof of Lemma 3.1 is complete. □

Remark 3.1. The above proof has been recently extended (December 1999) to the case where the coefficients of the principal part are variable in space [L-T-Y-Z.1], by using Bochner's techniques in Riemann geometry, in the style of [L-T-Y.1–3], [Y.1]. Extension of the present paper to that variable coefficient case is in progress. □

4. A basic pointwise inequality

We now make suitable choices in the functions $\ell(t,x)$ and $\psi(x)$ involved in Lemma 3.1.

Theorem 4.1. Let

$$(4.1) \quad w(t,x) \in C^2(\mathbb{R}_t \times \mathbb{R}_x^n); \quad d(x) \in C^3(\mathbb{R}_x^n), \quad \alpha(x) \in C^1(\mathbb{R}_x^n)$$

be three given functions [at this stage, w and d need not be the solution of Eqn. (1.1.1a), and the function provided by assumptions (A.1) and (A.2), respectively]. If $\tau > 0$ is a parameter, we introduce the functions

$$(4.2) \quad \ell(t,x) \equiv \tau\left[d(x) - c\left(t - \frac{T}{2}\right)^2\right] \equiv \tau\phi(t,x);$$

$$(4.3) \quad \psi(x) \equiv \tau\alpha(x); \quad \theta(t,x) = e^{\ell(t,x)} = e^{\tau\phi(t,x)},$$

where $\phi(t,x)$ is defined consistently with (1.1.8a), with a constant $0 < c < 1$ selected as in (1.1.8d). Then, with the above choices, Lemma 3.1 specializes as follows: setting $h = \nabla d$:

$$(4.4) \quad \ell_{x_i} = \tau d_{x_i}; \quad |\nabla \ell|^2 = \tau^2 |\nabla d|^2 = \tau^2 |h|^2; \quad \ell_{x_i x_j} = \tau d_{x_i x_j}; \quad \Delta \ell = \tau \Delta d;$$

$$(4.5) \quad \ell_t = -2c\tau\left(t - \frac{T}{2}\right); \quad \ell_{tt} = -2c\tau; \quad \ell_{tx_j} \equiv 0; \quad \psi_t = 0; \quad |\nabla \psi| = \tau|\nabla \alpha|,$$

so that the pointwise estimate (3.2) becomes

$$(4.6) \quad \theta^2(w_{tt} - \Delta w)^2 - \frac{\partial M}{\partial t} + \operatorname{div} V \geq 2\tau[\Delta d - 2c - \alpha]v_t^2$$

$$+ 2\tau\left[\alpha - \frac{\epsilon}{2\tau} - \Delta d - 2c\right]|\nabla v|^2 + 4\tau\left[\sum_{i,j=1}^{n} d_{x_i x_j} v_{x_i} v_{x_j}\right] + \theta^2 \tilde{B}w^2,$$

where M and $V = [V_1, \ldots, V_n]$ are given by (3.3) and (3.4b) respectively, as functions of w, as specialized via (4.4); while for A and $\tilde{B}$, we now obtain from (3.5), (3.6), via (4.3)–(4.5):

$$(4.7) \qquad A = \tau^2 \left[4c^2 \left(t - \frac{T}{2}\right)^2 - |\nabla d|^2 \right] + \tau[2c + \Delta d - \alpha]$$

$$(4.8) \qquad \tilde{B} = 2\tau^3 \left\{ [2c + \Delta d - \alpha]|\nabla d|^2 + 2\mathcal{H}_d \nabla d \cdot \nabla d \right.$$
$$\left. - (6c + \Delta d - \alpha) 4c^2 \left(t - \frac{T}{2}\right)^2 \right\} + \mathcal{O}(\tau^2).$$

Notice that it is the coefficient { } of the principal part of $\tilde{B}$ that justifies our interest in property $(p_3) = (1.1.15b)$.

Proof. The proof is a direct computation starting from Lemma 3.1 and using the choice of functions made in (4.2), (4.3). First, A in (3.5) becomes at once the expression (4.7) via (4.3)–(4.5). We then verify (4.8) for $\tilde{B}$. By recalling $\psi(x) = \tau\alpha(x)$ in (4.3), we obtain from (4.7), via (4.4)–(4.5):

$$(4.9) \qquad 2A\psi = 2\tau^3 \alpha \left[4c^2 \left(t - \frac{T}{2}\right)^2 - |\nabla d|^2 \right] + \mathcal{O}(\tau^2);$$

$$(4.10) \qquad (A + \psi)\ell_{x_j} = \tau^3 \left[4c^2 \left(t - \frac{T}{2}\right)^2 - |\nabla d|^2 \right] d_{x_j} + \tau^2 [\Delta d + 2c] d_{x_j};$$

$$(4.11) \quad \frac{\partial}{\partial x_j}[(A + \psi)\ell_{x_j}] = \tau^3 \left\{ \left[4c^2 \left(t - \frac{T}{2}\right)^2 - |\nabla d|^2 \right] d_{x_j x_j} \right.$$
$$\left. - \left(\frac{\partial}{\partial x_j}|\nabla d|^2\right) d_{x_j} \right\} + \mathcal{O}(\tau^2);$$

$$(4.12) \quad \sum_{j=1}^n \frac{\partial}{\partial x_j}[(A + \psi)\ell_{x_j}] = \tau^3 \left\{ \left[4c^2 \left(t - \frac{T}{2}\right)^2 - |\nabla d|^2 \right] \Delta d \right.$$
$$\left. - \nabla(|\nabla d|^2) \cdot \nabla d \right\} + \mathcal{O}(\tau^2);$$

$$(4.13) \qquad (A + \psi)\ell_t = \tau^2 \left[4c^2 \left(t - \frac{T}{2}\right)^2 - |\nabla d|^2 \right] \ell_t + \tau(\Delta d + 2c)\ell_t;$$

$$(4.14) \qquad \frac{\partial}{\partial t}[(A + \psi)\ell_t] = 2c\tau^3 \left[|\nabla d|^2 - 12c^2 \left(t - \frac{T}{2}\right)^2 \right] + \mathcal{O}(\tau^2).$$

Finally, one either verifies (or recalls from say [L-T.2]) the following identity, where $h = \nabla d$, and $J_h = \mathcal{H}_d$ are defined in (1.1.5):

(4.15) $$\nabla(|\nabla d|^2) \cdot \nabla d = \nabla(h \cdot \nabla d) \cdot \nabla d = J_h \nabla d \cdot \nabla d + \frac{1}{2}\nabla d \cdot \nabla(|\nabla d|^2),$$

hence

(4.16) $$\nabla(|\nabla d|^2) \cdot \nabla d = 2\mathcal{H}_d \nabla d \cdot \nabla d = 2 J_h h \cdot h.$$

Finally, first inserting (4.16) into (4.12), and next inserting the resulting (4.12) along with (4.9), (4.14), and (4.5) for ψ in the definition (3.6) of $\tilde{B}$, we readily obtain (4.8) for $\tilde{B}$, as desired. □

The pointwise estimate of interest in Corollary 4.2 below is then obtained for functions $d(x) \in C^3(\mathbb{R}^n_x)$, $\alpha(x) \in C^1(\mathbb{R}^n_x)$, such that the following three estimates hold true, for suitable positive constants $\rho > 0$, $\tilde{\beta} > 0$:

(H.1)

(4.17) $$\Delta d - 2c - \alpha \geq \rho > 0; \ \forall\, x \in \bar{\Omega};$$

(H.2)

(4.18) $$\begin{bmatrix} 2d_{x_1 x_1} + \gamma & 2d_{x_1 x_2} & \cdots & 2d_{x_1 x_n} \\ 2d_{x_2 x_1} & 2d_{x_2 x_2} + \gamma & \cdots & 2d_{x_2 x_n} \\ \vdots & & & \vdots \\ 2d_{x_n x_1} & 2d_{x_n x_2} & \cdots & 2d_{x_n x_n} + \gamma \end{bmatrix} \geq \rho I, \ \forall\, x \in \bar{\Omega},$$

where we have set $\gamma(x) = \alpha(x) - \Delta d(x) - 2c$;

(H.3)

(4.19) $$[2c + \Delta d - \alpha]|\nabla d|^2 + 2\mathcal{H}_d \nabla d \cdot \nabla d - (6c + \Delta d - \alpha)4c^2 \left(t - \frac{T}{2}\right)^2$$
$$\geq \tilde{\beta} > 0, \ \forall\, (t,x) \in Q^*(\sigma^*),$$

where $Q^*(\sigma^*)$ is the subset of $[0,T] \times \Omega$ defined in (1.1.16). But, as we have seen in Section 1, these three inequalities hold true, in particular, in the case of our interest where assumptions (A.1) and (A.2) hold true. Then (A.1) provides a strictly convex (positive potential) function $d(x)$, and we then choose $\alpha(x) = \Delta d(x) - 2c - 1 + k$ as in (1.1.11), to obtain (when $d(x)$ is, possibly, suitably rescaled, see Section 1.2) properties (p.1), (p.2), (p.3) listed in (1.1.12), (1.1.14), (1.1.15b), which then verify inequalities (4.17)–(4.19), respectively. We thus obtain from (4.6):

Corollary 4.2. With $0 < c < 1$ chosen in (1.1.8d), let $d(x) \in C^3(\mathbb{R}^n_x)$, and $\alpha(x) \in C^1(\mathbb{R}^n_x)$ be two functions such that inequalities (H.1) = (4.17), (H.2) = (4.18), (H.3) = (4.19) hold true. This is the case, in particular, if $d(x)$ is a (suitably rescaled, see Section 1.2) strictly convex function provided by assumptions (A.1) and (A.2), and then $\alpha(x) = \Delta d(x) - 2c - 1 + k$, as in (1.1.11) with k subject to

(1.1.14b). Let $w \in C^2(\mathbb{R}_t \times \mathbb{R}^n_x)$. Then, with such choices in (4.2), (4.3) for $\ell(t,x)$ and $\psi(x)$, respectively, Theorem 4.1, Eqn. (4.6), specializes to

(4.20)
$$\theta^2(w_{tt} - \Delta w)^2 - \frac{\partial M}{\partial t} + \text{div } V \geq 2\tau\rho[v_t^2 + |\nabla v|^2] + \tilde{B}v^2, \ 0 \leq t \leq T, \ x \in \bar{\Omega},$$

where, with the constant $\tilde{\beta} > 0$, we have via (4.8), (4.19):

(4.21) $\qquad \tilde{B}v^2 \geq [2\tau^3\tilde{\beta} + \mathcal{O}(\tau^2)]v^2, \quad \forall (t,x) \in Q^*(\sigma^*).$

Moreover, the scalar function M and the vector function V are given by (3.3) and (3.4), respectively, as functions of w, as specialized via (4.4). In particular, for future use below, we note that (3.4) yields on the boundary $\Gamma = \partial\Omega$, with outward unit normal $\nu = [\nu_1, \ldots, \nu_n]$, the following identity where $\nabla\ell = \tau\nabla d = \tau h$:

(4.22) on $\Gamma: V \cdot \nu = \sum_{j=1}^n V_j\nu_j = 2\theta^2 \Big\{ (w_t^2 - |\nabla w|^2)\nabla\ell \cdot \nu - 2\ell_t w_t \nabla w \cdot \nu$

$$+ 2(\nabla\ell \cdot \nabla w)\nabla w \cdot \nu + 2\left(|\nabla\ell|^2 - \ell_t^2 + \tau\frac{\alpha}{2}\right)w\nabla w \cdot \nu$$

$$+ (|\nabla\ell|^2 - \ell_t^2 - A)w^2 \nabla\ell \cdot \nu \Big\};$$

Moreover, via (4.5) for ℓ_t and (4.7) for A, we have:

(4.23) $\quad (|\nabla\ell|^2 - \ell_t^2 - A) = 2\tau^2\left[|h|^2 - 4c^2\left(t - \frac{T}{2}\right)^2\right] + \tau(\alpha - \Delta d - 2c). \qquad \square$

Notice that, through M in (3.3) and V in (3.4), the left-hand side of (4.20) is expressed in terms of w, while instead the right-hand side of (4.20) is still expressed in terms of $v = \theta w$, see (3.30). We no remedy this, and obtain a further corollary involving only w: it is then this corollary which, of course, will be used in the sequel.

Corollary 4.3. With $0 < c < 1$ chosen in (1.1.8d), let $d(x) \in C^3(\mathbb{R}^n_x)$, and $\alpha(x) \in C^1(\mathbb{R}^n_x)$ be two functions such that inequalities (H.1) - (4.17), (H.2) - (4.18), (H.3) = (4.19) hold true. This is the case, in particular, if $d(x)$ is any strictly convex function (suitably rescaled, see Section 1.2) provided by assumptions (A.1) and (A.2), and then $\alpha(x) \equiv \Delta d(x) - 2c - 1 + k$ as in (1.1.11), with k subject to (1.1.14b). Let $w \in C^2(\mathbb{R}_t \times \mathbb{R}^n_x)$.

Then, with such choices in (4.2), (4.3) for $\ell(t,x)$ and $\psi(x)$, respectively, Corollary 4.2 becomes: for any $1 > \epsilon > 0$, we obtain

(4.24) $\quad \theta^2(w_{tt} - \Delta w)^2 = \frac{\partial M}{\partial t} + \text{div } V \geq \epsilon\tau\rho\theta^2[w_t^2 + |\nabla w|^2] + \theta^2 B w^2,$

$$0 \leq t \leq T, \quad x \in \bar{\Omega},$$

where, recalling $\tilde{B}$ from (4.8), we have

(4.25a) $\qquad \begin{cases} B \equiv \tilde{B} - 2\epsilon\rho\tau^3(\phi_t^2 + |\nabla\phi|^2) \geq \tilde{B} - 2\epsilon\rho\tau^3 r; \\ \\ r = \max_{\bar{Q}}(\phi_t^2 + |\nabla\phi|^2). \end{cases}$

(4.25b)

Thus, for $\epsilon > 0$ suitably small, the cosntant $\beta \equiv \beta_\epsilon \equiv (\tilde{\beta} - \epsilon\rho r)$ is positive, via (4.19), and recalling (4.8), (4.19), (4.21), we obtain from (4.25):

(4.26) $$Bw^2 \geq [2\tau^3\beta + \mathcal{O}(\tau^2)]w^2, \quad \forall\, (t,x) \in Q^*(\sigma^*);$$

(4.27) $$\beta = \beta_\epsilon = (\tilde{\beta} - \epsilon\rho r) > 0;\ B = \mathcal{O}(\tau^3)\ \text{in}\ Q = (0,T] \times \Omega.$$

Proof. With $\ell_t = \tau\phi_t$ and $\ell_{x_j} = \tau\phi_{x_j} = \tau d_{x_j}$ from (4.2), we specialize v_t and v_x; in (3.30) and obtain from there that $\theta w_t = v_t - \theta\tau\phi_t w$; $\theta w_{x_j} = v_{x_j} - \theta\tau d_{x_j} w$, hence

(4.28) $$2v_t^2 \geq \theta^2 w_t^2 - 2\tau^2\phi_t^2 v^2;\quad 2|\nabla v|^2 \geq \theta^2|\nabla w|^2 - 2\tau^2|\nabla d|^2 v^2.$$

Returning to the right-hand side of (4.20), we then obtain via (4.28), for any $1 > \epsilon > 0$:

(4.29) $$2\tau\rho[v_t^2 + |\nabla v|^2] + \tilde{B}v^2 \geq \epsilon 2\tau\rho[v_t^2 + |\nabla v|^2] + \tilde{B}v^2$$

(4.30) $$\text{(by (4.28))} \geq \epsilon\tau\rho\theta^2[w_t^2 + |\nabla w|^2] + \tilde{B}v^2$$
$$- 2\epsilon\tau^3\rho[\phi_t^2 + |\nabla d|^2]v^2,$$

and (4.30) yields (4.24) as desired, via (4.25), as well as (4.26) and (4.27), as described below (4.25). □

5. Carleman estimates for smooth solutions of Eqn. (1.1a). First version

The next key result yields a Carleman-type estimate.

Theorem 5.1. With $0 < c < 1$ chosen in (1.1.8d), let $d(x) \in C^3(\bar{\Omega})$, $\alpha(x) \in C^1(\bar{\Omega})$ be two functions such that inequalities (H.1) = (4.17), (H.2) = (4.18), (H.3) = (4.19) hold true. This is the case, in particular, if $d(x)$ is a (suitably rescaled, see Section 1.2) strictly convex function provided by assumptions (A.1) and (A.2), and then $\alpha(x) = \Delta d(x) - 2c - 1 + k$, as in (1.1.11), with k subject to (1.1.14b). Let $\phi(x,t)$ be the pseudo-convex function defined by (1.1.8). Let $w \in C^2(\mathbb{R}_t \times \mathbb{R}_x^n)$ be a solution of Eqn. (1.1.1a) [and no B.C.], under the standing assumptions (1.1.2) for $F(w)$ and (1.1.3) for f. Then, the following one parameter family of estimates hold true, with $\rho > 0$, $\beta > 0$, as in (4.17)–(4.19); or (1.1.14a), (1.1.15b):

(i) for all $\tau > 0$ sufficiently large, and any $0 < \epsilon$ small:

(5.1) $$BT|_\Sigma + 2\int_0^T\!\!\int_\Omega e^{2\tau\phi}f^2 dQ + C_{1,T}e^{2\tau\sigma}\int_0^T\!\!\int_\Omega w^2 dQ$$

$$\geq (\tau\epsilon\rho - 2C_T)\int_0^T\!\!\int_\Omega e^{2\tau\phi}[w_t^2 + |\nabla w|^2]dQ$$

$$+ [2\tau^3\beta + \mathcal{O}(\tau^2) - 2C_T]\int_{Q(\sigma)} e^{2\tau\phi}w^2 dx\, dt$$

$$- c_T\tau^3 e^{-2\tau\delta}[E(0) + E(T)],$$

where $Q(\sigma)$ is the subset of $[0,T] \times \Omega \equiv Q$ defined by (1.1.19), where we recall from (4.26) that β depends on ϵ:

(ii) for all $\tau > 0$ sufficiently large and any $\epsilon > 0$ small,

$$(5.2) \quad BT|_\Sigma + 2\int_0^T \int_\Omega e^{2\tau\phi} f^2 dQ \geq (\tau\epsilon\rho - 2C_T)e^{2\tau\sigma} \int_{t_0}^{t_1} \int_\Omega [w_t^2 + |\nabla w|^2] d\Omega\, dt$$

$$- C_{1,T} e^{2\tau\sigma} \int_0^T E(t) dt - c_T \tau^3 e^{-2\tau\delta}[E(0) + E(T)].$$

Here $\delta > 0$, $\sigma > 0$, and $\sigma > -\delta$, are the constants in (1.1.9), (1.1.10), while $C_T > 0$ is a positive constant depending on T, as well as d. Moreover, the boundary terms $BT|_\Sigma$, $\Sigma = [0,T] \times \Gamma$, are defined by

$$(5.3) \quad BT|_\Sigma \equiv \int_0^T \int_\Omega \operatorname{div} V\, d\Omega\, dt = \int_0^T \int_\Gamma V \cdot \nu\, d\Gamma\, dt,$$

via the divergence theorem, and are explicitly given via (4.22), (4.23), $h \equiv \nabla d$ and $\nabla \ell = \tau \nabla d = \tau h$, ℓ_t in (4.5) and A in (4.7), and $\theta = \exp(\tau\phi)$ in (4.3), by:

$$(5.4) \quad BT|_\Sigma = 2\tau \int_0^T \int_{\Gamma_1} e^{2\tau\phi}(w_t^2 - |\nabla w|^2) h \cdot \nu\, d\Gamma\, dt$$

$$+ 8c\tau \int_0^T \int_\Gamma e^{2\tau\phi}\left(t - \frac{T}{2}\right) w_t \frac{\partial w}{\partial \nu}\, d\Gamma\, dt$$

$$+ 4\tau \int_0^T \int_\Gamma e^{2\tau\phi}(h \cdot \nabla w)\frac{\partial w}{\partial \nu}\, d\Gamma\, dt$$

$$+ 4\tau^2 \int_0^T \int_\Gamma e^{2\tau\phi}\left[|h|^2 - 4c^2\left(t - \frac{T}{2}\right)^2 + \frac{\alpha}{2\tau}\right] w \frac{\partial w}{\partial \nu}\, d\Gamma\, dt$$

$$+ 2\tau \int_0^T \int_{\Gamma_1} e^{2\tau\phi}\left[2\tau^2\left(|h|^2 - 4c^2\left(t - \frac{T}{2}\right)^2\right)\right.$$

$$\left. + \tau(\alpha - \Delta d - 2c)\right] w^2 h \cdot \nu\, d\Gamma\, dt,$$

since $h \cdot \nu = 0$ on Γ_0 by assumption (1.1.4). Moreover, as in (2.1.1), we have set

$$(5.5) \quad E(t) \equiv \int_\Omega [w_t^2(t,x) + |\nabla w(t,x)|^2 + w^2(t,x)] d\Omega.$$

(iii) The above inequality (5.2) may be then extended to all $w \in H^{2,2}(Q) = L_2(0,T; H^2(\Omega)) \cap H^2(0,T; L_2(\Omega))$. □

Proof. (i) **Step 1.** With $w \in C^2(\mathbb{R}_t \times \mathbb{R}_x^n)$, we return to inequality (4.24) of Corollary 4.3, supplemented by estimate (4.26) for Bw^2, and identity (4.22) for $V \cdot \nu$. Invoking the divergence theorem, we then obtain, with $\epsilon > 0$ fixed and small as in (4.27),

$$(5.6) \quad \int_0^T \int_\Omega \theta^2 (w_{tt} - \Delta w)^2 d\Omega\, dt - \left[\int_\Omega M\, d\Omega \right]_0^T + \int_0^T \int_\Gamma V \cdot \nu\, d\Gamma\, dt$$

$$\geq \epsilon \tau \rho \int_0^T \int_\Omega \theta^2 [w_t^2 + |\nabla w|^2] d\Omega\, dt + \int_0^T \int_\Omega \theta^2 B w^2 d\Omega\, dt.$$

Next, in view of estimate (4.26) for Bw^2 which holds true only on the subset $Q^*(\sigma^*)$ of Q defined by (1.1.16), split $Q \equiv [0,T] \times \Omega = Q^*(\sigma^*) \cup [Q^*(\sigma^*)]^c$ where $[\ ]^c$ denotes the complement in Q. See Figure 2. Thus, we then obtain by (4.26):

$$(5.7) \quad \int_0^T \int_\Omega \theta^2 Bw^2 d\Omega\, dt = \int_{Q^*(\sigma^*)} \theta^2 Bw^2 dx\, dt + \int_{[Q^*(\sigma^*)]^c} \theta^2 Bw^2 dx\, dt$$

$$\geq [2\tau^3 \beta + \mathcal{O}(\tau^2)] \int_{Q^*(\sigma^*)} \theta^2 w^2 dx\, dt + \int_{[Q^*(\sigma^*)]^c} \theta^2 Bw^2 dx\, dt.$$

Since the right side of (1.1a) is subject to estimate (1.2), we then obtain

$$(5.8) \quad \int_0^T \int_\Omega \theta^2 (w_{tt} - \Delta w)^2 d\Omega\, dt$$

$$\leq 2 C_T \left[\int_0^T \int_\Omega \theta^2 [w_t^2 + |\nabla w|^2 + w^2] d\Omega\, dt \right] + 2 \int_0^T \int_\Omega \theta^2 f^2 d\Omega\, dt.$$

As to the term $[\ ]_0^T$ at the time endpoints, if we recall M from (3.3), as well as (4.2)–(4.5), and A from (4.7), we then obtain (τ^3 comes from ℓ_t^3):

$$(5.9) \quad \left| \left[\int_\Omega M\, d\Omega \right]_0^T \right| \leq c_T \tau^3 \left[\int_\Omega e^{2\tau\phi}[w_t^2 + |\nabla w|^2 + w^2] d\Omega \right]_0^T$$

$$\text{(by (1.1.9))} \quad \leq c_T \tau^3 e^{-2\tau\delta} \left[\int_\Omega [w_t^2 + |\nabla w|^2 + w^2] d\Omega \right]_0^T$$

$$\text{(by (5.5))} \quad \leq c_T \tau^3 e^{-2\tau\delta} [E(0) + E(T)],$$

where in the last two steps we have recalled the critical property (1.1.9) for ϕ at $t = 0$ and $t = T$, as well as the definition of $E(t)$ in (5.5).

Next, via (4.22), (4.23), supplemented by (4.3)–(4.5), we obtain that the boundary terms $BT|_\Sigma$, as defined by (5.3), are explicitly given by (5.4).

Finally, we use (5.7), as well as (5.9), and (5.4) on the left side of inequality (5.6), and readily obtain for τ sufficiently large:

$$(5.10) \quad BT|_\Sigma + 2\int_0^T \int_\Omega e^{2\tau\phi} f^2 dQ - \int_{[Q^*(\sigma^*)]^c} e^{2\tau\phi} Bw^2 dx\, dt$$

$$\geq (\epsilon\tau\rho - 2C_T) \int_0^T \int_\Omega e^{2\tau\phi}[w_t^2 + |\nabla w|^2] dQ$$

$$+ [2\tau^3\beta + \mathcal{O}(\tau^2)] \int_{Q^*(\sigma^*)} e^{2\tau\phi} w^2 dx\, dt$$

$$- 2C_T \int_0^T \int_\Omega e^{2\tau\phi} w^2 dQ - c_T \tau^3 e^{-2\tau\delta}[E(0) + E(T)].$$

Step 2. By (4.27), we have $B = \mathcal{O}(\tau^3)$ on $[0,T] \times \Omega$. Moreover, we have that $\phi \leq \phi^* \leq \sigma^*$ on $[Q^*(\sigma^*)]^c$ by the very definition (1.1.16) and (1.1.18). Hence

$$(5.11) \quad -\int_{[Q^*(\sigma^*)]^c} e^{2\tau\phi} Bw^2 dx\, dt = \mathcal{O}(\tau^3) e^{2\tau\sigma^*} \int_{[Q^*(\sigma^*)]^c} w^2 dx\, dt.$$

Step 3. Recalling the subset $Q(\sigma)$ of $[0,T] \times \Omega \equiv Q$ in (1.1.19), we split $Q \equiv [0,T] \times \Omega = Q(\sigma) \cup [Q(\sigma)]^c$, where $[\quad]^c$ denotes complement in Q. Accordingly,

$$(5.12) \quad \int_0^T \int_\Omega e^{2\tau\phi} w^2 dQ = \int_{Q(\sigma)} e^{2\tau\phi} w^2 dx\, dt + \int_{[Q(\sigma)]^c} e^{2\tau\phi} w^2 dx\, dt.$$

Moreover, since $Q^*(\sigma^*) \supset Q(\sigma)$, see (1.1.20), we have via (5.12) for two right terms of (5.10):

$$(5.13) \quad [2\tau^3\beta + \mathcal{O}(\tau^2)] \int_{Q^*(\sigma^*)} e^{2\tau\phi} w^2 dx\, dt - 2C_T \int_0^T \int_\Omega e^{2\tau\phi} w^2 dQ$$

$$\geq [2\tau^3\beta + \mathcal{O}(\tau^2)] \int_{Q(\sigma)} e^{2\tau\phi} w^2 dx\, dt - 2C_T \int_{Q(\sigma)} e^{2\tau\phi} w^2 dx\, dt$$

$$- 2C_T \int_{[Q(\sigma)]^c} e^{2\tau\phi} w^2 dx\, dt.$$

Finally, in the last integral term in (5.13), we use that $\phi \leq \sigma$ in $[Q(\sigma)]^c$, by the very definition (1.1.19), so that

$$(5.14) \quad -2C_T \int_{[Q(\sigma)]^c} e^{2\tau\phi} w^2 dx\, dt \geq -2C_T e^{2\tau\sigma} \int_{[Q(\sigma)]^c} w^2 dx\, dt.$$

Using (5.14) in (5.13), we then conclude that

$$
(5.15) \quad [2\tau^3 \beta + \mathcal{O}(\tau^2)] \int_{Q^*(\sigma^*)} e^{2\tau\phi} w^2 \, dx \, dt - 2C_T \int_0^T \int_\Omega e^{2\tau\phi} w^2 \, dQ
$$

$$
\geq [2\tau^3 \beta + \mathcal{O}(\tau^2) - 2C_T] \int_{Q(\sigma)} e^{2\tau\phi} w^2 \, dx \, dt
$$

$$
- 2C_T e^{2\tau\sigma} \int_{[Q(\sigma)]^c} w^2 \, dx \, dt.
$$

Step 4. Using (5.11) and (5.15) in (5.10) yields

$$
(5.16) \quad BT|_\Sigma + 2 \int_0^T \int_\Omega e^{2\tau\phi} f^2 \, dQ
$$

$$
+ \mathcal{O}(\tau^3) e^{2\tau\sigma^*} \int_{[Q^*(\sigma^*)]^c} w^2 \, dx \, dt + 2C_T e^{2\tau\sigma} \int_{[Q(\sigma)]^c} w^2 \, dx \, dt
$$

$$
\geq (\epsilon\tau\rho - 2C_T) \int_0^T \int_\Omega e^{2\tau\phi} [w_t^2 + |\nabla w|^2] \, dQ
$$

$$
+ [2\tau^3 \beta + \mathcal{O}(\tau^2) - 2C_T] \int_{Q(\sigma)} e^{2\tau\phi} w^2 \, dx \, dt
$$

$$
- c_T \tau^3 e^{-2\tau\delta} [E(0) + E(T)].
$$

Finally, we use that, by construction, both $[Q^*(\sigma^*)]^c$ and $[Q(\sigma)]^c$ are subsets of $[0,T] \times \Omega$; and that, moreover, by the selection process in (1.1.16) through (1.1.20), we have chosen $0 < \sigma^* < \sigma$. Hence, these two facts yield

$$
(5.17) \quad \mathcal{O}(\tau^3) e^{2\tau\sigma^*} \int_{[Q^*(\sigma^*)]^c} w^2 \, dx \, dt + 2C_T e^{2\tau\sigma} \int_{[Q(\sigma)]^c} w^2 \, dx \, dt
$$

$$
\leq [\mathcal{O}(\tau^3) e^{2\tau\sigma^*} + 2C_T e^{2\tau\sigma}] \int_0^T \int_\Omega w^2 \, dQ
$$

$$
(5.18) \quad \leq C_{1,T} e^{2\tau\sigma} \int_0^T \int_\Omega w^2 \, dQ,
$$

for all τ sufficiently large. Inserting (5.18) into (5.16) yields (5.1), as desired.

(ii) We take τ sufficiently large so that, since $\beta > 0$ by assumption, see (4.27), we then have that the term $[2\tau^3 \beta + \mathcal{O}(\tau^2) - 2C_T]$ is positive, and we then drop the corresponding lower order *interior* term involving w^2 in (5.1). Moreover, we invoke the critical property (1.1.10) for ϕ on the first integral term on the right side of (5.1). Finally we majorize $\int_\Omega w^2(t) d\Omega$ by $E(t)$, see (5.5). This way, (5.1) readily yields (5.2). □

Remark 5.1. The statement and the proof of Theorem 5.1 use—as a consequence of assumptions (A.1) and (A.2)—that the principal part of the coefficient B acting on w^2 is *positive only on the set* $Q^*(\sigma^*)$; see (4.21), (4.27), and, ultimately, estimate (1.1.15b). If, instead, through possibly further rescaling of $d(x)$ [and consequent deterioration of the minimal time T_0, see Section 1.2], we assume, as in (1.2.1), that the principal part of the coefficient of B_1, hence of B, is positive *on the entire cylinder* $[0,T] \times \Omega$, then the proof and statement of Theorem 5.1 simplify. In particular: the term $e^{2\tau\sigma} \int_0^T \int_\Omega w^2 dQ$ is omitted in (5.1); thus the term $-e^{2\tau\sigma} \int_0^T E(t)dt$ is omitted in (5.2). The subsequent proof in Section 6 simplifies accordingly. □

6. Carleman estimate for smooth solutions of Eqn. (1.1a). Second version

A preliminary equivalence. Let $u \in H^1(\Omega)$. Then the following inequality holds true: there exist positive constants $0 < k_1 < k_2 < \infty$, independent of u, such that

$$(6.1) \quad k_1 \int_\Omega [u^2 + |\nabla u|^2] d\Omega \leq \int_\Omega |\nabla u|^2 d\Omega + \int_{\tilde{\Gamma}_1} u^2 d\Gamma \leq k_2 \int_\Omega [u^2 + |\nabla u|^2] d\Omega,$$

where $\tilde{\Gamma}_1$ is any (fixed) portion of the boundary Γ with positive measure. Inequality (6.1) is obtained by combining the following two inequalities:

$$(6.2) \quad \int_\Omega u^2 d\Omega \leq c_1 \left[\int_\Omega |\nabla u|^2 d\Omega + \int_{\tilde{\Gamma}_1} u^2 d\Gamma \right]; \quad \int_{\tilde{\Gamma}_1} u^2 d\Gamma \leq c_2 \int_\Omega [u^2 + |\nabla u|^2] d\Omega.$$

The inequality on the left of (6.2) [ ] replaces Poincaré's inequality, while the inequality on the right of (6.2) stems from (a conservative version of) trace theory. Thus, for $w \in C^2(\mathbb{R}_t \times \mathbb{R}_x^n)$, in fact, in $H^{2,2}(Q)$, if we introduce

$$(6.3) \quad \mathcal{E}(t) \equiv \int_\Omega [|\nabla w(t)|^2 + w_t^2(t)] d\Omega + \int_{\Gamma_1} w^2(t) d\Gamma_1,$$

with $\Gamma_1 = \Gamma \setminus \Gamma_0$, where Γ_0 is defined by (1.4), and recall $E(t)$ from (2.1.1) = (5.5), then (6.1) yields the equivalence

$$(6.4) \quad a E(t) \leq \mathcal{E}(t) \leq b E(t),$$

for some positive constant $a > 0$, $b > 0$.

We can now state the main result of the present section.

Theorem 6.1. With $0 < c < 1$ chosen in (1.1.8d), let $d(x) \in C^3(\bar{\Omega})$, $\alpha(x) \in C^1(\bar{\Omega})$ be two function such that inequalities (H.1) = (4.17), (H.2) = (4.18), (H.3) = (4.19) hold true. This is the case, in particular, if $d(x)$ is a (possibly, suitably rescaled, see Section 1.2) strictly convex function provided by assumptions (A.1) and (A.2), and then $\alpha(x) = \Delta d(x) - 2c - 1 + k$ as in (1.1.11), with k subject to (1.1.14b). Let $\phi(t,x)$ be the pseudo-convex function defined by (1.7) and define $\theta(t,x) = \exp(\tau\phi(t,x))$ as in (4.3). Finally, let $w \in H^{2,2}(Q)$ be a solution of Eqn. (1.1.1a) [and no B.C.], subject to the standing assumptions (1.1.2) on $F(w)$ and (1.1.3) on f. Then, the following one-parameter family of estimates hold true, for all τ sufficiently large, and any $\epsilon > 0$ small as in Corollary 4.3:

$$(6.5) \quad \overline{BT}|_\Sigma + 2\int_0^T \int_\Omega e^{2\tau\phi} f^2 dQ + \text{const}_\phi \int_0^T \int_\Omega f^2 dQ$$

$$\geq \left\{ \left[\frac{a}{2}(\epsilon\tau\rho - 2C_T)(t_1 - t_0)e^{-C_T T} - \frac{C_{1,T} b}{2a} Te^{C_T T}\right] e^{2\tau\sigma} \right.$$

$$\left. - c_T \tau^3 e^{-2\tau\delta} \right\} [E(0) + E(T)]$$

$$(6.6) \quad \geq k_\phi[E(0) + E(T)], \text{ for a constant } k_\phi > 0,$$

since $\sigma > -\delta$, see (1.1.10).

(a) Here, with $h = \nabla d = \nabla\phi$, the boundary terms $\overline{BT}|_\Sigma$ are given in terms of the boundary terms $BT|_\Sigma$ in (5.4) by

$$(6.7) \quad \overline{BT}|_\Sigma = BT|_\Sigma + \text{const}_\phi \left[\int_0^T \int_\Gamma \left|\frac{\partial w}{\partial\nu}\right| w_t\, d\Sigma \right.$$

$$\left. + \int_{t_0}^{t_1} \int_{\Gamma_1} w^2 d\Gamma_1 dt + \int_0^T \int_{\Gamma_1} |ww_t| d\Sigma_1 \right]$$

$$(6.8\text{a}) \quad (\text{by } (5.4)) = 2\tau \int_0^T \int_\Gamma e^{2\tau\phi}(w_t^2 - |\nabla w|^2) h \cdot \nu\, d\Gamma\, dt$$

$$+ 8c\tau \int_0^T \int_\Gamma e^{2\tau\phi}\left(t - \frac{T}{2}\right) w_t \frac{\partial w}{\partial\nu} d\Gamma\, dt$$

$$+ 4\tau \int_0^T \int_\Gamma e^{2\tau\phi}(h \cdot \nabla w) \frac{\partial w}{\partial\nu} d\Gamma\, dt$$

$$+ 4\tau^2 \int_0^T \int_\Gamma e^{2\tau\phi}\left[|h|^2 - 4c^2\left(t - \frac{T}{2}\right)^2 \right.$$

$$\left. + \frac{\alpha}{2\tau}\right] w \frac{\partial w}{\partial\nu} d\Gamma\, dt$$

$$+ 2\tau \int_0^T \int_\Gamma e^{2\tau\phi}\left[2\tau^2\left(|h|^2 - 4c^2\left(t - \frac{T}{2}\right)^2\right) \right.$$

$$\left. + \tau(\alpha - \Delta d - 2c)\right] w^2 h \cdot \nu\, d\Sigma$$

$$+ \text{const}_\phi \left\{\int_0^T \int_\Gamma \left|\frac{\partial w}{\partial\nu}\right| w_t\, d\Sigma \right.$$

$$\left. + \int_0^T \int_{\Gamma_1} |ww_t| d\Sigma_1 + \int_{t_0}^{t_1} \int_{\Gamma_1} w^2 d\Gamma_1 dt\right\}.$$

(b) Moreover, if in addition, w satisfies the pure Neumann B.C. $\frac{\partial w}{\partial \nu}|_\Sigma \equiv 0$ in (1.1.1c), so that $\nabla w = \nabla_{\tan} w$ (tangential gradient), then (6.8a) specializes (with $\alpha - \Delta d - 2c \equiv -4c - 1 + k$ by (1.1.13)) to:

$$
\text{(6.8b)} \quad \overline{BT}|_\Sigma = 2\tau \int_0^T \int_\Gamma e^{2\tau\phi}(w_t^2 - |\nabla_{\tan} w|^2) h \cdot \nu \, d\Sigma
$$

$$
+ 2\tau \int_0^T \int_\Gamma e^{2\tau\phi} \left[2\tau^2 \left(|h|^2 - 4c^2 \left(t - \frac{T}{2} \right)^2 \right) \right.
$$

$$
\left. - (4c + 1 - k)\tau \right] w^2 h \cdot \nu \, d\Sigma
$$

$$
+ \text{const}_\phi \left[\int_0^T \int_{\Gamma_1} |ww_t| d\Sigma_1 + \int_{t_0}^{t_1} \int_{\Gamma_1} w^2 d\Gamma_1 dt \right].
$$

Proof. Step 1. We return to estimate (5.2) of Theorem 5.1(ii), add the term $(\tau\epsilon\rho - 2C_T)e^{2\tau\sigma} \int_{t_0}^{t_1} \int_{\Gamma_1} w^2 d\Gamma_1 dt$ to both sides, recall (6.3) for $\mathcal{E}(t)$ and obtain

$$
\text{(6.9)} \quad BT|_\Sigma + (\tau\epsilon\rho - 2C_T)e^{2\tau\sigma} \int_{t_0}^{t_1} \int_{\Gamma_1} w^2 d\Gamma_1 dt + 2\int_0^T \int_\Omega e^{2\tau\phi} f^2 dQ
$$

$$
\geq (\tau\epsilon\rho - 2C_T)e^{2\tau\sigma} \int_{t_0}^{t_1} \mathcal{E}(t) dt - C_{1,T} e^{2\tau\sigma} \int_0^T E(t) dt
$$

$$
- c_T \tau^3 e^{-2\tau\delta} [E(0) + E(T)].
$$

Step 2. In a standard way, multiplying Eqn. (1.1a) by w_t and integrating over Ω yields, after an application of the first Green's identity

$$
\text{(6.10)} \quad \frac{1}{2} \frac{\partial}{\partial t} \left(\int_\Omega [w_t^2 + |\nabla w|^2] d\Omega + \int_{\Gamma_1} w^2 d\Gamma_1 \right)
$$

$$
= \int_\Gamma \frac{\partial w}{\partial \nu} w_t d\Gamma + \int_{\Gamma_1} ww_t d\Gamma_1 + \int_\Omega [F(w) + f] w_t d\Omega.
$$

Notice that on both sides of (6.10) we have added the term $\frac{1}{2} \frac{\partial}{\partial t} \int_{\Gamma_1} w^2 d\Gamma_1 = \int_{\Gamma_1} ww_t d\Gamma_1$. Recalling $\mathcal{E}(t)$ in (6.3), we integrate (6.10) over (s,t) and obtain

(6.11)
$$
\mathcal{E}(t) = \mathcal{E}(s) + 2\int_s^t \left[\int_\Gamma \frac{\partial w}{\partial \nu} w_t d\Gamma + \int_{\Gamma_1} ww_t d\Gamma_1 \right] dr + 2\int_s^t [F(w) + f] w_t d\Omega \, dr.
$$

We apply Schwarz inequality on $[F(w) + f]w_t$, recall estimate (1.2) for $F(w)$, invoke the left side $E(t) \leq \frac{1}{a}\mathcal{E}(t)$ of equivalence (6.4), and obtain

$$\mathcal{E}(t) \leq [\mathcal{E}(s) + N(T)] + C_T \int_s^t \mathcal{E}(r)dr; \tag{6.12}$$

$$\mathcal{E}(s) \leq [\mathcal{E}(t) + N(T)] + C_T \int_s^t \mathcal{E}(r)dr, \tag{6.13}$$

(C_T includes the constant $\frac{1}{a}$ of equivalence), where we have set

$$N(T) = \int_0^T \int_\Omega f^2 dQ + 2\int_0^T \int_\Gamma \left|\frac{\partial w}{\partial \nu} w_t\right| d\Sigma + 2\int_0^T \int_{\Gamma_1} |ww_t| d\Sigma_1. \tag{6.14}$$

Gronwall's inequality applied on (6.12), (6.13) then yields for $0 \leq s \leq t \leq T$,

$$\mathcal{E}(t) \leq [\mathcal{E}(s) + N(T)]e^{C_T(t-s)}; \quad \mathcal{E}(s) \leq [\mathcal{E}(t) + N(T)]e^{C_T(t-s)}. \tag{6.15}$$

Set $t = T$ and $s = t$ in the first (left) inequality of (6.15); and set $s = 0$ in the second (right) inequality of (6.15), to obtain

$$\mathcal{E}(T) \leq [\mathcal{E}(t) + N(T)]e^{C_T T}; \quad \mathcal{E}(0) \leq [\mathcal{E}(t) + N(T)]e^{C_T T}. \tag{6.16}$$

Summing up these two inequalites in (6.16) yields for $0 \leq t \leq T$,

$$\mathcal{E}(t) \geq \frac{\mathcal{E}(T) + \mathcal{E}(0)}{2} e^{-C_T T} - N(T) \tag{6.17}$$

$$\geq \frac{a}{2}[E(T) + E(0)]e^{-C_T T} - N(T), \tag{6.18}$$

after recalling the left side of the equivalence in (6.4). Similarly, summing up the left inequality of (6.15) for $s = 0$ and the right inequality of (6.15) for $s = t$ and $t = T$, and using the equivalence (6.4) yields for $0 \leq t \leq T$,

$$E(t) \leq \frac{1}{a}\left[\frac{b(E(0) + E(T))}{2} + N(T)\right]e^{C_T T}, \tag{6.19}$$

and hence, by (6.19),

$$-C_{1,T}e^{2\tau\sigma}\int_0^T E(t)dt \geq -\frac{C_{1,T}b}{2a} Te^{C_T T}e^{2\tau\sigma}[E(0) + E(T)] \tag{6.20}$$

$$-\frac{C_{1,T}T}{a} e^{C_T T}e^{2\tau\sigma}N(T).$$

Step 3. We insert (6.18) into the first integral on the right side of (6.9) and use (6.20) and readily obtain (6.5), (6.7), by invoking (6.14) for $N(T)$.

Finally, we recall the critical relation $\sigma > 0$, $\delta > 0$, $\sigma > -\delta$ from (1.1.9), (1.1.10), so that $[\epsilon\tau e^{2\tau\sigma} - \tau^3 e^{-2\tau\delta}]$ is positive for all τ large enough. Thus, (6.5) yields (6.6). □

7. A global uniqueness theorem with pure homogeneous Neumann B.C. on Σ

We consider the following over-determined problem with Γ_0 in (1.1.4) and $\Gamma_1 = \Gamma \setminus \Gamma_0$:

(7.1a) $\quad\quad\begin{cases} w_{tt} - \Delta w = F(w) & \text{in } (0,T] \times \Omega = Q; \\ \\ \left.\dfrac{\partial w}{\partial \nu}\right|_\Sigma \equiv 0 & \text{in } (0,T] \times \Gamma = \Sigma; \\ \\ w|_{\Sigma_1} \equiv 0 & \text{in } (0,T] \times \Gamma_1 = \Sigma_1. \end{cases}$

(7.1b)

(7.1c)

As a corollary of Theorem 6.1, we then obtain the following global uniqueness theorem.

Theorem 7.1. Assume hypotheses (A.1) and (A.2): thus there exists a strictly convex (possibly suitably rescaled, see Section 1.2) function $d(x)$, which along with the choice $\alpha(x) = \Delta d(x) - 2c - 1 + k$ in (1.1.11) [where $0 < c < 1$ chosen in (1.1.8d) and k subject to (1.1.14b)], satisfies properties $(p_1) = (1.1.12)$, $(p_2) = (1.1.14)$, $(p_3) = (1.1.15b)$, so that inequalities (H.1) = (4.17), (H.2) = (4.18), (H.3) = (4.19) hold true. Moreover, with $h = \nabla d$ as usual, we have that $h \cdot \nu = 0$ on Γ_0. Let $\Gamma_1 = \Gamma \setminus \Gamma_0$ as usual, and let $T > T_0$ in problem (7.1), with T_0 the constant in the definition (1.1.8b) of the pseudo-convex function $\phi(x,t)$ in (1.1.8a). Let $w \in H^{2,2}(Q)$ be a solution of problem (7.1a-b-c). Then, in fact, $w \equiv 0$ in Q; indeed, in $\mathbb{R}_t \times \Omega$.

Proof. Theorem 6.1 applies under the present assumptions (A.1) and (A.2). Thus estimate (6.6) holds true, where, because of the B.C. (7.1b), $\overline{BT}|_\Sigma$ is given by (6.8b). Moreover, we presently have two additional pieces of information: (i) $h \cdot \nu = 0$ on Γ_0 by assumption (A.1), and (ii) $w|_{\Sigma_1} \equiv 0$ by (7.1c); hence $w_t|_{\Sigma_1} \equiv |\nabla_{\tan} w|_{\Sigma_1} \equiv 0$, $\Sigma_1 = (0,T] \times \Gamma_1$, $\Gamma_0 \cup \Gamma_1 = \Gamma$. Thus, returning to (6.8b), we see that we now obtain

$$\overline{BT}|_\Sigma = 0, \text{ hence, by (6.6)}, \Rightarrow E(0) = 0, \text{ or } w_0 = w_1 = 0,$$

since $f = 0$ in (7.1). Then, as problem (7.1) is well-posed forward (and backward) in time as a s.c. group (F being a bounded operator: $\{w_1, w_2\} \in H^1(\Omega) \times L_2(\Omega) \to L_2(\Omega)$), we then obtain $w \equiv 0$ in Q; in fact, in $\mathbb{R}_t \times \Omega$. □

8. Extension of estimates to finite energy solutions

So far our estimates have been stated and proved only for $C^2(\mathbb{R}_t \times \mathbb{R}_x^n)$-solutions, hence $H^{2,2}(Q)$-solutions (Theorem 5.1(ii)), of Eqn. (1.1.1a), with $f \in L_2(Q)$ as in (1.1.3). In this section, we extend all our previous estimates to finite energy solutions of Eqn. (1.1.1a) in the following class

(8.1) $\quad\quad\begin{cases} w \in H^{1,1}(Q) = L_2(0,T; H^1(\Omega)) \cap H^1(0,T; L_2(\Omega)); \\ \\ w_t, \dfrac{\partial w}{\partial \nu} \in L_2(\Sigma) = L_2(0,T; L_2(\Gamma)). \end{cases}$

In order to achieve this goal, it suffices to extend the validity of estimate (5.1) of Theorem 5.1(i) from $H^{2,2}(Q)$-solutions to finite energy solutions defined by the class in (8.1). Here, the main difficulty is the fact that finite energy solutions subject to

Neumann B.C. do not produce (in dimension ≥ 2) H^1-traces on the boundary [L-T.5]. To overcome this difficulty, we shall invoke a regularizing procedure inspired from [La-Ta.1].

To this end, we shall make use of the following result.

Lemma 8.1. Let w be a solution of Eqn. (1.1a) in the class (8.1), with $f \in L_2(Q)$ as in (1.1.3). Then, in fact,

(i)

(8.2) $$\{w, w_t\} \in C([0,T]; H^1(\Omega) \times L_2(\Omega));$$

(ii) there is a constant $C_T > 0$, such that

(8.3) $$\int_0^T \int_\Gamma |\nabla_{\tan} w|^2 d\Gamma \, dt \leq C_T \left\{ \int_0^T \int_\Gamma \left[\left(\frac{\partial w}{\partial \nu}\right)^2 + w_t^2 \right] d\Sigma + \mathbb{E}(0) \right\},$$

where

(8.4) $$\mathbb{E}(0) = \|\{w(0, \cdot), w_t(0, \cdot)\}\|^2_{H^1(\Omega) \times L_2(\Omega)}.$$

Proof. (i) To Eqn. (1.1.1a) with $f \equiv 0$, we associate the B.C.

(8.5) $$\frac{\partial w}{\partial \nu} + w_t \equiv g \in L_2(\Sigma),$$

derived from (8.1b). Then, as is well-known [Las.1], [L-T.6, Chapter 7], for problem (1.1.1a) with $f \equiv 0$ and (8.5), the following regularity property holds true:

(8.6) the map : $f \equiv 0, g \to \{w, w_t\}$ is continuous :

$$L_2(\Sigma) \to C([0,T]; H^1(\Omega) \times L_2(\Omega)).$$

(This is so, because the presence of the boundary damping term w_t in the B.C. (8.5), which increases, when dim $\Omega \geq 2$, the interior regularity of $\{w, w_t\}$ to $H^1(\Omega) \times L_2(\Omega)$, over the case when such w_t is absent. [By (1.1.2), F facts as a bounded linear operator $\{w_1, w_2\} \in H^1(\Omega) \times L_2(\Omega) \to L_2(\Omega)$]. On the other hand, for Eqn. (1.1.1a) with $g = 0$ in (8.5), we have:

(8.7) the map : $g = 0, f \in L_2(Q)$

$$\to \{w, w_t\} \in C([0,T]; H^1(\Omega) \times L_2(\Omega)) \text{ is continuous}$$

(as seen, e.g., by using the variation of parameter formula, based on the corresponding s.c. contraction semigroup on $H^1(\Omega) \times L_2(\Omega)$, which describes the evolution of (1.1.1a) and (8.5) with $f = 0, g = 0$). Thus, (8.6), (8.7) yield (8.2).

(ii) Conclusion (8.3) follows from the by now classical identity, e.g. [Tr.1], [L-T.2], [Lio.1-2], obtained by the multiplier $m \cdot \nabla w$, where m is a $C^2(\bar{\Omega})$-vector field such that $m = \nu$ on Γ, so that $m \cdot \nu \equiv 1$, on Γ; i.e., by

(8.8) $$\int_\Sigma \frac{\partial w}{\partial \nu} m \cdot \nabla w \, d\Sigma + \frac{1}{2} \int_\Sigma \left\{ w_t^2 - \left[|\nabla_{\tan} w|^2 + \left(\frac{\partial w}{\partial \nu}\right)^2 \right] \right\} m \cdot \nu \, d\Sigma = \mathcal{O}(\mathbb{E}(0)).$$

Since $m = (m \cdot \nu)\nu + (m \cdot s)s$, s being a unit tangent vector on Γ, then

$$\nabla w \cdot m = (m \cdot \nu)\frac{\partial w}{\partial \nu} + (m \cdot s)\frac{\partial w}{\partial s}, \quad \nabla_{\tan} w = \frac{\partial w}{\partial s} s, \quad |\nabla_{\tan} w|^2 = \left(\frac{\partial w}{\partial s}\right)^2,$$

and (8.8) becomes with $m \cdot \nu = 1$,

$$(8.9) \quad \int_\Sigma \left(\frac{\partial w}{\partial s}\right)^2 d\Sigma = 2\int_\Sigma \left(\frac{\partial w}{\partial \nu}\right)\left(\frac{\partial w}{\partial s}\right) m \cdot s\, d\Sigma$$

$$+ \int_\Sigma \left[\left(\frac{\partial w}{\partial \nu}\right)^2 + w_t^2\right] d\Sigma + \mathcal{O}(\mathbb{E}(0)),$$

and (8.3) follows from (8.9) via $2ab \leq \epsilon a^2 + \frac{1}{\epsilon}b^2$, with $a = \frac{\partial w}{\partial s}$ and $b = \frac{\partial w}{\partial \nu}$. □

Theorem 8.2. Let $f \in L_2(Q)$. Let $w \in H^{2,2}(Q)$ be a solution of Eqn. (1.1.1a) for which inequality (5.1) holds true, at least as guaranteed by Theorem 5.1(iii). Let u be a solution of Eqn. (1.1.1a) in the class defined by (8.1): i.e., $u \in H^{1,1}(Q)$; $\frac{\partial u}{\partial \nu}, u_t \in L_2(\Sigma)$. Then, estimate (5.1) is satisfied by such solution u as well.

Proof. Step 1. Let u be a solution of Eqn. (1.1a) in the class (8.1), and define accordingly the boundary function $g = \frac{\partial u}{\partial \nu} + u_t \in L_2(\Sigma)$ as in (8.5) and the interior function $\hat{f} = Fu + f \in L_2(Q)$. Let $u_0 \in H^1(\Omega)$ and $u_1 \in L_2(\Omega)$ be the initial conditions for such solution u. Given these data, there exist sequences $\{\hat{f}^n\}_{n=0}^\infty$, $\{g^n\}_{n=0}^\infty$, $\{u_0^n\}_{n=0}^\infty$, $\{u_1^n\}_{n=0}^\infty$, such that

$$(8.10) \quad \hat{f}^n \in H^{1,1}(Q), \text{ and } \hat{f}^n \to \hat{f} \equiv Fu + f \text{ in } L_2(Q);$$

$$(8.11) \quad g^n \in H^{1,1}(\Sigma), \text{ and } g^n \to g \equiv \frac{\partial u}{\partial \nu} + u_t \text{ in } L_2(\Sigma);$$

$$(8.12) \quad u_0^n \in H^2(\Omega), \text{ and } u_0^n \to u_0 \text{ in } H^1(\Omega);$$

$$(8.13) \quad u_1^n \in H^1(\Omega), \text{ and } u_1^n \to u_1 \text{ in } L_2(\Omega),$$

subject to the Compatibility Relation (C.R.): $\frac{\partial u_0^n}{\partial \nu} + u_1^n = g^n(0)$ on Γ.

Next, we consider the problem corresponding to these smooth data

$$(8.14a) \quad \begin{cases} w_{tt}^n = \Delta w^n + \hat{f}^n & \text{in } Q; \\ (8.14b) \quad w^n(0, \cdot) = u_0^n;\ w_t^n(0, \cdot) = u_1^n & \text{in } \Omega; \\ (8.14c) \quad \dfrac{\partial w^n}{\partial \nu} + w_t^n = g^n & \text{in } \Sigma, \end{cases}$$

Step 2. It follows *a-fortiori* from the given regularity of the data in (8.10)–(8.13) (left), that problem (8.14) admits the following regularity:

$$(8.15) \quad \{w^n, w_t^n, w_{tt}^n\} \in C([0,T]; H^2(\Omega) \times H^1(\Omega) \times L_2(\Omega)).$$

Indeed, $g^n \in H^{1,1}(\Sigma)$ and $\hat{f}^n \in H^{1,1}(Q)$ imply *a-fortiori* [L-M.1, p. 19] that

(8.16)
$$g^n \in H^1(0,T; L_2(\Gamma)) \cap C([0,T]; H^{\frac{1}{2}}(\Gamma)), \quad \hat{f}^n \in H^1(0,T; L_2(\Omega)) \cap C([0,T]; H^{\frac{1}{2}}(\Omega)).$$

It will suffice to let $u_0^n = 0$; $u_1^n = 0$. Then, the problem

(8.17a) $$(w_t^n)_{tt} - \Delta w_t^n = \hat{f}_t^n \in L_2(0,T; L_2(\Omega));$$

(8.17b) $$\frac{\partial(w_t^n)}{\partial \nu} + (w_t^n)_t = g_t^n \in L_2(0,T; L_2(\Gamma)),$$

obtained from differentiating (8.14a,c) in t yields the regularity stated in (8.15) for $\{w_t^n, w_{tt}^n\}$, by simply invoking for problem (8.17) the regularity maps (8.6), (8.7), *mutatis mutandis*. This preliminary regularity of $\{w_t^n, w_{tt}^n\} \in C([0,T]; H^1(\Omega) \times L_2(\Omega))$, hence $w_t^n \in C([0,T]; H^{\frac{1}{2}}(\Gamma))$ by trace theory, is then used in (8.14) for the resulting elliptic problem at each t. Elliptic regularity [L-M.1, p. 188] with $\hat{f}^n$, g^n as in (8.16) (right yields $w^n \in H^2(\Omega)$ at each t, as desired. Thus, (8.15) is established.

Step 3. Next, multiplying problem (8.14a) by w_t^n and integrating by parts yields

(8.18) $$\int_\Omega [|w_t^n(t)|^2 + |\nabla w^n(t)|^2]d\Omega + \int_0^t \int_\Gamma |w_t^n|^2 d\Gamma \, dr$$
$$= \int_\Omega [|u_1^n|^2 + |\nabla u_0^n|^2]d\Omega + \int_0^t \int_\Gamma g^n w_t^n d\Gamma \, dr + \int_0^t \int_\Omega \hat{f}^n w_t^n d\Omega \, dr.$$

This, combined with estimate (8.3) then yields

(8.19) $$\int_0^t \int_\Omega [|w_t^n(t)|^2 + |\nabla w^n(t)|^2]d\Omega \, dr + \int_0^t \int_\Gamma \left(\frac{\partial w^n}{\partial \nu}\right)^2 d\Gamma \, dr$$
$$+ \int_0^t \int_\Gamma |\nabla_{\tan} w^n|^2 d\Gamma \, dr + \int_0^t \int_\Gamma |w_t^n|^2 d\Gamma \, dr$$
$$\leq C_T \left\{ \|\{u_0^n, u_1^n\}\|_{H^1(\Omega) \times L_2(\Omega)}^2 + \int_0^T \int_\Gamma |g^n|^2 d\Sigma \right.$$
$$\left. + \int_0^T \int_\Omega |\hat{f}^n|^2 dQ \right\}, \quad 0 \leq t \leq T.$$

In fact, integrating (8.18) over $\int_0^t$ and majorizing its right-hand side yields

(8.20) $$\left(1 - \frac{\epsilon}{2}\right) \int_0^t \int_\Omega (w_t^n) d\Omega \, dr$$
$$+ \int_0^t \int_\Omega |\nabla w^n|^2 d\Omega \, dr + \left(1 - \frac{\epsilon}{2}\right) \int_0^t \int_\Gamma (w_t^n) d\Gamma \, dr$$
$$\leq T \left\{ \|\{w_0^n, w_1^n\}\|_{H^1(\Omega) \times L_2(\Omega)}^2 \right.$$
$$\left. + \frac{1}{2\epsilon} \int_0^T \int_\Gamma (g^n)^2 d\Sigma + \frac{1}{2\epsilon} \int_0^T \int_\Omega (\hat{f}^n)^2 dQ \right\}.$$

Next, adding to both sides of (8.20) the quantity

$$\int_0^t \int_\Gamma |\nabla_{\tan} w^n|^2 d\Gamma \, dr + \int_0^t \int_\Gamma \left(\frac{\partial w^n}{\partial \nu}\right)^2 d\Gamma \, dr,$$

and using first inequality (8.3), next (8.5) for $\frac{\partial w^n}{\partial \nu}$, and finally the bound (8.20) for $\int_0^t \int_\Gamma (w_t^n)^2 d\Gamma \, dr$, yields readily (8.19), as desired.

Step 4. The above estimate (8.19), when applied to Cauchy sequences, allows one to pass to the limit and obtain the following convergence relations at the interior and at the boundary

(8.21a) $$w^n \to w^* \text{ in } H^1(Q).$$

(8.21b) $$\frac{\partial w^n}{\partial \nu} \to \frac{\partial w^*}{\partial \nu} \text{ in } L_2(\Sigma); \quad \frac{\partial w^n}{\partial s} \to \frac{\partial w^*}{\partial s} \text{ in } L_2(\Sigma); \quad w_t^n \to w_t^* \text{ in } L_2(\Sigma).$$

From (8.10)–(8.13) and (8.21), passing to the limit on problem (8.14), we obtain that the limit w^* obtained above in (8.21a) satisfies

(8.22a) $$\begin{cases} w_{tt}^* = \Delta w^* + F(u) + f & \text{on } Q; \\ w^*(0, \cdot) = u_0, \, w_t^*(0, \cdot) = u_1 & \text{in } \Omega; \\ \frac{\partial w^*}{\partial \nu} + w_t^* = g & \text{in } \Sigma. \end{cases}$$
(8.22b)
(8.22c)

Comparing problem (8.22) with the problem satisfied by u, by its very definition

(8.23a) $$\begin{cases} u_{tt} = \Delta u + F(u) + f & \text{in } Q; \\ u(0, \cdot) = u_0, \, u_t(0, \cdot) = u_1 & \text{in } \Omega; \\ \frac{\partial u}{\partial \nu} + u_t = g & \text{in } \Sigma, \end{cases}$$
(8.23b)
(8.23c)

we see that the difference $\hat{w} = w^* - u$ satisfies

(8.24) $$\hat{w}_{tt} = \Delta \hat{w} \text{ in } Q; \quad \hat{w}(0, \cdot) = 0, \, \hat{w}_t(0, \cdot) = 0 \text{ in } \Omega; \quad \frac{\partial \hat{w}}{\partial \nu} + \hat{w}_t = 0 \text{ in } \Sigma,$$

and hence

(8.25) $$\hat{w} = w^* - u \equiv 0 \text{ in } Q.$$

Step 5. Since the solution w^n of problem (8.14) is in $H^{2,2}(Q)$, by Step 1, then w^n satisfies estimate (5.1) of Theorem 5.1(ii) with f there replaced by $\hat{f} = Fu + f$ now. By the limit properties (8.21), we obtain that the limit $w^* = u$ satisfies estimate (5.1), with f there replaced by $\hat{f} = Fu + f$ now. Finally, recalling estimate (1.1.2) for F, we obtain that estimate (5.1) holds true for the postulated finite energy solution u in the class (8.1) as well. The proof of Theorem 8.2 is complete. □

Remark 8.1. As a consequence of Theorem 8.2, estimate (5.2) of Theorem 5.1(ii) and all subsequent estimates through Section 7 can be extended from $H^{2,2}(Q)$-solutions of Eqn. (1.1.1a) to finite energy solutions the class (8.1).

9. Continuous observability without geometrical conditions on Γ_1. Non-explicit constant

Key to the elimination of geometrical conditions on the (controlled or observed) portion Γ_1 of the boundary Γ, is the following result from [L-T.3, Section 7.2].

Lemma 9.1. Let w be a solution of Eqn. (1.1.1a) in the class (8.1). Given $\epsilon > 0$, $\epsilon_0 > 0$ arbitrary, given $T > 0$ there exists a constant $C_{\epsilon,\epsilon_0,T} > 0$ such that

$$(9.1) \quad \int_\epsilon^{T-\epsilon} \int_\Gamma |\nabla_{\tan} w|^2 d\Sigma \leq C_{\epsilon,\epsilon_0,T}\left\{\int_0^T \int_\Gamma \left[\left(\frac{\partial w}{\partial \nu}\right)^2 + w_t^2\right] d\Sigma \right.$$

$$\left. + \|w\|^2_{L_2(0,T;H^{\frac{1}{2}+\epsilon_0}(\Omega))} + \|f\|^2_{H^{-\frac{1}{2}+\epsilon_0}(Q)}\right\}. \quad \square$$

Using Lemma 9.1, we shall establish the sought-after continuous observability inequality.

Remark 9.1. We remark that estimate (9.1) is much sharper than (8.3) in that—unlike (8.3)—it is given in terms of a lower-order term, *below* energy level, while (8.3) contains an energy term $\mathbb{E}(0)$. The argument needed to prove (9.1) is much more subtle than the argument for (8.3). The 'loss' of ϵ in the time interval in (9.1) is not critical, as seen in the proof of Theorem 9.2.

Theorem 9.2. Assume hypotheses (A.1) and (A.2). Let $w \in H^{1,1}(Q)$ be a solution of problem (1.1.1a-b-c) with $f \equiv 0$. Then, the following continuous observability inequality holds true for $T > T_0$, with T_0 given by (1.1.8b): there exists a constant $C_T > 0$ such that

$$(9.2) \quad \int_0^T \int_{\Gamma_1} [w^2 + w_t^2] d\Sigma_1 \geq C_T E(0).$$

Proof. Step 1. Lemma 9.3. Under hypotheses (A.1) and (A.2), we first establish the weaker conclusion

$$(9.3) \quad \int_0^T \int_{\Gamma_1} [w^2 + w_t^2] d\Sigma + \|w\|^2_{L_2(0,T;H^{\frac{1}{2}+\epsilon_0}(\Omega))} \geq C_T E(0),$$

which is the desired inequality (9.2) polluted by the interior l.o.t $\|w\|$.

Proof of (9.3). To this end, we invoke Theorem 6.1—which holds true under the present assumptions (A.1) and (A.2), also for $H^{1,1}(Q)$-solutions of class (8.1), by virtue of the extension Theorem 8.2 and consequent Remark 8.1. We then apply estimate (6.6), except on the interval $[\epsilon, T - \epsilon]$, rather than on $[0, T]$ as in (6.6). Thus, we obtain since $f \equiv 0$:

$$(9.4) \quad \overline{BT}|_{[\epsilon, T-\epsilon] \times \Gamma} \geq k_\phi E(\epsilon),$$

where $\overline{BT}|_{[\epsilon,T-\epsilon]\times\Gamma}$ is the counterpart of (6.8b) since $\frac{\partial w}{\partial \nu}|_\Sigma \equiv 0$ by (1.1c), with the additional information that $h \cdot \nu = 0$ on Γ_0 by (A.1); i.e., with $\Sigma_1^\epsilon = (\epsilon, T - \epsilon) \times \Gamma_1$:

$$
(9.5) \quad \overline{BT}|_{[\epsilon,T-\epsilon]\times\Gamma} = 2\tau \int_\epsilon^{T-\epsilon} \int_{\Gamma_1} e^{2\tau\phi}(w_t^2 - |\nabla_{\tan}w|^2) h \cdot \nu \, d\Sigma_1
$$

$$
+ 2\tau \int_\epsilon^{T-\epsilon} \int_{\Gamma_1} e^{2\tau\phi}\left[2\tau^2\left(|h|^2 - 4c^2\left(t - \frac{T}{2}\right)^2\right)\right.
$$

$$
\left. - (4c + 1 - k)\tau\right] w^2 h \cdot \nu \, d\Sigma_1^\epsilon
$$

$$
+ \text{const}_\phi \left[\int_\epsilon^{T-\epsilon} \int_{\Gamma_1} |w w_t| d\Sigma_1^\epsilon + \int_{t_0}^{t_1} \int_{\Gamma_1} w^2 d\Gamma_1 dt\right].
$$

Next, by the right side of equivalences (6.4) and (6.18), we obtain

$$
(9.6) \quad E(\epsilon) \geq \frac{\mathcal{E}(\epsilon)}{b} \geq \frac{a}{2b} E(0) e^{-C_T T} - 2 \int_0^T \int_{\Gamma_1} |w w_t| d\Sigma_1,
$$

recalling $N(T)$ in (6.14) via (1.1.1c). We use (9.6) in (9.4). Finally, we invoke estimate (9.1) of Lemma 9.1 on the first integral term of (9.5) and recall that $\frac{\partial w}{\partial \nu}|_\Sigma = 0$ by (1.1.1c). This way, we readily obtain (9.3). □

Step 2. To eliminate the interior l.o.t. in estimate (9.3), we apply the by now standard compactness/uniqueness argument [Lit.1], [L-T.2], [Lio.1-2]. To this end, we need to invoke the global uniqueness Theorem 7.1, which has been extended to $H^{1,1}(Q)$-solutions by virtue of Theorem 8.2 and consequence Remark 8.1. □

10. Replacement of assumption (A.2) = (1.1.6) by virtue of two vector fields

Orientation. As the examples in Appendix A persuasively illustrate, assumption (A.2) = (1.1.6) introduces undesirable limitations, when enforced along with assumption (A.1) Accordingly, the purpose of this section is to replace this assumption suitably. This will be done by using the following strategy. First, from the given domain $\Omega \in \mathbb{R}^n$, we extract two overlapping subdomains Ω_1 and Ω_2, to which we can apply, separately, the setting of Section 1 (including the counterpart version of assumption (A.2) = (1.1.6)), in correspondence to two postulated strictly convex functions $d_1(x)$ and $d_2(x)$. This leads to the estimate of Corollary 4.3— Eqns. (4.24), (4.26)—to each subproblem on Ω_i: see estimates (10.2.16), (10.2.17) below of Proposition 10.2.1. Next, it is then a delicate matter to combine these estimates (10.2.16) for each separate subproblem to finally obtain the desired global estimate (10.5.1), hence (10.6.3), for the original problem, this time, however, having dispensed with assumption (A.2) = (1.1.6) on all of Ω. This latter step requires the introduction of a suitable cut-off function $\chi_i(t,x)$ [in (10.2.10) below] for each subproblem on Ω_i. In particular, each $\chi_i(t,x)$ is *space-independent* on a small layer of the boundary $\Gamma = \partial \Omega$, so that the cut-off solutions w_i corresponding to each subdomain Ω_i, $i = 1, 2$, and the original solution w have same traces on Γ, see (10.2.14) below, except for a multiplicative time-dependent function. In short: we replace assumptions (A.1), (A.2) in (1.1.4), (1.1.5), (1.1.6) of Section 1 with assumptions (A.1i), (A.2i) in (10.1.2), (10.1.3), (10.1.4) below. The latter are weaker than the former in many cases, e.g., when the unobserved boundary Γ_0 is *flat*: see Appendix. The advantages of removing assumption (A.2) = (1.1.6) on all of Ω are multiple. This is illustrated in the Appendix.

10.1. Basic setting using two conservative vector fields of the same class as in Section 1. Statement of main result. Continuous observability. Global uniqueness. Postulated setting. We divide the original Ω into two overlapping subdomains Ω_1 and Ω_2:

$$(10.1.1) \qquad \Omega = \Omega_1 \cup \Omega_2, \qquad \Omega_1 \cap \Omega_2 = \text{non-empty},$$

chosen (in infinitely many ways) as to fulfill the following conditions (following Section 1): there exist two strictly convex functions $d_i : \bar{\Omega} \to \mathbb{R}$ of class C^3, $i = 1, 2$, such that for the corresponding (conservative) vector fields $h_i(x) = [h_{i;1}(x), \ldots, h_{i;n}(x)] \equiv \nabla d_i(x)$, $x \in \Omega$, the following three properties hold true for $i = 1, 2$:

(A.1i) (a)

$$(10.1.2) \qquad \frac{\partial d_i}{\partial \nu} = \nabla d_i \cdot \nu = h_i \cdot \nu = 0 \quad \text{on } \Gamma_0, \ h_i = \nabla d_i;$$

(b) the Hessian matrix $\mathcal{H}_{d_i}$ of $d_i(x)$ [i.e., the Jacobian matrix J_{h_i} of $h_i(x)$] is strictly positive definite on $\bar{\Omega}_i$: there exists a constant $\rho_0 > 0$, such that for all $x \in \bar{\Omega}_i$, we have

$$(10.1.3) \quad \mathcal{H}_{d_i}(x) = J_{h_i}(x) = \begin{bmatrix} d_{i;x_1 x_1}, & \cdots, & d_{i;x_1 x_n} \\ \vdots & & \vdots \\ d_{i;x_n x_1}, & \cdots, & d_{i;x_n x_n} \end{bmatrix}$$

$$= \begin{bmatrix} \frac{\partial h_{i;1}}{\partial x_1}, & \cdots, & \frac{\partial h_{i;1}}{\partial x_n} \\ \vdots & & \vdots \\ \frac{\partial h_{i;n}}{\partial x_1}, & \cdots, & \frac{\partial h_{i;n}}{\partial x_n} \end{bmatrix} \geq \rho_0 I. \quad i = 1, 2.$$

Indeed, after rescaling if necessary (Section 1.2), we may and shall require that $\rho_0 \geq 2$;

(A.2i)

$$(10.1.4) \qquad \inf_{x \in \Omega_i} |h_i(x)| \equiv \inf_{x \in \Omega_i} |\nabla d_i(x)| \geq p > 0, \quad i = 1, 2.$$

The validity of the above setting, comprising hypotheses (A.1i), (A.2i), for large classes of triples $\{\Omega, \Gamma_0, \Gamma_1\}$ is established in Theorem C.1 of Appendix C. This culminates the analysis in the appendices, carried our by virtue of several different approaches, aimed at verifying the required geometrical assumptions of the setting of Section 1, as well as of the setting of the present Section 10.

Pseudo-convex functions. Let $d_i(x) : \bar{\Omega} \to \mathbb{R}$ be the C^3-functions provided by the above setting, $i = 1, 2$, and satisfying assumptions (A.1i) and (A.2i). Without loss of generality, we shall require—after, possibly, translation as in Section 1—that: $\min_{\bar{\Omega}_i} d_i(x) \equiv m > 0$. We then define

(10.1.5) $$T_{0,i}^2 = 4 \max_{x \in \bar{\Omega}_i} d_i(x), \quad i = 1, 2.$$

We next define the pseudo-convex functions

(10.1.6a) $$\phi_i(x,t) = d_i(x) - c\left(t - \frac{T}{2}\right)^2, \quad x \in \Omega, \ 0 \le t \le T,$$

where $T > T_{0,i}$, $i = 1, 2$, and the constant $0 < c < 1$ is selected as follows. If $T > T_{0,i}$, there exists a constant $\delta > 0$ such that [as in (1.1.8c)]

(10.1.6b) $$T^2 > 4 \max_{x \in \bar{\Omega}_i} d_i(x) + 4\delta, \quad i = 1, 2.$$

For such $\delta > 0$, there exists c, $0 < c < 1$, such that

(10.1.6c) $$cT^2 > 4 \max_{x \in \bar{\Omega}_i} d_i(x) + 4\delta, \quad i = 1, 2.$$

Henceforth, let $\phi_i(x,t)$ be defined by (10.1.6a) with T and c chosen above, unless otherwise explicitly noted. Such constant $0 < c < 1$, close to 1, may be taken independent of rescaling of $d_i(x)$, see Section 1.2. Such functions $\phi_i(x,t)$ have the following properties (as in Section 1):

(a) for the constant $\delta > 0$ fixed in (10.1.6b), we have

(10.1.7) $\phi_i(x,0) = \phi_i(x,T) = d_i(x) - c\dfrac{T^2}{4} \le -\delta$, uniformly in $x \in \Omega_i$, $i = 1, 2$;

(b) there are t_0, t_1, with $0 < t_0 < \frac{T}{2} < t_1 < T$, say symmetric about $\frac{T}{2}$, such that

(10.1.8) $\quad \phi_i(x,t) \ge \sigma > 0, \quad 0 < \sigma < m$, for all $(t,x) \in [t_0, t_1] \times \Omega_i$, $i = 1, 2$,

since $\phi_i(x, \frac{T}{2}) = d_i(x) \ge m > 0$ in Ω_i: indeed, we take

$$m - c\left(t_1 - \frac{T}{2}\right)^2 \equiv \sigma > 0 \text{ and } t_1 - \frac{T}{2} < \sqrt{\frac{m-\sigma}{c}},$$

as in Section 1.1.

Consequences of above setting. [Assumptions (A.1i), (A.2i)] Let $d_i(x) : \bar{\Omega}_i \to \mathbb{R}$, be the C^3-functions satisfying assumptions (A.1i), (A.2i) and the rescaling choice $\rho_0 \ge 2$. Then, as in Section 1, it follows that: there exist functions $\alpha_i(x) \in C^1(\bar{\Omega})$, in fact, take

(10.1.9) $$\alpha_i(x) \equiv \Delta d_i(x) - 2c - 1 + k \in C^1(\bar{\Omega})$$

for a constant $1 + 4c - 2\rho_0 < k < 1$, to be selected below, such that the following properties $(p_{1;i})$, $(p_{2;i})$ hold true:

$(p_{1;i})$

(10.1.10) $$\Delta d_i(x) - 2c - \alpha_i(x) \equiv 1 - k > 0, \quad x \in \bar{\Omega},$$

$(p_{2;i})$

(10.1.11) $$\gamma \equiv \alpha_i(x) - 2c - \Delta d_i(x) \equiv -4c - 1 + k,$$

and the Hessian matrix $\mathcal{H}_{d_i}$ of $d_i(x)$ satisfies the following inequality for all $x \in \bar{\Omega}_i$:

(10.1.12) $2\mathcal{H}_{d_i}(x) + [\alpha_i(x) - 2c - \Delta d_i(x)]I = 2\mathcal{H}_{d_i}(x) + \gamma I$

$$= \begin{bmatrix} 2d_{i;x_1x_1} + \gamma & 2d_{i;x_1x_2} & \cdots & 2d_{i;x_1x_n} \\ 2d_{i;x_2x_1} & 2d_{i;x_2x_2} + \gamma & \cdots & 2d_{i;x_2x_n} \\ \vdots & \vdots & \ddots & \vdots \\ 2d_{i;x_nx_1} & 2d_{i;x_nx_2} & \cdots & 2d_{i;x_nx_n} + \gamma \end{bmatrix}$$

$$\geq \rho I, \quad \forall\, x \in \Omega_i,\; i = 1, 2,$$

for some constant $\rho = 2\rho_0 + \gamma = 2\rho_0 - 4c - 1 + k > 0$, see (10.1.3), for a constant k chosen as to satisfy: $1 + 4c - 2\rho_0 < k < 1$, which is possible since $\rho_0 > 2c$, $0 < c < 1$, due to the choice $\rho \geq 2$. In addition, by additional rescaling, if necessary [at worst by imposing the rescaling condition: $|\nabla d_i(x)|^2 - 4d_i(x) \geq 0, \quad \forall\, x \in \Omega_i,\; i = 1, 2$ as in (1.2.3) of Section 1.2], we may require that such $d_i(x)$ fulfill, in addition, also the following property $(p_{3;i})$:

$(p_{3;i})$ Noting via (10.1.9) that

(10.1.13) $\qquad 6c + \Delta d_i(x) - \alpha_i(x) \equiv 8c + 1 - k, \quad x \in \bar{\Omega},$

we have that the following inequalities hold true, by virtue of assumption (A.2i) = (10.1.4):

(10.1.14) $\begin{cases} (2c + \Delta d_i - \alpha_i)|\nabla d_i|^2 + 2\mathcal{H}_{d_i}\nabla d_i \cdot \nabla d_i \\ \quad -(6c + \Delta d_i - \alpha_i)4c^2\left(t - \frac{T}{2}\right)^2 \\ \equiv (4c + 1 - k)|\nabla d_i|^2 + 2\mathcal{H}_{d_i}\nabla d_i \cdot \nabla d_i \\ \quad -(8c + 1 - k)4c^2\left(t - \frac{T}{2}\right)^2 \geq \tilde{\beta} > 0, \quad \forall\,(x,t) \in \text{set } Q_i^*(\sigma^*), \end{cases}$

for a constant $\tilde{\beta} > 0$, where the set $Q_i^*(\sigma^*)$ is defined (as in Section 1.1) by

(10.1.15) $\qquad Q_i^*(\sigma^*) \equiv \{(x,t) : x \in \Omega,\; 0 \leq t \leq T,\; \phi_i^*(x,t) \geq \sigma^* > 0\},$

for a constant σ^* chosen as to satisfy $0 < \sigma^* < \sigma$, see (10.1.8), where the functions ϕ_i^* are in turn defined by

(10.1.16) $\qquad \phi_i^*(x,t) \equiv d_i(x) - c^2\left(t - \frac{T}{2}\right)^2, \quad x \in \Omega,\; 0 \leq t \leq T.$

Since $0 < c < 1$, we obtain, by (10.1.6a) and (10.1.16), that

(10.1.17) $\qquad \phi_i^*(x,t) \geq \phi_i(x,t), \quad x \in \Omega,\; 0 \leq t \leq T.$

Thus, if we define, in agreement with (10.1.15), the set $Q_i(\sigma)$ by

(10.1.18) $\qquad Q_i(\sigma) = \{(x,t) : x \in \Omega,\; 0 \leq t \leq T,\; \phi_i(x,t) \geq \sigma > 0\},$

we see, since $0 < \sigma^* < \sigma$, and by (10.1.8), that:

(10.1.19)
$$\begin{cases} [t_0, t_1] \times \Omega_i \subset Q_i(\sigma) \subset Q_i^*(\sigma^*) \subset [0,T] \times \Omega, \text{ properly} \\ \text{by (10.1.7), at } t = 0 \text{ and } t = T \text{: no point } x \in \Omega_i \text{ belongs to } Q_i(\sigma). \end{cases}$$

The main result of this paper is the following Theorem 10.1.1, which in many cases (e.g., when Γ_0 is flat) extends Theorem 2.1.1, by replacing assumptions (A.1), (A.2) with assumptions (A.1i), (A.2i), $i = 1, 2$.

Theorem 10.1.1. Let the above setting of Section 10.1 based on assumptions (A.1i) and (A.2i) be in force for a given triple $\{\Omega, \Gamma_0, \Gamma_1\}$. Let $T_{0,i} > 0$ be the constants defined by (10.1.5) and let $T > T_{0,i}$. Then

(a) for all $T > T_{0,i}$, the following continuous observability inequality holds true for $H^{1,1}(Q)$-solutions of problem (1.1.1), with F satisfying (1.1.2): there is a constant $C_T > 0$ such that

(10.1.20) $$C_T E(0) \leq \int_0^T \int_{\Gamma_1} [w_t^2 + w^2] d\Sigma_1 + \int_0^T \int_\Omega f^2 dQ.$$

The constant C_T is *explicit* if in addition we assume geometrical star-shaped conditions on Γ_1:

(10.1.21) $$h_i \cdot \nu \geq 0, \quad i = 1, 2 \quad \text{on } \Gamma_1.$$

Such constant C_T is of the order of Ce^{Cr^2}, where C is a generic constant and r is the norm in (1.1.2c) on the involved coefficients.

(b) A-fortiori, the following global uniqueness result holds true: let $T > T_{0,i}$ and let w be an $H^{1,1}(Q)$-solution of problem (1.1.1a) with $f \equiv 0$, along with the B.C.

(10.1.22) $$\left.\frac{\partial w}{\partial \nu}\right|_\Sigma = 0 \quad \text{and} \quad w|_{\Sigma_1} = 0 \text{ where } h_i \cdot \nu = 0 \text{ on } \Gamma_0, \ i = 1, 2,$$

as in the assumed Eqn. (10.1.2). Then, in fact, $w \equiv 0$, in Q (in fact, $w \equiv 0$ in $\mathbb{R}_t \times \Omega$). □

The same comments as those below Theorem 2.1.1 apply. In effect we shall first prove the uniqueness statement of part (b) in Section 10.6 [as a direct consequence of the Carleman estimates of Theorem 10.5.1 for $H^{2,2}(Q)$-solutions]. Next, part (b) will be used to establish part (a) in Section 10.7.

A discussion was given in Section 1 on several topics including: (i) the role of the set $Q_i^*(\sigma^*)$ in relationship to the set $Q_i(\sigma)$; (ii) that in the case where Γ_0 is flat, whereby then $d_i(x) = |x - z_i|^2$, z_i just outside Ω on the hyperplane containing Γ_0 (radial vector field case), properties ($p_{1;i}$), ($p_{2;i}$), and ($p_{3;i}$) automatically hold true, with no rescaling needed; (iii) the issue of possibly rescaling $d_i(x)$ as to satisfy properties ($p_{1;i}$), ($p_{2;i}$), and ($p_{3;i}$), the latter one even on the cylinder $[0,T] \times \Omega_i$, at the price of deteriorating the minimal time $T_{0,i}$ of observability, by imposing the rescaling condition (as in (1.2.3)):

(10.1.23) $$|\nabla d_i(x)|^2 - 4d_i(x) \geq 0, \quad \forall\, x \in \Omega_i, \ i = 1, 2.$$

10.2. The cut-off functions $\chi_i(t,x)$ and corresponding subproblems for $w_i \equiv \chi_i w$. Preliminary estimate. Definition of cut-off functions χ_i.
Step 1. First, we recall the time-space sets $Q_i(\sigma) \subset Q_i^*(\sigma^*)$ (properly), defined in (10.1.15), (10.1.18) for $0 < \sigma^* < \sigma$, $i = 1, 2$, which are proper subsets of the basic cylinder $Q \equiv (0, T] \times \Omega$. Accordingly, we may introduce C^∞-functions $m_i(t,x)$, $0 \leq m_i(t,x) \leq 1$, on $[0,T] \times \Omega$, such that

(10.2.1) $\qquad m_i(t,x) \equiv 1$ on $Q_i(\sigma)$; $\quad$ supp $m_i \subset Q_i^*(\sigma^*)$,

so that $m_i \equiv 0$ on $[Q_i^*(\sigma^*)]^c$, the complement of $Q_i^*(\sigma^*)$ with respect to $[0,T] \times \Omega$.

Step 2. Next, for $x \in \mathbb{R}^n$, we define

(10.2.2) $\qquad \rho(x) = \inf_{y \in \Gamma} |x - y|, \qquad \Gamma = \partial\Omega,$

so that the set $\{x \in \Omega : \rho(x) < \epsilon\}$ is an ϵ-internal layer of the boundary $\Gamma = \partial\Omega$. For any $\epsilon > 0$, let μ_ϵ be the usual mollifier [Kes.1, p. 4]
(10.2.3a)
$$\mu_\epsilon(x) = \begin{cases} k\epsilon^{-n} \exp\left(-\frac{\epsilon^2}{\epsilon^2 - |x|^2}\right), & |x| < \epsilon \\ 0, & |x| \geq \epsilon \end{cases}, \quad k^{-1} = \int_{|x| \leq 1} \exp\left(-\frac{1}{1 - |x|^2}\right) dx,$$

and then let

(10.2.3b) $\qquad \rho^\epsilon(x) = (\mu_\epsilon * \rho)(x) \in C^\infty(\mathbb{R}^n), \qquad x \in \mathbb{R}^n.$

It is well known [Kes.1, p.] that, given $\epsilon_1 > 0$, then for all $\epsilon > 0$ sufficiently small, we have

(10.2.4) $\qquad \sup_{\bar{\Omega}} |\rho^\epsilon(x) - \rho(x)| < \frac{\epsilon_1}{4}.$

Step 3. Introduce the following two subsets of $[0,T]$ on the t-axis:

(10.2.5) $E_i(\sigma) \equiv$ orthogonal projection of the set $Q_i(\sigma)$ onto the t-axis;

(10.2.6) $E_i^*(\sigma^*) \equiv$ orthogonal projection of the set $Q_i^*(\sigma^*)$ onto the t-axis,

see Figure 3. Thus, properties (10.1.19) yield

(10.2.7) $\qquad [t_0, t_1] \subset E_i(\sigma) \subset E_i^*(\sigma^*) \subset [0,T]$ properly.

We then introduce a function $b_i(t) \in C_0^\infty(\mathbb{R}; [0,1])$ by setting

(10.2.8) $\qquad b_i(t) = \begin{cases} 1 & \text{for } t \in E_i(\sigma); \\ 0 & \text{for } t \in \mathbb{R} \setminus E_i^*(\sigma^*). \end{cases} \quad i = 1, 2.$

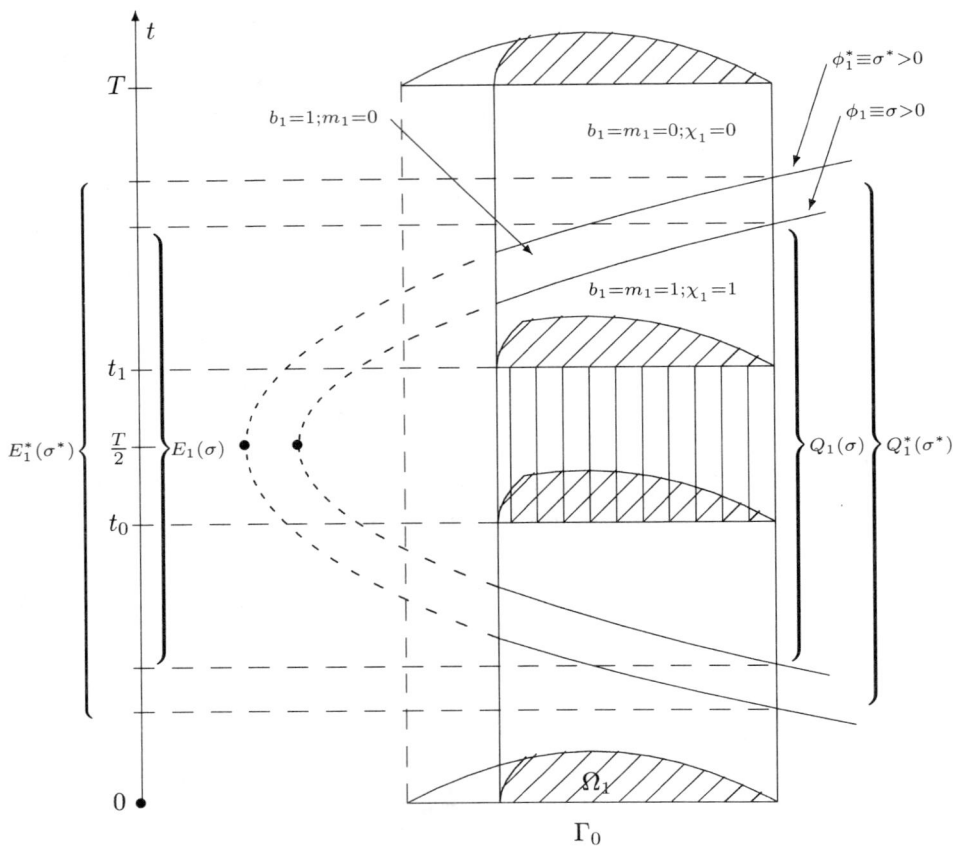

FIGURE 3

Step 4. Let $g(x)$ be a $C^\infty(\mathbb{R}; [0,1])$-function such that

(10.2.9) $$g(s) = \begin{cases} 1 & \text{for } s > \epsilon_1; \\ 0 & \text{for } s < \dfrac{\epsilon_1}{2}. \end{cases}$$

Finally, we define the following C^∞-function by setting

(10.2.10) $\quad \chi_i(t,x) = m_i(t,x)g(\rho^\epsilon(x)) + [1 - g(\rho^\epsilon(x))]b_i(t), \quad t, x \in [0,T] \times \Omega,$

with $\epsilon, \epsilon_1 > 0$ as in (10.2.4).

Properties of $\chi_i(t,x)$. First, let $x \in \Omega$ satisfy $\rho^\epsilon(x) < \frac{\epsilon_1}{2}$, i.e., $\rho(x) < \frac{\epsilon_1}{4}$ by (10.2.4), so that x lies in the $\frac{\epsilon_1}{4}$-internal layer of Γ. Then, $g(\rho^\epsilon(x)) = 0$ by (10.2.9), and then $\chi_i(t,x) \equiv b_i(t)$ for all $t \in [0,T]$.

Next, let $x \in \Omega$ satisfy $\rho^\epsilon(x) > \epsilon_1$, i.e., $\rho(x) > \frac{3}{4}\epsilon_1$. Then, $g(\rho^\epsilon(x)) = 1$ by (10.2.9), and then $\chi_i(t,x) = m_i(t,x)$ on $[0,T]$.

Also, if $(x,t) \in Q_i(\sigma)$, then: $m_i(t,x) = 1$ by (10.2.1), and $b_i(t) = 1$ by (10.2.8) and (10.2.5), and thus $\chi_i(t,x) \equiv m_i(t,x) \equiv 1$ by (10.2.10).

Let $t \in [0,T] \setminus E_i^*(\sigma^*)$: then $b_i(t) = 0$ by (10.2.8) and $m_i(t,x) = 0$ for all $x \in \Omega$ by (10.2.1) and (10.2.6). Hence $\chi_i(t,x) = 0$. We summarize (see Figure 3):

$$(10.2.11) \quad \chi_i(t,x) = \begin{cases} b_i(t) & \forall\, x \in \Omega \text{ s.t. } \rho^\epsilon(x) < \frac{\epsilon_1}{2}, \text{ or } \rho(x) < \frac{\epsilon_1}{4}; \\ m_i(t,x) & \forall\, x \in \Omega \text{ s.t. } \rho^\epsilon(x) > \epsilon_1, \text{ or } \rho(x) > \frac{3}{4}\epsilon_1; \end{cases}$$

$$(10.2.12) \quad \begin{cases} \chi_i(t,x) = m_i(t,x) = 1 \text{ on } Q_i(\sigma); \\ \chi_i(t,x) = 0, \quad \forall\, t \in \mathbb{R} \setminus E_i^*(\sigma^*),\ \forall x \in \Omega. \end{cases}$$

Remark 10.2.1. We can now explain our goal in this section. We sought cut-off functions which, among other features, are *only* time dependent (but *not* space dependent) on a small interior layer of the boundary Γ. This goal is dictated by the fact that we are dealing with Neumann B.C., and it would not be necessary if we were dealing instead with Dirichlet B.C. Once the above goal is achieved in (10.2.11), then the Neumann B.C. $\frac{\partial w_i}{\partial \nu}$ of the corresponding subproblems are readily expressed in terms of $\frac{\partial w}{\partial \nu}$ as in (10.2.14f). If, by contrast, χ_i were also space dependent near Γ, it would then pollute $\frac{\partial w_i}{\partial \nu}$.

Dynamical systems for $w_i \equiv \chi_i w$. Let $w \in C^2(\mathbb{R}_t \times \mathbb{R}_x^n)$ be a solution of Eqn. (1.1.1a). We then introduce new variables on $[0,T] \times \Omega$:

$$(10.2.13) \quad w_i(t,x) \equiv \chi_i(t,x) w(t,x); \quad f_i(t,x) \equiv \chi_i(t,x) w(t,x), \quad i = 1, 2.$$

Then, we see that each term w_i satisfies the following problem:

$$(10.2.14a) \quad \begin{cases} w_{i,tt} - \Delta w_i = F(w_i) + f_i + [D_t^2 - \Delta - F, \chi_i] w, \quad i = 1, 2; \\ w_i(0, \cdot) = w_{i,0},\ w_{i,t}(0, \cdot) = w_{i,1}, \quad \text{in } \Omega; \end{cases}$$
$(10.2.14b)$

$(10.2.14c) \quad \kappa_i \equiv [D_t^2 - \Delta - F, \chi_i]$ commutator active only on $(\operatorname{supp} \chi_i)$;

$(10.2.14d) \quad w_{i,0} = \chi_i(0, \cdot) w(0, \cdot);\ w_{i,1} = \chi_{i,t}(0, \cdot) w(0, \cdot) + \chi_i(0, \cdot) w_t(0, \cdot);$

$$(10.2.14e) \quad \begin{cases} w_i(t,x) = b_i(t) w(t,x),\ \forall\, x \text{ in an } \frac{\epsilon_1}{4}\text{-internal layer of } \Gamma, \\ \text{hence} \end{cases}$$
$$(10.2.14f) \quad w_i(t,x) = b(t) w(t,x);\quad \frac{\partial w_i}{\partial \nu} = b_i(t) \frac{\partial w}{\partial \nu} \quad \text{on } [0,T] \times \Gamma.$$

In (10.2.14a), $D_t = \frac{\partial}{\partial t}$, while $[\ ,\ ]$ denote the corresponding commutator of order 1 in time and space. Moreover, the key relation in (10.2.14e), hence the key trace properties in (10.2.14f), are a consequence of (10.2.11a): $\chi_i = b(t)$ on an $\frac{\epsilon_1}{4}$-internal layer of Γ.

Since χ_i is smooth and the commutator in (10.2.14a) is of order 1 in time and space, we then obtain via (10.2.14a) and (1.2) on F:

$$(10.2.15)$$
$$(w_{i,tt} - \Delta w_i)^2 \leq C_T \{ [w_{i,t}^2 + |\nabla w_i|^2 + w_i^2 + f_i^2] + [w_t^2 + |\nabla w|^2 + w^2] \},$$
$$(t,x) \in [0,T] \times \Omega.$$

Preliminary estimate: Counterpart of Corollary 4.3. As constructed above, each problem w_i in (10.2.14), $i = 1, 2$, satisfies the setting of assumptions (A.1i) and (A.2i), i.e., the setting of Section 1. As a result, each problem (10.2.14) satisfies the counterpart of Theorem 4.1/Corollary 4.2/Corollary 4.3, Eqns. (4.24)–(4.27), in particular, we recall (4.8). We take this result as our present starting point.

Proposition 10.2.1. Let $w \in C^2(\mathbb{R}_t \times \mathbb{R}_x^n)$ be a solution of Eqn. (1.1.1a). Let the setting of Section 10.1 based on assumptions (A.1i) and (A.2i) be in force. Then, each problem (10.2.14), $i = 1, 2$, satisfies the following pointwise inequality for $\epsilon > 0$ small:

$$(10.2.16) \quad \theta_i^2 (w_{i,tt} - \Delta w_i)^2 - \frac{\partial M_i}{\partial t} + \operatorname{div} V_i$$
$$\geq \epsilon\tau \theta_i^2 \rho[w_{i,t}^2 + |\nabla w_i|^2] + \theta_i^2 B_i w_i^2, \quad 0 \leq t \leq T, \ x \in \Omega,$$

see (4.24), where $\rho > 0$ is a constant, and where for $i = 1, 2$ and (10.1.14), we have recalling (4.8), (4.19), (4.25)–(4.27):

$$(10.2.17) \quad B_i \equiv \tilde{B}_i - 2\epsilon\rho\tau^3(\phi_{i,t}^2 + |\nabla d_i|^2)$$
$$\geq 2\tau^3 \Big\{ [2c + \Delta d_i - \alpha_i]|\nabla d_i|^2 + 2\mathcal{H}_{d_i}\nabla d_i \cdot \nabla d_i$$
$$- (6c + \Delta d_i - \alpha_i)4c^2 \left(t - \frac{T}{2}\right)^2 \Big\}$$
$$+ \mathcal{O}(\tau^2) - 2\epsilon\rho\tau^3 r$$
$$\geq 2\tau^3 \beta + \mathcal{O}(\tau^2), \quad \forall\, (x,t) \in \operatorname{set} Q_i^*(\sigma^*),$$

$$(10.2.18) \quad \beta \equiv \beta_\epsilon \equiv \tilde{\beta} - \epsilon\rho r > 0; \ r = \max_i \max_Q (\phi_{i,t}^2 + |\nabla d_i|^2); \ B_i = \mathcal{O}(\tau^3),$$

where the set $Q_i^*(\sigma^*)$ is defined in (10.1.15), and $\beta > 0$ is a constant depending on $\epsilon > 0$. Above, $\theta_i = e^{\tau\phi_i}$; ϕ_i as in (10.1.6). Moreover, M_i, V_i are obtained from M and V in (3.3), (3.4) in the present case: i.e., by replacing $\ell = \tau d$, $\psi = \tau\alpha$, etc., with $\ell_i = \tau d_i$, $\psi_i = \tau\alpha_i$, where d_i is given by assumption (A.1i) and α_i is defined by (10.1.9) so that $\tilde{B}_i$ on the left of (10.2.17) is the counterpart of (4.8), while the estimate in (10.2.17) is due to (10.1.14). $\square$

10.3. Proof of Theorem: Carleman estimate for the w_i-problem. Building up on Proposition 10.2.1, we obtain the counterpart of Theorem 5.1 (Carleman estimate, first version) for the w_i-problems.

Proposition 10.3.1. Let $w \in C^2(\mathbb{R}_t \times \mathbb{R}_x^n)$ be a solution of Eqn. (1.1.1a). Let the setting of Section 10.1 based on assumptions (A.1i) and (A.2i) be in force. Let $w_i = \chi_i w$ as in (10.2.13). Let $E(t)$ be defined by (5.5), or (2.1.1), as usual. Then, for $\epsilon > 0$ small as in (10.2.18), and for all τ sufficiently large (with $\epsilon\tau$ large with respect to C_T in (1.1.2)), the following estimate holds true:

$$(10.3.1) \quad (BT)_{w_i}|_\Sigma + C_{1,T}e^{2\tau\sigma}\int_0^T E(t)dt + C_{2,T}\int_0^T\int_\Omega f_i^2 d\Omega\, dt$$

$$\geq (\epsilon\tau\rho - 2C_T)\int_0^T\int_\Omega e^{2\tau\phi_i}[w_{i,t}^2 + |\nabla w_i|^2]d\Omega\, dt$$

$$+ [2\tau^3\beta + \mathcal{O}(\tau^2) - 2C_T]\int_{Q_i(\sigma)} e^{2\tau\phi_i}w^2 dx\, dt$$

$$- c_T\tau^3 e^{-2\tau\delta}[E(0) + E(T)],$$

where $Q_i(\sigma)$ is the subset of $[0,T]\times\Omega \equiv Q$ defined by (10.1.18); moreover, the constants $\sigma > 0$, $\delta > 0$ are defined in (10.1.8) and (10.1.7), while the constant $\beta > 0$ is defined via (10.1.14) for $\tilde{\beta}$ and (10.2.18). In (10.3.1), the boundary term $(BT)_{w_i}|_\Sigma$ is defined (counterpart of (5.3)) by

$$(10.3.2) \quad (BT)_{w_i}|_\Sigma \equiv \int_0^T\int_\Omega \operatorname{div} V_i d\Omega\, dt = \int_0^T\int_\Gamma V_i\cdot\nu\, d\Gamma\, dt, \quad i = 1,2,$$

and is explicitly given below (as in (5.4)), recalling $w_i = b_i w$, $\frac{\partial w_i}{\partial\nu} = b_i\frac{\partial w}{\partial\nu}$ on Γ, by (10.2.14f); as well as $h_i\cdot\nu = 0$ on Γ_0 by (10.1.21)):

$$(10.3.3) \quad (BT)_{w_i}|_\Sigma \equiv 2\tau\int_0^T\int_{\Gamma_1} e^{2\tau\phi_i}(w_{1,t}^2 - |\nabla w|^2 b_i^2(t))h_i\cdot\nu\, d\Gamma_1 dt$$

$$+ 8c\tau\int_0^T\int_\Gamma e^{2\tau\phi_i}\left(t - \frac{T}{2}\right)w_{i,t}\frac{\partial w}{\partial\nu}b_i^2(t)d\Gamma\, dt$$

$$+ 4\tau\int_0^T\int_\Gamma e^{2\tau\phi_i}(h_i\cdot\nabla w)\frac{\partial w}{\partial\nu}b_i^2(t)d\Gamma\, dt$$

$$+ 4\tau^2\int_0^T\int_\Gamma e^{2\tau\phi_i}\left[|h_i|^2 - 4c^2\left(t - \frac{T}{2}\right)^2\right.$$

$$\left. + \frac{\alpha_i}{2\tau}\right]w\frac{\partial w}{\partial\nu}b_i^2(t)d\Gamma\, dt$$

$$+ 2\tau\int_0^T\int_{\Gamma_1} e^{2\tau\phi_i}\left[2\tau^2\left(|h_i|^2 - 4c^2\left(t - \frac{T}{2}\right)^2\right)\right.$$

$$\left. + \tau(\alpha_i - \Delta d_i - 2c)\right]w^2 b_i^2(t)h_i\cdot\nu\, d\Gamma_1 dt.$$

(ii) The above inequality (10.3.3) may be extended to $H^{2,2}(Q)$-solutions.

Remark 10.3.1. For τ sufficiently large as to obtain $[2\tau^3\beta + \mathcal{O}(\tau^2) - C_T] > 0$, see (10.2.18), we reach one of our goals and drop the integral term involving w^2, accordingly, from inequality (10.3.1). □

Proof. Step 1. We return to Eqn. (10.2.16) of Proposition 10.2.1, which we now integrate over $[0, T] \times \Omega$. By use of the divergence theorem, we obtain the counterpart of (5.6) for $\epsilon > 0$:

$$(10.3.4) \quad \int_0^T \int_\Omega \theta_i^2 (w_{i,tt} - \Delta w_i)^2 d\Omega\, dt - \left[\int_\Omega M_i d\Omega\right]_0^T + \int_0^T \int_\Gamma V_i \cdot \nu\, d\Gamma\, dt$$

$$\geq \epsilon\tau\rho \int_0^T \int_\Omega \theta_i^2 [w_{i,t}^2 + |\nabla w_i|^2] d\Omega\, dt + \int_0^T \int_\Omega \theta_i^2 B_i w_i^2 \, d\Omega\, dt,$$

with $\rho > 0$. Moreover, M_i and V_i are the counterpart of Eqns. (3.3) and (3.4), as specialized to the two cases $i = 1, 2$. With reference to (10.3.4), we now define $(BT)_{w_i}|_\Sigma$ by (10.3.2), and then obtain, via the counterpart of (5.4), that $(BT)_{w_i}|_\Sigma$ coincides with the expression given by (10.3.3) [which is the counterpart of (5.4)]. In doing so, we use two ingredients: (a) that $h_i \cdot \nu \equiv 0$ on Γ_0, $i = 1, 2$ by assumption (10.1.2), so that integration where $h_i \cdot \nu$ occurs is restricted to Γ_1 only; (b) that by (10.2.14f) $w_i \equiv b_i(t)w$, $\frac{\partial w_i}{\partial \nu} \equiv b_i(t) \frac{\partial w}{\partial \nu}$ on an interior strip of Γ and for all $0 \leq t \leq T$, where $b_i(t)$ is defined by (10.2.8).

Step 2. We next estimate the first term on the left side of inequality (10.3.4). We shall prove that

$$(10.3.5)$$

$$\int_0^T \int_\Omega \theta_i^2 (w_{i,tt} - \Delta w_i)^2 d\Omega\, dt \leq C_T \Bigg\{ \int_0^T \int_\Omega e^{2\tau\phi_i} [w_{i,t}^2 + |\nabla w_i|^2] d\Omega\, dt$$

$$+ \int_{Q_i(\sigma)} e^{2\tau\phi_i} w_i^2 dx\, dt + e^{2\tau\sigma} \int_0^T E(t) dt + \int_0^T \int_\Omega f_i^2 d\Omega\, dt \Bigg\},$$

where the subset $Q_i(\sigma)$ of $[0, T] \times \Omega$ is defined in (10.1.18). In fact, to prove (10.3.5), we introduce the following simplified notation on $[0, T] \times \Omega$:

$$(10.3.6) \qquad e_i(t, x) \equiv w_{i,t}^2(t, x) + |\nabla w_i(t, x)|^2 + w_i^2(t, x);$$

$$(10.3.7) \qquad e(t, x) \equiv w_t^2(t, x) + |\nabla w(t, x)|^2 + w^2(t, x);$$

$$(10.3.8) \qquad E_i(t) \equiv \int_\Omega e_i(t, x) d\Omega; \quad E(t) \equiv \int_\Omega e(t, x) d\Omega.$$

As in the proof of Theorem 5.1, Step 3, we further split $Q \equiv [0, T] \times \Omega = Q_i(\sigma) \cup [Q_i(\sigma)]^c$, where $[\ ]^c$ denotes complement with respect to Q, and the set $Q_i(\sigma)$ is defined by (10.1.18). Moreover, we have

$$(10.3.9) \quad \begin{cases} \phi_i(t, x) \leq \sigma \text{ on } [Q_i(\sigma)]^c \text{ by (10.1.18)}; \ e_i(t, x) \leq \text{const } e(t, x) \\ \quad \text{in particular on } [Q_i(\sigma)]^c, \text{ by (10.2.10), (10.2.13)}; \\ e_i(t, x) \equiv e(t, x) \text{ on } Q_i(\sigma) \text{ by (10.2.12), i.e., } \chi_i \equiv 1 \text{ on } Q_i(\sigma). \end{cases}$$

We next invoke inequality (10.2.15), as well as (10.3.6)–(10.3.9), recall $\theta_i = e^{\tau\phi_i}$, and estimate

$$(10.3.10) \quad \int_0^T \int_\Omega \theta_i^2 (w_{i,tt} - \Delta w_i)^2 d\Omega\, dt$$

$$\leq C_T \left\{ \int_0^T \int_\Omega \theta_i^2 [e_i(t,x) + f_i^2 + e(t,x)] d\Omega\, dt \right\}$$

$$(10.3.11) \quad \leq C_T \left\{ \int_{Q_i(\sigma)} e^{2\tau\phi_i} [e_i(t,x) + e(t,x)] dx\, dt \right.$$

$$\left. + \int_{[Q_i(\sigma)]^c} e^{2\tau\phi_i} [e_i(t,x) + e(t,x)] dx\, dt + \int_0^T \int_\Omega f_i^2 d\Omega\, dt \right\}$$

(10.3.12) (by (10.3.9))

$$\leq C_T \left\{ \int_{Q_i(\sigma)} e^{2\tau\phi_i} 2 e_i(t,x) dx\, dt \right.$$

$$\left. + \text{const} \int_{[Q_i(\sigma)]^c} e^{2\tau\sigma} e(t,x) dx\, dt + \int_0^T \int_\Omega f_i^2 d\Omega\, dt \right\}$$

$$(10.3.13) \quad \leq C_T \left\{ 2 \int_{Q_i(\sigma)} e^{2\tau\phi_i} e_i(t,x) dx\, dt \right.$$

$$\left. + \text{const } e^{2\tau\sigma} \int_0^T \int_\Omega e(t,x) d\Omega\, dt + \int_0^T \int_\Omega f_i^2 d\Omega\, dt \right\}.$$

Finally, we estimate, since $Q_i(\sigma)$ is a subset of $[0,T] \times \Omega$, via (10.3.6):

$$(10.3.14) \quad \int_{Q_i(\sigma)} e^{2\tau\phi_i} e_i(t,x) dx\, dt$$

$$= \int_{Q_i(\sigma)} e^{2\tau\phi_i} [w_i^2(t,x) + w_{i,t}^2(x,t) + |\nabla w_i(x,t)|^2] dx\, dt$$

$$\leq \int_{Q_i(\sigma)} e^{2\tau\phi_i} w_i^2 dx\, dt + \int_0^T \int_\Omega e^{2\tau\phi_i} [w_{i,t}^2 + |\nabla w_i|^2] d\Omega\, dt,$$

by majorizing the integral term on $Q_i(\sigma)$ but only for the integrand $[w_{i,t}^2 + |\nabla w_i|^2]$. Finally, we insert (10.3.14) into the right side of (10.3.13) and obtain (10.3.5) by invoking (10.3.8).

Step 3. Next, we use the counterpart of inequality (5.9) in the present case, to obtain, recalling (10.1.7) for ϕ_i at $t = 0$ and $t = T$:

$$(10.3.15) \quad \left| \left[\int_{\Omega_i} M_i d\Omega_i \right]_0^T \right| \leq c_T \tau^3 \left[\int_{\Omega_i} e^{2\tau\phi_i} [w_{i,t}^2 + |\nabla w_i|^2 + w_i^2] d\Omega_i \right]_0^T$$

$$(10.3.16) \quad \text{(by (10.1.7))} \quad \leq c_T \tau^3 e^{-2\tau\delta} [E_i(0) + E_i(T)]$$

$$(10.3.17) \quad \text{(by (10.2.13))} \quad \leq c_T \tau^3 e^{-2\tau\delta} [E(0) + E(T)],$$

where $E_i(t)$ and $E(t)$ are defined by (10.3.8), and where in the last step we have recalled $w_i = \chi_i w$ from (10.2.13).

Step 4. Thus, inequalities (10.3.5) and (10.3.17) used on the left side of estimate (10.3.4), yield

$$(10.3.18) \quad C_T e^{2\tau\sigma} \int_0^T E(t) dt + C_T \int_0^T \int_\Omega f_i^2 d\Omega \, dt + (BT)_{w_i}|_\Sigma$$

$$\geq (\epsilon\tau\rho - C_T) \int_0^T \int_\Omega e^{2\tau\phi_i} [w_{i,t}^2 + |\nabla w_i|^2] d\Omega \, dt$$

$$+ \int_0^T \int_\Omega e^{2\tau\phi_i} B_i w_i^2 d\Omega \, dt$$

$$- C_T \int_{Q_i(\sigma)} e^{2\tau\phi_i} w_i^2 dx \, dt - c_T \tau^3 e^{-2\tau\delta} [E(0) + E(T)].$$

Step 5. In this step, we refine (10.3.18) to arrive at (10.3.1) by estimating the integral term containing B_i in (10.3.18). To this end, we proceed as in the proof of Theorem 5.1, Step 1. We split now $[0,T] \times \Omega = Q_i^*(\sigma^*) \cup [Q_i^*(\sigma^*)]^c$, where $[Q_i^*(\sigma^*)]^c$ is the complement in $[0,T] \times \Omega$ of the set $Q_i^*(\sigma^*)$ defined by (10.1.15). It is here that we use the critical property that B_i is strictly positive on $Q_i^*(\sigma^*)$ noted in (10.2.17) via (10.1.14). We compute

$$(10.3.19) \quad \int_0^T \int_\Omega e^{2\tau\phi_i} B_i w_i^2 d\Omega \, dt$$

$$= \int_{Q_i^*(\sigma^*)} e^{2\tau\phi_i} B_i w_i^2 dx \, dt + \int_{[Q_i^*(\sigma^*)]^c} e^{2\tau\phi_i} B_i w_i^2 dx \, dt$$

$$(10.3.20) \quad \text{(by (10.2.17))}$$

$$\geq [2\tau^3 \beta + \mathcal{O}(\tau^2)] \int_{Q_i^*(\sigma^*)} e^{2\tau\phi_i} w_i^2 dx \, dt + \int_{[Q_i^*(\sigma^*)]^c} e^{2\tau\phi_i} B_i w_i^2 dx \, dt.$$

As to the last term on the right of (10.3.20), we estimate as in the proof of Theorem 5.1, Step 2, recalling from its definition in (10.2.18) that B_i is $\mathcal{O}(\tau^3)$, and that $\phi_i \leq \phi_i^* \leq \sigma^*$ in $[Q_i^*(\sigma^*)]^c$, by the very definition of (10.1.15) and by (10.1.17),

$$(10.3.21) \quad -\int_{[Q_i^*(\sigma^*)]^c} e^{2\tau\phi_i} B_i w_i^2 \, dx \, dt \leq C\tau^3 \int_{[Q_i^*(\sigma^*)]^c} e^{2\tau\phi_i} w_i^2 \, dx \, dt$$

$$(10.3.22) \quad \leq C\tau^3 \int_{[Q_i^*(\sigma^*)]^c} e^{2\tau\sigma^*} w_i^2 \, dx \, dt \leq C\tau^3 e^{2\tau\sigma^*} \int_0^T \int_\Omega w^2 \, dx \, dt,$$

majorizing w_i^2 by const w^2, and $[Q_i^*(\sigma^*)]^c$ by $[0,T] \times \Omega$. Next, we insert (10.3.20) in the right side of (10.3.18), move the last term of (10.3.20) on $[Q_i^*(\sigma^*)]^c$ to the left side of (10.3.18), and apply for it estimate (10.3.22). We obtain

$$(10.3.23) \quad C_1 \tau^3 e^{2\tau\sigma^*} \int_0^T \int_\Omega w^2 \, d\Omega \, dt + C_T e^{2\tau\sigma} \int_0^T E(t) \, dt$$

$$+ C_T \int_0^T \int_\Omega f_i^2 \, d\Omega \, dt + (BT)_{w_i}|_\Sigma$$

$$\geq (\epsilon\tau\rho - C_T) \int_0^T \int_\Omega e^{2\tau\phi_i}[w_{i,t}^2 + |\nabla w_i|^2] \, d\Omega \, dt$$

$$+ [2\tau^3\beta + \mathcal{O}(\tau^2)] \int_{Q_i^*(\sigma^*)} e^{2\tau\phi_i} w_i^2 \, dx \, dt$$

$$- C_T \int_{Q_i(\sigma)} e^{2\tau\phi_i} w_i^2 \, dx \, dt - c_T \tau^3 e^{-2\tau\delta}[E(0) + E(T)].$$

Finally, we use the following estimates:

(i) Regarding the first two terms on the left of (10.3.23), we use that, by selection, $0 < \sigma^* < \sigma$, see (10.1.15)–(10.1.19). Thus, for all τ sufficiently large, the first term is absorbed by the second, and we obtain

$$(10.3.24) \quad C\tau^3 e^{2\tau\sigma^*} \int_0^T \int_\Omega w^2 \, d\Omega \, dt \leq C_{1,T} e^{2\tau\sigma} \int_0^T E(t) \, dt.$$

(ii) Regarding the integral terms on $Q_i^*(\sigma^*)$, $Q_i(\sigma)$, we see that since $Q_i^*(\sigma^*) \supset Q_i(\sigma)$, we have that for all τ sufficiently large, as in Step 3 in the proof of Theorem 5.1,

$$(10.3.25) \quad [2\tau^3\beta + \mathcal{O}(\tau^2)] \int_{Q_i^*(\sigma^*)} e^{2\tau\phi_i} w_i^2 \, dx \, dt - C_T \int_{Q_i(\sigma)} e^{2\tau\phi_i} w_i^2 \, dx \, dt$$

$$\geq [2\tau^3\beta + \mathcal{O}(\tau^2)] \int_{Q_i(\sigma)} e^{2\tau\phi_i} w_i^2 \, dx \, dt - C_T \int_{Q_i(\sigma)} e^{2\tau\phi_i} w_i^2 \, dx \, dt$$

$$(10.3.26) \quad = [2\tau^3\beta + \mathcal{O}(\tau^2) - C_T] \int_{Q_i(\sigma)} e^{2\tau\phi_i} w^2 \, dx \, dt,$$

since, on $Q_i(\sigma)$, we have that $\chi_i \equiv 1$, by (10.2.12), hence $w_i \equiv w$ by (10.2.13). Thus, using (10.3.24) and (10.3.25) in (10.3.23), we arrive at the derived estimate (10.3.1). The proof of Proposition 11.3.1 is complete.

10.4. Carleman estimate, first version, for the w-problem.

The counterpart of Theorem 5.1(ii) is now:

Theorem 10.4.1. Let $w \in H^{2,2}(Q)$ be a solution of Eqn. (1.1.1a). Let the setting of Section 10.1 based on assumptions (A.1i) and (A.2i) be in force. Then

(i) for all $\tau > 0$ sufficiently large (with $\epsilon\tau$ large, as compared to the constant C_T in (1.1.2)), the following one-parameter family of inequalities holds true:

$$(10.4.1) \qquad BT|_\Sigma + 2C_T \int_0^T \int_\Omega f^2 d\Omega\, dt \geq (\epsilon\tau\rho - 2C_T)e^{2\tau\sigma} \int_{t_0}^{t_1} \mathcal{E}(t)dt$$

$$-2C_{1,T}e^{2\tau\sigma} \int_0^T E(t)dt - 2c_T\tau^3 e^{-2\tau\delta}[E(0) + E(T)],$$

where

$$(10.4.2) \qquad BT|_\Sigma = \sum_{i=1}^{2}(BT)_{w_i}|_\Sigma + 2(\epsilon\tau\rho - 2C_T)e^{2\tau\sigma} \int_{t_0}^{t_1} \int_{\Gamma_1} w^2 d\Gamma_1 dt,$$

with $(BT)_{w_i}|_\Sigma$ is defined in (10.3.2), (10.3.3); while $\mathcal{E}(t)$ and $E(t)$ are defined by (6.3) and (5.5), respectively.

(ii) By virtue of Theorem 8.2 and Remark 8.1, the above estimate (10.4.1) may be extended to $H^{1,1}(Q)$-solutions of (1.1.1a) in the class (8.1).

Proof. Step 1. We recall property (10.1.8): $\phi_i \geq \sigma > 0$ on $[t_0, t_1] \times \Omega_i$; finally, that $[t_0, t_1] \times \Omega_i \subset Q_i(\sigma)$ by (10.1.19), where $w_i \equiv w$ on $Q_i(\sigma)$ by $\chi_i \equiv 1$ on $Q_i(\sigma)$, see (10.2.12) and (10.2.13). Thus, we estimate the first terms on the right side of (10.3.1) for $i = 1, 2$:

$$(10.4.3) \qquad \sum_{i=1}^{2} \int_0^T \int_\Omega e^{2\tau\phi_i}[w_{i,t}^2 + |\nabla w_i|^2]d\Omega\, dt$$

$$\text{(by (10.1.8))} \quad \geq \sum_{i=1}^{2} \int_{t_0}^{t_1} \int_{\Omega_i} e^{2\tau\sigma}[w_{i,t}^2 + |\nabla w_i|^2]d\Omega_i\, dt$$

$$\text{(as } w_i \equiv w \text{ on } Q_i(\sigma) \supset [t_0,t_1] \times \Omega_i)$$

$$= e^{2\tau\sigma} \sum_{i=1}^{2} \int_{t_0}^{t_1} \int_{\Omega_i} [w_t^2 + |\nabla w|^2] d\Omega_i\, dt$$

$$\text{(by (10.1.1))} \quad \geq e^{2\tau\sigma} \int_{t_0}^{t_1} \int_\Omega [w_t^2 + |\nabla w|^2] d\Omega\, dt,$$

where the last step follows, since the integral terms over Ω_i, $i = 1, 2$, collects also contributions on the non-empty portion $\Omega_1 \cap \Omega_2$, see (10.1.1).

Step 2. With τ sufficiently large as to have the coefficients $[2\tau^3\beta + \mathcal{O}(\tau^2) - C_T] > 0$, so that Remark 10.3.1 applies, we sum up Eqn. (10.3.1) of Proposition

10.3.1 for $i = 1, 2$, and obtain also by virtue of (10.4.3),

$$(10.4.4) \quad 2C_{1,T}e^{2\tau\sigma}\int_0^T E(t)dt + C_{2,T}\sum_i \int_0^T \int_\Omega f_i^2 d\Omega\, dt + \sum_i (BT)_{w_i}|_\Sigma$$

$$\geq 2(\epsilon\tau\rho - 2C_T)e^{2\tau\sigma}\int_{t_0}^{t_1}\int_\Omega [w_t^2 + |\nabla w|^2]d\Omega\, dt$$

$$- 2c_T\tau^3 e^{-2\tau\delta}[E(0) + E(T)].$$

Step 3. Finally, adding to both sides of inequality (10.4.4) the term

$$\left[2(\epsilon\tau\rho - 2C_T)e^{2\tau\sigma}\int_{t_0}^{t_1}\int_{\Gamma_1} w^2 d\Gamma_1 dt\right],$$

and invoking the definition of $\mathcal{E}(t)$ in (6.3), we readily obtain the desired inequality (10.4.1), with $BT|_\Sigma$ as in (10.4.2). □

10.5. Carleman estimate, second version, for the w-problem. Our main final step in establishing Theorem 10.1.1 is the following counterpart of Theorem 6.1, Eqn. (6.6).

Theorem 10.5.1. Let w be a $H^{2,2}(Q)$-solution of problem (1.1.1), including the Neumann B.C. on Σ. Let the setting of Section 10.1, based on assumptions (A.1i) and (A.2i), be in force. Then:

(i) for $\epsilon > 0$ small as in (10.2.18), and for all $\tau > 0$ sufficiently large (with $\epsilon\tau$ large as compared to the constant C_T in (1.1.2)), the following one-parameter family of inequality holds true: there is a constant $k_{1\phi} > 0$, depending on the pseudo-convex functions, such that

$$(10.5.1) \quad \overline{BT}|_\Sigma + \int_0^T \int_\Omega f^2 d\Omega\, dt \geq k_{1\phi}[E(0) + E(T)],$$

where, recalling (10.4.2) for $BT|_\Sigma$, we have

$$(10.5.2a) \quad \overline{BT}|_\Sigma = BT|_\Sigma + C_\phi \int_0^T \int_{\Gamma_1} |ww_t|d\Gamma_1 dt$$

$$(10.5.2b) \quad \leq \text{const}\left\{\int_0^T \int_{\Gamma_1}[w^2 + w_t^2]d\Gamma_1 dt\right.$$

$$\left.+ \sum_i \int_0^T \int_{\Gamma_1} e^{2\tau\phi_i}[w_t^2 - |\nabla_{\tan}w|^2]b_i^2 h_i \cdot \nu\, d\Gamma_1 dt\right\}.$$

Proof. Same as the proof of Theorem 6.1. First we note that, with the Neumann B.C. (1.1c), $\frac{\partial w}{\partial \nu} = 0$ on Σ in force now, we have that the term $N(T)$ in (6.14) specializes to

$$(10.5.3) \quad N(T) = \int_0^T \int_\Omega f^2 dQ + 2\int_0^T \int_{\Gamma_1} |ww_t|d\Sigma_1.$$

Thus, with reference to estimate (10.4.1) of Theorem 10.4.1, and recalling (6.18), we obtain

$$(10.5.4) \quad 2(\epsilon\tau\rho - 2C_T)e^{2\tau\sigma}\int_{t_0}^{t_1}\mathcal{E}(t)dt$$

$$\geq 2(\epsilon\tau\rho - 2C_T)e^{2\tau\sigma}(t_1 - t_0)\left\{\frac{a}{2}[E(T) + E(0)]e^{-C_TT} - N(T)\right\}.$$

Moreover, recalling (6.20), we obtain

$$(10.5.5) \quad -2C_{1,T}e^{2\tau\sigma}\int_0^T E(t)dt \geq -C_{1,T}e^{2\tau\sigma}\frac{Te^{C_TT}}{a}\{b[E(T) + E(0)] - 2N(T)\}.$$

Adding up (10.5.4)–(10.5.6), we can estimate the right-hand side (RHS) of (10.4.1) as follows:

(10.5.6)

$$\text{RHS of (10.4.1)} = 2(\epsilon\tau\rho - 2C_T)e^{2\tau\sigma}\int_{t_0}^{t_1}\mathcal{E}(t)dt$$

$$- 2C_{1,T}e^{2\tau\sigma}\int_0^T E(t)dt - 2c_T\tau^3 e^{-2\tau\delta}[E(0) + E(T)]$$

$$(10.5.7) \quad \geq \left\{\left[(\epsilon\tau\rho - 2C_T)(t_1 - t_0)ae^{-C_TT} - C_{1,T}\frac{b}{a}Te^{C_TT}\right]e^{2\tau\sigma}\right.$$

$$\left. - 2c_T\tau^3 e^{-2\tau\delta}\right\}[E(T) + E(0)] - \text{const}_\phi N(T).$$

We now use critically that $0 < \sigma$, $\delta > 0$, see (10.1.8), (10.1.6), that, for large τ, $[\tau e^{2\tau\sigma} - \tau^3 e^{-2\tau\delta}]$ is positive. We then obtain that: there exists a critically positive constant $k_{1\phi} > 0$, and a constant $k_{2\phi}$ (also positive, but this is not critical) depending on the pseudo-convex functions ϕ_i, such that for τ sufficiently large we have:

$$(10.5.8) \quad \text{RHS of (10.4.1)} \geq k_{1,\phi}[E(T) + E(0)] - k_{2,\phi}N(T).$$

Then, using (10.5.8) in (10.4.1), moving $N(T)$ on the left side of (10.4.1) and invoking (10.5.3) for it, we readily find estimate (10.5.1). Moreover, $\overline{BT}|_\Sigma$ is given by (10.5.2a).

We now establish (10.5.2b): for this, we return to (10.4.2) for $BT|_\Sigma$ and (10.3.3) for $(BT)_{w_i}|_\Sigma$, with $\frac{\partial w}{\partial \nu} \equiv 0$ on $(0,T] \times \Gamma$: thus, in (10.3.3), we see that the 3 terms (2nd, 3rd, and 4th) which involve $\frac{\partial w}{\partial \nu}$ on $[0,T] \times \Gamma$ vanish. That is, for $\frac{\partial w}{\partial \nu} \equiv 0$ on Σ, we obtain from (10.3.3),

$$(10.5.9) \quad (BT)_{w_i}|_\Sigma = 2\tau\int_0^T\int_{\Gamma_1} e^{2\tau\phi_i}(w_{i,t}^2 - b_i^2(t)|\nabla w|^2)h_i \cdot \nu\, d\Gamma_1\, dt$$

$$+ 2\tau\int_0^T\int_{\Gamma_1} e^{2\tau\phi_i}\left[2\tau^2\left(|h_i|^2 - 4c^2\left(t - \frac{T}{2}\right)^2\right)\right.$$

$$\left. + \tau(\alpha_i - \Delta d_i - 2c)\right]w^2 b_i^2(t)h_i \cdot \nu\, d\Gamma_1\, dt,$$

where $|\nabla_{\tan} w| = |\nabla w|$ since $\frac{\partial w}{\partial \nu} \equiv 0$ on Γ, and where, by (10.2.14e), near Γ we have

$$(10.5.10) \qquad w_{i,t}^2 = \left[\frac{\partial}{\partial t}(b_i(t)w)\right]^2 = (b_i'(t))^2 w^2 + b_i^2(t) w_t^2 + 2b_i(t) b_i'(t) w w_t,$$

in an $\frac{\epsilon_1}{4}$-internal layer of Γ. Thus, substituting (10.5.10) into the first integral of (10.5.9) readily yields (10.5.2b) by recalling also (10.4.2). The proof of Theorem 10.5.1 is complete. □

10.6. Global uniqueness: Theorem 10.1.1(b). Let $w \in H^{1,1}(Q)$ be a solution of Eqn. (1.1.1a) with $f \equiv 0$, satisfying $\frac{\partial w}{\partial \nu} \equiv 0$ on Σ, as in (1.1.1c), and, in addition, $w|_{\Sigma_1} \equiv 0$, where $h_i \cdot \nu = 0$ in Γ_0, $i = 1,2$, as in (10.1.22). Then, $\nabla_{\tan} w \equiv 0$ on Σ_1 as well, and thus $\overline{BT}|_\Sigma = 0$ by (10.5.2b) of Theorem 10.5.1(ii). Thus, estimate (10.5.1) with $f \equiv 0$ yields $E(0) = 0$, hence $w \equiv 0$ in Q, in fact in $\mathbb{R}_t \times \Omega$, as in the proof of Theorem 7.1. □

10.7. Continuous observability: Theorem 10.1.1(a). To complete the proof of Theorem 10.1.1(a) and thus obtain the final estimate (10.1.20) from estimate (10.5.1) already proved with $\overline{BT}|_\Sigma$ satisfying (10.5.2b), we proceed as in the proof of Theorem 9.2. that is, we invoke Lemma 9.1 on the interval $[\epsilon, T - \epsilon]$, $\epsilon > 0$ small, see (9.4) and ff. In particular, we make use of the global uniqueness already established, in Section 10.6, as stated in Theorem 10.1.1(b), to absorb the interior l.o.t., as in Step 2 in the proof of Theorem 9.2. This way the constant C_T in (10.1.20) is *not* explicit. However, if we assume also (10.1.21) that is

$$(10.7.1) \qquad h_i \cdot \nu \geq 0 \qquad \text{on } \Gamma_1,$$

then we may drop the terms

$$(10.7.2) \qquad \int_0^T \int_{\Gamma_1} e^{2\tau \phi_i}(-|\nabla_{\tan} w|^2) h_i \cdot \nu \, d\Sigma_1 \leq 0, \quad i = 1, 2,$$

in (10.5.2b) and obtain

$$(10.7.3) \qquad BT|_\Sigma \leq \text{const} \int_0^T \int_{\Gamma_1} [w^2 + w_t^2] d\Gamma_1 dt,$$

as desired, with an explicit constant C_T of the order Ce^{Cr^2}, where r is defined by (1.1.2c). The proof of Theorem 10.1.1(a) is complete. □

Appendix A: Classes of triples $\{\Omega, \Gamma_0, \Gamma_1\}$, $\partial \Omega = \Gamma = \Gamma_0 \cup \Gamma_1$ satisfying assumptions (A.1) and (A.2). Setting of Section 1.

Orientation. Let $\Omega \subset \mathbb{R}^n$ be an open, bounded domain with boundary $\partial \Omega = \Gamma = \overline{\Gamma_0 \cup \Gamma_1}$, $\Gamma_0 \cap \Gamma_1 = \emptyset$; Γ_0 is the uncontrolled or unobserved part of the boundary, while Γ_1 is the controlled or observed part of the boundary. The purpose of Appendices A through C is to illustrate the claim that: assumptions (A.1) and (A.2) of Section 1, or their counterpart version (A.1i) and (A.2i) of Section 10, hold true for large classes of triples $\{\Omega, \Gamma_0, \Gamma_1\}$. Moreover, in some subcases—typically when Γ_0 is *flat*—it is possible to satisfy the additional assumption (10.1.21): $h_i \cdot \nu \geq 0$ on Γ_1, $i = 1, 2$, in the setting of Section 10, although in many cases this requirement $h \cdot \nu \geq 0$ on Γ_1 in addition to the setting of Section 1 is incompatible with the requirement (A.2) = (1.1.6) that the inf $|h(x)|$ over $\bar{\Omega}$ be positive.

We shall provide a few approaches addressing these assumptions and present a few rather general results illustrated by canonical examples. Several other illustrative examples may be given where the unobserved boundary Γ_0 is given analytically by the common elementary functions. However, an exhaustive analysis of these geometrical assumptions, or even a presentation of the most general results within each approach, is beyond the scope of the present Appendices.

A.1 The case where Γ_0 is flat: Explicit construction of $h(x)$ for the setting of Section 1 and the setting of Section 10. In this special but important case where Γ_0 is flat, we can readily consider the setting of Section 1 and the setting of Section 10 in one shot. Another approach for the setting of Section 10 will be given in Appendix C.

Setting of Section 1. Let the triple $\{\Omega, \Gamma_0, \Gamma_1\}$ be given, $\Omega \subset \mathbb{R}^n$, where Γ_0 is assumed flat. Then, we take any point x_0 on the hyperplane containing Γ_0, with $x_0 \notin \bar{\Omega}$, and define the radial field $h(x) = 2(x - x_0) = 2\nabla d(x)$ where $d(x) = \|x - x_0\|^2$. See Fig. A.1. Then, the Jacobian matrix $J_h(x)$ of h is twice the identity matrix; $h \cdot \nu = 0$ on Γ_0; and $|h(x)| \geq p > 0$, $\forall x \in \Omega$. Accordingly, see (1.1.21b), the time $T_0 = 2\times$ (diameter of Ω) is then optimal in this case. Thus, assumptions (A.1) and (A.2) hold true for this large, yet special class. However, the additional requirement $h \cdot \nu \geq 0$ on Γ_1 is incompatible, in this case, with the condition that $\inf|h(x)|$ over Ω be positive.

Setting of Section 10. Now we decompose $\Omega = \Omega_1 \cup \Omega_2$, $\Omega_1 \cap \Omega_2 \neq \emptyset$. Let x_1 and x_2 be two points in common to $\bar{\Gamma}_0$ and $\bar{\Gamma}_1$, with x_1 at a finite distance from Ω_1 and x_2 at a finite distance from Ω_2. See Fig. A.2. We then define $h_i(x) = 2(x - x_i) = 2\nabla d_i(x)$, $d_i(x) = \|x - x_i\|^2$, $i = 1, 2$. Then, assumptions (A.1i) and (A.2i) of Section 10 are satisfied: $h_i \cdot \nu = 0$ on Γ_0, since each x_i lies on the hyperplane containing the flat Γ_0; $J_{h_i} = 2(\text{Identity})$, and $\|h_i(x)\| \geq p > 0$, $\forall x \in \bar{\Omega}_i$. Moreover, in this case, we may also satisfy assumption (10.1.21): $h_i \cdot \nu \geq 0$ on Γ_1, $i = 1, 2$.

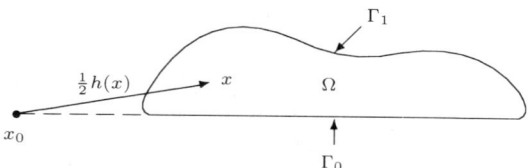

FIG. A.1: SETTING OF SECTION 1: THE REQUIRED VECTOR FIELD $h = 2(x - x_0)$ SATISFYING (A.1), (A.2) WHEN Γ_0 IS FLAT

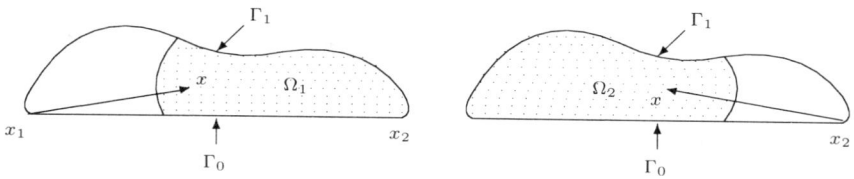

FIG. A.2: SETTING OF SECTION 10: THE REQUIRED VECTOR FIELDS $h_i = 2(x - x_i)$ SATISFYING (A.1i), (A.2i), AND (10.1.21): $h_i \cdot \nu \geq 0$ ON Γ_1

A.2 A first approach where Γ_0 is curved: Explicit construction of $h(x,y)$, 2-dimensional case. Setting of Section 1. Orientation.

In this section we consider the two-dimensional case, dim $\Omega = 2$, where the construction of the required vector field $h(x)$ is most transparent. However, the given treatment admits a natural generalization to higher dimensions as well. See Appendix B.2.

In the two-dimensional case, our approach will, *in particular*, encompass the following result.

Theorem A.2.1. Let the uncontrolled (or unobserved) boundary Γ_0 satisfy the following assumptions:

(i) Γ_0 is described by the graph

(A.2.1)
$$y = \begin{cases} f_1(x) & x_0 \leq x \leq x_1, \ y \geq 0; \\ f_2(x) & x_0 \leq x \leq x_1, \ y < 0, \end{cases}$$
(A.2.2)

where the functions $f_i(x)$ are of class C^3 and satisfy

(A.2.3)
$$\begin{cases} f_1(x_1) = f_2(x_1) = 0; \ f_1'(x_1) = -\infty; \ f_2'(x_1) = \infty; \\ f_2''(x_1) = -\infty; \ f_2''(x_1) = +\infty; \ f_i'(x) \neq 0, \ x_0 \leq x \leq x_1, \ i=1,2; \end{cases}$$

so that the graph of Γ_0 does nowhere have horizontal tangent, while it has vertical tangent at x_1.

(ii) Both $f_1(x)$ and $f_2(x)$ are *logarithmic concave* on $x_0 < x < x_1$; equivalently (Remark A.2.1), they satisfy the following conditions:

(A.2.4) $$\frac{d}{dx}\left(\frac{f_1(x)}{f_1'(x)}\right) > 0 \quad \text{and} \quad \frac{d}{dx}\left(\frac{f_2(x)}{f_2'(x)}\right) > 0, \text{ on } x_0 \leq x < x_1.$$

Then, there exists a conservative vector field $h(x,y) = \{h_1, h_2\}$, constructively defined in a neighborhood of $\Gamma_0 \setminus \{x_1, 0\}$ [the boundary Γ_0 with the point $(x_1, 0)$ removed] by:

(A.2.5a) $$\begin{cases} h_1(x) = \dfrac{f_1(x)}{f_1'(x)}, \ h_2(y) = y, \text{ for } x_0 \leq x \leq x_1; \ y \geq -\epsilon(x - x_1); \\ h_1(x) = \dfrac{f_2(x)}{f_2'(x)}, \ h_2(y) = y, \text{ for } x_0 \leq x \leq x_1; \ y \leq \epsilon(x - x_1), \end{cases}$$
(A.2.5b)

for a sufficiently small $\epsilon > 0$, such that on its domain of definition specified above, we have:

(a)

(A.2.6) $$h \cdot \nu = 0 \text{ on } \Gamma_0; \ J_h(x,y) = \begin{bmatrix} \dfrac{d}{dx}\left(\dfrac{f_i(x)}{f_i'(x)}\right) & 0 \\ 0 & 1 \end{bmatrix} > 0,$$

where $i=1$ for $y \geq 0$, and $i=2$ for $y < 0$; and moreover,

(b)

(A.2.7) $$h = \nabla d, \text{ where } d(x,y) = \int^x h_1(\xi)d\xi + \frac{y^2}{2} + C.$$

If the graph of Γ_0 is symmetric with respect to the x-axis, i.e., if $f_1(x) = -f_2(x)$, then we may take $\epsilon = 0$ in (A.2.5).

(c) Finally, the above vector field $h(x,y)$ can be extended by continuity in the triangular region: $\epsilon(x-x_1) < y < -\epsilon(x-x_1)$; $x_0 \leq x < x_1$, as well as for $x < x_0$, so that the condition $J_h(x,y) > 0$ is preserved. □

Remark A.2.1. Let $f \in C^2$, either $f(x) > 0$ or $f(x) < 0$. Define

(A.2.8) $$G(x) \equiv \ln|f(x)|, \text{ so that } G'(x) = \frac{f'(x)}{f(x)} \equiv \frac{1}{F(x)}.$$

Then [concerning assumptions (A.2.4)] we have that

(A.2.9) $$G''(x) = \frac{f''(x)f(x) - [f'(x)]^2}{f^2(x)} = -\frac{F'(x)}{F^2(x)} < 0 \iff F'(x) = \frac{d}{dx}\left(\frac{f(x)}{f'(x)}\right) > 0,$$

in which case $G(x)$ is strictly concave and so $f(x)$ is logarithmic concave.

If $f''(x) \leq 0$ (f is concave) and, without loss of generality modulo a translation, $f(x) \geq 0$, then condition (A.2.9) is satisfied. Similarly, if $f''(x) \geq 0$ (f is convex) and, without loss of generality modulo a translation, $f(x) \leq 0$, then condition (A.2.9) is satisfied as well.

For the purposes of Figure A.3, which illustrates a possible graph Γ_0 covered by Theorem A.2.1, we note that $f(x) = \sin x$, $-\frac{\pi}{2} < x < \frac{\pi}{2}$, and $f(x) = \cos x$, $0 < x < \pi$ are logarithmic concave; i.e., satisfy (A.2.9). But they are neither convex nor concave. □

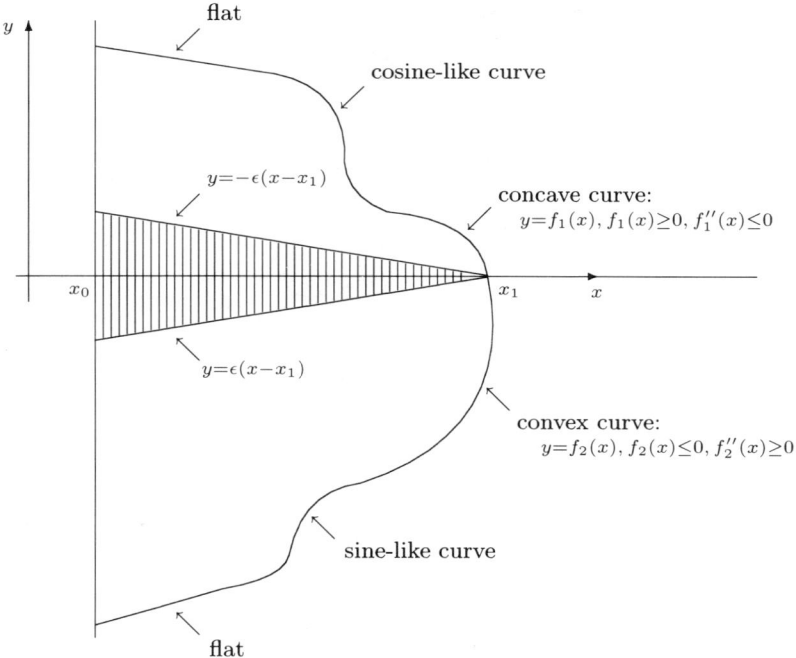

FIG. A.3. A POSSIBLE CONFIGURATION OF Γ_0 COVERED BY THEOREM A.2.1. Ω MAY BE ON EITHER SIDE OF Γ_0.

Geometrical description of the approach encompassing Theorem A.2.1.
Let $\{\Omega, \Gamma_0, \Gamma_1\}$, $\partial \Omega = \Gamma = \overline{\Gamma_0 \cup \Gamma_1}$, Γ_0 and Γ_1 relatively open in Γ, $\Gamma_0 \cap \Gamma_1 = \emptyset$. Let the (non-intersecting) curve Γ_0 be given explicitly as a level set by the equation

(A.2.10) $$\Gamma_0 = \{(x,y) \in \mathbb{R}^2 : \ell(x,y) = 0\}, \quad \ell \in C^3.$$

with $|\nabla \ell| \neq 0$ on Γ_0, for a suitable domain in (x,y).

Assumption (A.1) of Section 1. As exemplified by the statement of Theorem A.2.1, we shall explicitly construct a conservative vector field $h = \{h_1, h_2\}$, such that the following conditions are satisfied:

(A.2.11) $$h \cdot \nu \equiv 0 \text{ on } \Gamma_0; \quad J_h(x,y) > 0 \text{ near } \Gamma_0;$$

J_h being the Jacobian matrix of h. Moreover, it will be possible to extend smoothly such $h(x,y)$ so that $J_h(x,y) > 0$ on $\bar{\Omega}$, as well.

This way, *condition (A.1) of Section 1 is satisfied*. In fact, the constructed vector field will be, near Γ_0, of the form $h_1 = h_1(x)$, $h_2 = h_2(y)$, i.e., with first (second) component depending only on the first (second) coordinate, as in the statement of Theorem A.2.1.

Assumption (A.2) of Section 1. Depending upon the given unobserved/ uncontrolled boundary Γ_0, the constructed vector field h may vanish at one point P_0 of Γ_0. This is the case if Γ_0 has a U-turn, as in the case of Theorem A.2.1, where $h(x_1, 0) = 0$. Then, the present setting of Section 1, assumption (A.2) can be satisfied only by restricting Γ_0; that is, by removing from Γ_0 a small neighborhood, on Γ_0, of the pathological point $P_0 \in \Gamma_0$, and assigning it instead to the observed/controlled boundary Γ_1. This way, neither the resulting, new Γ_0 nor the resulting, new observed boundary Γ_1 are connected. See also Remark 1.1.3. In Appendix C below, we will be able, in many cases, to avoid the above situation and overcome the related difficulty. To do this it will be necessary to rely on the 2-vector field setting of Section 10. This way, we will be able to keep, as unobserved boundary, the original portion Γ_0, even if it has a U-turn, and still fit the setting of Section 10. See Theorem C.1 in Appendix C.

First setting: Nowhere horizontal and nowhere vertical tangent to Γ_0.
We shall at first introduce our geometrical construction of h in the case where Γ_0, the graph of ℓ, *has nowhere horizontal and nowhere vertical tangent* in the (x,y)-plane. Thus, the slope of Γ_0 is finite and either always positive, or else always negative. Locally the domain Ω may lie on either side of Γ_0. In this case, the construction of the required vector field h goes through the following two steps:

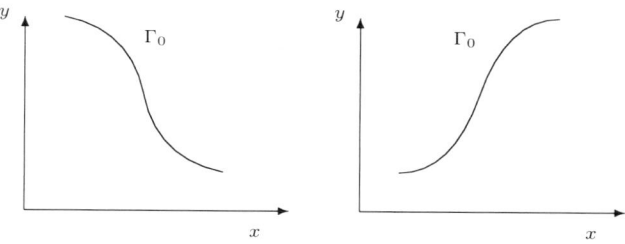

FIG. A.4: FIRST SETTING FOR Γ_0

Step 1: Definition of h on Γ_0. Let $P = (x, y) \in \Gamma_0$. Then, $\{\ell_x, \ell_y\}$ is a vector orthogonal to Γ_0 at P, and thus $\{-\ell_y, \ell_x\}$ is a vector tangent to Γ_0 at P. Multiplying this tangent vector either by $\frac{y}{\ell_x}$, or else by $\frac{-x}{\ell_y}$ respectively, we obtain two vectors

(A.2.12)
$$h = \overrightarrow{XP} = \left\{-y\frac{\ell_y}{\ell_x}, y\right\}, \text{ and respectively, } h = \overrightarrow{YP} = \left\{x, -x\frac{\ell_x}{\ell_y}\right\}, \quad (x, y) \in \Gamma_0,$$

which are also tangent to Γ_0 at P. Geometrically, the points

(A.2.13) $$X = \left\{x + y\frac{\ell_y}{\ell_x}, 0\right\}, \text{ and } Y = \left\{0, y + x\frac{\ell_x}{\ell_y}\right\},$$

are the intersection points of the tangent line $\frac{\xi - x}{-\ell_y} = \frac{\eta - y}{\ell_x}$ to Γ_0 at the point P with the x-axis and the y-axis, respectively. In many cases, the following assumption is satisfied by $\ell(x, y)$, and this is surely the case, at least locally, by the Implicit Function Theorem, which is valid under our assumptions:

(A.2.14)
$$\text{for } (x, y) \in \Gamma_0 : \begin{cases} \text{either the term } \left(-y\frac{\ell_y}{\ell_x}\right)(x) \text{ is only a function of } x; \\ \text{or else the term } \left(-x\frac{\ell_x}{\ell_y}\right)(y) \text{ is only a function of } y. \end{cases}$$

Assumption (A.2.14) holds true in particular (but not exclusively, see examples A.2.1 and A.2.2 below) when

(A.2.15)
either $\Gamma_0: \ell(x, y) = y - f(x) \equiv 0$, $a \leq x \leq b$; or else $\Gamma_0: \ell(x, y) = x - g(y) \equiv 0$, $c \leq y \leq d$, in which case, we obtain, respectively,

(A.2.16) $\quad \left(-y\dfrac{\ell_y}{\ell_x}\right)(x) = \dfrac{f(x)}{f'(x)}; \quad \left(-x\dfrac{\ell_x}{\ell_y}\right)(y) = \dfrac{g(y)}{g'(y)}, \quad (x, y) \in \Gamma_0,$

with finite $f'(x) \neq 0$, or finite $g'(y) \neq 0$, respectively, as assumed.

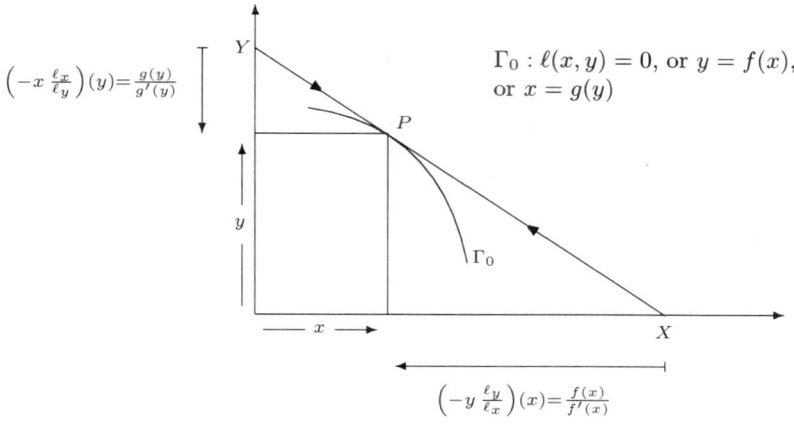

FIG. A.5: DEFINITION OF $h = \overrightarrow{XP}$ OR $h = \overrightarrow{YP}$ FOR $P \in \Gamma_0$

Step 2: Extension of h near Γ_0. We now extend h, defined by (A.2.12) on Γ_0, to any point $P = (x, y) \in \Omega$ near the graph of Γ_0 of ℓ. The horizontal and vertical lines through P meet Γ_0 at points P_h and P_v, respectively, on Γ_0. Let X_h and X_v [respectively, Y_h and Y_v] be the intersections of the tangent lines to Γ_0 at P_h and P_v with the x-axis [respectively, with the y-axis]

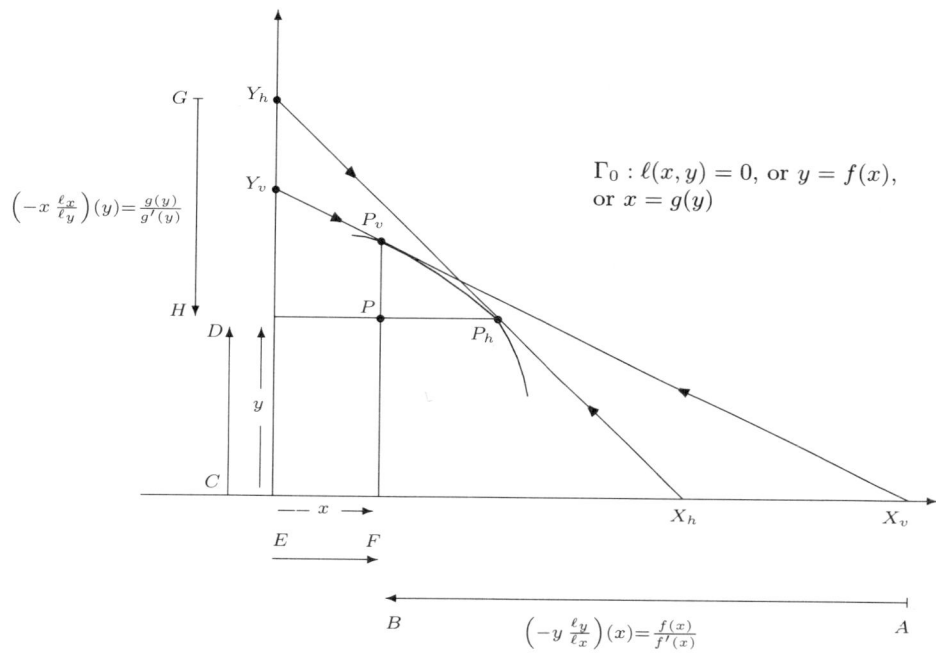

FIG. A.6: DEFINITION OF $h = \{h_1 = \overrightarrow{AB},\ h_2 = \overrightarrow{CD}\}$ FOR P NEAR Γ_0: $\overrightarrow{AB}$ = HORIZONTAL COMPONENT OF $\overrightarrow{X_v P_v}$; $\overrightarrow{CD}$ = VERTICAL COMPONENT OF $\overrightarrow{X_h P_h}$. OR ELSE, DEFINITION OF $h = \{h_1 = \overrightarrow{EF},\ h_2 = \overrightarrow{GH}\}$, $\overrightarrow{EF}$ = HORIZONTAL COMPONENT OF $\overrightarrow{Y_v P_v}$; $\overrightarrow{GH}$ = VERTICAL COMPONENT OF $\overrightarrow{Y_h P_h}$

At such nearby point $P = (x, y)$, we define the vector field $h = \{h_1, h_2\}$ by either

(A.2.17)
$$\begin{cases} h_1 = \text{horizontal component of } \overrightarrow{X_v P_v} = \left(-y\dfrac{\ell_y}{\ell_x}\right)(x), \\ \qquad \text{in particular, } \dfrac{f(x)}{f'(x)},\ a \leq x \leq b; \\ h_2 = \text{vertical component of } \overrightarrow{X_h P_h} = y; \end{cases}$$

or, respectively, by

(A.2.18)
$$\begin{cases} h_1 = \text{horizontal component of } \overrightarrow{Y_v P_v} = x; \\ h_2 = \text{vertical component of } \overrightarrow{Y_h P_h} = \left(-x\dfrac{\ell_x}{\ell_y}\right)(y), \\ \qquad \text{in particular, } \dfrac{g(y)}{g'(y)},\ c \leq y \leq d. \end{cases}$$

The Jacobian matrix of h is then either
(A.2.19)
$$J_h = \begin{bmatrix} \dfrac{d}{dx}\left(-y\dfrac{\ell_y}{\ell_x}\right)(x) & 0 \\ 0 & 1 \end{bmatrix}, \text{ in particular, } J_h = \begin{bmatrix} \dfrac{d}{dx}\left(\dfrac{f(x)}{f'(x)}\right) & 0 \\ 0 & 1 \end{bmatrix},$$
for (A.2.17), or else
(A.2.20)
$$J_h = \begin{bmatrix} 1 & 0 \\ 0 & \dfrac{d}{dy}\left(-x\dfrac{\ell_x}{\ell_y}\right)(y) \end{bmatrix}, \text{ in particular, } J_h = \begin{bmatrix} 1 & 0 \\ 0 & \dfrac{d}{dy}\left(\dfrac{g(y)}{g'(y)}\right) \end{bmatrix},$$
for (A.2.18), respectively.

Lemma A.2.2. Let $\ell \in C^3$ and assume hypothesis (A.2.14) with $\ell_x \neq 0$, or $\ell_y \neq 0$, respectively, unless a cancellation occurs, in particular, let either one of the situations in (A.2.15) hold true, with $f \in C^3$, $f'(x) \neq 0$; or $g \in C^3$, $g'(y) \neq 0$, respectively, unless a cancellation occurs.

(a) Assume, moreover, that either
(A.2.21)
$$\frac{d}{dx}\left(-y\frac{\ell_y}{\ell_x}\right)(x) > 0, \text{ in particular } \frac{d}{dx}\left(\frac{f(x)}{f'(x)}\right) \equiv 1 - \frac{f(x)f''(x)}{[f'(x)]^2} > 0, \ a \leq x \leq b,$$
i.e., $f(x)$ is logarithmic concave, see Remark A.2.1; or else
(A.2.22)
$$\frac{d}{dy}\left(-x\frac{\ell_x}{\ell_y}\right)(y) > 0, \text{ in particular } \frac{d}{dy}\left(\frac{g(y)}{g'(y)}\right) \equiv 1 - \frac{g(y)g''(y)}{[g'(y)]^2} > 0, \ c \leq y \leq d,$$
respectively, i.e., $g(y)$ is logarithmic concave, see Remark A.2.1. Then:

(i) the Jacobian matrix J_h is positive definite near Γ_0, in either case;

(ii) the vector field $h = \{h_1, h_2\}$ defined in (A.2.17), or respectively (A.2.18), satisfies $h = \overrightarrow{XP}$, or $h = \overrightarrow{YP}$ for $(x,y) = P \in \Gamma_0$, respectively, and thus $h \cdot \nu \equiv 0$ on Γ_0;

(iii) the function $d(x,y) \in C^3$ defined near Γ_0 either by
(A.2.23)
$$d(x,y) = \int \left(-y\frac{\ell_y}{\ell_x}\right)(x)dx + \frac{y^2}{2}, \text{ or else by } d(x,y) = \frac{x^2}{2} + \int \left(-x\frac{\ell_x}{\ell_y}\right)(y)dy,$$
satisfies the first two conditions (i) = (1.1.4) and (ii) = (1.1.5) of assumption (A.1) in Section 1.

(b) The conclusion of part (a) applies, in particular, if either (b_1) $f(x)$ is convex: $f''(x) \geq 0$, $a \leq x \leq b$; or else (b_2) $f(x)$ is concave: $f''(x) \leq 0$, $a \leq x \leq b$. Similarly, for $g(y)$.

Proof. (a) The proof of (a) is contained in the construction of h.

(b) To prove (b), we notice that $f(x)$ can be translated without loss of generality. Thus, if $f(x)$ is convex [respectively, concave] we can always assume that $f(x) \leq 0$ [respectively, $f(x) \geq 0$] so that in either case we have: $f(x)f''(x) \leq 0$, $a \leq x \leq b$, and condition (A.2.21) holds true. □

Examples. Lemma A.2.2 can be used to construct many examples, in fact even more general than those admitted by Lemma A.2.2 itself, in the sense that (as in some of the examples below) the slope of Γ_0 may be infinite. These are given in Appendix B.

Second setting: The case of Theorem A.2.1. We next generalize the construction leading to Lemma A.2.2 to the case where the boundary Γ_0 has a U-turn (such as in the situation covered by Theorem A.2.1, and such as in the case of examples B.1.1, B.2.2 below). In particular, Γ_0 (locally) bounds either a convex, or a concave domain. After an appropriate choice of axes, we are let to the situation depicted in Fig. A.7, where Γ_0 has nowhere horizontal tangent.

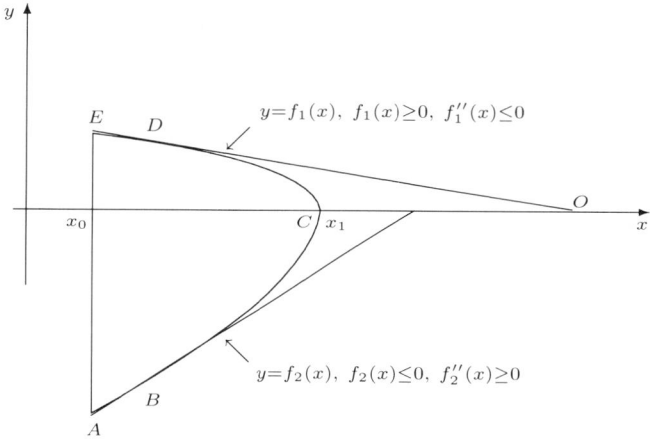

FIG. A.7: Ω MAY BE ON EITHER SIDE OF Γ_0: $ABCDE$, WITH ABC A CONVEX CURVE AND CDE A CONCAVE CURVE. AB AND DE ARE FLAT.

We notice that the key assumption, beside (A.2.4), in particular convexity/concavity, that Γ_0 has nowhere a horizontal tangent can be rephrased as follows: *there exists a point O*—in the setting of (A.2.1), (A.2.3), on the x-axis, in fact, see Fig. A.7—such that the *radial field* $\overrightarrow{OP}$ *is entering or exiting Ω (locally) through* Γ_0: $\overrightarrow{OP} \cdot \nu \leq 0$ on Γ_0, or else $\overrightarrow{OP} \cdot \nu \geq 0$ on Γ_0. The tangent line at x_1 is vertical. In the present case we define a vector field h in Ω locally near Γ_0, following the geometric ideas and construction leading to Lemma A.2.2, suitably adapted, thus establishing Theorem A.2.1.

Proof of Theorem A.2.1. Case 1. (symmetric case) If $f_2(x) = -f_1(x)$, $x_0 \leq x \leq x_1$, then the construction leading to Lemma A.2.2 can be applied, to obtain a vector field $h = \{h_1, h_2\}$,

(A.2.24) $$h_1(x) = \frac{f_1(x)}{f_1'(x)} = \frac{f_2(x)}{f_2'(x)}, \quad x_0 \leq x < x_1, \ h_1(x_1) = 0, \ h_2(y) = y.$$

The vector field h in (A.2.24) is tangent to Γ_0. We next verify that the positivity condition of the Jacobian matrix:

(A.2.25)
$$J_h > 0 \text{ on } x_0 \leq x < x_1, \text{ where } \frac{d}{dx}\left(\frac{f_1(x)}{f_1'(x)}\right) = 1 - \frac{f_1(x)f_1''(x)}{[f_1'(x)]^2} > 0, \quad x_0 \leq x < x_1,$$

can be extended up to $x = x_1$. At $x = x_1$, we have $f_1(x_1) = 0$, $f_1'(x_1) = -\infty$, $f_1''(x_1) = -\infty$, by (A.2.3). Near x_1, setting $x = g_1(y)$ to be the inverse of $y = f_1(x)$, so that $x \equiv g_1(f_1(x))$, then the chain rule shows that

(A.2.26)
$$\frac{f_1''(x)}{[f_1'(x)]^2} = -\frac{g_1''(y)}{g_1'(y)}, \quad x < x_1, \ x \text{ near } x_1.$$

From here, one can prove that

(A.2.27)
$$\lim_{x \to x_1} \frac{f_1(x)f_1''(x)}{[f_1'(x)]^2} = -\lim_{y \to 0} \frac{y g_1''(y)}{g_1'(y)} = -1,$$

thus extending the validity of the positivity condition $J_h > 0$ in (A.2.25) up to $x = x_1$, as required. To show (A.2.27), we simply use the Taylor formula with remainder in the form of Lagrange to the functions $g_1'(y)$, $g_1''(y)$ where $g_1'(0) = \frac{1}{f_1'(x_1)} = 0$, and for some η_y and ζ_y comprised between 0 and y, we have:

(A.2.28)
$$g_1'(y) = g_1''(0)y + g_1'''(\eta_y)\frac{y^2}{2}; \qquad g_1''(y) = g_1''(0) + g_1'''(\zeta_y)y.$$

Using (A.2.28), one readily shows

(A.2.29)
$$\lim_{y \to 0} \frac{y g_1''(y)}{g_1'(y)} = \frac{g_1''(0)}{g_1''(0)} = 1,$$

and (A.2.27) (right) follows.

Case 2. (Local symmetry near the vertex point x_1) If $f_1(x) = -f_2(x)$ only near $x = x_1$, say, $x_1 - \delta \leq x \leq x_1$, for some small $\delta > 0$, we define h on the set $x_1 - \delta \leq x \leq x_1$, $f_2(x) \leq y \leq f_1(x)$, as in Case 1. Instead, for $x_0 \leq x < x_1 - \delta$, where $f_1(x) \neq -f_2(x)$, we define h, consistently, as follows:

(A.2.30)
(A.2.31)
$$\begin{cases} h_1(x) = \dfrac{f_1(x)}{f_1'(x)}, & h_2(y) = y, \ x_0 \leq x \leq x_1 - \delta, \ y \geq f_1(x_1 - \delta); \\ h_2(x) = \dfrac{f_2(x)}{f_2'(x)}, & h_2(y) = y, \ x_0 \leq x \leq x_1 - \delta, \ y \leq f_2(x_1 - \delta). \end{cases}$$

Then, in the strip: $\{x_0 \leq x < x_1 - \delta, \ f_2(x_1 - \delta) < y < f_1(x_1 - \delta)\}$ we extend the original $h_1(x)$ in (A.2.30) and (A.2.31) smoothly (as done in Example B.1.1) to obtain a global C^2-function.

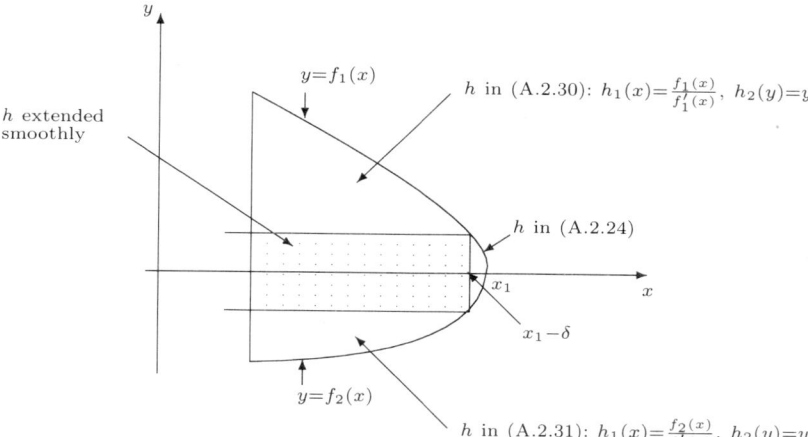

FIG. A.8: DEFINITION OF h IN THE CASE OF LOCALLY SYMMETRIC Γ_0 NEAR x_1: $f_1(x) = -f_2(x)$, $x_1 - \delta \leq x \leq x_1$.

Case 3. (general case) If generally $f_1(x) \neq -f_2(x)$, we first consider a small triangular domain $\mathcal{T}$ with one vertex at x_1: $\mathcal{T} = \{x_0 \leq x \leq x_1,\ \epsilon(x - x_1) \leq y \leq -\epsilon(x - x_1)\}$ for a small $\epsilon > 0$.

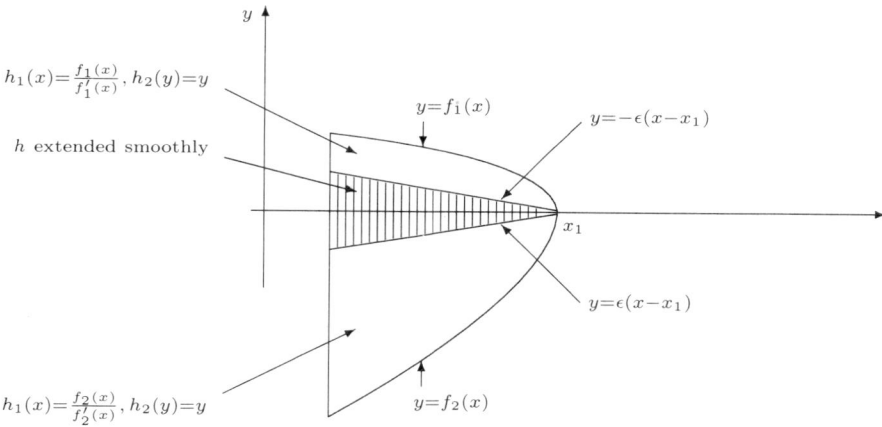

FIG. A.9: DEFINITION OF h IN THE GENERAL CASE.

We then define h first outside the triangular domain that is for: $x_0 \leq x \leq x_1$, and either $y > -\epsilon(x - x_1)$ or else $y < \epsilon(x - x_1)$, as in (A.2.30), (A.2.31). Next, we extend h_1 smoothly across the triangular domain. This is possible since $h_1(x_1) = h_2(0) = 0$ at $(x_1, 0)$. □

A.3 A second approach where Γ_0 is curved: Conformal mapping methods, 2-dimensional case. Setting of Section 1. Assumption (A.1) in (1.1.4), (1.1.5) of Section 1. In this subsection we point out the possibility of using conformal mapping methods to obtain, in the 2-dimensional case, a smooth *conservative* vector vield $h(x,y)$ such that the two conditions in (A.2.11) [i.e., assumption (A.1) of Section 1] are satisfied. The approach presented here is based on the following well-known result.

Theorem A.3.1. [B-C, p. 294] Suppose that a transformation

(A.3.1) $$w = f(z) = u(x,y) + iv(x,y), \qquad z = x + iy$$

is conformal on a smooth arc C_0, and let Γ_0 be the image of C_0 under the transformation: $\Gamma_0 = f(C_0)$. Assume that there exists a scalar function $d(u,v) \in C^3$, $h = \nabla d \neq 0$, which along Γ_0 in the (u,v)-plane, satisfies

(A.3.2) $$\left.\frac{\partial d}{\partial \nu}\right|_{\Gamma_0} = \nabla d \cdot \nu = h \cdot \nu = 0 \quad \text{on } \Gamma_0,$$

ν = unit normal vector to Γ_0. Then, the function defined by

(A.3.3) $$\delta(x,y) \equiv d(u(x,y), v(x,y)) \in C^3,$$

along C_0 in the (x,y)-plane, satisfies

(A.3.4) $$\left.\frac{\partial \delta}{\partial \bar{n}}\right|_{C_0} = \nabla \delta \cdot \bar{n} = 0 \text{ on } C_0,$$

$\bar{n}$ = unit normal vector to C_0. □

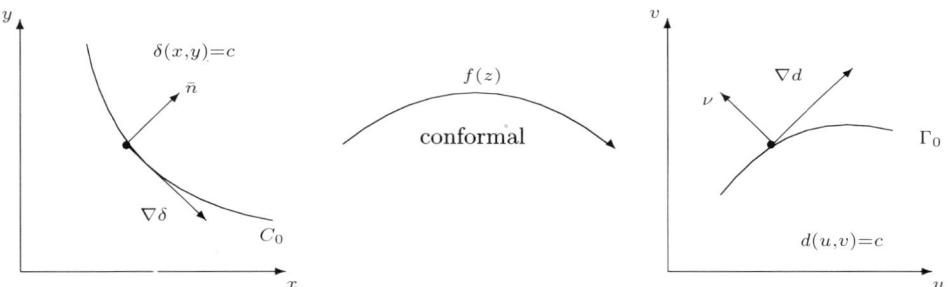

FIG. A.10: CONFORMAL MAPPING FROM C_0 TO Γ_0

Thus, the idea of the present approach is as follows:

Step 1. In the (u,v)-plane, select a curve Γ_0 for which a scalar function $d(u,v) \in C^3$ is known to exist, such that assumption (A.1) of Section 1 is satisfied: this means that such $d(u,v)$ satisfies the boundary condition (A.3.2), as well as the positivity condition

(A.3.5) $$\mathcal{H}_d(\Gamma_0) = J_h(\Gamma_0) = \begin{bmatrix} d_{uu} & d_{uv} \\ d_{uv} & d_{vv} \end{bmatrix}(\Gamma_0) > 0.$$

Here, $\mathcal{H}_d(\Gamma_0)$ is the Hessian matrix of $d(u,v)$, evaluated on Γ_0, $h = \nabla d$. To this end, we may use Sections A.1 and A.2. In particular, we may take, in the simplest case, Γ_0 = line segment, the case of Section A.1; or else Γ_0 = part of a circumference as in Section A.2 (Example B.1.1), etc.

Step 2. If $f(z)$ is any conformal mapping from C_0 in the (x,y)-plane onto Γ_0 in the (u,v)-plane, Theorem A.3.1 already yields a function $\delta(x,y)$ which fulfills half of assumption (A.1) of Section 1, concerning the arc C_0, namely condition (A.3.4). Therefore, it *remains to select the conformal mapping $f(z)$* [or the arc C_0] *such that the resulting function $\delta(x,y)$ constructed via (A.3.3) satisfies also the positivity condition*

$$(A.3.6) \qquad \mathcal{H}_\delta(C_0) = J_{\nabla\delta}(C_0) = \begin{bmatrix} \delta_{xx} & \delta_{xy} \\ \delta_{xy} & \delta_{yy} \end{bmatrix}(C_0) > 0$$

on C_0. Once this is established, then, by continuity, $\mathcal{H}_\delta > 0$ in a neighborhood of the arc C_0 as well, and $\delta(x,y)$ can then be extended smoothly to all of Ω while preserving positivity of the Hessian matrix $\mathcal{H}_\delta$.

The positivity condition (A.3.6) may, in turn, be tested according to any of the following well-known *equivalent* characterizations:

(a) the principal minors have positive determinant: $\delta_{xx} > 0$ and $\det \mathcal{H}_\delta > 0$ on C_0;

(b) the eigenvalues λ_1 and λ_2 of the matrix $\mathcal{H}_\delta(C_0)$ are both positive: $\lambda_1, \lambda_2 > 0$;

(c) the determinant and the trace of the matrix $\mathcal{H}_\delta(C_0)$ are both positive:

$$(A.3.7) \quad \det \mathcal{H}_\delta(C_0) = \lambda_1 \lambda_2 > 0; \quad \operatorname{tr} \mathcal{H}_\delta(C_0) = \lambda_1 + \lambda_2 = [\delta_{xx} + \delta_{yy}](C_0) > 0.$$

Test (c) is the most useful here, in view of the well-known identity [B-C.1, p. 298],

$$(A.3.8) \qquad \delta_{xx}(x,y) + \delta_{yy}(x,y) = [d_{uu}(u,v) + d_{vv}(u,v)]|f'(z)|^2,$$

which can be easily derived from (A.3.3), where, moreover, $[d_{uu} + d_{vv}] > 0$ on Γ_0, by assumption (A.3.5) in Step 1, via test (c) applied this time to the matrix $\mathcal{H}_d(\Gamma_0)$. Hence, by conformality, the trace $\operatorname{tr} \mathcal{H}_\delta(C_0)$ in (A.3.8) is always positive on C_0, as desired in (A.3.7). In view of this, we see then that the positivity condition (A.3.6) for the matrix $\mathcal{H}_\delta(C_0)$ holds true if and only if $\det \mathcal{H}_\delta(C_0) > 0$. Overall, the above argument has thus shown the following result.

Theorem A.3.2. *Let Γ_0 be a curve (in the (u,v)-plane) for which assumption (A.1) of Section 1 holds true* [*this means that both (A.3.2) and (A.3.5) are fulfilled*]. *Let $f(z)$ be a conformal mapping of an arc C_0 in the (x,y)-plane onto Γ_0.*

Then C_0 satisfies assumption (A.1) of Section 1 as well [*i.e., both conditions (A.3.4) and (A.3.6)*] *if and only if:* $\det \mathcal{H}_\delta(C_0) > 0$.

A full exploration of the conformal mapping approach here proposed, and related Riemann mapping theorem, remains to be done; in particular, it would be desirable to characterize explicitly classes of conformal mappings as well as classes of arcs C_0 satisfying Theorem A.3.2 and mapping C_0 onto elementary curves Γ_0 (straight segments, portions of circumferences, etc.) for which assumption (A.1) of Section 1 is satisfied [as in Sections A.1 and A.2]. Here, we confine ourselves to analyzing the simplest case. Specific examples in Appendix B.3 show positive features (Example B.3.2) as well as limitations (Example B.3.4) of this approach over the method of Appendix A.2.

The case where Γ_0 is a straight segment. Let Γ_0 be a segment in the (u,v)-plane, say

$$(A.3.9) \qquad \Gamma_0 : u \equiv u_0 > 0; \quad v_0 \le v \le v_1.$$

Step 1. From Section A.1, we know that the function

$$(A.3.10) \qquad d(u,v) = \frac{1}{2}\|(u,v) - (u_0, 0)\|^2 = \frac{1}{2}[(u-u_0)^2 + v^2]$$

satisfies assumption (A.1) of Section 1 on Γ_0: i.e., (A.3.2) as well as (A.3.5).

Step 2. Let $f(z)$ as in (A.3.1) be a conformal mapping as required by Theorem A.3.2. According to (A.3.3), we define the function $\delta(x, y)$ via (A.3.8) by

$$(A.3.11) \qquad 2\delta(x,y) = (u(x,y) - u_0)^2 + v^2(x, y).$$

We obtain the relevant partial derivatives

$$(A.3.12) \qquad \delta_x = (u - u_0)u_x + vv_x; \quad \delta_y = (u - u_0)u_y + vv_y;$$

(A.3.13)
$$\delta_{xx} = u_x^2 + v_x^2 + (u - u_0)u_{xx} + vv_{xx}; \quad \delta_{yy} = u_y^2 + v_y^2 + (u - u_0)u_{yy} + vv_{yy};$$

$$(A.3.14) \qquad \delta_{xy} = \delta_{yx} = u_x u_y + (u - u_0)u_{xy} + v_x v_y + vv_{xy}.$$

By the Cauchy-Riemann equations:

$$(A.3.15) \quad u_x = v_y, \ u_y = -v_x \text{ and hence } |f'(z)|^2 = u_x^2 + v_x^2 = u_y^2 + v_y^2 \neq 0,$$

by conformality. We next evaluate (A.3.13), (A.3.14) on a curve C_0 which is mapped into Γ_0, i.e., such that $u(x, y) \equiv u_0 > 0$. Then (A.3.13), (A.3.14) specialize to:
on C_0:

$$(A.3.16) \qquad \begin{cases} \delta_{xx} = |f'(z)|^2 + vv_{xx}; \quad \delta_{yy} = |f'(z)|^2 + vv_{yy} \\ \\ (A.3.17) \qquad \delta_{xy} = \delta_{yx} = vv_{xy}; \\ \\ (A.3.18) \qquad \delta_{xx}\delta_{yy} - \delta_{xy}^2 = |f'(z)|^4 - v^2[v_{xx}^2 + v_{xy}^2] \end{cases}$$

Eqn. (A.3.17) is obtained from (A.3.14) after a cancellation $u_x u_y + v_x v_y \equiv 0$ by the Cauchy-Riemann equations. Moreover, Eqn. (A.3.18) is obtained after using (twice) that v is harmonic: $v_{xx} + v_{yy} \equiv 0$.

A specialization of the general Theorem A.3.2 to this case where Γ_0 is given by (A.3.9) is given next.

Corollary A.3.3. Assume that $f(z)$ is a conformal mapping of C_0 onto Γ_0, where Γ_0 is the straight segment in (A.3.9). With reference to (A.3.18), we have that the curve C_0 satisfies assumption (A.1) of Section 1 [i.e., both Eqns. (A.3.4) and (A.3.6)] if and only if

$$(A.3.19) \ \det \mathcal{H}_\delta(C_0) = \delta_{xx}\delta_{yy} - \delta_{xy}^2 = |f'(z)|^4 - v^2(v_{xx}^2 + v_{xy}^2) > 0 \text{ on } C_0. \qquad \square$$

With reference to test (a) above, we remark in passing that condition (A.3.19) clearly implies $\delta_{xx} > 0$, see (A.3.16), via

$$(A.3.20) \qquad [|f'(z)|^2 - vv_{xx}][|f'(z)|^2 + vv_{xx}] > v^2 v_{xy}^2 \geq 0,$$

where the only option possible is for both terms in the square brackets to be positive. Illustrations are given in Appendix B.3.

Assumption (A.2) = (1.1.6) of Section 1. In preparation for our discussion on the setting of Section 10, we now identify the point(s) c on the boundary C_0, if any, where the conservative vector field $\nabla\delta$ constructed by (A.3.3) in Theorem A.3.1 [or its specialization (A.3.11) of Corollary A.3.3] vanishes: $\nabla\delta(c) = 0$, and hence violates assumption (A.2) = (1.1.6). This will then allow us, in the subsequent Appendix C, to remedy the situation by falling into the setting of Section 10. To this end, we recall the following well-known relationship [B-C.1, p. 296],

$$(A.3.21) \qquad |\nabla\delta(x,y)| = |\nabla d(u,v)|\,|f'(z)|$$

between the gradients of the function $\delta(x,y)$ and $d(u,v)$ respectively [which can be readily shown from (A.3.3) via the Cauchy-Riemann equations]. From (A.3.21) since $f(z)$ is conformal, we obtain

Corollary A.3.4. With $f(z)$ a conformal mapping of the curve C_0 in the (x,y)-plane onto the curve Γ_0 in the (u,v)-plane, we have that the boundary point $z = (x,y) \in C_0$ is a critical point of the function δ, i.e., $\nabla\delta(x,y) = 0$ if and only if the point $w = f(z) = (u,v) \in \Gamma_0$ is a critical point of the function d, i.e., $\nabla d(u,v) = 0$. □

In the setting of Section A.2 [as well as its 3-dimensional generalization B.2] and of the subsequent Section A.4 (see Corollary A.4.2), we have that: *there is at most one critical point for $d(u,v)$ on Γ_0.*

A.4 A third approach where Γ_0 is convex or concave: Multidimensional case. Setting of Section 1. Assumption (A.1) in (1.1.4), (1.1.5) of Section 1. The present section gives a result which shows, in the n-dimensional space, that assumption (A.1) of Section 1 holds true, in the case where: (i) the set Ω is convex (respectively, concave) near Γ_0, and (ii) a radial vector field exists which is entering (respectively, exiting) through Γ_0.

From our inquiries within geometric circles at the conference at the University of Colorado, Boulder, and elsewhere, it appears that a version of the statement of Theorem A.4.1 may perhaps be known. However, we were neither given, nor were we able to find, specific references to it. In a related context, the perturbation formula (A.4.6), (A.4.7) was communicated to the first author by D. Tataru, and it appears in his unpublished manuscript, where an argument is given based on Poisson brackets to verify the pseudo-convex property. By contrast, our direct proof and computations below verify the positivity condition on the Hessian matrix under perturbation.

Theorem A.4.1. Consider the triple $\{\Omega, \Gamma_0, \Gamma_1\}$, $\Omega \subset \mathbb{R}^n$, $\Gamma = \overline{\Gamma_0 \cup \Gamma_1}$, $\Gamma_0 \cap \Gamma_1 = \phi$, where the surface Γ_0 is given explicitly, as a level set, as in (A.2.10), by the equation,

$$(A.4.1) \qquad \Gamma_0 = \{x = (x_1, \ldots, x_n) \in \mathbb{R}^n : \ell(x) = 0\},\ \ell \in C^3,$$

with $\nabla\ell \neq 0$ on Γ_0 (for a suitable domain in $(x_1, \ldots, x_n)$). Assume that:

(i) the Hessian matrix $\mathcal{H}_\ell$ of ℓ is non-negative definite on Γ_0:

$$(A.4.2) \qquad \mathcal{H}_\ell(\Gamma_0) \equiv \begin{bmatrix} \ell_{x_1 x_1}, \ldots, \ell_{x_1 x_n} \\ \vdots \qquad \vdots \\ \ell_{x_n x_1}, \ldots, \ell_{x_n x_n} \end{bmatrix}(\Gamma_0) \geq 0$$

which is a characterization for the surface $z = \ell(x)$ to be convex [Fl.1, Theorem 36, p. 114] or having convex epigraph, or for the set Ω being a convex set near Γ_0 [Fl.1, Proposition 3.5, p. 108], so that $\ell(x) \leq 0$ for $x \in \Omega$ near Γ_0. Moreover, the gradient $\nabla \ell$ points toward the exterior of Ω.

(ii) there exists a point $x_0 \in \mathbb{R}^n$, outside of Ω, such that

$$\text{(A.4.3)} \quad \left.\frac{\partial d_0}{\partial \nu}\right|_{\Gamma_0} = \nabla d_0 \cdot \nu = (x - x_0) \cdot \nu \leq 0 \text{ on } \Gamma_0;$$

$$\text{where } d_0(x) = \frac{1}{2}\|x - x_0\|^2, \ \nabla d_0(x) = x - x_0; \ \mathcal{H}_{d_0} = I,$$

where ν is the unit outward normal vector to Γ_0, thus pointing in the same direction of $\nabla \ell(x)$.

Alternatively, in place of (A.4.2) and (A.4.3), respectively, assume $\mathcal{H}_\ell(\Gamma_0) \leq 0$ and $\nabla d_0 \cdot \nu \geq 0$ on Γ_0 with $\nu(x)$ still outward in the same direction of $\nabla \ell(x)$, but now with $(x - x_0)$ making an acute or right angle with $\nu(x)$, for $x \in \Gamma_0$.

[The proof below works for the product $(\nabla d_0 \cdot \nu)\mathcal{H}_\ell(\Gamma_0) \geq 0$, see (A.4.31) below.]

Then, there exists a scalar function $d(x)$, defined explicitly below in (A.4.6)–(A.4.7) in a layer (collar) of Γ_0, such that the following two conditions are satisfied:

(a)

$$\text{(A.4.4)} \quad \left.\frac{\partial d}{\partial \nu}\right|_{\Gamma_0} = \nabla d \cdot \nu = 0 \text{ on } \Gamma_0;$$

(b) the Hessian matrix of d, evaluated on Γ_0, $\mathcal{H}_d(\Gamma_0)$, is positive definite:

$$\text{(A.4.5)} \quad \mathcal{H}_d(\Gamma_0) = \begin{bmatrix} d_{x_1 x_1}, \ldots, d_{x_1 x_n} \\ \vdots \quad\quad \vdots \\ d_{x_n x_1}, \ldots, d_{x_n x_n} \end{bmatrix} (\Gamma_0) > 0; \text{ in fact, } \mathcal{H}_d(\Gamma_0) \geq (1 - \epsilon)I.$$

Thus, assumption (A.1) of Section 1 holds true for $d(x)$. The scalar function $d(x)$ is defined on Ω, near Γ_0, as a 'perturbation' of $d_0(x)$ as follows:

$$\text{(A.4.6)} \quad \begin{cases} d(x) \equiv d_0(x) + z(x), \\ z(x) \equiv -\left(\dfrac{\partial d_0}{\partial \nu}\right)\ell k + \lambda \ell^2, \ k \equiv \dfrac{1}{|\nabla \ell|}, \end{cases}$$
(A.4.7)

where λ is a sufficiently large parameter, while $\left(\frac{\partial d_0}{\partial \nu}\right)$ denotes an extension of $\left.\left(\frac{\partial d_0}{\partial \nu}\right)\right|_{\Gamma_0}$ from $\Gamma_0 : \ell = 0$ to a layer (collar) of Γ_0 within Ω, defined by:

$$\text{(A.4.8)} \quad \text{near } \Gamma_0: \ \frac{\partial d}{\partial \nu} \equiv \nabla d \cdot \nu = \nabla d \cdot (k \nabla \ell), \ \text{ where } \nu \equiv \frac{\nabla \ell}{|\nabla \ell|}, \ k = \frac{1}{|\nabla \ell|}.$$

Remark A.4.1. Figure A.10 illustrates two typical cases covered under Theorem A.4.1. On the other hand, the assumptions of Theorem A.4.1 exclude the case where Ω is an annulus, say $1 \leq \|x\| \leq 2$, with Γ_0 the internal sphere $\|x\| = 1$, and x_0 the origin. $\square$

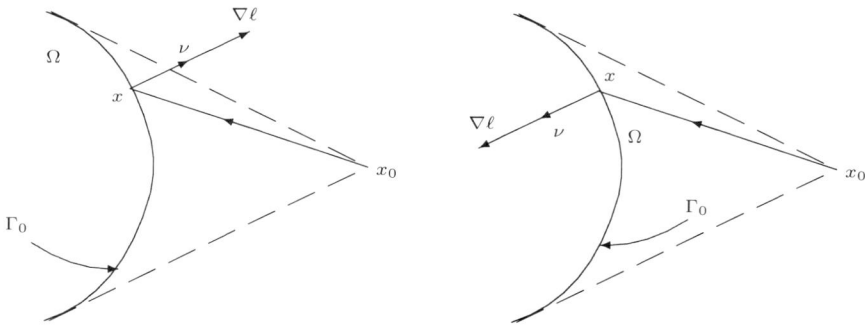

FIG. A.10: THEOREM A.4.1: CASE 1: $\ell = x^2 + y^2 - 1$; $\mathcal{H}_\ell(\Gamma_0) \geq 0$; $(x - x_0) \cdot \nu(x) \leq 0$; CASE 2: $\ell = 1 - x^2 - y^2$; $\mathcal{H}_\ell(\Gamma_0) \leq 0$; $(x - x_0) \cdot \nu(x) \geq 0$

Proof. (a) We first establish (A.4.4). Direct computations show that: on Γ_0:

(A.4.9)
$$\left. \frac{\partial(\ell k)}{\partial \nu} \right|_{\Gamma_0} = \ell \frac{\partial k}{\partial \nu} + k \frac{\partial \ell}{\partial \nu} = k \nabla \ell \cdot k \nabla \ell = 1,$$

(A.4.10)
$$\left. \frac{\partial(\ell^2)}{\partial \nu} \right|_{\Gamma_0} = 2\ell \frac{\partial \ell}{\partial \nu} = 0,$$

since, on $\Gamma_0 : \nu = k \nabla \ell$ (by (A.4.8); and $\ell = 0$. Then, returning to (A.4.6), (A.4.7), we obtain, by virtue of (A.4.9), (A.4.10):

(A.4.11) on Γ_0:
$$\left. \frac{\partial d}{\partial \nu} \right|_{\Gamma_0} = \left[\frac{\partial d_0}{\partial \nu} + \frac{\partial z}{\partial \nu} \right]_{\Gamma_0}$$

$$= \frac{\partial d_0}{\partial \nu} - \left(\frac{\partial^2 d_0}{\partial \nu^2} \right)(\ell k) - \frac{\partial d_0}{\partial \nu} \frac{\partial(\ell k)}{\partial \nu} + \lambda \frac{\partial(\ell^2)}{\partial \nu}$$

(A.4.12)
$$= \left[\frac{\partial d_0}{\partial \nu} - \frac{\partial d_0}{\partial \nu} \right]_{\Gamma_0} = 0,$$

and (A.4.4) is established.

(ii) We now prove (A.4.5). First, for convenience, set

(A.4.13)   near $\Gamma_0 : p \equiv -\frac{\partial d_0}{\partial \nu} k$, so that, by (A.4.7), $z \equiv p\ell + \lambda \ell^2$ near Γ_0.

We shall now use the diadatic product notation

(A.4.14)
$$A \otimes B \equiv \begin{bmatrix} a_1 \\ \vdots \\ a_n \end{bmatrix} [b_1, \ldots, b_n] = \begin{bmatrix} a_1 b_1 & a_1 b_2 & \cdots & a_1 b_n \\ a_2 b_1 & a_2 b_2 & \cdots & a_2 b_n \\ \cdot & \cdot & & \cdot \\ a_n b_1 & a_n b_2 & \cdots & a_n b_n \end{bmatrix}.$$

Step 1. Proposition A.4.2. In the notation of (A.4.14), and with reference to z in (A.4.13), the Hessian matrix $\mathcal{H}_z$ of z is given by

(a)
(A.4.15)
$$\text{near } \Gamma_0 : \mathcal{H}_z = p\mathcal{H}_\ell + \ell\mathcal{H}_p + \nabla\ell \otimes \nabla p + \nabla p \otimes \nabla\ell + 2\lambda\ell\mathcal{H}_\ell + 2\lambda\nabla\ell \otimes \nabla\ell;$$

(b)
(A.4.16)
$$\text{on } \Gamma_0 : \mathcal{H}_z = p\mathcal{H}_\ell + \nabla\ell \otimes \nabla p + \nabla p \otimes \nabla\ell + 2\lambda\nabla\ell \otimes \nabla\ell.$$

Proof of Proposition A.4.2. In the notation of (A.4.14), we can rewrite the Hessian matrix $\mathcal{H}_z$ of z as follows:

(A.4.17)
$$\mathcal{H}_z = \nabla \otimes \nabla z = \begin{bmatrix} \frac{\partial}{\partial x_1} \\ \vdots \\ \frac{\partial}{\partial x_n} \end{bmatrix} \begin{bmatrix} \frac{\partial z}{\partial x_1}, \ldots, \frac{\partial z}{\partial x_n} \end{bmatrix} = \begin{bmatrix} z_{x_1 x_1}, \ldots z_{x_1 x_n} \\ \vdots \quad \vdots \\ z_{x_n x_1}, \ldots z_{x_n x_n} \end{bmatrix}.$$

From (A.4.13) we have

(A.4.18)
$$\nabla z = \nabla(p\ell + \lambda\ell^2) = p\nabla\ell + \ell\nabla p + 2\lambda\ell\nabla\ell,$$

and hence, using (A.4.18) in (A.4.17), we find

(A.4.19)
$$\mathcal{H}_z = \nabla \otimes [\ell\nabla p + p\nabla\ell + 2\lambda\ell\nabla\ell].$$

Next, we verify that

(A.4.20)
$$\nabla \otimes (\ell\nabla p) = \nabla\ell \otimes \nabla p + \ell\mathcal{H}_p,$$

and hence that

(A.4.21)
$$\nabla \otimes (p\nabla\ell) = \nabla p \otimes \nabla\ell + p\mathcal{H}_\ell; \quad \nabla \otimes (\ell\nabla\ell) = \nabla\ell \otimes \nabla\ell + \ell\mathcal{H}_\ell.$$

Then (A.4.20), (A.4.21), used in (A.4.19), yield (A.4.15), from which (A.4.16) follows upon setting $\ell = 0$ on Γ_0. We now verify (A.4.20): we compute

(A.4.22)
$$\nabla \otimes (\ell\nabla p) = \begin{bmatrix} \frac{\partial}{\partial x_1} \\ \vdots \\ \frac{\partial}{\partial x_n} \end{bmatrix} [\ell p_{x_1}, \ldots, \ell p_{x_n}]$$

$$= \begin{bmatrix} \ell_{x_1} p_{x_1} + \ell p_{x_1 x_1}, \ldots, \ell_{x_1} p_{x_n} + \ell p_{x_1 x_n} \\ \vdots \quad \vdots \\ \ell_{x_n} p_{x_1} + \ell p_{x_1 x_n}, \ldots, \ell_{x_n} p_{x_n} + \ell p_{x_n x_n} \end{bmatrix}$$

(A.4.23)
$$= \begin{bmatrix} \ell_{x_1} \\ \vdots \\ \ell_{x_n} \end{bmatrix} [p_{x_1}, \ldots, p_{x_n}] + \ell\mathcal{H}_p,$$

and (A.4.20) follows from (A.4.23). Proposition A.4.2 is proved. □

Step 2. Lemma A.4.3. For $x \in \mathbb{R}^n$, and with reference to (A.4.16), we have

$$\text{(A.4.24)} \quad \text{on } \Gamma_0 : (\mathcal{H}_z x, x)_{R^n} = p(\mathcal{H}_\ell x, x)_{R^n} + 2(\nabla \ell \cdot x)(\nabla p \cdot x) + 2\lambda(\nabla \ell \cdot x)^2,$$

where $(\ ,\ )_{R^n}$ is the inner product in $\mathbb{R}^n$, and "·" denotes the usual dot product.

Proof. We first note that for any two vectors A and B in R^n, we have

$$\text{(A.4.25)} \quad ((A \otimes B)x, x)_{R^n} = (A \cdot x)(B \cdot x), \quad \forall\, x \in R^n.$$

This can be readily verified from the definition (A.4.14). Thus, specializing (A.4.25), we obtain

$$\text{(A.4.26)} \quad \begin{cases} ((\nabla \ell \otimes \nabla p)x, x)_{R^n} = ((\nabla p \otimes \nabla \ell)x, x)_{R^n} = (\nabla \ell \cdot x)(\nabla p \cdot x) \\ ((\nabla \ell \otimes \nabla \ell)x, x)_{R^n} = (\nabla \ell \cdot x)^2. \end{cases}$$
(A.4.27)

Then, using (A.4.26), (A.4.27) in (A.4.16) yields (A.4.24), as desired. □

Step 3. Proposition A.4.4. Assume hypotheses (A.4.2) and (A.4.3) on Γ_0: $\frac{\partial d_0}{\partial \nu} \leq 0$ and $\mathcal{H}_d \geq 0$ on Γ_0 (hence $p \geq 0$ on Γ_0 by (A.4.13) since $k > 0$ by (A.4.8)). Then, with reference to (A.4.28), for any $\epsilon_0 > 0$, there exists $\lambda_{\epsilon_0} > 0$, in fact $\lambda_{\epsilon_0} = \frac{1}{2\epsilon_0}$, such that, for all $\lambda > \lambda_{\epsilon_0}$ and all $x \in R^n$, we have:

$$\text{(A.4.28)} \quad \text{on } \Gamma_0 : (\mathcal{H}_z x, x)_{R^n} \geq -\epsilon_0 \| |\nabla p| \|^2 \|x\|^2 + \left[2\lambda - \frac{1}{\epsilon_0}\right](\nabla \ell \cdot x)^2$$

$$\geq -\epsilon \|x\|^2, \quad \epsilon = \epsilon_0 \max_{\Gamma_0} \| |\nabla p| \|.$$

Proof. We preliminarily have

$$\text{(A.4.29)} \quad 2(\nabla \ell \cdot x)(\nabla p \cdot x) \geq -\epsilon_0 (\nabla p \cdot x)^2 - \frac{1}{\epsilon_0}(\nabla \ell \cdot x)^2,$$

and hence

$$\text{(A.4.30)} \quad 2(\nabla \ell \cdot x)(\nabla p \cdot x) + 2\lambda(\nabla \ell \cdot x)^2 \geq \left[2\lambda - \frac{1}{\epsilon_0}\right](\nabla \ell \cdot x)^2 - \epsilon_0(\nabla p \cdot x)^2.$$

Returning to (A.4.24) on Γ_0, where $p \geq 0$ and $\mathcal{H}_\ell \geq 0$ on Γ_0, or else $p \leq 0$ and $\mathcal{H}_\ell \leq 0$, respectively; or more generally, $p\mathcal{H}_\ell \geq 0$, by assumption, we finally obtain via (A.4.30),

$$\text{(A.4.31)} \quad \text{on } \Gamma_0 : (\mathcal{H}_z x, x)_{R^n} \geq -\epsilon_0(\nabla p \cdot x)^2 + \left[2\lambda - \frac{1}{\epsilon_0}\right](\nabla \ell \cdot x)^2$$

$$\text{(A.4.32)} \quad \geq -\epsilon_0 \| |\nabla p| \|^2 \|x\|^2 \geq -\epsilon \|x\|^2,$$

by the Schwarz inequality, for all $\lambda > \frac{1}{2\epsilon_0}$. Thus, (A.4.32) establishes (A.4.28), as desired. □

Step 4. We return to (A.4.6) and obtain that, under the given assumptions (A.4.2), (A.4.3), Proposition A.4.4 holds true, and then (A.4.28) yields, since $\mathcal{H}_{d_0} \equiv I$ (identity):

(A.4.33) $\qquad\qquad\qquad$ on $\Gamma_0 : \mathcal{H}_d = \mathcal{H}_{d_0} + \mathcal{H}_z \geq I - \epsilon I.$

Then, (A.4.32) proves property (ii) in (A.4.5). Theorem A.4.1 is established. $\square$

Assumption (A.2) = (1.1.6) of Section 1. In preparation for our discussion on the setting of Section 10, we now identify the point(s) on the boundary Γ_0, if any, where the conservative vector field ∇d constructed in Theorem A.4.1 violates assumption (A.2) = (1.1.6). This will then allow us, in the subsequent Appendix C, to remedy the situation by devising a strategy [based on splitting Ω as the union of two overlapping subdomains Ω_1, Ω_2: $\Omega = \Omega_1 \cup \Omega_2$ and the employment of two vector fields], which will then fit the setting of Section 10.

Corollary A.4.2. *Under the assumptions of Theorem A.4.1, the scalar function $d(x)$ there constructed has the following third property: its gradient $\nabla d|_{\Gamma_0}$, once restricted on the boundary Γ_0, vanishes at the unique point $x \in \Gamma_0$, if such exists on Γ_0, where the vector field $\nabla d_0(x) = x - x_0$ (see (A.4.2)) is orthogonal to Γ_0. In symbols: for $x \in \Gamma_0$,*

$$\nabla d(x) = 0 \iff (x - x_0) \text{ parallel to normal } \nu(x) \text{ at } x \in \Gamma_0,$$

(A.4.34) $\qquad\qquad\qquad\qquad\qquad$ *i.e., orthogonal to Γ_0 at x,*

and such point $x \in \Gamma_0$ is unique, if it exists on Γ_0.

Proof. The proof is a direct computation, starting from the definition of the function $d(x)$ in (A.4.6), (A.4.7). The gradient ∇d of d is:

(A.4.35) $\qquad \nabla d = \nabla d_0 + \nabla z = \nabla d_0 - \left(\dfrac{\partial d_0}{\partial \nu}\right) k \nabla \ell - \ell \nabla \left(-\dfrac{\partial d_0}{\partial \nu} k\right) - 2\lambda \ell \nabla \ell.$

Its restriction on Γ_0, where $\ell \equiv 0$, is then, recalling k from (A.4.7):

(A.4.36) $\qquad$ on $\Gamma_0 : \nabla d|_{\Gamma_0} = \nabla d_0|_{\Gamma_0} - \dfrac{\partial d_0}{\partial \nu}\bigg|_{\Gamma_0} \nu, \quad \left(\dfrac{\nabla \ell}{|\nabla \ell|}\right)_{\Gamma_0} = \nu.$

Thus, if $x \in \Gamma_0$, since $\nabla d_0(x) = x - x_0$ (see (A.4.2)), we obtain

(A.4.37) $\qquad$ for $x \in \Gamma_0 : \nabla d(x) = (x - x_0) - [(x - x_0) \cdot \nu(x)]\nu(x).$

Thus, for $x \in \Gamma_0$, we obtain $\nabla d(x) = 0$ in (A.4.36) if and only if (A.4.34) holds true. Such point $x \in \Gamma_0$ in (A.4.34) is unique, if it exists on Γ_0, since Γ_0 is convex (concave).

In fact, if Γ_0 is described, say, by the level set $\ell(x,y) = f(x) - y = 0$, with $f(x)$ a convex function defined on a convex set K of $\mathbb{R}^{n-1}$, choose the axes so that: the origin 0 is the point $f(0) = 0$; $f(x) \geq 0$; on Ω near Γ_0 we have $\ell(x,y) = f(x) - y \leq 0$. Let $Y_0 : (0, y_0)$, with $y_0 < 0$, be a point outside of Ω such that $\overrightarrow{Y_0 O}$ is orthogonal to Γ_0. Pick any other point $P : (x, f(x))$ on Γ_0, with normal $\bar{n} = \{\nabla f(x), -1\}$. We claim that $\overrightarrow{Y_0 P} = \{x, f(x) - y_0\}$ cannot be orthogonal to Γ_0; i.e., that $\overrightarrow{Y_0 P}$ cannot be parallel to $\bar{n}$. This is so since the last coordinates, -1 and $[f(x) - y_0]$, of $\bar{n}$ and

$\overrightarrow{Y_0P}$ have opposite sign, while $\nabla f(x) \cdot x \geq f(x) \geq 0$, since, by convexity of $f(x)$, we have: $f(0) - f(x) \geq \nabla f(x) \cdot (0 - x)$ [Fl.1, Prop. 3.6a, p. 111]. □

Appendix B: Illustrations satisfying the setting of Section 1

B.1 Illustrations of approach of Section A.2. In this subsection, we provide illustrations of the approach of Section A.2 and its extension to the 3-dimensional case.

Two-dimensional illustrations. **Example B.1.1.** Let Ω be the two-dimensional ellipsoidal region: $\frac{x^2}{a^2} + \frac{y^2}{b^2} < 1$ in the (x,y)-plane, surrounded by the ellipse $\Gamma : \frac{x^2}{a^2} + \frac{y^2}{b^2} = 1$. Define

(B.1.1) $\qquad \Gamma_0 = \{(x,y) \in \Gamma : 0 < x_1 \leq x \leq x_2 < a\}; \ \Gamma_1 = \Gamma \setminus \Gamma_0.$

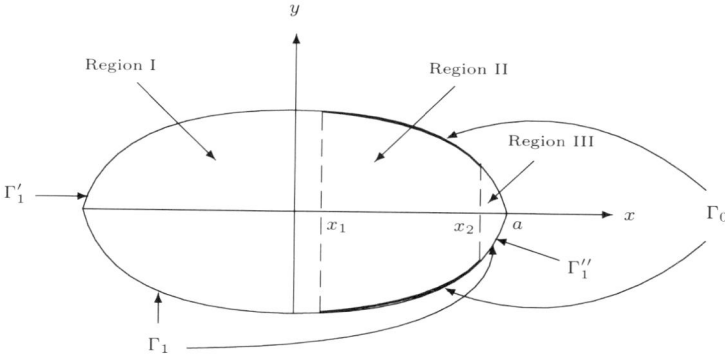

FIG. B.1: UNCONTROLLED PART Γ_0 AND CONTROLLED PART Γ_1, $0 < x_1 < x_2 < a$

Here, x_1 and x_2 are fixed points, arbitrarily close to 0 and a, respectively. It is assumption (A.2) = (1.1.6) of Section 1 that imposes the constraint $x_2 < a$: indeed, the vector field h constructed below in (B.1.5) vanishes at the point $\{x = a, y = 0\}$, thus violating (A.2) =(1.1.6) of Section 1. It is then necessary (under the present or similar constructions) based on the one-vector setting of Section 1 to 'cut off' an arbitrarily small portion of the boundary Γ around the point $\{a, 0\}$ to assign to Γ_1, in order to achieve condition (A.2) = (1.1.6): see (B.1.9) below. Thus, Γ_0 is still 'almost' $\frac{1}{2}$ of the boundary Γ, in line with known control theoretic results for second-order hyperbolic equations, but Γ_0 must miss a small arc around $(a, 0)$ in our present construction, based only on Section 1.

In Section C below we shall refine the present analysis by having Γ_0 connected and almost $\frac{1}{2}$ of the boundary Γ; i.e., the present portion Γ_1'' will be dispensed with. Here we have, recalling (A.2.12).

(B.1.2) $\qquad -y \frac{\ell_y}{\ell_x} = \frac{-y^2}{b^2} \frac{a^2}{x} = \frac{a^2}{x} \left(\frac{x^2}{a^2} - 1 \right) = x - \frac{a^2}{x}, \ 0 \leq x \leq a, \ (x, y) \in \Gamma_0.$

This yields the expression $h_1^{II}(x)$ of the vector field h in (B.1.6b) below, where the additional assumption (A.2) = (1.1.6) of Section 1 [not guaranteed by Lemma A.2.1] forces the constraint: $x_2 < a$; thus, the need to extend smoothly and suitably $h_1^{II}(x)$. This is done below.

Lemma B.1.1. The following function $d(x,y) \in C^3$

$$d(x,y) = \begin{cases} d^I(x) + \dfrac{y^2}{2}, & -a \leq x \leq x_1; \quad \text{(B.1.3a)} \\ d^{II}(x) + \dfrac{y^2}{2}, & 0 < x_1 \leq x \leq x_2 < a; \quad \text{(B.1.3b)} \\ d^{III}(x) + \dfrac{y^2}{2}, & x_2 \leq x \leq a, \quad \text{(B.1.3c)} \end{cases}$$

satisfies assumption (A.1) of Section 1, where

(B.1.4a)
$$d^I(x) = -\frac{a^2(x-x_1)^3}{3x_1^3} + \left(1 + \frac{a^2}{x_1^2}\right)\frac{(x-x_1)^2}{2}$$
$$+ \left(x_1 - \frac{a^2}{x_1}\right)x + 1 - \frac{x_1^2}{2} - a^2 \ln x_1, \quad -1 \leq x \leq x_1;$$

(B.1.4b)
$$d^{II}(x) = \frac{x^2}{2} - a^2 \ln x \qquad x_1 \leq x \leq x_2;$$

(B.1.4c)
$$d^{III}(x) = \frac{x^2}{2} - a^2 \ln x - \frac{\epsilon(x-x_2)^4}{4} \qquad x_2 \leq x \leq a,$$

for $0 < \epsilon$ sufficiently small, say $3\epsilon(a-x_2)^2 < \frac{1}{2}$, as to obtain (B.1.8c) below.

Proof. We use the ideas leading to Lemma A.2.1 to construct $d(x,y)$, starting from (B.1.2) to obtain the component $h_1^I(x)$ below and then extend smoothly. Here we verify the requirements of assumptions (A.1) and (A.2):

(a) $d(x,y) \in C^3$: using also the formulas below in (B.1.6), (B.1.8), one checks directly that

$$d^I(x_1) = d^{II}(x_1); \quad (d^I)'(x_1) = (d^{II})'(x_1); \quad (d^I)''(x_1) = (d^{II})''(x_2);$$

$$d^{II}(x_2) = d^{III}(x_2); \quad (d^{II})'(x_2) = (d^{III})'(x_2); \quad (d^{II})''(x_2) = (d^{III})''(x_2).$$

(b) The gradient $h(x,y) = (h_1, h_2) = \nabla d(x,y)$ is given by

$$h_1(x) = \begin{cases} h_1^I(x) & -a \leq x \leq x_1 \quad \text{(B.1.5a)} \\ h_1^{II}(x) & x_1 \leq x \leq x_2; \quad \text{(B.1.5b)} \\ h_1^{III}(x) & x_2 \leq x \leq a \quad \text{(B.1.5c)} \end{cases} \qquad h_2(y) = y;$$

(B.1.6a)
$$h_1^I(x) = -\frac{a^2}{x_1^3}(x-x_1)^2$$
$$+ \left(1 + \frac{a^2}{x_1^2}\right)(x-x_1) + x_1 - \frac{a^2}{x_1} < 0, \quad -a \leq x \leq x_1;$$

(B.1.6b)
$$h_1^{II}(x) = x - \frac{a^2}{x} < 0 \qquad x_1 \leq x \leq x_2;$$

(B.1.6c)
$$h_1^{III}(x) = x - \frac{a^2}{x} - \epsilon(x-x_2)^3 < 0 \qquad x_2 \leq x \leq a.$$

The Jacobian matrix J_h of h is given by

$$(B.1.7) \quad J_h(x,y) = \begin{bmatrix} h_1'(x) & 0 \\ 0 & 1 \end{bmatrix} = \text{positive definite on } \bar{\Omega},$$

where

$$(B.1.8a) \quad (h_1^I)'(x) = -\frac{2a^2}{x_1^3}(x-x_1) + \left(1+\frac{a^2}{x_1^2}\right) > 0, \quad -a \leq x \leq x_1;$$

$$(B.1.8b) \quad h_1'(x) = \begin{cases} (h_1^{II})'(x) = 1 + \frac{a^2}{x^2} > 0, & x_1 \leq x \leq x_2; \end{cases}$$

$$(B.1.8c) \quad (h_1^{III})'(x) = 1 + \frac{a^2}{x^2} - 3\epsilon(x-x_2)^2 > 0, \quad x_2 \leq x \leq a,$$

for $3\epsilon(a-x_2)^2 < \frac{1}{2}$, as desired.

(c) $h \cdot \nu \equiv 0$ on Γ_0: indeed, for $(x,y) \in \Gamma_0$, see (B.1.2), we have that $h_1^{II}(x)$ in (B.1.6b) is the active component and then $\left(x - \frac{a^2}{x}, y\right) \cdot \left(\frac{x}{a^2}, \frac{y}{b^2}\right) \equiv \frac{x^2}{a^2} - 1 + \frac{y^2}{b^2} \equiv 1$, as required.

(d) For all $(x,y) \in \bar{\Omega}$, we have by (B.1.6),

$$(B.1.9) \quad |h(x,y)| \geq |h_1(x)| \geq p > 0.$$

[Notice that each term in (B.1.6a-b-c) is non-positive on its specific range.]

Remark B.1.1. As pointed out at the outset, the first two requirements (i) = (1.1.4) and (ii) = (1.1.5) of assumption (A.1) are satisfied also by taking $x_2 = a$ in which case the region III vanishes and Γ_0 is connected.

Remark B.1.2. We remark explicitly that, in the present example, we have $h \cdot \nu \geq 0$ on $\Gamma_1' = \{(x,y) \in \Gamma_1 : -a \leq x \leq x_1\}$, but not on $\Gamma_1'' \equiv \Gamma_1 \setminus \Gamma_1' \equiv \{(x,y) \in \Gamma_1 : x_2 \leq x \leq a\}$. Indeed, condition $h \cdot \nu \geq 0$ on all of Γ_1 and condition (A.2) = (1.1.6) appear to be incompatible. □

Example B.1.2. (non-convex Ω) We now let Ω be a region exterior to the critical part of the boundary $\Gamma_0 \cup \Gamma_1''$ of Example B.1.1: see Fig. B.2.

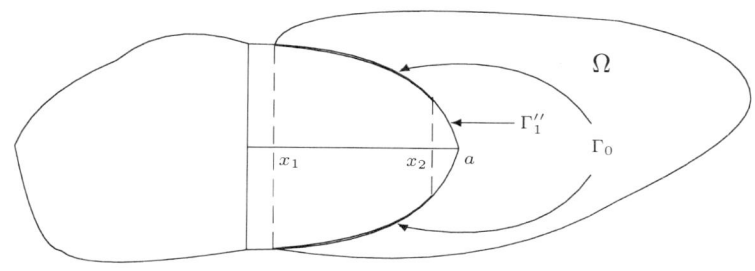

FIG. B.2: EXAMPLE B.1.2

This case is actually simpler than Example B.1.1: here the counterpart of Lemma B.1.1 is that: the function $d(x, y) \in C^3$, with gradient $\nabla d = h = (h_1, h_2)$: (B.1.10a)

(B.1.10b)
$$d(x,y) = \begin{cases} d^{II}(x) + \dfrac{y^2}{2} \\ d^{III}(x) + \dfrac{y^2}{2} \end{cases} ; \quad h_1(x) = \begin{cases} h_1^{II}(x) & 0 < x_1 \leq x \leq x_2 < a; \\ h_1^{III}(x) & x_2 \leq x \leq x_3. \end{cases}$$

$h_2(y) = y$, with $d^{II}(x)$, $d^{III}(x)$ defined by (B.1.4b-c), hence $h_1^{II}(x)$, $h_1^{III}(x)$ defined by (B.1.6b-c), except that now $3\epsilon(x_3 - x_2)^2 < \frac{1}{2}$, *satisfies assumption (A.1) of Section 1*. Again, the first two requirements (i) = (1.1.4) and (ii) = (1.1.5) of assumption (A.1) are fulfilled also with $x_2 = a$, in which case $h_1(x) = x - \frac{a^2}{x}$, $0 < x_1 \leq x \leq x_3$; $h_2(y) = y \in \mathbb{R}$ does the job. $\square$

Example B.1.3. (parabola) Here $\ell(x, y) = x - 1 + y^2 \equiv 0$ and then $-y\frac{\ell_y}{\ell_x} = -2y^2 = 2x - 2$, $(x, y) \in \Gamma_0$. Then

(B.1.11) $$h_1(x) = 2x - 2, \quad h_2(y) = y$$

provides the required vector field $h(x, y)$, $h_1'(x) = 2 > 0$, which satisfies assumption (A.1) everywhere within, or without the parabolic sector. However, to satisfy also assumption (A.2) we must exclude the vertex point $\{x = 1, y = 0\}$, as in Example B.1.1.

Example B.1.4. (hyperbola) Here $\ell(x, y) = x^2 - y^2 - k^2 \equiv 0$ and then $-y\frac{\ell_y}{\ell_x} = x - \frac{k^2}{x}$, $(x, y) \in \Gamma_0$. Then

(B.1.12) $$h_1(x) = x - \frac{k^2}{x}, \quad h_2(y) = y, \quad |x| \geq \epsilon > 0$$

provides the required vector field $h(x, y)$, $h_1'(x) = 1 + k^2/x^2 > 0$, which satisifies assumption (A.1) everywhere within, or without, the hyperbolic sector, away from the y-axis.

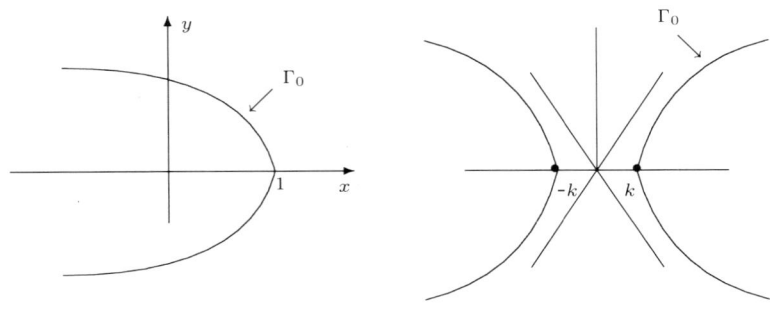

FIG. B.3: EXAMPLES B.1.3 (PARABOLA) AND B.1.4 (HYPERBOLA)

Example B.1.5. (logarithm) Here $f(x) = \ln x$, so that

(B.1.13) $$h_1(x) = \frac{f(x)}{f'(x)} = x \ln x, \quad h_2(y) = y$$

provides the required vector field, however, under the constraint $x > \frac{1}{e}$, whereby then $h'_1(x) = 1 + \ln x > 0$, as required. We note, however, that if the *same* curve is viewed as $y = f(x) = e^x$, then the test of Lemma A.2.2 [which requires a suitable choice of the coordinate axes] fails in x [but the test of Lemma A.2.2 works in y, as seen from the analysis above for $x = \ln y$], as now the vector $\{1, y\}$, $\frac{f(x)}{f'(x)} \equiv 1$, suggested by Lemma A.2.2 does not satisfy the positivity condition of its Jacobian matrix. However, that test can be easily modified to yield a positive conclusion. For $(x, y) \in \Gamma_0$, described by $y = e^x$, we take the tangential vector $h = \{e^x \cdot 1, e^x \cdot y\}$ [instead of $\{1, y\}$], which we then extend to all (x, y), $y > 0$, as $h = \{e^x, y^2\}$. Then this vector field h satisfies assumption (A.1) of Section 1 for the exponential curve $y = e^x$ on any finite interval in x.

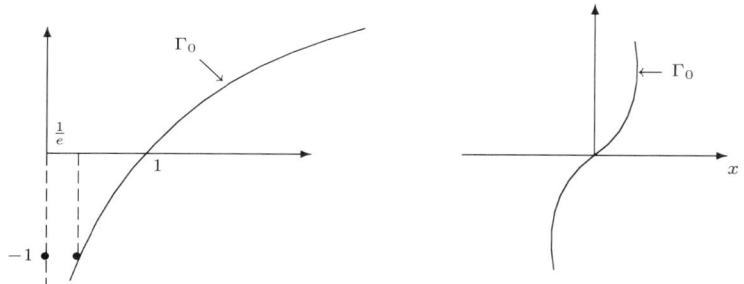

FIG. B.4: EXAMPLE B.1.5 (LOGARITHM) AND B.1.6 (CUBIC)

Thus, $h(x)$ satisfies assumption (A.1) for $x > \frac{1}{e}$.

Example B.1.6. (cubic) Here $f(x) = x^3$ so that

(B.1.14) $$h_1(x) = \frac{f(x)}{f'(x)} = \frac{1}{3}x, \quad h_2(y) = y$$

provides the required vector field $h(x, y)$, as $h'_1(x) \equiv \frac{1}{3} > 0$.

Example B.1.7. (sine and cosine) Here $f(x) = \sin x$, $-\frac{\pi}{2} < x < \frac{\pi}{2}$; or else $f(x) = \cos x$, $0 < x < \pi$. Then the vector field

(B.1.15) $$h_1(x) = \frac{f(x)}{f'(x)} = \begin{cases} \tan x \\ -\cot x \end{cases} \quad h_2(y) = y$$

provides the required vector field so that $h'_1(x) = 1/\cos^2 x$, or else $h'_1(x) = 1/\sin^2 x$, respectively, are positive in the indicated intervals

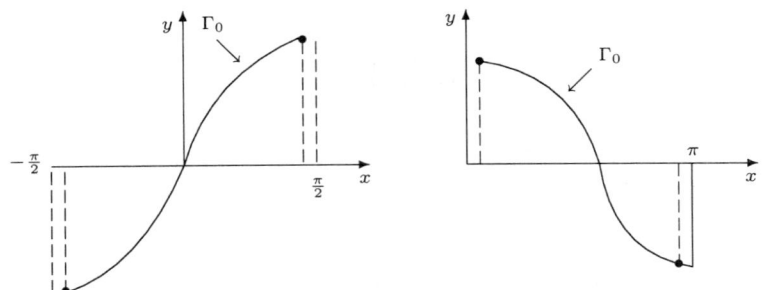

FIG. B.5: EXAMPLE B.1.7 (SINE AND COSINE)

B.2 Three-dimensional analysis: Extension of Section A.2. In this subsection we extend the analysis of Section A.2 from the 2-dimensional to the 3-dimensional case. Let Γ_0 be a 3-dimensional surface described as a level set:

(B.2.1) $$\Gamma_0 = \{(x,y,z) \in R^3 : \ell(x,y,z) = 0\}, \quad \ell \in C^3,$$

with $|\nabla \ell| \neq 0$ on Γ_0. Let $P = (x,y,z) \in \Gamma_0$. Assume that Γ_0 displays symmetry with respect to the x-axis [or else to the y-axis; or else to the z-axis, respectively]. Then, we seek the point X [or else the point Y; or else the point Z, respectively] of intersection between the tangent plane at P and the x-axis [or else the y-axis; or else the z-axis, respectively]. We then consider the vector fields

(B.2.2) $$\overrightarrow{XP} = \left\{-\frac{y\ell_y + z\ell_z}{\ell_x}, y, z\right\}, \quad P = (x,y,z) \in \Gamma_0, \ \ell_x \neq 0;$$

(B.2.3) $$\overrightarrow{YP} = \left\{x, -\frac{x\ell_x + z\ell_z}{\ell_y}, z\right\}, \quad P = (x,y,z) \in \Gamma_0, \ \ell_y \neq 0;$$

(B.2.4) $$\overrightarrow{ZP} = \left\{x, y, -\frac{x\ell_x + y\ell_y}{\ell_z}\right\}, \quad P = (x,y,z) \in \Gamma_0, \ \ell_x \neq 0,$$

respectively. Thus, in each case, the corresponding vector $\overrightarrow{XP}, \overrightarrow{YP}, \overrightarrow{ZP}$ is tangent to Γ_0 at P.

In many cases, at least one of the following conditions is satisfied by $\ell(x,y,z)$: for $(x,y,z) \in \Gamma_0$, then

(B.2.5) $$\begin{cases} \text{either the term } \left(-\frac{y\ell_y + z\ell_z}{\ell_x}\right)(x) \text{ is only a function of } x; \ \ell_x \neq 0; \\ \text{or else the term } \left(-\frac{x\ell_x + z\ell_z}{\ell_y}\right)(y) \text{ is only a function of } y; \ \ell_y \neq 0; \\ \text{or else the term } \left(-\frac{x\ell_x + y\ell_y}{\ell_z}\right)(z) \text{ is only a function of } z; \ \ell_z \neq 0. \end{cases}$$

Then, in either case, we extend the boundary vectors $\overrightarrow{XP}, \overrightarrow{YP}, \overrightarrow{ZP}$ in (B.2.2)–(B.2.4) to all points (x,y,z) near Γ_0 as follows: define a vector field $h(x,y,z) = \{h_1, h_2, h_3\}$, where

(B.2.6) $$h_1(x) = \left(-\frac{y\ell_y + z\ell_z}{\ell_x}\right)(x), \ h_2(y) = y; \ h_3(z) = z,$$

or else

(B.2.7) $\quad h_1(x) = x; \quad h_2(y) = \left(-\dfrac{x\ell_x + z\ell_z}{\ell_y}\right)(y), \quad h_3(z) = z,$

or else

(B.2.8) $\quad h_1(x) = x; \quad h_2(y) = y; \quad h_3(z) = \left(-\dfrac{x\ell_x + y\ell_y}{\ell_z}\right)(z),$

respectively. An extension of Lemma A.2.2 is then:

Lemma B.2.1. Let $\ell \in C^3$ and assume hypothesis (B.2.5) with either $\ell_x \neq 0$, or $\ell_y \neq 0$, or $\ell_z \neq 0$, respectively, unless a cancellation occurs. Assume, moreover, that on Γ_0, either

(B.2.9) $\quad \begin{cases} \dfrac{d}{dx}\left[-\dfrac{y\ell_x + z\ell_z}{\ell_x}\right](x) > 0; \quad \text{or} \quad \dfrac{d}{dy}\left[-\dfrac{x\ell_x + z\ell_z}{\ell_y}\right](y) > 0; \\ \text{or} \quad \dfrac{d}{dz}\left[-\dfrac{x\ell_x + y\ell_y}{\ell_z}\right](z) > 0, \end{cases}$

respectively. Then:

(i) the Jacobian matrix J_h of the vector field h defined in (B.2.6)–(B.2.8), respectively, is positive definite on, and near, Γ_0;

(ii) the vector field $h = \{h_1, h_2, h_3\}$ defined in (B.2.6)–(B.2.8) respectively, satisfies $h = \overrightarrow{XP}$, or $h = \overrightarrow{YP}$, or $h = \overrightarrow{ZP}$ for $P \in \Gamma_0$, respectively, and thus $h \cdot \nu = 0$ on Γ_0;

(iii) the vector field $h(x,y,z)$ is conservative, $h = \nabla d$, where the function $d(x,y,z) \in \ell^3$ defined near Γ_0 by

(B.2.10)
(B.2.11) $\quad d(x,y,z) = \begin{cases} \displaystyle\int\left(-\dfrac{y\ell_y + z\ell_z}{\ell_x}\right)(x)dx + \dfrac{y^2}{2} + \dfrac{z^2}{2}, & \text{for } h \text{ in (B.2.6)}; \\[2mm] \dfrac{x^2}{2} + \displaystyle\int\left(-\dfrac{x\ell_x + z\ell_z}{\ell_x}\right)(y)dy + \dfrac{z^2}{2}, & \text{for } h \text{ in (B.2.7)}; \\[2mm] \dfrac{x^2}{2} + \dfrac{y^2}{2} + \displaystyle\int\left(-\dfrac{x\ell_x + y\ell_y}{\ell_z}\right)(z)dz, & \text{for } h \text{ in (B.2.8)}; \end{cases}$
(B.2.12)

satisfies the first two conditions (i) = (1.1.4) and (ii) = (1.1.5) of assumption (A.1) of Section 1. □

Illustrations. Example B.2.1. (ellipsoid) Here $\Gamma_0 : \ell(x,y,z) = \dfrac{x^2}{a^2} + \dfrac{y^2}{b^2} + \dfrac{z^2}{z^2} - 1 = 0$. Then, for $(x,y,z) \in \Gamma_0$, we find

(B.2.13) $\quad -\dfrac{y\ell_y + z\ell_z}{\ell_x} = -\dfrac{a^2}{x}\left(\dfrac{y^2}{b^2} + \dfrac{z^2}{c^2}\right) = x - \dfrac{a^2}{x}, \quad 0 < x \leq a,$

a function only of x, and the first condition in (B.2.5) is satisfied. Then, according to Lemma B.2.1,

(B.2.14) $\quad h_1(x) = x - \dfrac{a^2}{x}, \quad h_2(y) = y, \quad h_3(z) = z, \quad J_h = \begin{bmatrix} 1 + \dfrac{a^2}{x^2} & & \\ & 1 & \\ & & 1 \end{bmatrix}$

provides the required vector field, $h'_1(x) = 1 + \frac{a^2}{x^2} > 0$, $0 < x \le a$, which satisfies assumption (A.1) of Section 1, for any $0 < \rho \le x \le a$. An analysis as in Example B.1.1 could be carried out, extending the corresponding functionn $d(x, y, z)$ to the entire ellipsoidal region.

Example B.2.2. (elliptic paraboloid) Here $\ell(x, y, z) = \frac{x^2}{a^2} + \frac{y^2}{b^2} - \frac{z}{c} = 0$ with symmetry with respect to the z-axis. Accordingly, we test the third condition in (B.2.5). For $(x, y, z) \in \Gamma_0$, we find

(B.2.15) $$-\frac{x\ell_x + y\ell_y}{\ell_z} = 2c\left(\frac{x^2}{a^2} + \frac{y^2}{b^2}\right) = 2z,$$

a function only of z. Thus, the first condition of (B.2.5) is satisfied. Then, according to Lemma B.2.1,

(B.2.16) $$h_1(x) = x, \quad h_2(y) = y, \quad h_3(z) = 2z, \quad J_h = \begin{bmatrix} 1 & & \\ & 1 & \\ & & 2 \end{bmatrix}$$

provides the required vector field, $h'_3(z) = 2 > 0$, which satisfies assumption (A.1) of Section 1.

Example B.2.3. (hyperbolic paraboloid) Here $\ell(x, y, z) = \frac{x^2}{a^2} - \frac{y^2}{b^2} - \frac{z}{c} = 0$, with symmetry with respect to the z-axis. Accordingly, we test the third condition in (B.2.5). For $(x, y, z) \in \Gamma_0$, we find

(B.2.17) $$-\frac{x\ell_x + y\ell_y}{\ell_z} = 2c\left(\frac{x^2}{a^2} - \frac{y^2}{b^2}\right) = 2z,$$

a function only of z. Thus, the same definition as in (B.2.16) provides the required vector field, which satisfies assumption (A.1) of Section 1.

Example B.2.4. (hyperboloid of one sheet) Here, $\ell(x, y, z) = \frac{x^2}{a^2} + \frac{y^2}{b^2} - \frac{z^2}{c^2} - 1 = 0$, with symmetry with respect to the z-axis. Accordingly,

(B.2.18) $$-\frac{x\ell_x + y\ell_y}{\ell_z} = \frac{c^2}{z}\left(\frac{x^2}{a^2} + \frac{y^2}{b^2}\right) = z + \frac{c^2}{z}, \quad 0 < z,$$

a function only of z. The third condition of (B.2.5) is satisfied. Then, according to Lemma B.2.1,

(B.2.19) $$h_1(x) = x, \quad h_2(y) = y, \quad h_3(z) = z + \frac{c^2}{z}; \quad J_h = \begin{bmatrix} 1 & & \\ & 1 & \\ & & 1 - \frac{c^2}{z^2} \end{bmatrix}$$

provides the required vector field, which satisfies assumption (A.1) of Section 1, for $|z| > |c|$.

Example B.2.5. (hyperboloid of two sheets) Here $\ell(x, y, z) = \frac{x^2}{a^2} + \frac{y^2}{b^2} - \frac{z^2}{c^2} + 1 = 0$, with symmetry with respect to the z-axis. Accordingly,

(B.2.20) $$-\frac{x\ell_x + y\ell_y}{\ell_z} = \frac{c^2}{z}\left(\frac{x^2}{a^2} + \frac{y^2}{b^2}\right) = z - \frac{c^2}{z}, \quad 0 < z,$$

a function only of z. Thus, the third condition of (B.2.5) is satisfied. Then, according to Lemma B.2.1,

$$(B.2.21) \quad h_1(x) = x, \quad h_2(y) = y, \quad h_3(z) = z - \frac{c^2}{z}; \quad J_h = \begin{bmatrix} 1 & & \\ & 1 & \\ & & 1 + \frac{c^2}{z^2} \end{bmatrix}$$

provides the required vector field, which satisfies assumption (A.1) of Section 1, for $z > 0$. □

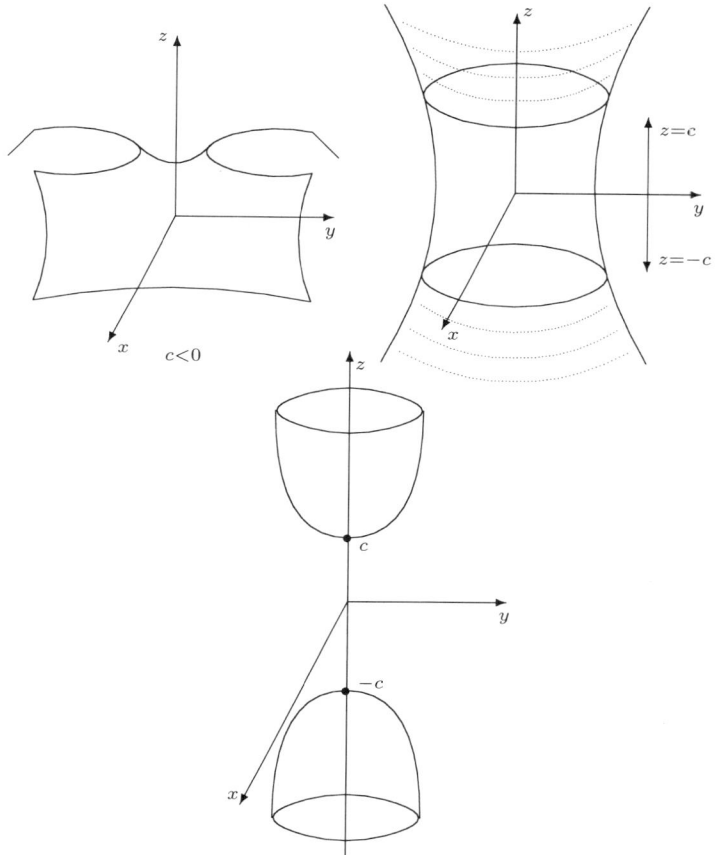

FIG. B.6: EXAMPLES B.2.3 (HYPERBOLIC PARABOLOID), B.2.4 (HYPERBOLOID OF ONE SHEET), B.2.5 (HYPERBOLOID OF TWO SHEETS)

B.3 Illustrations of the conformal mapping approach of Section A.3.
In this subsection we provide a few illustrations of the conformal mapping approach presented in Section A.3 to obtain 2-dimensional curves C_0, where assumption (A.1) of Section 1 holds true.

Example B.3.1. We take the conformal mapping

$$(B.3.1) \quad f(z) = z^2 = u + iv, \quad u(x,y) = x^2 - y^2; \quad v(x,y) = 2xy,$$

which maps the hyperbola C_0:

(B.3.2) $\quad C_0 : x^2 - y^2 = u_0, \quad u_0 > 0$ onto the line $\Gamma_0 : u \equiv u_0 > 0$.

We test condition (A.3.19) of Corollary A.3.3, with $u_x = 2x$, $v_x = 2y$, $v_{xx} \equiv 0$, $v_{xy} = 2$, on C_0:

(i)

(B.3.3) $\quad |f'(z)|^4 - v^2(v_{xx}^2 + v_{xy}^2)$

$$= [4(x^2 + y^2)]^2 - 4xy \cdot 4 = 16(x^4 + y^4 + 16x^2y^2)$$

$$= 16[(u_0 + y^2)^2 + y^4 + 16(u_0 + y^2)y^2 > 0.$$

Thus, the assumption of Corollary A.3.3 is satisfied. Then, Corollary A.3.3 yields the following: *any finite portion of the hyperbola $x^2 - y^2 = u_0$, $u_0 > 0$ satisfies assumption (A.1) of Section 1*, thus re-proving by conformal mapping methods the result of Example B.1.4. More precisely, the function

(B.3.4) $\quad \delta(x,y) = \frac{1}{2}[(u(x,y) - u_0)^2 + v^2(x,y)] = \frac{1}{2}[(x^2 - y^2 - u_0)^2 + 4x^2y^2]$

[see (A.3.11) plus (B.3.1)] satisfies assumption (A.1) of Section 1 for any portion C_0 of the hyperbola, with corresponding vector field $h = \nabla d$ given by (see (A.3.12) in a different notation $\delta(x,y)$]

(B.3.5) $\quad \begin{cases} h_1 = \delta_x = (u - u_0)u_x + vv_x = 2(x^3 + 2xy^2 - xy^2 - u_0 x); \\ h_2 = \delta_y = (u - u_0)u_y + vv_y = 2(2x^2 y - x^2 y + y^3). \end{cases}$

Example B.3.2. Here we take the conformal mapping

(B.3.6) $\quad f(x) = e^z = u + iv, \quad u(x,y) = e^x \cos y, \quad v(x,y) = e^x \sin y,$

which maps the curve

(B.3.7) $\quad C_0 : e^x \cos y \equiv u_0 \neq 0$ onto the line $\Gamma_0 : u \equiv u_0$.

We test condition (A.3.19) of Corollary A.3.3, with $u_x = u$, $v_x = v_{xx} = v$, $v_{xy} = e^x \cos y$, on C_0:

(i)

(B.3.8) $\quad |f'(z)|^4 - v^2(v_{xx} + v_{xy}^2) = e^{4x} - e^{2x} \sin^2 y(e^{2x})$

$$= e^{4x} \cos^2 y = e^{2x} u_0^2 > 0, \quad u_0 \neq 0.$$

Thus, the assumption of Corollary A.3.3 is satisfied. Then, Corollary A.3.3 yields the following: *any finite portion of the curve $C_0 : e^x \cos y = u_0$, or $x = \ln(u_0/\cos y)$ satisfies assumption (A.1) of Section 1*, with function (see (A.3.11)),

(B.3.9) $\quad \delta(x,y) = \frac{1}{2}[(u(x,y) - u_0)^2 + v^2(x,y)] = \frac{1}{2}(e^{2x} - 2u_0 e^x \cos y + u_0^2),$

and vector field $h = \nabla \delta$ given by (see (A.3.12)),

(B.3.10) $\quad h_1 = \delta_x = e^{2x} - u_0 e^x \cos y; \quad h_2 = \delta_y = u_0 e^x \sin y.$

Remark B.3.1. By contrast, if we apply the test of Lemma A.2.2 to the curve $C_0 : \ell(x,y) = e^x \cos y - u_0 = 0$, or $x = -\ln \cos y = g(y)$ for say, $u_0 = 1$, we find a worse conclusion. Indeed, we find that on the curve C_0, where $\cos y = e^{-x}$, we have

(B.3.11) $$\left. \frac{d}{dy}\left(\frac{g(y)}{g'(y)}\right) \right|_{C_0} = 1 - x - e^{-2x} = 1 + \ln(\cos y) - \cos^2 y,$$

and thus

(B.3.12)
(B.3.13) $$\left. \frac{d}{dy}\left(\frac{g(y)}{g'(y)}\right) \right|_{C_0} \begin{cases} > 0 & \text{for } 0 \leq |y| < y_1; \\ \leq 0 & \text{for } y_1 \leq |y| \leq \frac{\pi}{2}, \end{cases}$$

for some point y_1, where $e^{-x_1} = \cos y_1 < \frac{1}{e}$, $x_1 = 0.79681\ldots$.

This is a result *less precise* than the one obtained above by conformal mapping.

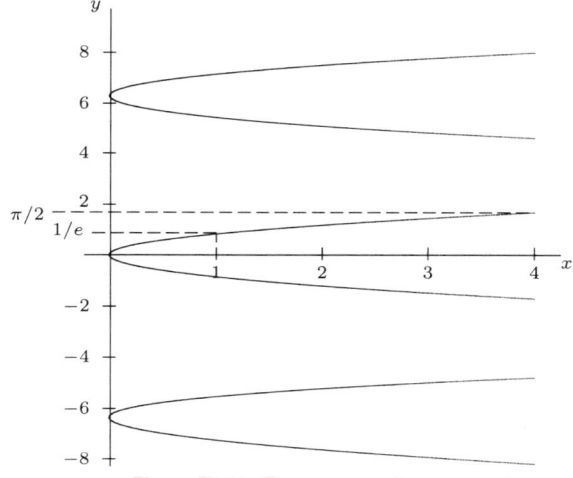

FIG. B.7: PLOT OF $e^x \cos y = 1$

Example B.3.3. Again we take the conformal mapping $f(z) = e^z$ in (B.3.7), which this time we view as a mapping from the family of vertical lines in the (x,y)-plane onto the family of circles centered at the origin in the (u,v)-plane:

(B.3.14) $$C_0 : x \equiv x_0 \qquad \text{onto } \Gamma_0 : u^2 + v^2 = e^{2x_0}.$$

Specialize to $x_0 = 0$, so that for the unit circle Γ_0 we make take, according to Example B.1.1, Eqn. (B.1.2),

(B.3.15) $$d(u,v) = \frac{1}{2}(u^2 + v^2) - \ln u, \quad \nabla d = h = \left\{u - \frac{1}{u}, v\right\}$$

to satisfy assumption (A.1) of Section 1, for $0 < u \leq 1$. Then, according to Section A.3, Theorem A.3.2, we take

(B.3.16) $$\delta(x,y) = \frac{1}{2}[u^2(x,y) + v^2(x,y)] - \ln u(x,y) = \frac{1}{2}e^{2x} - \ln(e^x \cos y),$$

whose gradient vector field is then

(B.3.17) $$\tilde{h} = \nabla \delta = \{\delta_x, \delta_y\} = \{e^{2x} - 1, \tan y\}.$$

On C_0, where $x \equiv 0$, this vector field satisfies (in agreement with Example B.1.1 and Theorem A.3.2, the orthogonality condition

(B.3.18) $\qquad$ on $C_0 : \tilde{h}|_{C_0} = \{0, \tan y\}$, so that $\tilde{h} \cdot \nu = 0$ on C_0.

Moreover, the Hessian matrix $\mathcal{H}_\delta$ is (since $\delta_{xx} = 2e^{2x}$, $\delta_{xy} \equiv 0$, $\delta_{yy} = 1/\cos^2 y$,

(B.3.19) $\qquad \mathcal{H}_\delta(C_0) = \begin{bmatrix} 2e^{2x} & 0 \\ 0 & \dfrac{1}{\cos^2 y} \end{bmatrix}(C_0) = \begin{bmatrix} 2 & 0 \\ 0 & \dfrac{1}{\cos^2 y} \end{bmatrix} > 0$,

provided $-\frac{\pi}{2} < y < \frac{\pi}{2}$ in which case Γ_0 is arbitrarily closed to a half-circumference (in agreement with Example B.1.1, while C_0 is arbitrarily close to the straight segment: $x = 0$, $-\frac{\pi}{2} < y < \frac{\pi}{2}$.

Example B.3.4. Here, we take the conformal mapping

(B.3.20) $\qquad f(z) = \dfrac{1}{z} = u + iv, \quad u(x,y) = \dfrac{x}{x^2 + y^2}, \quad v(x,y) = \dfrac{-y}{x^2 + y^2}$,

which maps the unit circle C_0

(B.3.21) $\qquad C_0 : (x-1)^2 + y^2 = 1$ onto the straight line $\Gamma_0 : u \equiv \dfrac{1}{2}$.

We test condition (A.3.19) of Corollary A.3.3 and after straightforward computations we obtain that on C_0:

(B.3.22) $\quad |f'(z)|^4 - v(v_{xx}^2 + v_{xy}^2) > 0 \quad \text{iff} \quad 16x^4(2x - 3) > 0$, i.e., iff $x > \dfrac{3}{2}$.

Thus, only the portion of the circle $C_0 : (x,y) \in C_0$ with $\frac{3}{2} < x \leq 2$ satisfies the test. Thus, in this case, this mapping $f(z)$ does *not* provide an optimal result. As we know, "the optimal case" is when C_0 is arbitrarily close to half-circle, which would require $x > 1$ on C_0. Thus, in this example, we may take in the (x,y)-plane *any* finite segment $u \equiv \frac{1}{2}$, $v_0 \leq v \leq v_1$, while in the (u,v)-plane the arc C_0 is *limited* by $\frac{3}{2} < x \leq 2$.

Appendix C: Illustrations of the setting of Section 10: Assumptions (A.1i) and (A.2i)

Sections A.1 through B.2 have provided large classes of triples $\{\Omega, \Gamma_0, \Gamma_1\}$ where assumption (A.1) of Section 1 is satisfied. Throughout those illustrations, we have also noted, however, that the additional requirement of fulfilling also assumption (A.2) = (1.1.6) of Section 1 imposes geometrical limitations on the allowed triples $\{\Omega, \Gamma_0, \Gamma_1\}$, which are covered by Theorem 2.1.1, under hypotheses (A.1) and (A.2). The setting of Section 10 is meant to relax these geometrical restrictions, by mitigating the impact of assumption (A.2).

The multi-faced treatment of the entire Appendix (A, B, C) then culminates with the following result, which shows that the setting of Section 10 does apply to large classes of triples $\{\Omega, \Gamma_0, \Gamma_1\}$, constructed by various techniques.

Theorem C.1. Let $\{\Omega, \Gamma_0, \Gamma_1\}$ be given with $\overline{\Gamma_0 \cup \Gamma_1} = \Gamma$, $\Gamma_0 \cap \Gamma_1 = \emptyset$. Assume that it satisfies: (i) either the setting of Sections A.1 and A.2 (or its 3-dimensional generalization in Section B.2); in particular, Theorem A.2.1, Lemma A.2.2, or Lemma B.2.1; (ii) or else the setting of Section A.4, in particular, Theorem A.4.1; (iii) or else the setting of Section A.3, in particular, Theorem A.3.2. Thus, in all these cases there exists (constructively) a scalar function $d \in C^3$ (called δ in Section A.3) such that the conservative vector field $h = \nabla d$ satisfies assumption (A.1) [i.e., both conditions (1.1.4) and (1.1.5)] of Section 1. Moreover, in all these cases, there is at most one point $P \in \Gamma_0$ (called C_0 in Section A.3) such that: $h(P) = \nabla d(P) = 0$; so that assumption (A.2) = (1.1.6) of Section 1 fails.

Then, it is possible to split Ω as the union of two overlapping subsets Ω_1, Ω_2 as in (10.1.1) of Section 10, such that both assumptions (A.1i) [i.e., both conditions (10.1.2) and (10.1.3)] and (A.2i) = (10.1.4) hold true.

Proof. The case where Γ_0 is flat. This case was already treated (by the simplest method) and illustrated in Section A.1, thus providing a special but important subclass of triples $\{\Omega, \Gamma_0, \Gamma_1\}$ where the setting of Section 10 is fulfilled, and thus Theorem 10.1.1 holds true.

Another method which employs the technique of Section A.4 is illustrated in Fig. C.1.

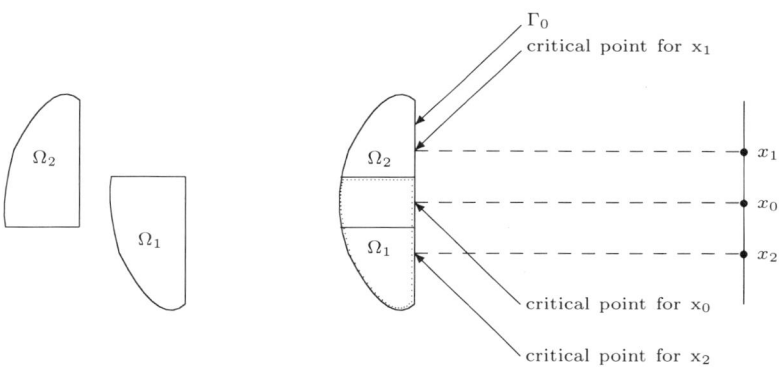

FIG. C.1: DECOMPOSITION OF Ω WITH FLAT Γ_0, AS REQUIRED BY THE SETTING OF SECTION 10, ACCORDING TO THE CONSTRUCTION OF TWO VECTOR FIELDS PROVIDED BY THE METHOD OF SECTION A.4.

The case where Γ_0 has a U-turn with one critical point for $d(x)$ on Γ_0 (according to Section A.2 paragraph below (A.2.11); Corollary A.3.4; Corollary A.4.2). We begin by illustrating the idea in the case where Γ_0 is arbitrarily close to a half-circumference.

The case where Γ_0 is arbitrarily close to a half-circumference. We now show how it is possible to take the unobserved (or uncontrolled) portion Γ_0 of the boundary of the unit disk to be arbitrarily close to a half-circumference. To this end, we need the setting of Section 10. This analysis, therefore, improves upon that of Example B.1.1, which was only based on the setting of Section 1, whereby the sub-boundary Γ_0 obtained there was non-connected.

Sections A.2 and A.4 have provided two distinct geometrically intrinsic ways of constructing conservative vector fields $h = \nabla d$, which both satisfy Assumption (A.1) of Section 1, but have a critical point for d on the mid-point of Γ_0 (point D in Fig C.3).

Method of Section A.2. With reference to Fig. C.2, let Ω_1 be the circular sector OAB, and let Ω_2 be the circular sector $OA'B'$. Thus, Ω_2 is the mirror image of Ω_1 with respect to the x-axis, and we shall then confine on the description of Ω_1. The axis OA makes an angle ϵ with the positive y-axis. The positive η-axis makes an angle $\frac{\epsilon}{2}$ with the positive y-axis, and an angle $\frac{\pi}{2}$ with the positive ξ-axis. The axis OB makes an angle $\frac{\epsilon}{4}$ with the x-axis and an angle $\frac{\epsilon}{4}$ with the positive ξ-axis. We have that $\Omega = \Omega_1 \cup \Omega_2$, the circular sector $OAB'BA'$, which is arbitrarily close to the half-disk, while $\Omega_1 \cap \Omega_2 \neq \emptyset$, as required. We claim that we can take $\Gamma_0 =$ arc $AB'BA'$ as the unobserved or uncontrolled portion of the boundary. We now verify assumptions (A.1i) in (10.1.2), (10.1.3) and (A.2i) = (10.1.4), $i = 1, 2$.

Regarding Ω_1. Given a point $P = (\xi, \eta)$ [with respect to the orthogonal system (ξ, η)] on the arc AB (= boundary of Ω_1), by the analysis of Section A.2 and its specialization in Example B.1.1, we begin by taking the boundary vector field $h_1 = \{h_{1,1}, h_{1,2}\} = \overrightarrow{XP}$, where X is the point of intersection of the tangent line to Γ_0 at P with the ξ-axis. Next, as in Section A.2, Eqn. (A.2.17), we extend h_1 to all of the circular sector OAN containing Ω_1, as follows: let $P' = (\xi', \eta')$ [with respect to the orthogonal system (ξ, η)] be in Ω_1. Then, take $h_1 = \{h_{1,1}, h_{1,2}\}$, where [still with respect to the (ξ, η) axes]:

$$\begin{cases} h_{1,1} &= \text{horizontal component } XF \text{ of } \overrightarrow{XP}; \\ h_{1,2} &= \text{vertical component } FP' = \eta'. \end{cases}$$

Then, as seen in Section A.2, this vector field h_1, which is defined geometrically, satisfies assumption (A.1$_1$) in (10.1.2), (10.1.3). Moreover, such a vector field $\overrightarrow{XP} = h_1$ vanishes only at the boundary point N of intersection between the circumference and the ξ-axis; which is not a point in Ω_1. Thus, for all $P \in$ arc $AB =$ boundary of Ω_1, the vector field $\overrightarrow{XP}$ has a length which is bounded away from zero. Thus, h_1 satisfies assumption (A.2$_1$) = (10.1.4), as well.

The analysis for Ω_2 is symmetric. In conclusion: *Theorem C.1 and hence Theorem 10.1.1 hold true with $\Gamma_0 =$ arc $AB'BA'$ arbitrarily close to the half-circumference.*

Method of Section A.4. This is illustrated in Fig. C.3. The circular region $\Omega = ABCDEFGHLA$ with boundary $\Gamma_0 = ABCDEFG$ is split as the union of two overlapping subdomains $\Omega_1 = ABCDEHA$ and $\Omega_2 = CDEFGLC$, as required by (10.1.1). The mid-point D of Γ_0 is the original critical point for the scalar function $d(x)$, *constructed through Eqns. (A.4.6), (A.4.7), with respect to the original external point x_0* (see Corollary A.4.2).

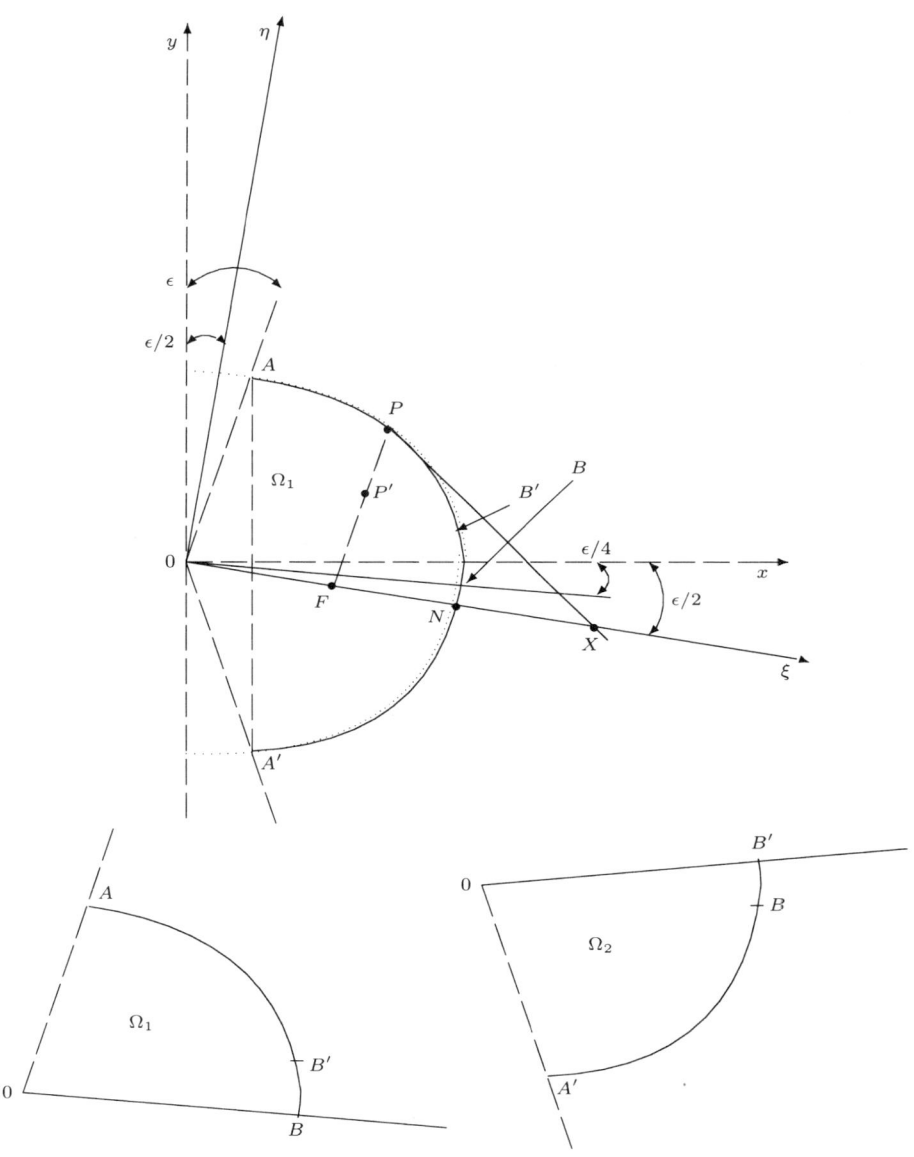

Fig. C.2: Decomposition of a circular region with Γ_0 arbitrarily close to a half-circle, as required by the setting of Section 10, according to the construction of two vector fields provided by the method of Section A.2.

Next, we replace x_0 with x_1 [respectively, x_0 with x_2] and produce, accordingly, a function $d_1(x)$ [respectively, a function $d_2(x)$], defined by the counterpart of Eqns. (A.4.6), (A.4.7), where this time $d_0(x)$ is defined as $d_{0,1}(x) = \frac{1}{2}\|x - x_1\|^2$ in the first case, and as $d_{0,2}(x) = \frac{1}{2}\|x - x_2\|^2$ in the second case. We claim that:

the vectors required by the setting of Section 10 are $h_1(x) = \nabla d_1(x)$ and $h_2(x) = \nabla d_2(x)$.

Indeed, they satisfy assumption (A.1i) [i.e., both conditions (10.1.2) and (10.1.3)] by Theorem A.4.1.

Moreover, the critical point for $d_1(x)$ is $F \in \Gamma_0$ (see Corollary A.4.2), where $F \notin \Omega_1$, and similarly the critical point for $d_2(x)$ is $B \in \Gamma_0$, where $B \notin \Omega_2$. Thus, $|h_1(x)| = |\nabla d_1(x)| \geq p_1 > 0$ for all $x \in \Omega_1$; and similarly, $|h_2(x)| = |\nabla d_2(x)| \geq p_2 > 0$ for all $x \in \Omega_2$. Thus, $h_i(x)$ satisfy also assumption (A.2i) = (10.1.4). Our claim is established. Thus, Theorem C.1 is proved in this case, and then Theorem 10.1.1 holds true.

Thus, *Theorem C.1 and Theorem 10.1.1 hold true with* $\Gamma_0 = ABFG$ *arbitrarily close to a half-circle.*

The general case where Γ_0 has a U-turn and satisfies either Theorem A.2.1 or Lemma B.2.1, or else Theorem A.4.1 requires only minor modifications over the case of Γ_0 being close to a half-circumference. □

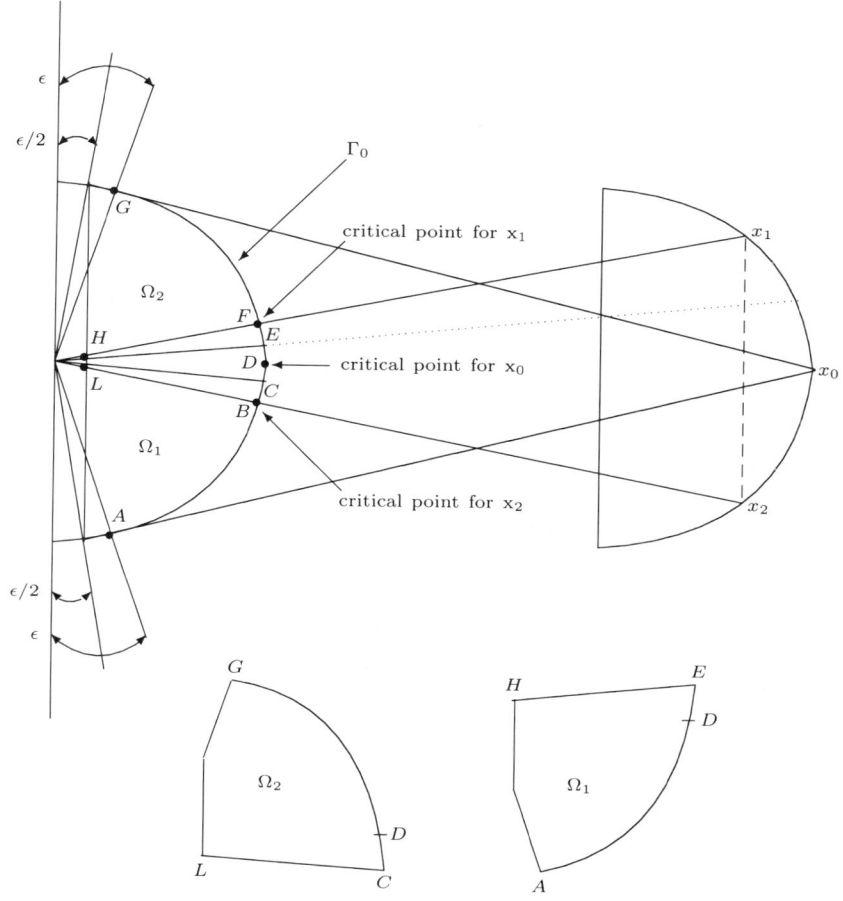

FIG. C.3: DECOMPOSITION OF A CIRCULAR REGION WITH Γ_0 ARBITRARILY CLOSE TO A HALF-CIRCLE, AS REQUIRED BY THE SETTING OF SECTION 10, ACCORDING TO THE CONSTRUCTION OF TWO VECTOR FIELDS PROVIDED BY THE METHOD OF SECTION A.4.

References

[B-L-R.1] C. Bardos, G. Lebeau, and J. Rauch, Sharp sufficient conditions for the observation, control, and stabilization of waves from the boundary, *SIAM J. Control & Optimiz.* **30** (1992), 1024–1065.

[B-C.1] J. W. Brown and R. V. Churchill, *Complex Variables and Applications*, 6th ed., McGraw-Hill, 1995.

[C.1] G. Chen, Energy decay estimates and exact boundary controllability for the wave equation in a bounded domain, *J. Math. Pures et Appl.* **9** (8) (1979), 240–274.

[Fl.1] W. Fleming, *Functions of Several Variables*, Springer-Verlag, 2nd ed., 1977.

[F-I.1] A. Fursikov and O. Yu. Imanuvilov, *Controllability of Evolution Equations*, Lecture Notes, Vol. 34, Research Institute of Mathematics, Seoul National University, Seoul, Korea.

[H.1] L. F. Ho, Observabilite frontiere de l'equation des ondes, *C. R. Acad. Sci. Paris, Ser I Math* **302** (1986), 443–446.

[Ho.1] L. Hörmander, *Linear Partial Differential Operators*, Springer-Verlag, 1969.

[Ho.2] L. Hörmander, *The Analysis of Linear Partial Differential Operators*, Vol. III, springer-Verlag, Berlin/New York, 1985.

[Ho.3] L. Hörmander, On the uniqueness of the Cauchy problem under partial analyticity assumptions, in "Geometrical Optics and Related Topics" (F. Colombini and N. Lerner, Eds.), Birkhäuser, Boston, 1997.

[I.1] V. Isakov, *Inverse Problem for Partial Differential Equations*, Springer-Verlag, Berlin, 1998.

[I-Y.1] V. Isakov and M. Yamamoto, Carleman estimate with Neumann boundary condition and its application to the observability inequality and inverse hyperbolic problems, this volume of *Contemporary Mathematics*.

[K-K.1] M. Kazemi and M. V. Klibanov, Stability estimates for ill-posed Cauchy problems involving hyperbolic equations and inequalities, *Applicable Analysis* **50** (1993), 93–102.

[Ke.1] S. Kesavan, *Topics in Functional Analysis and Applications*, John Wiley & Sones, 1989.

[Ko.1] V. Komornik, *Exact Controllability and Stabilization: The Multiplier Method*, Masson, Paris, 1994.

[Lag.1] J. Lagnese, Decay of solutions of the wave equation in a bounded region with boundary dissipation, *J. Diff. Eqns.* **50** (1983), 163–182.

[Lag.2] J. Lagnese, Note on boundary stabilization of wave equations, *SIAM J. Contr. & Optimiz.* **26** (1988), 1250–1256.

[Las.1] I. Lasiecka, Stabilization of hyperbolic and parabolic systems of non-linearly perturbed boundary conditions, *J. Diff. Eqns.* **75** (1988), 53–87.

[La-Ta.1] I. Lasiecka and D. Tataru, Uniform boundary stabilization of semilinear wave equations with nonlinear boundary conditions, *J. Diff. & Int. Eqn.* 6(3) (1993), 507–533.

[L-T.1] I. Lasiecka and R. Triggiani, Uniform exponential energy decay of wave equations in a bounded region with $L_2(0,\infty; L_2(\Gamma))$-feedback control in the Dirichlet boundary conditions, *J. Diff. Eqns.* **66** (1987), 340–390.

[L-T.2] I. Lasiecka and R. Triggiani, Exact controllability of the wave equation with Neumann boundary control, *Appl. Math. & Optimiz.* **19** (1989), 243–290. PReliminary version appeared in

[L-T.3] I. Lasiecka and R. Triggiani, Uniform stabilization of the wave equation with Dirichlet or Neumann feedback control without geometrical conditions, *Appl. Math. & Optimiz.* **25** (1992), 189–244.

[L-T.4] I. Lasiecka and R. Triggiani, Carleman estimates and exact boundary controllability for a system of coupled, nonconservative second order hyperbolic equations, in Partial Differential Equations Methods in Control and Shape Analysis, *Lecture Notes in Pure and Applied Mathematics*, Marcel Dekker, New York, Vol. 188, 215–243.

[L-T.5] I. Lasiecka and R. Triggiani, Sharp regularity for mixed second-order hyperbolic equations of Neumann type, Part I: The L_2-boundary case, *Annali di Matematica Pura e Applicata* (IV), **157** (1990), 285–367.

[L-T.6] I. Lasiecka and R. Triggiani, *Control Theory for Partial Differential Equations: Continuous and Approximation Theories*, Encyclopedia of Mathematics and its Applications, Vols. 72 and 73, 1999.

[L-T-Y.1] I. Lasiecka, R. Triggiani, and P. F. Yao, Exact controllability for second-order hyperbolic equations with variable coefficient principal part and first-order terms, *Non-Linear Analysis, Theory, Methods & Applications* **30** (1997), 111–122.

[L-T-Y.2] I. Lasiecka, R. Triggiani, and P. F. Yao, Inverse/observability estimates for second order hyperbolic equations with variable coefficients, *J. Math. Anal. & Appl.* **235** (1999), 13–57.

[L-T-Y.3] I. Lasiecka, R. Triggiani, and P. F. Yao, An observability estimate in $L_2(\Omega) \times H^{-1}(\Omega)$ for second order hyperbolic equations with variable coefficients, in "Control of Distributed Parameter and Stochastic Systems" (S. Chen, X. Li, J. Yong, and X. Y. Zhou, Eds.), pp. 71–78, Kluwer Academic Publishers, 1999.

[L-T-Y-Z.1] I. Lasiecka, R. Triggiani, P. F. Yao, and X. Zhang, Work in progress on the variable coefficient principal part case.

[L-R-S.1] M. M. Lavrentev, V. G. Romanov, and S. P. Shishataskii, *Ill-Posed Problems of Mathematical Physics and Analysis*, Amer. Math. Soc., Vol. 64 (1986).

[Lio.1] J. L. Lions, Exact controllability, stabilization and perturbations for distributed systems, *SIAM Review* 30 (1988), 1–68.

[Lio.2] J. L. Lions, *Controllabilité Exacte, Perturbations et Stabilisation de Systemes Distribues*, Vol. 1, Masson, Paris, 1988.

[L-M.1] J. L. Lions and E. Magenes, *Non-homogeneous Boundary Value Problems and Applications*, Vol. I, Springer-Verlag, Berlin, 1972.

[Lit.1] W. Littman, Near optimal time boundary controllability for a class of hyperbolic equations, *Lecture Notes in Control and Information*, Vol. 97, pp. 307–312, Springer-Verlag, Berlin/New York, 1987.

[Lit.2] W. Littman, Boundary controllability for polyhedral domains, *Lecture Notes in Control and Information*, Vol. 178, pp. 272–284, Springer-Verlag, Berlin/New York, 1992.

[Lit.3] W. Littman, Remarks on global uniqueness for partial differential equations, this volume of *Contemporary Mathematics*.

[Li-Ta.1] W. Littman and S. Taylor, Smoothing evolution equations and boundary control theory, *J. d'Analyse Mathematique* 59 (1992), 117–131.

[R.1] D. L. Russell, Exact boundary controllability theorems for wave and heat processes in star complemented regions, in *Differential Games and Control Theory*, E. Roxin, Lin, & Steinberg, eds., Marcel Dekker 1974, Lectures in Pure and Applied Mathematics.

[R.2] D. Russell, Controllability and stabilization theoryf or linear partial differential equations: recent progress and open questions, *SIAM Review* 20 (1978), 639–431.

[Ta.1] D. Tataru, Boundary controllability for conservative PDE's, *Appl. Math. & Optimiz.* 31 (1995), 257–295.

[Ta.2] D. Tataru, Carleman estimates and unique continuation for solutions to boundary value problems, *J. Math. Pures et Appl.* 75 (1996), 367–408.

[Ta.3] D. Tataru, Unique continuation for solutions to PDEs: Between Hörmander theorem and Holmgren theorem, *Comm. Part. Diff. Eqns.* 20 (5 & 6) (1995), 855–884.

[Tr.1] R. Triggiani, Exact boundary controllability on $L_2(\Omega) \times H^{-1}(\Omega)$ of the wave equation with Dirichlet boundary control acting on a portion of the boundary, *AMO* 18 (1988), 241–277.

[Tr.2] R. Triggiani, Wave equation on a bounded domain with boundary dissipation: An operator approach, *JMAA*, Vol. 137 (1989).

[Y.1] P. F. Yao, On the observability inequalities for exact controllability of wave equations with variable coefficients, *SIAM J. Control & Optimiz.*

[Z.1] X. Zhang, Rapid exact controllability of the semi-linear wave equation, *Chin. Ann. of Math.* 20B: 3 (1999), 377-384.

[Z.2] X. Zhang, Explicit observability estimate for the wave equation with potential and its application, *Proc. R. Soc. London A*, 456 (2000), 1101–1115.

[Z.3] X. Zhang, Explicit observability inequalities fro the wave equation with lower order terms by means of Carleman inequalities, *SIAM J. Contr. & Optimiz.*, to appear.

Department of Mathematics, University of Virginia, Charlottesville, VA 22903
E-mail address: il2v@virginia.edu, rt7u@virginia.edu

Mathematics Department, Fudan University, Shanghai 200433 PRC and Sichuan University, Chengdu 610064 PRC
E-mail address: xuzhang@fudan.edu

Uniform Stability of a Coupled Structural Acoustic System with Thermoelastic Effects and Weak Structural Damping

Catherine Lebiedzik

ABSTRACT. We consider a coupled PDE system which arises in problems dealing with the active control of structural acoustic systems. An acoustic chamber is surrounded by an active plate or wall, and the pressure in the chamber is controlled via piezoceramic actuators. We propose and study a model that includes thermal effects in the wall and the possibility of structural damping. Our main result is the uniform stabilization of the system as a consequence of thermal effects. The decay rates obtained are independent of the amount of structural damping within an L_∞ bound.

1. Introduction

In recent years there has been a great deal of interest in the active control of structural acoustic systems. Several contributions in the present Volume [**Ava00, CJ00**] confirm that this is lively area of research. The motivation for studying these problems comes from a variety of situations where noise control is desired – for example reducing the level of pressure in helicopter cabins, or lessening the interior cabin noise in an aircraft. These applications have been well documented in the engineering literature [**CdL87, FGS90, BSS94, BSYW95**]. We consider a structural acoustic model consisting of an acoustic chamber with a combination of flexible (vibrating) and hard walls. The goal of this paper is to address the problem by using a model that includes thermal effects in the wall and allows for the possibility of structural damping effects [**CR81, CT89, CT88**] as well.

It is known that either thermal effects [**Leb00**] or the presence of full structural damping [**Ava96**], along with a linear boundary damping imposed on the wave component, are enough to stabilize the system. However, since structural damping is rather rare, and its modeling is poorly understood (as is well known and is acknowledged in the engineering literature), it is less desirable to rely on it to produce the stabilizing effect needed. Nevertheless, there is often some sort of structural damping inherent in the system under study. For this reason, we introduce a 'weak' structural damping term into our model. By this we mean that the space coefficient α in (1.2a) below is only assumed to be non-negative (and in L_∞). Since it

1991 *Mathematics Subject Classification.* 35B35, 35B37.

can vanish in parts (or all) of the domain Γ_0, it cannot serve as the sole source of dissipation in the model. In addition, it is a higher order operator and thus it may in principle contribute to the mathematical complexity of the problem. Our conclusion is that the stability of the thermoelastic model is robust with respect to the coefficient α, i.e. to any form of structural damping which may be present in the model, as a result of thermal effects alone.

1.1. Statement of the Problem. Let $\Omega \in \mathbb{R}^n, n = 2$ or 3 be an open bounded domain with boundary Γ. The boundary Γ consists of two connected regions: the active wall Γ_0, which is flat and is modeled by a thermoelastic plate; and the 'hard' wall Γ_1, which is convex. The precise assumptions required on Γ will be collected below in Hypotheses 1.1 and 1.2 with appropriate comments, after introducing the mathematical model.

1.1.1. *The Model.* The PDE model considered consists of the wave equation in the variable z

(1.1a) $\qquad z_{tt} = c^2 \Delta z \quad \text{in} \quad (0, T] \times \Omega$

(1.1b) $\qquad \dfrac{\partial}{\partial \nu} z = 0 \quad \text{on} \quad (0, T] \times \Gamma_1$

(1.1c) $\qquad \dfrac{\partial}{\partial \nu} z = -d\, z_t + w_t \quad \text{on} \quad (0, T] \times \Gamma_0$

(1.1d) $\qquad z(0, \cdot) = z_0 \in H^1_{\Gamma_1}(\Omega), \qquad z_t(0, \cdot) = z_1 \in L^2(\Omega)$

and the elastic equation representing the displacement w of the wall Γ_0 subject to thermal effects [see, e.g., [**Lag89**]]:

(1.2a) $\quad w_{tt} - \gamma \Delta w_{tt} + \Delta^2 w + \Delta\bigl(\alpha(x)\Delta w_t\bigr) = -\Delta \theta - \rho z_t \quad \text{on} \quad (0, T] \times \Gamma_0$

(1.2b) $\quad \theta_t - \Delta \theta = \Delta w_t \quad \text{on} \quad (0, T] \times \Gamma_0$

(1.2c) $\quad w = \Delta w = 0; \quad \theta = 0 \quad \text{on} \quad (0, T] \times \partial \Gamma_0$

Here, the quantity ρz_t is the acoustic pressure, ρ is the density of the fluid, θ is the temperature, c^2 is the speed of sound as usual, $d > 0$ is boundary damping on the active wall, and the constant $\gamma \geq 0$ accounts for rotational forces. The function $\alpha(x)$ accounts for the possible existence of structural damping in the plate. Accordingly, we will assume that

(1.3) $\qquad\qquad 0 \leq \alpha(x) \leq \alpha_0 \text{ a.e. } \text{ for } x \in \Gamma_0, \quad \alpha(x) \in L_\infty(\Gamma_0)$

Here x is a 1-dimensional ($n = 2$) or 2-dimensional ($n = 3$) space variable restricted to Γ_0. The boundary conditions given in (1.2c) are those for a hinged plate, though clamped boundary conditions can be considered as well with no increase in complexity. For convenience, in what follows we will choose $c = \rho = 1$.

1.1.2. *Assumptions on the triple* $\{\Omega, \Gamma_0, \Gamma_1\}$. . We collect here the assumptions needed on the chamber Ω and its boundary Γ in connection with the main result of this paper: the stabilization result in Theorem 1.4. The following geometrical conditions are required:

HYPOTHESIS 1.1. *Ω is a 2- or 3-dimensional open bounded domain with boundary $\Gamma = \overline{\Gamma_0 \cup \Gamma_1}$, $\Gamma_0 \cap \Gamma_1 = \emptyset$; Γ_0 is flat and Γ_1 is convex and smooth (say of class C^1); so that Ω is a convex set.*

HYPOTHESIS 1.2. *There exists a smooth vector field $h(x)$, such that (*[**Tri88**], *p. 270)*

(1.4a)   • h *is coercive:* $H(x)v \cdot v \geq c_0|v|^2$, $c_0 > 0$

(1.4b)   • $h(x) \cdot \nu(x) \equiv 0$ $\forall x \in \Gamma_1$

where $H(x)$ is the transpose of the Jacobian matrix of $h(x)$ and ν is the unit outward normal vector on Γ_1.

We shall invoke Hypothesis 1.1 for the well-posedness Theorem 1.3, and both Hypotheses for the uniform stabilization result of Theorem 1.4.

REMARK 1.1. *Suitability of the geometric assumptions.* The above assumptions warrant some comments regarding their nature, their validity, and their implications on the problem considered. The presence of the apparently strong condition (1.4b) (that the vector field h be parallel along Γ_1), is due to the presence of the Neumann (rather than Dirichlet) homogeneous boundary condition (1.1b) on the *uncontrolled* portion of the boundary. It was first pointed out in [**Tri88**] (Eqns. (5.18), (5.19), p. 273), that in this case the (by now) classical multiplier method appears to require the condition (1.4b) in order to annihilate two terms which do not appear to compensate each other. The problem arises regardless of whether or not the control action exercised on the complementary portion of the boundary Γ_0 is of Dirichlet or Neumann type. Thus, a relevant question is: how large is the class of triples $\{\Omega, \Gamma_0, \Gamma_1\}$ which satisfy both the parallelism condition (1.4b) and the usual coercive condition (1.4a)? Several classes are illustrated in [**Tri88**], but a systematic study was not given there.

It was precisely the structural acoustic problem – which calls for the Neumann boundary condition (1.1b) as being more realistic than the corresponding Dirichlet condition – that has resurrected [**LL99**] interest in this non-classical case. Two papers in this Volume are in fact devoted to it [**IY00, LTZ00**], and the characterization of the class of triples $\{\Omega, \Gamma_0, \Gamma_1\}$ which satisfy (1.4a),(1.4b) is relevant in them. Several approaches giving checkable sufficient conditions are given in the Appendix of [**LTZ00**]. One such *sufficient* criterion, which is relevant to our problem here, and which is in effect valid for any dimension, is as follows([**LTZ00**], Theorem A.4.1, Appendix A.4; see also [**LL99**]):

PROPOSITION 1.1. *Sufficient conditions for* (1.4a),(1.4b):

(i) *Let the level set $l(x) = 0$, $l \in C^3(\mathbb{R}^n)$ describing Γ_1 be convex (i.e. the Hessian of l is non-negative in Γ_1).*
(ii) *Let there exist a point $x_0 \in \mathbb{R}^n$ such that $(x - x_0) \cdot \nu \leq 0$ for $x \in \Gamma_1$.*

Then there exists (by construction) a conservative vector field $h(x)$ such that the assumptions (1.4a),(1.4b) *hold true.*

REMARK 1.2. *Implications on regularity of solutions and multiplier method.* If we apply the sufficient criterion given in Proposition 1.1 to guarantee the validity of assumptions (1.4a),(1.4b) on Γ_1, and we further take into account that Γ_0 is flat by assumption, it follows that Ω must have corners at the intersection $\bar{\Gamma}_0 \cap \bar{\Gamma}_1$. This raises the potential problem of whether the solution of (1.1),(1.2) (particularly z) has sufficient regularity to justify the energy (multiplier) method employed. For

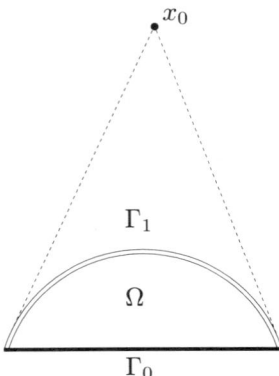

FIGURE 1. Possible cross section of the domain Ω.

this to succeed, one needs to start with initial conditions $\{z_0, z_1, w_0, w_1, \theta_0\}$ in the domain $\mathcal{D}(A_\gamma)$ of the generator asserted by Theorem 1.3, and run the multiplier method for the corresponding solution. These steps, however, require that such a z_0 has $H^2(\Omega)$ smoothness, an elliptic problem essentially. This required regularity is guaranteed in our case by [**Gri85**], p.139-140, 149 see also [**Gri89**].

1.1.3. *Purpose of this paper.* Our goal is to show the uniform stability in the appropriate topology of the coupled PDE system (1.1)-(1.2). To accomplish this we shall use differential multipliers developed in the context of sharp regularity [**LLT86**]/exact controllability [**Lio88, LT89, Tri88**]/ stability analysis for the wave equation together with the operator multiplier method introduced for thermoelastic problems in [**AL97**]. Our task will be to combine these coupled estimates for the variables z, w, θ, to obtain the inequality (2.23) from which then it routinely follows that there exists some $T > 0$ such that the overall energy is *strictly* less than the original energy, see (2.26). Standard semigroup theory will then yield, as usual, the desired exponential decay in the corresponding uniform operator topology, as in (1.9) below.

1.2. Statement of Main Results. We begin with a preliminary result that the system is well-posed in the semigroup sense. To this end, we introduce the following function space:

(1.5) $$X_\gamma \equiv H^1_{\Gamma_1}(\Omega) \times L^2(\Omega) \times H^2_0(\Gamma_0) \times H^1_{0,\gamma}(\Gamma_0) \times L^2(\Gamma_0)$$

Here the function spaces used are defined as

(1.6) $$H^1_{\Gamma_1}(\Omega) = \{f \in H^1(\Omega) \text{ with } f|_{\Gamma_1} = 0\}$$

as usual, while $H^1_{0,\gamma}(\Gamma_0)$ is the space $H^1_0(\Gamma_0)$ with inner product

(1.7) $$(\omega_1, \omega_2)_{H^1_{0,\gamma}(\Gamma_0)} \equiv (\omega_1, \omega_2)_{L^2(\Gamma_0)} + \gamma(\nabla \omega_1, \nabla \omega_2)_{L^2(\Gamma_0)} \quad \forall \omega_1, \omega_2 \in H^1_0(\Gamma_0)$$

Next we define the subspace Y_γ as X_γ under an appropriate compatibility condition.

(1.8) $$Y_\gamma = \left\{ y = [z_0, z_1, w_0, w_1, \theta] \in X_\gamma \Big| \int_\Omega z_1 \, d\Omega = \int_{\Gamma_0} (w_0 - d\, z_0) \, d\Gamma_0 \right\}$$

THEOREM 1.3. *(well-posedness) Let Ω be a bounded open domain in $\mathbb{R}^n$, $n = 2$ or 3 with boundary Γ satisfying the geometrical conditions described by Hypothesis 1.1. For all initial data $y_0 \in Y_\gamma$, where Y_γ is defined in (1.8), the map $y_0 \to y(t) = [z(t), z_t(t), w(t), w_t(t), \theta(t)] = e^{A_\gamma t} y_0 \in C([0,T]; Y_\gamma)$ corresponding to the the problem (1.1) - (1.2) generates a s.c. contraction semigroup $e^{A_\gamma t}$ in Y_γ*

PROOF. The proof follows by a standard application of the Lumer-Phillips theorem after rewriting the problem (1.1)-(1.2) in the abstract form $y_t = A_\gamma y$. Indeed, as in other similar cases, e.g. [**Leb00**], the problem under consideration is maximal dissipative. We note also that subspace Y_γ as defined in (1.8) is closed and invariant under the generator A_γ (we omit the proof because of space considerations). A compatibility condition such as (1.8) is typical in these problems, as it eliminates the possibility of a zero eigenvalue (see [**Tri88**],[**Cam99**] and [**CJ00**], Lemma 2.1 for a different structural acoustic model). □

Our main result is the following.

THEOREM 1.4. *(uniform stability) Let Ω be a bounded open domain in $\mathbb{R}^n$, $n = 2$ or 3 with boundary Γ satisfying the geometrical conditions described by Hypotheses 1.1 and 1.2, for which sufficient conditions are listed in Proposition 1.1. Let $\gamma_0 \geq \gamma \geq 0$ in (1.2a). Then, problem (1.1)-(1.2) – which is well-posed by virtue of Theorem 1.3 – is exponentially stable in the uniform operator topology of $\mathcal{L}(Y_\gamma)$, where Y_γ is defined in (1.8). That is, there exist constants $\delta > 0$ and $M_\delta \geq 1$ (independent of γ_0 and α_0), such that any semigroup solution $y(t) = [z(t), z_t(t), w(t), w_t(t), \theta(t)] = e^{A_\gamma t} y_0$ satisfies*

(1.9) $\quad \|y(t)\|_{Y_\gamma} \leq M_\delta e^{-\delta t} \|y_0\|_{Y_\gamma}, \text{ or } \|e^{A_\gamma t}\|_{\mathcal{L}(Y_\gamma)} \leq M_\delta e^{-\delta t}, \quad t \geq 0$

The proof will be given in Section 2.

REMARK 1.5. The structural damping (variable) coefficient $\alpha \geq 0$ *increases* the dissipativity of the corresponding generator over the case $\alpha \equiv 0$. (The explicit form of these generators, which demonstrates this claim, can be given along the lines of the past literature, but is omitted here because of space constraints.) Accordingly, we may as whether the above Theorem 1.4 may be reduced to the case $\alpha \equiv 0$. However, this does not appear to be the case. It is known [**Tri92**] that under certain conditions, if A is the generator of a s.c. semigroup and e^{At} is uniformly stable, then a 'more dissipative' generator B will share the same stability properties. This positive result was shown in [**Tri92**] to be true under the assumption that $\|e^{At}\| \leq M_\delta e^{-\delta t}$ with $M_\delta \equiv 1$. Furthermore, the condition that $M_\delta \equiv 1$ was shown there to be sharp by means of counterexamples. In our case, when $\alpha \equiv 0$, we do not have the case that $M_\delta \equiv 1$, and thus we cannot apply the result from [**Tri92**]. However, the energy method in Section 2 below shows the close relationship between the present case of 'weak' structural damping as defined in (1.3), and the case $\alpha \equiv 0$, the former being a refinement of the latter.

1.3. Previous Literature. Structural acoustic systems modeled by a system of two coupled partial differential equations have existed for a long time (see, e.g. [**MI68**]). Recently, there has been an emergence of interest in such models because of the applications to control systems mentioned earlier. The properties of the overall system depend on the type of coupling between the wave equation within the chamber Ω and the particular model of the 'elastic' wall Γ_0. A few models of the latter have been proposed.

Hyperbolic/parabolic model. The original structural acoustic model assumed the active wall Γ_0 to be a 'structurally damped' beam [**CR81**] or plate: in this case, the equation governing its behavior is parabolic in nature, in the sense that the dynamics of the structurally damped wall generate an analytic semigroup[**CT88, CT89**]. In this case, the overall system couples a hyperbolic component with a parabolic component. It is, then, natural to ask whether the overall structural acoustic system is stable as well. In fact, this is shown in [**Ava96**].

Hyperbolic/hyperbolic model. In a physically more accurate description, the vibrating wall Γ_0 is modeled by an elastic Kirchoff equation with $\gamma > 0$, which is therefore hyperbolic. Thus, the overall system now couples two hyperbolic components: the wave equation within the chamber Ω and the Kirchoff equation over Γ_0. Its well-posedness, as an open loop system, is now more delicate and is given in [**CT99**]. Moreover, when damped by suitable boundary dissipation, the overall closed loop has been shown to be uniformly stable in both the case of two dimensions [**Cam99**] and three dimensions [**CJ00**], by combining energy methods for the two components.

The model here considered includes thermal effects by modeling Γ_0 as a thermoelastic plate, with either $\gamma = 0$ (the analytic case [**LR95**]; this property continues to hold true under *all* boundary conditions, see [**LT99**], Chapter 3, and Appendices E-I.) or $\gamma > 0$ (the hyperbolic-dominated case [**LT00**]). In this sense, our present note is a companion paper to our previous work [**Leb00**]. In both cases, here and in [**Leb00**], our goal is to dispense with the structural damping as in [**Ava96**] as well as with the boundary damping on the active wall used in [**Cam99, CJ00**], and rely instead on thermal effects alone to stabilize the overall system. In contrast with [**Leb00**], we consider here the physically more realistic Neumann boundary condition $\frac{\partial}{\partial \nu} z \equiv 0$ on the rigid wall Γ_1, whereas [**Leb00**] imposed, instead, the homogeneous Dirichlet condition $z \equiv 0$ on Γ_1. This difference accounts for the more demanding geometrical conditions (1.4a),(1.4b) as noted. In both papers, here and in [**Leb00**], where the active wall is modeled by a thermoelastic plate, the thermal effects in the plate will produce exponential decay of the energy, since they serve to propagate dissipation from the heat component to the elastic component.

In addition to selecting the physically more appealing Neumann boundary condition (1.1b) on Γ_1, the present paper intends to also account for the possible presence of some 'structural damping' [**CR81**] included in the model. However, it may be very small or vanish in certain areas, so we cannot rely on this term alone to stabilize the structure. We need to show that the decay rates obtained for the model are uniform with respect to the bound α_0 in (1.3) on the parameter representing structural damping. Thus, in the limit case we recover the case of no structural damping at all.

2. Uniform Stabilization

Our goal is to show the uniform stability of the coupled PDE system (1.1)-(1.2). Our strategy is to study the thermoelastic equations on Γ_0 and the wave equation on Ω separately and combine the results. In the case of the thermoelastic plate, we will run the multiplier $A_D^{-1}\theta$ (introduced in [**AL97**]) on the elastic equation to yield an estimate of the plate energy. For the wave equation, we run the (by now classical) multipliers z, $z\,\mathrm{div}\,h$, and $h\cdot\nabla z$. This leads to an estimate of energy plus lower-order terms, which are then absorbed via a standard compactness/uniqueness argument. The goal of these methods is to yield the inequality

$$E_\gamma(T) < E_\gamma(0)$$

for some $T > 0$, where $E_\gamma(t)$ is the energy of the system at time t, given by (2.2). Standard semigroup theory will then yield, as usual, the desired exponential decay in the corresponding uniform operator topology, as in (1.9).

We note that our decay rates are independent of $\gamma \geq 0$. As has been previously stated, the dynamical properties of the overall system are very different in the cases of $\gamma = 0$ and $\gamma > 0$. Nevertheless, our stability results are 'robust' and they do not depend on these characteristics. This is in contrast to other papers dealing with thermoelasticity in which the decay rates blow up as $\gamma \to 0$ [**BBPMZ98**]. In addition, the decay rates are uniform in α. They do not depend on the profile of $\alpha(x)$ but instead depend only on the bound α_0 in (1.3). This dependence is such that in the limit $\alpha = 0$, we recover the decay rates derived in [**Leb00**] (where the thermoelastic system without the coefficient α was considered). This shows that the stability of the thermoelastic model is indeed 'robust' with respect to other forms of damping possibly existing in the system.

One could consider other boundary conditions for the plate component (such as, for example, clamped boundary conditions). This will not affect the analysis, though it demands the use of an additional trace regularity result[**AL97**]. The special case of 'free' boundary conditions, however, *is* more complicated and will be the subject of future work.

We shall adopt the following notation:

$$|w|_{s,\Omega} \equiv |w|_{H^s(\Omega)}; \quad (u,v)_\Omega \equiv \int_\Omega uv\,d\Omega$$

The same notation will be used with Ω replaced by Γ etc.

We begin with a preliminary, standard energy identity which illustrates the fact that the system is dissipative.

PROPOSITION 2.1. *With respect to the system of equations* (1.1)-(1.2), *the following energy equality holds for all* $T > 0$:

(2.1)
$$E_\gamma(0) = E_\gamma(T) + 2d\int_0^T |z_t|^2_{0,\Gamma_0}\,dt + 2\int_0^T |\alpha^{\frac{1}{2}}\Delta w_t|^2_{0,\Gamma_0}\,dt + 2\int_0^T |\nabla\theta|^2_{0,\Gamma_0}\,dt$$

where the 'energy' $E_\gamma(t)$ *is defined by*

(2.2) $\quad E_\gamma(t) \equiv |\nabla z|^2_{0,\Omega} + |z_t|^2_{0,\Omega} + |w_t|^2_{0,\Gamma_0} + \gamma|\nabla w_t|^2_{0,\Gamma_0} + |\Delta w|^2_{0,\Gamma_0} + |\theta|^2_{0,\Gamma_0}$

PROOF. By running the multipliers z_t on the wave equation, w_t on the elastic equation, θ on the thermal equation, and then integrating by parts, we obtain the above equality for smooth solutions. A density argument allows us to extend this equality to all solutions of finite energy. Note that this result implies that $\theta \in L_2(0,T;H^1(\Gamma_0))$. □

2.1. Thermoelastic Equations.

LEMMA 2.1. *Let the plate energy $E_{w,\gamma}(t)$ be defined as*

$$(2.3) \qquad E_{w,\gamma}(t) = |w_t(t)|^2_{0,\Gamma_0} + \gamma|\nabla w_t(t)|^2_{0,\Gamma_0} + |\Delta w(t)|^2_{0,\Gamma_0} + |\theta(t)|^2_{0,\Gamma_0}$$

With respect to the thermoelastic component of the model (1.2), the following inequality holds:

$$(2.4) \quad \int_0^T \left[|\Delta w|^2_{0,\Gamma_0} + |w_t|^2_{0,\Gamma_0} + \gamma|\nabla w_t|^2_{0,\Gamma_0}\right] dt$$

$$\leq C\left[E_{w,\gamma}(0) + E_{w,\gamma}(T) + \int_0^T \left(|\theta|^2_{1,\Gamma_0} + |\alpha\Delta w_t|^2_{0,\Gamma_0} + |z_t|^2_{0,\Gamma_0}\right) dt\right]$$

where the constant C does not depend on γ and α.

PROOF. We define $A_D : L^2(\Gamma_0) \supset \mathcal{D}(A_D) \to L^2(\Gamma_0)$ to be the positive self-adjoint operator $A_D = -\Delta$ with Dirichlet boundary conditions,

$$(2.5) \qquad A_D f = -\Delta f, \quad \mathcal{D}(A_D) = \left\{f \in H^2(\Gamma_0) : f|_{\partial\Gamma_0} = 0\right\}$$

We multiply the first equation in (1.1) by $A_D^{-1}\theta$ [**AL97**] and integrate from 0 to T to obtain

$$(2.6) \qquad \int_0^T (w_{tt} - \gamma\Delta w_{tt} + \Delta^2 w + \Delta(\alpha\Delta w_t) + \Delta\theta + z_t, A_D^{-1}\theta)_{0,\Gamma_0} dt = 0.$$

We deal with each part of (2.6) separately:

(1) Using integration by parts, substitution of boundary conditions and (1.2b), and the fact that A_D^{-1} is smoothing gives (detailed calculations are in [**AL97**])

$$(2.7)$$
$$\left|\int_0^T (w_{tt} - \gamma\Delta w_{tt}, A_D^{-1}\theta)_{L^2(\Gamma_0)} dt - \int_0^T \left[|w_t|^2_{0,\Gamma_0} + \gamma|\nabla w_t|^2_{0,\Gamma_0}\right] dt\right|$$

$$\leq C\left[E_{w,\gamma}(0) + E_{w,\gamma}(T)\right] + \varepsilon\int_0^T \left[|w_t|^2_{0,\Gamma_0} + \gamma|\nabla w_t|^2_{0,\Gamma_0}\right] dt + C_\varepsilon \int_0^T |\theta|^2_{1,\Gamma_0}$$

where neither of the constants C and C_ε depend on T or γ.

(2) The next step involves the new term α. Another integration by parts, application of boundary conditions, and the fact that $A_D^{-1}\theta \in \mathcal{D}(A_D)$ by (2.1) gives

$$(2.8) \qquad \int_0^T (\Delta^2 w + \Delta(\alpha\Delta w_t), A_D^{-1}\theta)_{0,\Gamma_0} dt = \int_0^T (\Delta w + \alpha\Delta w_t, \Delta A_D^{-1}\theta)_{0,\Gamma_0} dt$$

We estimate the right hand side of equation (2.8) using the fact that since $A_D^{-1}\theta \in L_2(0,T;\mathcal{D}(A_D))$ a.e., $\Delta A_D^{-1}\theta = -A_D A_D^{-1}\theta = -\theta \in L_2(0,T;\mathcal{D}(A_D))$.

$$(2.9) \quad \left| \int_0^T \left(\Delta^2 w + \Delta(\alpha \Delta w_t), A_D^{-1}\theta \right)_{L^2(\Gamma_0)} dt \right|$$
$$\leq C_\varepsilon \int_0^T |\theta|_{1,\Gamma_0}^2 \, dt + \varepsilon \int_0^T |\Delta w|_{0,\Gamma_0}^2 \, dt + \varepsilon \int_0^T |\alpha \Delta w_t|_{0,\Gamma}^2$$

(3) For the next term in (2.6) we just use Green's second theorem and the boundary conditions on θ:

$$(2.10) \quad \int_0^T \left(\Delta\theta, A_D^{-1}\theta \right)_{L^2(\Gamma_0)} dt + \int_0^T (\theta, \Delta A_D^{-1}\theta)_{\Gamma_0} \, dt \leq C \int_0^T |\theta|_{0,\Gamma_0}^2$$

(4) Finally, for the last term in (2.6), using (2.1):

$$(2.11) \quad \left| \int_0^T \left(z_t, A_D^{-1}\theta \right)_{L^2(\Gamma_0)} \right| dt \leq C_\varepsilon \int_0^T |z_t|_{0,\Gamma_0}^2 \, dt + \varepsilon \int_0^T |\theta|_{1,\Gamma_0}^2 \, dt$$

Combining equations (2.7) - (2.11) results in the fact that for ε small enough, there exists a constant $C > 0$ so that

$$(2.12) \quad (1 - 2\varepsilon) \int_0^T \left[|w_t|_{0,\Gamma_0}^2 + \gamma |\nabla w_t|_{0,\Gamma_0}^2 \right] dt \leq C \left[E_{w,\gamma}(0) + E_{w,\gamma}(T) \right]$$
$$+ C \int_0^T \left[|\theta|_{1,\Gamma_0}^2 + |z_t|_{0,\Gamma_0}^2 + |\alpha \Delta w_t|_{0,\Gamma_0}^2 \right] dt + \varepsilon \int_0^T |\Delta w|_{0,\Gamma_0}^2 \, dt$$

where the non-crucial dependence of C on ε has not been noted.

Next, we multiply (1.2a) by w and integrate from 0 to T, using Green's Theorem and the boundary conditions on (1.2a) to obtain

$$(w_t, w)_{0,\Gamma_0} \Big|_0^T + \gamma(\nabla w_t, \nabla w)_{0,\Gamma_0} \Big|_0^T - \int_0^T \left[|w_t|_{0,\Gamma_0}^2 + \gamma|\nabla w_t|_{0,\Gamma_0}^2 \right] dt$$
$$= - \int_0^T |\Delta w|_{0,\Gamma_0}^2 \, dt - \int_0^T (\alpha \Delta w_t, \Delta w)_{0,\Gamma_0} \, dt$$
$$+ \int_0^T (\nabla\theta, \nabla w)_{0,\Gamma_0} \, dt - \int_0^T (z_t, w)_{0,\Gamma_0} \, dt$$

Taking norms and using the trace theorem and the definition (2.3) gives that

(2.13)
$$\int_0^T |\Delta w|_{0,\Gamma_0}^2 dt \leq C \left[E_{w,\gamma}(0) + E_{w,\gamma}(T) \right] + C_1 \int_0^T \left[|w_t|_{0,\Gamma_0}^2 + \gamma|\nabla w_t|_{0,\Gamma_0}^2 \right] dt$$
$$+ 2\varepsilon \int_0^T |\Delta w|_{0,\Gamma_0}^2 \, dt + C_\varepsilon \int_0^T \left[|\theta|_{1,\Gamma_0}^2 + |z_t|_{0,\Gamma_0}^2 + |\alpha \Delta w_t|_{0,\Gamma_0}^2 \right] dt$$

Thus, we have that there exists a constant $C > 0$ such that for $\varepsilon > 0$ small enough,

(2.14)
$$(1-2\varepsilon)\int_0^T |\Delta w|^2_{0,\Gamma_0} dt \leq C[E_{w,\gamma}(0)+E_{w,\gamma}(T)]+C_1\int_0^T \left[|w_t|^2_{0,\Gamma_0}+\gamma|\nabla w_t|^2_{0,\Gamma_0}\right] dt$$
$$+ C_\varepsilon \int_0^T \left[|\theta|^2_{1,\Gamma_0} + |z_t|^2_{0,\Gamma_0} + |\alpha\Delta w_t|^2_{0,\Gamma_0}\right] dt$$

If the ε of equations (2.12) and (2.14) is small enough, they can be combined to produce the inequality (2.4), which is the desired result. □

2.2. Wave Equation.

LEMMA 2.2. *Consider the wave equation (1.1). Let $E_z(t)$ be the energy defined by*

(2.15) $$E_z(t) = |z_t(t)|^2_{0,\Omega} + |\nabla z(t)|^2_{0,\Omega}$$

Then, for any $\beta < T/2$, there exist constants $C_1, C_2,$ and C_3 such that

(2.16) $$\int_\beta^{T-\beta} E_z(t)\, dt \leq C_1\left[E_z(\beta) + E_z(T-\beta)\right] + C_2 \int_0^T \left(|z|^2_{0,\Gamma_0} + |z|^2_{0,\Omega}\right) dt$$
$$+ C_3 \int_0^T \left(|w_t|^2_{0,\Gamma_0} + |z_t|_{0,\Gamma_0}\right) dt + C_T\, lot(z)$$

where $lot(z) \leq C \int_0^T \left(|z|^2_{1-\delta,\Omega} + |z_t|^2_{-\delta,\Omega}\right) dt;\quad \delta > 0$

PROOF. The method shown here for stabilization of the wave equation is standard by now so we present only a few key inequalities. Detailed calculations can be seen in [**LT89**]. Assume that h is a smooth vector field on Ω satisfying the geometrical conditions (1.4), sufficient conditions for which are presented in Proposition 1.1. We perform standard multiplier calculations on equation (1.1), i.e. multiplying equation (1.1) by $z\, div\, h$, $z \cdot \nabla h$, and z, respectively, integrating from s to T, and estimate. The use of the geometrical conditions (1.4) is necessary at this point in order to dispatch two terms which appear in the case the Neumann homogeneous condition ([**Tri88**], Section 5). This leads to

(2.17) $$\int_s^T E_z(t)\, dt \leq C_1[E_z(s) + E_z(T)] + C_2 \int_s^T \left(|z_t|^2_{0,\Gamma_0} + |w_t|^2_{0,\Gamma_0}\right) dt$$
$$+ C_3 \int_s^T \left(|z|^2_{0,\Omega} + |z|^2_{0,\Gamma_0} + \left|\frac{\partial}{\partial\tau}z\right|^2_{0,\Gamma_0}\right) dt$$

The existence of Neumann conditions on Γ gives rise to the tangential derivative $\frac{\partial}{\partial\tau}$ on the right hand side of equation (2.17). In order to estimate this term need the following trace regularity result valid for the wave equation.

PROPOSITION 2.2. *Let z be a solution to (1.1) with the interior regularity $z \in C(0,T;H^1(\Omega)) \cap C^1(0,T;L^2(\Omega))$ and boundary regularity*

$$\frac{\partial}{\partial\nu}z, z_t \in L^2((0,T) \times \Gamma_0)$$

Let $T > 0$ be arbitrary and let β be an arbitrarily small constant such that $\beta < T/2$. Then we have that

(2.18) $$\int_\beta^{T-\beta} \left|\frac{\partial}{\partial \tau} z\right|^2_{0,\Gamma_0} dt \le C_{T,\beta} \left[\int_0^T \left[\left|\frac{\partial}{\partial \nu} z\right|^2_{0,\Gamma} + |z_t|^2_{0,\Gamma_0}\right] dt + lot(z)\right]$$

The inequality above does not follow from standard trace theory; it is an independent trace regularity result proved in [**LT92**]. We combine this inequality with the estimate (2.17) with s replaced by β and T replaced by $T - \beta$ to derive (2.16). □

2.2.1. *Uniform Stability Analysis.* In the final analysis, we will combine the energy estimates on plate and wave equations, and then absorb the lower order terms by means of a standard compactness/uniqueness argument.

PROPOSITION 2.3. *With respect to the coupled PDE system* (1.1)-(1.2), *the following estimate holds:*

(2.19)
$$\int E_\gamma(t)\, dt \le C_1[E_\gamma(0) + E_\gamma(T)] + C_2 \int_0^T \left(|z_t|^2_{0,\Gamma_0} + |\theta|^2_{1,\Gamma_0} + |\alpha\Delta w_t|^2_{0,\Gamma_0}\right) dt$$
$$+ C_3 \int_0^T \left(|z|^2_{0,\Gamma_0} + |z|^2_{0,\Omega}\right) dt + C_T lot(z)$$

Here the energy $E_\gamma(t)$ is defined as in (2.2) and the constants $C_1, C_2,$ and C_3 are independent of the parameters γ and α.

PROOF. The inequality follows from equations (2.4) and (2.16). First, we have added (2.4) and (2.16) after multiplying (2.4) by a suitable constant in order to consolidate the $|w_t|^2_{0,\Gamma_0}$ terms. Then, terms involving β were eliminated using the dissipation equality (2.1) and the inequality

$$\left(\int_0^\beta + \int_{(T-\beta)}^T\right) E_\gamma(t)\, dt \le 2\beta E_\gamma(0)$$

which is due to (2.1) □

PROPOSITION 2.4. *With respect to the coupled PDE system* (1.1)-(1.2), *there exists a constant $C_T > 0$ such that*

(2.20)
$$\int_0^T \left(|z|^2_{0,\Gamma_0} + |z|^2_{0,\Omega}\right) dt + C_T lot(z) \le C_T \int_0^T \left(|z_t|^2_{0,\Gamma_0} + |\theta|^2_{1,\Gamma_0} + |\alpha\Delta w_t|^2_{0,\Gamma_0}\right) dt$$

PROOF. The conclusion follows by contradiction via the usual compactness and uniqueness argument. Since this argument is standard, we shall only point out the main steps. The compactness of the lower order terms in z with respect to the topology induced by the energy E follows from trace theory and the compact imbeddings $H^1(\Omega) \hookrightarrow L^2(\Omega)$, $H^{\frac{1}{2}}(\Gamma) \hookrightarrow L^2(\Gamma)$.

As far as uniqueness, we deal with the following overdetermined system:

(2.21)
$$\begin{aligned}
\tilde{z}_{tt} &= \Delta \tilde{z} &&\text{on } (0,T] \times \Omega \\
\tilde{z}_t &= 0 &&\text{on } (0,T] \times \Gamma \\
\tilde{z} &= 0 &&\text{on } (0,T] \times \Gamma_1 \\
\frac{\partial \tilde{z}}{\partial \nu} &= \tilde{w}_t &&\text{on } (0,T] \times \Gamma_0
\end{aligned}$$

$$\begin{aligned}
\tilde{w}_{tt} - \gamma \Delta \tilde{w}_{tt} + \Delta^2 \tilde{w} + \Delta(\alpha \Delta \tilde{w}_t) &= 0 &&\text{on } (0,T] \times \Gamma_0 \\
\tilde{\theta} &\equiv 0 &&\text{on } (0,T] \times \Gamma_0 \\
\tilde{w} = \Delta \tilde{w} &= 0 &&\text{on } (0,T] \times \partial\Gamma_0
\end{aligned}$$

$$\{\tilde{z}(0), \tilde{z}_t(0), \tilde{w}(0), \tilde{w}_t(0), \tilde{\theta}(0)\} = \{\tilde{z}_0, \tilde{z}_1, \tilde{w}_0, \tilde{w}_1, \tilde{\theta}_0\} \in Y_\gamma$$

Since $\tilde{\theta} \equiv 0$, (1.2b) tells us that $\Delta \tilde{w}_t = 0$. However, we have that $\tilde{w}_t = 0$ on $\partial\Gamma_0$. Thus, by elliptic theory $\tilde{w}_t \equiv 0$. Substituting this into (2.21) gives the following system:

(2.22)
$$\begin{aligned}
\tilde{z}_{tt} &= \Delta \tilde{z} &&\text{on } (0,T] \times \Omega \\
\tilde{z}_t &= 0 &&\text{on } (0,T] \times \Gamma \\
\tilde{z} &= 0 &&\text{on } (0,T] \times \Gamma_1 \\
\frac{\partial \tilde{z}}{\partial \nu} &= 0 &&\text{on } (0,T] \times \Gamma_0
\end{aligned}$$

$$\begin{aligned}
\Delta^2 \tilde{w} &= 0 &&\text{on } (0,T] \times \Gamma_0 \\
\tilde{\theta} &\equiv 0 &&\text{on } (0,T] \times \Gamma_0 \\
\tilde{w} = \Delta \tilde{w} &= 0 &&\text{on } (0,T] \times \partial\Gamma_0
\end{aligned}$$

$$\{\tilde{z}(0), \tilde{z}_t(0), \tilde{w}(0), \tilde{w}_t(0), \tilde{\theta}(0)\} = \{\tilde{z}_0, \tilde{z}_1, \tilde{w}_0, \tilde{w}_1, \tilde{\theta}_0\} \in Y_\gamma$$

Again by elliptic theory, $\tilde{w} \equiv 0$, and for $t > 0$ large enough, the solution of (2.22) is exactly zero. Thus by Holmgren-John's uniqueness theorem, $\{\tilde{z}, \tilde{w}, \tilde{\theta}\} \equiv 0$, for all $t > t_0 > 0$. This allows us to assert (2.20) as desired. $\square$

To finish off the proof of Theorem 1.4, we use the inequality (2.20) to combine terms on the right hand side of (2.19) and note that $\alpha \leq \alpha_0$ and by the dissipation property (2.1), $TE_\gamma(T) \leq \int_0^T E_\gamma(t)\,dt$. Thus,

$$TE_\gamma(T) \leq C_1 [E_\gamma(0) + E_\gamma(T)]$$
$$+ C_T \int_0^T \left[d|z_t|^2_{0,\Gamma_0} + |\theta|^2_{1,\Gamma_0}\right] dt + C_T \alpha_0 \int_0^T |\alpha^{\frac{1}{2}} \Delta w_t|^2_{0,\Gamma}\,dt$$

Substituting in the dissipation equality and noticing that $E_\gamma(T) \leq E_\gamma(0)$ gives that

(2.23) $$C'_T E_\gamma(T) \leq \int_0^T \left(d|z_t|^2_{0,\Gamma_0} + |\theta|^2_{1,\Gamma_0}\right) dt + \alpha_0 \int_0^T |\alpha^{\frac{1}{2}} \Delta w_t|^2_{0,\Gamma}\,dt$$

where $C'_T = (T - 2C_1)/(C_1 + 2C_T)$ is strictly positive for $T > 2C_1$. Again, we can apply the dissipation relation (2.1) separately to both terms of the right hand side of (2.23) to give

(2.24) $$C'_T E_\gamma(T) \leq \frac{1}{2}[E_\gamma(0) - E_\gamma(T)] + \frac{\alpha_0}{2}[E_\gamma(0) - E_\gamma(T)]$$

or

(2.25) $$(2C'_T + 1 + \alpha_0)E_\gamma(T) \leq (1 + \alpha_0)E_\gamma(0)$$

or

(2.26) $$E_\gamma(T) < E_\gamma(0)$$

for $T > 2C_1$.

References

[AL97] G. Avalos and I. Lasiecka, *Exponential stability of a thermoelastic system without mechanical dissipation. Dedicated to the memory of Pierre Grisvard*, Rend. Istit. Mat. Univ. Trieste **28** (1997), no. suppl., 1–27.

[Ava96] G. Avalos, *The exponential stability of a coupled hyperbolic/parabolic system arising in structural acoustics*, Abstr. Appl. Anal. **1** (1996), no. 2, 203–217.

[Ava00] G. Avalos, *Well-posedness of a canonical structural acoustic model under point boundary control*, Contemporary Mathematics **this volume** (2000).

[BBPMZ98] E. Bisognin, V. Bisognin, G. Perla Menzala, and E. Zuazua, *On the exponential stability for Von Kármán equations in the presence of thermal effects*, Math. Methods Appl. Sci. **21** (1998), no. 5, 393–416.

[BSS94] H. T. Banks, R. J. Silcox, and R. C. Smith, *The modeling and control of acoustic/structure interaction problems via piezoceramic actuators: 2-d numerical examples*, ASME Journal of Vibration and Acoustics **116** (1994), no. 3, 386–396.

[BSYW95] H. T. Banks, R. C. Smith, and Y. Y. Wang, *The modeling of piezoceramic patch interactions with shells, plates, and beams*, Quarterly of Applied Mathematics **53** (1995), no. 2, 353–381.

[Cam99] M. Camurdan, *Uniform stability of a coupled structural acoustic system by boundary dissipation*, Abstract and Applied Analysis (1999).

[CdL87] E. F. Crawley and J. de Luis, *Use of piezoceramic actuators as elements of intelligent structures*, AIAA Journal **25** (1987), no. 10, 1373–1385.

[CJ00] M. Camurdan and G. Ji, *A noise reduction problem arising in structural acoustics: a three-dimensional solution*, Contemporary Mathematics **this volume** (2000).

[CR81] S. Chen and D. Russell, *A mathematical model for linear elastic systems with structural damping*, Quart. Appl. Math. **39** (1981), no. 4, 433–454.

[CT88] S. Chen and R. Triggiani, *Proof of two conjectures of G. Chen and D. Russell in structural damping for elastic systems: the case $\alpha = 1/2$*, Lecture Notes in Mathematics, vol. 1354, Springer-Verlag, New York, 1988, pp. 234–256.

[CT89] S. Chen and R. Triggiani, *Proof of extensions of two conjectures on structural damping for elastic systems*, Pacific J. Math. **136** (1989), no. 1, 15–55.

[CT99] M. Camurdan and R. Triggiani, *Sharp regularity of a coupled system of a wave equation and a Kirchoff equation with point control arising in noise reduction*, Diff. Int. Eqns. **12** (1999), 101–118.

[FGS90] C. R. Fuller, G. P. Gibbs, and R. J. Silcox, *Simultaneous active control of flexural and extensional power flow in beams*, Journal of Intelligent Materials, Systems and Structures **1** (1990).

[Gri85] P. Grisvard, *Elliptic problems in nonsmooth domains*, Monographs and Studies in Mathematics, 24, Pitman Publishing Ltd., London, 1985.

[Gri89] P. Grisvard, *Controlabilité exacte des solutions de l'equation des ondes en présence de singularités*, J. Math. Pures et appl. **68** (1989), 215–259.

[IY00] V. Isakov and M. Yamamoto, *Carleman estimate with the Neumann boundary condition and its application to observability inequality and inverse hyperbolic problems*, Contemporary Mathematics **this volume** (2000).

[Lag89] J. Lagnese, *Boundary stabilization of thin plates*, SIAM, Philadelphia, PA, 1989.

[Leb00] C. Lebiedzik, *Uniform stability of a coupled structural acoustic system with thermoelastic effects*, Dynamics of Continuous, Discrete, and Impulsive Systems **to appear** (2000).

[Lio88] J. L. Lions, *Exact controllability, stabilization, and pertubation*, SIAM Rev. **30** (1988), 1–68.

[LL99]   I. Lasiecka and C. Lebiedzik, *Uniform stability in structural acoustic systems with thermal effects and nonlinear boundary damping*, Control and Cybernetics **28** (1999), no. 3, 557–581.

[LLT86]   I. Lasiecka, J. L. Lions, and R. Triggiani, *Nonhomogeneous boundary value problems for second-order hyperbolic operators*, J. Math. pures et appl. **65** (1986), 149–192.

[LR95]   Z. Lui and M. Renardy, *A note on the equations of a thermoelastic plate*, Appl. Math. Lett. **8** (1995), no. 3, 1–6.

[LT89]   I. Lasiecka and R. Triggiani, *Exact controllability of the wave equation with Neumann boundary control*, Appl. Math. Optim. **19** (1989), 243–290.

[LT92]   I. Lasiecka and R. Triggiani, *Uniform stabilization of the wave equation with Dirichlet or Neumann feedback control without geometric conditions*, Appl. Math. Optim. **25** (1992), no. 2, 189–224.

[LT99]   I. Lasiecka and R. Triggiani, *Control theory for partial differential equations*, Encyclopedia of Mathematics and its Applications, vol. I and II, Cambridge University Press, 1999.

[LT00]   I. Lasiecka and R. Triggiani, *Structural decomposition of thermoelastic semigroups with rotational forces*, Semigroup Forum **60** (2000), 16–66.

[LTZ00]   I. Lasiecka, R. Triggiani, and X. Zhang, *Nonconservative wave equations with unobserved Neumann boundary conditions: global uniqueness and observability in one shot*, Contemporary Mathematics **this volume** (2000).

[MI68]   P.M. Morse and K.U. Ingard, *Theoretical acoustics*, McGraw-Hill, 1968.

[Tri88]   R. Triggiani, *Exact controllability on $L_2(\Omega) \times H^{-1}(\Omega)$ of the wave equation with Dirichlet boundary control*, Appl. Math. Optim. **18** (1988), 241–277.

[Tri92]   R. Triggiani, *Counterexamples to some stability questions for dissipative generators*, J. Math. Anal. Appl. **170** (1992), no. 1, 49–64.

DEPARTMENT OF MATHEMATICS, UNIVERSITY OF VIRGINIA, CHARLOTTESVILLE, VA 22902

TOPOLOGICAL DERIVATIVE FOR NUCLEATION OF NON-CIRCULAR VOIDS. THE NEUMANN PROBLEM.

Tomasz Lewiński and Jan Sokołowski

ABSTRACT. The hitherto existing literature concerning the topological derivative of shape functionals concerned perturbations of domains caused by introduction of circular or ball openings. In the present study the notion of the topological derivative is generalized to the case of openings of arbitrary shape. The paper is concerned with energy functionals related to the $2D$ Neumann problem.

1. Introduction

Let us consider the following model problem of shape optimization. We are given a plane domain Ω, called a feasible domain, a part of which, Ω_μ, should be filled up with a material whose physical properties are characterized by a modulus μ, $\mu > 0$. A part Γ of the boundary $\partial\Omega$ is subjected to a loading $p = p(x)$, $x \in \Gamma$, which means the Neumann boundary data, satisfying

$$\int_\Gamma p\,ds = 0\,.$$

We demand that $\Gamma \subset \partial\Omega_\mu$. The given amount of the material is smaller than the area of Ω, or $|\Omega_\mu| = A < |\Omega|$, $A > 0$, being a given constant. We assume that $\Omega_\mu = \Omega\backslash K$, K being a compact subset of Ω. Just K is the removed part of Ω, while the remaining part, Ω_μ, forms a body, see Fig. 1.1.

1991 *Mathematics Subject Classification.* Primary 49B22, 73K15, 93C20; Secondary 35J50, 65K10, 73L99.

Key words and phrases. Shape optimization, shape derivative, topological derivative, compound asymptotics, Polya-Szegö mass matrix, $2D$ elasticity, Neumann's problem.

Partially supported by the Institute Elie Cartan, University Nancy 1 and by the French–Polish research programme POLONIUM between INRIA–Lorraine and Systems Research Institute of the Polish Academy of Sciences .

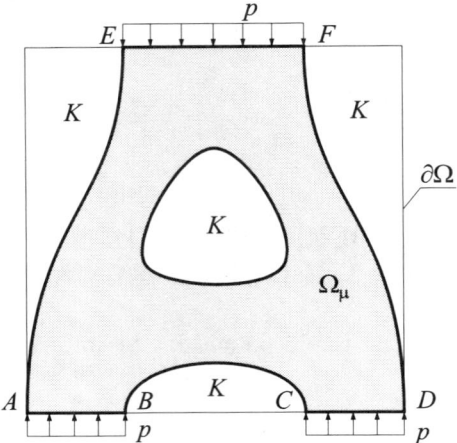

Fig. 1.1. Here Ω is a rectangular domain, while Γ consists of three intervals: AB, CD, EF

For the field u satisfying

$$(P_\mu) \left| \begin{array}{l} \Delta u = 0 \quad \text{in } \Omega_\mu \\ \mu \dfrac{\partial u}{\partial n} = p \quad \text{on } \Gamma, \end{array} \right.$$

where n is a unit vector, outward normal to Ω, we define the shape functional

$$J(\Omega_\mu) = \int_\Gamma pu\,dx\,.$$

The necessary conditions of optimum for the layout problem:

$$J(\Omega^*) = \inf\{J(\Omega_\mu) \,:\, |\Omega_\mu| = A < |\Omega|\}$$

concern the admissible variations of the shape functional resulting from

- the change of shape of Ω, see Sokołowski and Zolesio (1992), due to boundary variations; such optimality conditions are derived e.g. in the framework of the so-called material derivative method;
- the topological changes within Ω, see Sokołowski and Żochowski (1999a-c) for the definition of topological derivative of an arbitrary shape functional; the topological derivative of J should be nonnegative in the interior of an optimal domain.

The topological derivative determines an increment of $J(\Omega_\mu)$ implied by appearing (or nucleation) of a small opening (void) within Ω. The definition of the topological derivative depends on the shape of the appearing of small openings. The case of appearing of circular openings was considered by Sokołowski and Żochowski (1999a). In the present paper the notion of topological derivative is extended to the case of appearing of openings of arbitrary shape.

To make this generalization a new technique of finding the increment of $J(\Omega_\mu)$ is proposed. This technique makes use of the compound asymptotic expansions of solutions to partial differential equations of elliptic type developed e.g. by Maz'ya and Nazarov (1987) and Mazja et al. (1991). We refer the reader to the recent survey paper by Nazarov (1999) for complete description of this technique for elliptic systems which satisfy the so-called polynomial property. Such a technique is currently used in applications to optimum design in structural mechanics.

In the years 1994-1995 Eschenauer et al. (1994) and Schumacher (1995) put forward a new numerical iterative scheme for solving the layout problem mentioned above and the analogous problems of minimum compliance of elastic bodies. According to this scheme, in the subsequent iteration steps, new openings within Ω are predicted at the points, where the so-called characteristic function assumes minima. Lewiński and Sokołowski (1999) reported that in the case of nucleating circular voids in plane elastic bodies, this characteristic function differs in a factor from the expression representing the topological derivative of the functional of the elastic compliance. This equivalence shows a practical role of the notion of topological derivative.

We show in the present paper that the topological derivative of the energy functional of the Neumann problem is a positive definite quadratic form (with matrix $G = (G^{\alpha\beta})$) in the case of nucleation of star-shaped openings. To this end we apply the sensitivity analysis methods of Sokołowski and Zolesio (1992). A stronger result has been reported by Maz'ya and Nazarov (1997) who proved that the energy increment due to appearing of an opening is a positive definite quadratic form, independently of the fact if the opening is star-shaped or not. The proof is based on the property of the negative definiteness of the mass matrix m of Polya-Szegö, see Schiffer and Szegö (1949) and Appendix A. In the present paper we prove that the matrices m and G are identical if the opening is an ellipse.

The paper is organized as follows. A notion of the topological derivative is introduced in Sec. 2. The Neumann problem for a domain with a small opening is analyzed in Sec. 3. Its solution is represented by the compound asymptotic expansion (Eq. (3.8)). The first three terms of this expansion are explicity constructed. The third term depends on the Polya-Szegö mass matrix m. This matrix is defined by the expansion (3.16). However, its components can be evaluated by integral representations. They are collected and justified in Appendix A.

The main result of the present paper, Theorem 4.1, is given in Sec. 4. This theorem says that the topological derivative of the energy functional is expressed by the quadratic form (4.5) with components forming a matrix G, Eq. (4.6). The proof uses the tools of shape sensitivity methods along with the compound asymptotic expansions.

A link between our result (Theorems 4.1 and 4.2) and that of Maz'ya and Nazarov (1987, Theorem 4.8), reported in Appendix B, is provided in Theorem 4.3, according to which the identity $m = G$ holds for openings of elliptical shapes.

The notation of the present paper complies with that used in the books by Mazja et al. (1991), Movchan and Movchan (1995) and Sokołowski and Zolesio (1992). To avoid misunderstandings the summation convention is not adopted. Small Greek indices: α, β, ... (except for ε) run over 1,2. The symbol ε means a small, positive parameter. Two coordinate systems are used: (x_1, x_2) and (y_1, y_2), the gradient and the Laplace differential operators with respect to the both systems are denoted by (∇, Δ) and (∇_y, Δ_y), respectively.

2. The notion of topological derivative

Let $\Omega \subset \mathbb{R}^2$ be an open domain. This domain is parametrized by the cartesian coordinate system (x_1, x_2). It is assumed that $O = (0,0) \in \Omega$. Let us form a family of domains ω_ε around O according to the formula

$$\omega_\varepsilon = \left\{ x = (x_1, x_2) \in \mathbb{R}^2 \, \bigg| \, \frac{x}{\varepsilon} \in \omega \right\}, \qquad (2.1)$$

where ω is a domain in $\mathbb{R}^2$; ε is a positive, small parameter. The domain ω will be parametrized by the cartesian coordinate system (y_1, y_2); its central point (the origin) $O \in \omega$, see Fig. 2.1, thus $O \in \omega_\varepsilon$. We say that a small domain (or a void) ω_ε nucleates at the point O of the virgin domain Ω.

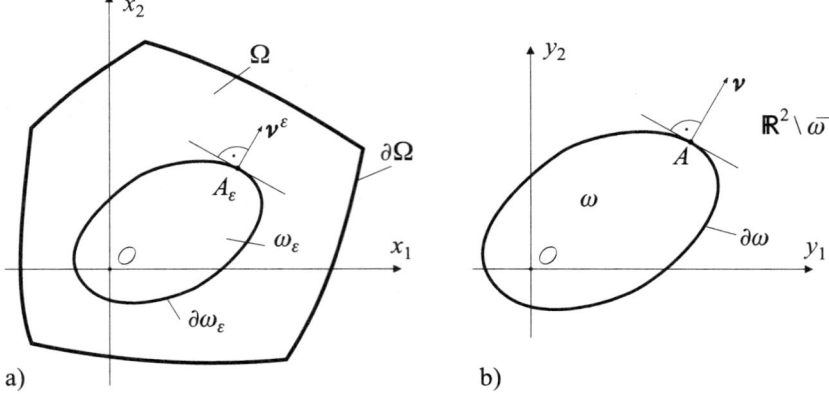

Fig. 2.1. a) Domains Ω and ω_ε; b) The rescaled domain ω

If $A = (y_1, y_2) \in \partial\omega$, then $A_\varepsilon := (\varepsilon y_1, \varepsilon y_2) \in \partial\omega_\varepsilon$. The normal vectors: $\boldsymbol{\nu} = (\nu_1, \nu_2)$ at A and $\boldsymbol{\nu}^\varepsilon = (\nu_1^\varepsilon, \nu_2^\varepsilon)$ at A_ε are directed outward to the domains ω and ω_ε, respectively. Let us note that $\nu_\alpha^\varepsilon = \nu_\alpha$ for every $\varepsilon > O$. The areas of ω_ε and ω are linked by the following relation $|\omega_\varepsilon| = \varepsilon^2 |\omega|$. Let $K \subset \overline{\Omega}$ be a compact subset. For a given shape functional:

$$J : \Omega \setminus K \to \mathbb{R}, \tag{2.2}$$

assume that the following limit exists

$$\mathfrak{T}_\omega(O) = \lim_{\varepsilon \searrow 0} \frac{J(\Omega \setminus \overline{\omega}_\varepsilon) - J(\Omega)}{|\overline{\omega}_\varepsilon|}. \tag{2.3}$$

This limit will be called the topological derivative of the functional J, computed at the point $O \in \Omega$, for nucleation of voids of the shape ω. The value $\mathfrak{T}_\omega(x_0)$ is defined in the same way at any other point $x_0 = (x_{01}, x_{02}) \in \Omega$. Note that

$$\mathfrak{T}_\omega(O) = |\omega|^{-1} \lim_{\varepsilon \searrow 0} \frac{J(\Omega \setminus \overline{\omega}_\varepsilon) - J(\Omega)}{\varepsilon^2}. \tag{2.4}$$

In the case when ω is a circle the index ω in (2.3) will be omitted. The topological derivative defined in this way was introduced by Sokołowski and Żochowski (1997, 1999a, b) for the shape functionals in $\mathbb{R}^n$; ω is assumed there to be an open ball in $\mathbb{R}^n$.

3. The Neumann problem posed on a domain with a small opening

A solution of the Neumann problem posed on a plane domain with a small void can be constructed by the compound asymptotic method, cf. Maz'ya and Nazarov (1987). The main term of the asymptotic expansion turns out to be the solution v referred to the virgin domain (or a domain without a void), while the subsequent term depends linearly on ∇v measured at the point at which a small void nucleates. This result will be used further to assess the relevant change of energy.

Let us consider the following Neumann problem posed on the domain $\Omega_\varepsilon = \Omega \setminus \overline{\omega}_\varepsilon$:

$$(P_\varepsilon) \quad \begin{vmatrix} \text{find } u_\varepsilon \in H^{2+k}(\Omega_\varepsilon) \; k \geq 2, \text{ such that} \\[4pt] \Delta u_\varepsilon = 0 \quad \text{in} \quad \Omega_\varepsilon\,, \hfill (3.1) \\[4pt] \dfrac{\partial u_\varepsilon}{\partial \nu^\varepsilon} = 0 \quad \text{on} \quad \partial \omega_\varepsilon\,, \hfill (3.2) \\[4pt] \mu \dfrac{\partial u_\varepsilon}{\partial n} = p \quad \text{on} \quad \partial \Omega\,. \hfill (3.3) \end{vmatrix}$$

The Laplace operator is defined by $\Delta = \dfrac{\partial^2}{\partial (x_1)^2} + \dfrac{\partial^2}{\partial (x_2)^2}$ and n represents a unit vector outward normal to $\partial \omega$. Moreover, μ is a positive constant.

The given loading $p \in H^{\frac{1}{2}+k}(\partial \Omega)$ satisfies the compatibility condition

$$\int_{\partial \Omega} p\, ds = 0\,. \tag{3.4}$$

We assume that the domain Ω_ε satisfies the known conditions of regularity such that $u_\varepsilon \in C^2(\overline{\Omega}_\varepsilon)$ and u_ε is determined up to an additive constant.

In order to derive the asymptotic expansion for u_ε we start with the Neumann problem for the virgin domain Ω:

$$(P_0) \quad \begin{vmatrix} \text{find } v \in C^2(\overline{\Omega}) \text{ such that} \\[4pt] \Delta v = 0 \quad \text{in} \quad \Omega\,, \hfill (3.5) \\[4pt] \mu \dfrac{\partial v}{\partial n} = p \quad \text{on} \quad \partial \Omega\,. \hfill (3.6) \end{vmatrix}$$

The first derivatives of v at $x = \boldsymbol{O}$ are finite. Let us introduce the notation

$$\epsilon_\alpha(x) = \frac{\partial v}{\partial x_\alpha}(x), \quad \epsilon_\alpha^0 = \frac{\partial v}{\partial x_\alpha}(\boldsymbol{O}) \tag{3.7}$$

The asymptotic expansion of the solution u_ε was derived by Maz'ya and Nazarov (1987, Sec. 4.4). It has the form

$$u_\varepsilon(x) = v(x) + \varepsilon u^{(1)}\left(\frac{x}{\varepsilon}\right) + \varepsilon^2 u^{(2)}(x) + \varepsilon^2 O\left(\frac{\varepsilon}{\|x\|}\right)\,. \tag{3.8}$$

Here we denote $\|x\| = [(x_1)^2 + (x_2)^2]^{\frac{1}{2}}$. The function $u^{(1)}$ defined in $\mathbb{R}^2 \setminus \overline{\omega}$ is represented as

$$u^{(1)}(y) = \epsilon_1^0 w_1(y) + \epsilon_2^0 w_2(y)\,, \quad y \in \mathbb{R}^2 \setminus \overline{\omega}\,. \tag{3.9}$$

The functions $w_\alpha(y)$ are solutions to the following problems for $\ell \in (0,1)$,

$$(P_\omega) \quad \begin{vmatrix} \text{find } w_\alpha = w_\alpha(y) \in V^2_{2,\ell}(\mathbb{R}^2 \setminus \overline{\omega}) \quad \text{such that} \\[4pt] \Delta_y w_\alpha = 0 \hfill \text{in} \quad \mathbb{R}^2 \setminus \overline{\omega}\,, \hfill (3.10) \\[4pt] \dfrac{\partial w_\alpha}{\partial \nu} = -\nu_\alpha \hfill \text{on} \quad \partial \omega\,, \hfill (3.11) \\[4pt] w_\alpha(y) \to 0 \hfill \text{if} \quad \|y\| \to \infty\,, \hfill (3.12) \end{vmatrix}$$

where $\Delta_y = \dfrac{\partial^2}{\partial (y_1)^2} + \dfrac{\partial^2}{\partial (y_2)^2}$.

The weight function space $\mathbb{V}^2_{2,\ell}(\mathbb{R}^2 \setminus \overline{\omega})$ is defined as the closure

$$\mathbb{V}^2_{2,\ell}(\mathbb{R}^2 \setminus \overline{\omega}) = \overline{C^\infty_0(\mathbb{R}^2 \setminus \overline{\omega})}$$

with respect to the norm

$$\|u\|^2 = \sum_{\alpha,\gamma \leq 2} \int_{\mathbb{R}^2 \setminus \overline{\omega}} |y|^{2(\ell+\alpha+\gamma-1)} \left| \frac{\partial^{\alpha+\gamma} u(y)}{\partial (y_1)^\alpha \partial (y_2)^\gamma} \right|^2 dy$$

The problems (P_ω) are uniquely solvable in the weight spaces, we refer e.g. to Mazya et al. (1991, Sec. 1.5.2) for a proof.

To determine $u^{(2)}(x)$ we must first recall the definition of the *mass matrix*. Let us denote by $G(x,y) = S(x-y)$ the Green function for the Laplace equation in $\mathbb{R}^2$. The function $S(y) = G(y,0)$ satisfies the equation

$$\mu \Delta_y S + \delta(y - \boldsymbol{O}) = 0 \qquad (3.13)$$

with δ being Dirac's distribution, and the condition

$$\left\| \frac{\partial S}{\partial y_\alpha} \right\| = 0\left(\frac{1}{\|y\|}\right), \quad \text{if} \quad \|y\| \to \infty. \qquad (3.14)$$

We note that

$$S(y) = -\frac{1}{2\pi\mu} \ln \|y\|. \qquad (3.15)$$

The functions w_α are harmonic, singular at $y = \boldsymbol{O}$ and vanish at infinity. Thus they are linear combinations of $\dfrac{\partial S}{\partial y_\alpha}, \dfrac{\partial^2 S}{\partial y_\alpha \partial y_\beta}, \ldots$, see Schiffer (1959). In the neighbourhood of $y = \infty$ the functions w_α can be expanded as follows

$$w_\alpha(y) = \sum_{\beta=1}^{2} m_{\alpha\beta} \frac{\partial S}{\partial y_\beta} + O\left(\frac{1}{\|y\|^2}\right), \qquad (3.16)$$

where $(m_{\alpha\beta})$ are certain constant coefficients constituting a symmetric *mass matrix* – such a definition was proposed by Polya, see Schiffer and Szegö (1949).

Now we are ready to determine the function $u^{(2)}(x)$. It is represented by

$$u^{(2)}(x) = \sum_{\alpha=1}^{2} \sum_{\beta=1}^{2} m_{\beta\alpha} \epsilon^0_\beta z_\alpha(x), \qquad (3.17)$$

where the functions z_α satisfy

$$\left(\widehat{P}\right) \quad \left|\begin{array}{ll} \Delta z_\alpha = 0 & \text{in } \Omega, \qquad (3.18) \\ \mu \dfrac{\partial z_\alpha}{\partial n} = \dfrac{1}{2\pi} \dfrac{\partial}{\partial n}\left(\dfrac{x_\alpha}{\|x\|^2}\right) & \text{on } \partial\Omega. \qquad (3.19) \end{array}\right.$$

The subsequent terms of the expansion (3.8) have been formed according to the *compound asymptotic method* developed in Maz'ya and Nazarov (1987) and Mazja et al. (1991). The idea of this method is as follows. The first term of (3.8) satisfies the conditions (3.1) and (3.3), but violates the condition (3.2). The second term of (3.8) introduces a correction that makes the condition (3.2) satisfied, but introduces new errors into the boundary condition (3.3). To correct the latter mismatch the third term of (3.8) is introduced, which inevitably leads to an incompatibility in the condition (3.2). This iterative process can be continued, its convergence having been proved e.g. in Nazarov (1999, §7). Moreover, the expansion (3.8) can be differentiated term by term, which gives

$$\frac{\partial u_\varepsilon}{\partial x_\alpha} = \frac{\partial v}{\partial x_\alpha} + \left.\frac{\partial u^{(1)}(y)}{\partial y_\alpha}\right|_{y=\frac{x}{\varepsilon}} + \varepsilon^2 \frac{\partial u^{(2)}}{\partial x_\alpha} + o(\varepsilon) \qquad (3.20)$$

or, taking into account (3.7) and (3.9), one finds

$$\frac{\partial u_\varepsilon}{\partial x_\alpha} = \epsilon_\alpha(x) + \sum_{\beta=1}^{2} \epsilon_\beta^0 \left.\frac{\partial w_\beta}{\partial y_\alpha}\right|_{y=\frac{x}{\varepsilon}} + \varepsilon^2 \frac{\partial u^{(2)}}{\partial x_\alpha} + o(\varepsilon). \qquad (3.21)$$

In the neighbourhood of $\partial \omega_\varepsilon$ we have $\epsilon_\alpha(x) = \epsilon_\alpha^0 + 0(\varepsilon)$, see (3.7). Thus, in this neighbourhood the expansion (3.21) can be put in the form

$$\frac{\partial u_\varepsilon}{\partial x_\alpha} = \sum_{\beta=1}^{2} \epsilon_\beta^0 \left.\frac{\partial W_\beta(y)}{\partial y_\alpha}\right|_{y=\frac{x}{\varepsilon}} + O(\varepsilon) \qquad (3.22)$$

with

$$W_\beta(y) = w_\beta(y) + y_\beta. \qquad (3.23)$$

Let us note that the functions $W_\beta(y)$ are solutions to the following problem

$$\left(\widehat{P}_\omega\right) \quad \left|\begin{array}{ll} \Delta_y W_\beta = 0 & \text{in } \mathbb{R}^2 \setminus \overline{\omega}, \\ \dfrac{\partial W_\beta}{\partial \nu} = 0 & \text{on } \partial \omega, \\ W_\beta \to y_\beta & \text{if } \|y\| \to \infty. \end{array}\right. \qquad \begin{array}{l}(3.24)\\ (3.25)\\ (3.26)\end{array}$$

Additional explanations concerning the compound asymptotic method are included in the Appendices A and B.

4. Topological derivative of the energy functional

Let us consider the shape functional

$$J(\Omega \setminus \overline{\omega}_\varepsilon) = \frac{1}{2} \int_{\Omega \setminus \overline{\omega}_\varepsilon} \mu |\nabla u_\varepsilon|^2 \, dx - \int_{\partial \Omega} p u_\varepsilon \, ds, \qquad (4.1)$$

where

$$\nabla u_\varepsilon = \left(\frac{\partial u_\varepsilon}{\partial x_1}, \frac{\partial u_\varepsilon}{x_2}\right) \qquad (4.2)$$

and u_ε is the solution to the problem (P_ε) of Sec. 3. Let $j(\varepsilon) := J(\Omega \setminus \overline{\omega}_\varepsilon)$. We shall prove the following result on the asymptotic expansion of the function $j(\varepsilon)$.

Theorem 4.1.
Under the conditions concerning the domains Ω and ω and the loading p on $\partial \Omega$, introduced in Sec. 3, the function $j(\varepsilon)$ can be expanded as follows

$$j(\varepsilon_0 + \varepsilon) = j(\varepsilon_0) + j'(\varepsilon_0)\varepsilon + \frac{1}{2}j''(\varepsilon_0)\varepsilon^2 + o\left((\varepsilon_0 + \varepsilon)^2\right), \qquad (4.3)$$

where

$$j'(\varepsilon_0) \to j'(0^+) = 0, \quad \text{if} \quad \varepsilon_0 \searrow 0^+, \qquad (4.4)$$

$$j''(\varepsilon_0) \to j''(0^+) = \sum_{\alpha,\beta=1}^{2} \epsilon_\beta^0 G_{\alpha\beta} \epsilon_\beta^0, \quad \text{if} \quad \varepsilon_0 \searrow 0^+ \tag{4.5}$$

and the matrix $(G_{\alpha\beta})$ is expressed by

$$G_{\alpha\beta} = -\mu[\delta_{\alpha\beta}|\omega| + \int_{\partial\omega} \gamma_{\alpha\beta}(\boldsymbol{w}) y \cdot \nu \, ds] . \tag{4.6}$$

Here $y \cdot \nu = y_1\nu_1 + y_2\nu_2$, $\boldsymbol{w} = (w_1, w_2)$ and

$$\gamma_{\alpha\beta}(\boldsymbol{w}) = \frac{1}{2}\left(\frac{\partial w_\alpha}{\partial y_\beta} + \frac{\partial w_\beta}{\partial y_\alpha} + \nabla_y w_\alpha \cdot \nabla_y w_\beta\right), \tag{4.7}$$

where $\nabla_y f = \left(\dfrac{\partial f}{\partial y_1}, \dfrac{\partial f}{\partial y_2}\right)$.

The functions w_1, w_2 are solutions to the problems (P_ω). The quantity $\frac{1}{2}j''(0^+)$ determines the topological derivative of the shape functional $J(\Omega \setminus \overline{\omega}_\varepsilon) = J(\Omega_\varepsilon)$.

Proof.
We shall draw upon the following lemma, concerning sensitivity of the solutions of the Neumann problem. Consider the following family of Neumann's problem in the domain Ω_t of boundaries γ_t and $\partial\Omega$, see Fig. 4.1:

$$(P_t) \quad \left| \begin{array}{l} \text{find } u_t \in H_1(\Omega_t)/\mathbb{R} \text{ such that} \\ \mu \displaystyle\int_{\Omega_t} \nabla u_t \cdot \nabla \varphi \, dx = \int_{\partial\Omega} p\varphi \, ds \quad \forall \varphi \in H^1(\Omega_t) \end{array} \right. \tag{4.8}$$

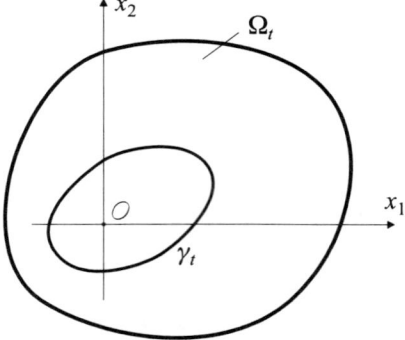

Fig. 4.1. Domain Ω_t with the varying internal boundary γ_t

The domain Ω_t is given in the form of an image of the domain Ω

$$\Omega_t = T_t(\boldsymbol{V})(\Omega) . \tag{4.9}$$

The mapping T_t depends on the velocity field $\boldsymbol{V}$ such that $\operatorname{supp} \boldsymbol{V} \cap \partial\Omega = \emptyset$ and

$$\gamma_t = T_t(\boldsymbol{V})(\gamma) . \tag{4.10}$$

Then the shape derivative of u_t, denoted by u'_t (cf. Sokołowski and Zolesio (1992)) is given by the unique solution to the following problem

$$(\overline{P}_t) \quad \left| \begin{array}{l} \text{find } u'_t \in H^1(\Omega_t) \backslash \mathbb{R} \text{ such that} \\ \displaystyle\int_{\Omega_t} \nabla u'_t \cdot \nabla \varphi \, dx + \int_{\gamma_t} \nabla u_t \cdot \nabla \varphi \, \boldsymbol{V} \cdot \boldsymbol{\nu}_t \, ds_t = 0 \quad \forall \varphi \in H^2(\Omega_t) \,. \end{array} \right. \quad (4.11)$$

The latter problem can be reformulated as follows, cf. Sokołowski and Zolesio (1992, Prop. 3.2, p. 120)

$$\left| \begin{array}{l} \Delta u'_t = 0 \quad \text{in } \Omega_t \\ \nabla u'_t \cdot \nu_t = \text{div}_{\gamma_t}(\boldsymbol{V} \cdot \nu_t \nabla_{\gamma_t} u_t) \quad \text{on } \gamma_t \,. \end{array} \right.$$

Using this result it is easy to see that the derivative of the energy functional associated with the Neumann problem equals,

$$\frac{dJ(\Omega_t)}{dt} = -\frac{1}{2}\int_{\partial\Omega} p u'_t \, ds = -\frac{1}{2}\mu \int_{\Omega_t} \nabla u'_t \cdot \nabla u_t \, dx$$

hence

$$\frac{dJ(\Omega_t)}{dt} = \frac{1}{2}\mu \int_{\gamma_t} |\nabla u_t|^2 \boldsymbol{V} \cdot \boldsymbol{\nu}_t \, ds_t \quad (4.12)$$

for

$$J(\Omega_t) = \frac{1}{2}\mu \int_{\Omega_t} |\nabla u_t|^2 \, dx - \int_{\partial\Omega} p u_t \, ds = -\frac{1}{2}\int_{\partial\Omega} p u_t \, ds \,. \quad (4.13)$$

To analyse the behaviour of the function $j(\varepsilon)$ at the point 0^+ we introduce two small parameters: $\varepsilon_0 > 0$ and ε being sufficiently small. The latter parameter will play the role of the parameter t. The following perturbation of the domain will be considered, cf. Fig. 4.2

$$\Omega_{\varepsilon_0+\varepsilon} = \Omega \setminus \overline{\omega}_{\varepsilon_0+\varepsilon} \,. \quad (4.14)$$

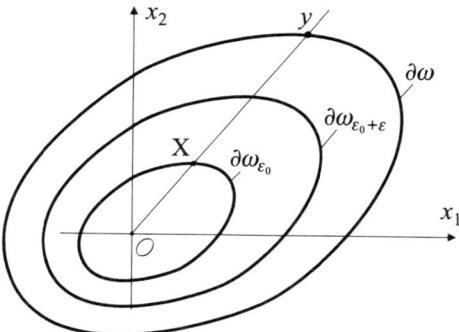

Fig. 4.2. Variation of the boundary: from $\partial\omega_{\varepsilon_0}$ to $\partial\omega_{\varepsilon_0+\varepsilon}$

For any $\varepsilon_0 > 0$, ε_0 small enough and $|\varepsilon| < \dfrac{\varepsilon_0}{2}$, we define the one-to-one mapping

$$\widetilde{T}_\varepsilon \,:\, \partial\omega_{\varepsilon_0} \to \partial\omega_{\varepsilon_0+\varepsilon} \quad (4.15)$$

of the form
$$\tilde{x}(\varepsilon) = \tilde{T}_\varepsilon(X) = \left(1 + \frac{\varepsilon}{\varepsilon_0}\right) X, \qquad X \in \partial\omega_{\varepsilon_0}. \tag{4.16}$$

The mapping $\tilde{T}_\varepsilon$ is extended to a small neighbourhood of $\partial\omega_{\varepsilon_0}$. Let $\varepsilon_0 > 0$ be fixed and let $\eta \in C_0^\infty(\Omega), 0 \leq \eta(x) \leq 1$ be a function such that $\eta(x) = 0$ for $|x| < r_0$ and $|x| > r_1$, for some $r_1 > r_0 > 0$ which depend on ε_0, $\eta(x) = 1$ for $x \in \omega_{\varepsilon_0+\varepsilon}$ and for all $\varepsilon \in [0, \varepsilon_0/4]$. Introduce the vector field

$$\boldsymbol{V}(\varepsilon, x) = \eta(x)\frac{x}{\varepsilon_0 + \varepsilon}, \qquad x \in \Omega, \quad |\varepsilon| < \frac{\varepsilon_0}{2} \tag{4.17}$$

and denote by $x(\varepsilon) = T_\varepsilon(\boldsymbol{V})(X)$ a solution to the system

$$\frac{dx(\varepsilon)}{d\varepsilon} = \boldsymbol{V}(\varepsilon, x(\varepsilon)) \tag{4.18}$$

$$x(0) = X. \tag{4.19}$$

Then $T_\varepsilon : \Omega \to \Omega$ is an extension of $\tilde{T}_\varepsilon$ which equals to the identity in the ball $B_{r_0}(\boldsymbol{O})$ and in the exterior of the ball $B_{r_1}(\boldsymbol{O})$. Here we assume that r_1 is chosen in such a way that $\partial\omega_{\varepsilon_0+\varepsilon} \subset B_{r_1}(\boldsymbol{O})$ for all $\varepsilon \in [0, \varepsilon_0/4]$. Using the field $\boldsymbol{V}$ and the material derivative method we can evaluate the shape derivatives with respect to the parameter ε, at $\varepsilon = 0$, of the energy functional $j(\varepsilon_0 + \varepsilon) = J(\Omega_{\varepsilon_0+\varepsilon})$ defined by (4.20).

The energy functional is given by

$$J(\Omega_{\varepsilon_0+\varepsilon}) = J(\Omega\setminus\overline{\omega}_{\varepsilon_0+\varepsilon}) = \frac{1}{2}\mu \int_{\Omega_{\varepsilon_0+\varepsilon}} |\nabla u_{\varepsilon_0+\varepsilon}|^2 \, dx - \int_{\partial\Omega} p u_{\varepsilon_0+\varepsilon} \, ds. \tag{4.20}$$

Now we can apply the general formula (4.12) to find the topological derivative. For ε sufficiently small we find

$$\frac{dJ(\Omega_{\varepsilon_0+\varepsilon})}{d\varepsilon} = j'(\varepsilon_0 + \varepsilon) \tag{4.21}$$

and

$$j'(\varepsilon_0 + \varepsilon) = -\frac{1}{2}\mu \int_{\partial\omega_{\varepsilon_0+\varepsilon}} |\nabla u_{\varepsilon_0+\varepsilon}|^2 \frac{1}{\varepsilon_0 + \varepsilon} x(\varepsilon) \cdot \nu(x(\varepsilon)) \, ds_{\varepsilon_0+\varepsilon}, \tag{4.22}$$

where $ds_{\varepsilon_0+\varepsilon} = (\varepsilon_0 + \varepsilon) ds$. According to (3.22) we have

$$|\nabla u_{\varepsilon_0+\varepsilon}|^2 = \sum_{\alpha,\beta=1}^{2} \epsilon_\beta^0 \left(\nabla_y W_\beta|_{y=\frac{x}{\varepsilon_0+\varepsilon}}\right) \cdot \left(\nabla_y W_\alpha|_{y=\frac{x}{\varepsilon_0+\varepsilon}}\right) \epsilon_\alpha^0 + O(\varepsilon_0 + \varepsilon). \tag{4.23}$$

We change the integration domain from $\partial\omega_{\varepsilon_0+\varepsilon}$ to $\partial\omega$. Using (4.18) we find

$$j'(\varepsilon_0 + \varepsilon) = -\frac{1}{2}(\varepsilon_0 + \varepsilon)\mu \left(\sum_{\alpha,\beta=1}^{2} \int_{\partial\omega} (\nabla_y W_\beta \cdot \nabla_y W_\alpha) y \cdot \nu \, ds\right) \epsilon_\alpha^0 \epsilon_\beta^0 + O\left((\varepsilon_0 + \varepsilon)^2\right). \tag{4.24}$$

Let us note that $j'(0^+) = 0$.

Direct differentiation of (4.24) gives

$$j''(\varepsilon_0) = -\frac{1}{2}\mu \sum_{\alpha,\beta=1}^{2} \left(\int_{\partial\omega} (\nabla_y W_\beta \cdot \nabla_y W_\alpha) y \cdot \nu \, ds\right) \epsilon_\alpha^0 \epsilon_\beta^0 + o(\varepsilon_0). \tag{4.25}$$

hence one arrives at

$$j''(0^+) = \sum_{\alpha,\beta=1}^{2} \epsilon_\beta^0 G_{\beta\alpha} \epsilon_\alpha^0 , \qquad (4.26)$$

with

$$G_{\beta\alpha} = -\frac{1}{2}\mu \int_{\partial\omega} (\nabla_y W_\beta \cdot \nabla_y W_\alpha)\, y \cdot \nu\, ds . \qquad (4.27)$$

To rearrange the matrix $(G_{\beta\alpha})$ to the form (4.6) one should make use of (3.23) and of the formula:

$$\delta_{\beta\alpha}|\omega| = \int_\omega (\nabla_y y_\beta) \cdot (\nabla_y y_\alpha)\, dy = \int_{\partial\omega} y_\beta \frac{\partial y_\alpha}{\partial \nu}\, ds = \int_{\partial\omega} y_\beta \nu_\alpha\, ds . \qquad (4.28)$$

Hence

$$\int_{\partial\omega} y \cdot \nu\, ds = 2|\omega| , \qquad (4.29)$$

which confirms the equation (4.6).

Definition 4.1.
Let the boundary $\partial\omega$ be piecewise smooth. The domain ω will be called star-shaped if one can place the origin O of the coordinate system (y_1, y_2) such that $O \in \omega$ and

$$y(s) \cdot \nu(s) > 0 \qquad \text{a.e.} \quad \text{on } \partial\omega , \qquad (4.30)$$

where s is a natural parameter of $\partial\omega$.

The convex domains are star-shaped. The star-shaped domains are not in general convex.

Let us consider the functional

$$E(\Omega_\varepsilon) = \mu \int_{\Omega_\varepsilon} |\nabla u_\varepsilon|^2\, dx \qquad (4.31)$$

or $E(\Omega_\varepsilon) = -2J(\Omega_\varepsilon)$. Passing to the limit $\varepsilon_0 \to 0^+$ in the expansion (4.3) gives

$$E(\Omega_\varepsilon) = E(\Omega) - \varepsilon^2 \sum_{\alpha,\beta=1}^{2} \epsilon_\alpha^0 G_{\alpha\beta} \epsilon_\beta^0 + o(\varepsilon^2) , \qquad (4.32)$$

where

$$E(\Omega) = \mu \int_\Omega |\nabla v|^2\, dx . \qquad (4.33)$$

We establish the following property of star-shaped openings.

Theorem 4.2.
Assume that the domain ω is star-shaped. Then, nucleation of a small opening ω_ε increases the energy:

$$E(\Omega_\varepsilon) > E(\Omega) . \qquad (4.34)$$

Proof.
We have

$$E(\Omega_\varepsilon) - E(\Omega) = \frac{1}{2}\varepsilon^2 \mu \int_{\partial\omega} |\nabla_y W|^2 y \cdot \nu\, ds + o(\varepsilon^2) , \qquad (4.35)$$

where $W = \epsilon_1^0 W_1 + \epsilon_2^0 W_2$. Taking into account the boundary condition (3.32) one finds

$$\frac{\partial W_\beta}{\partial y_\alpha} = \frac{\partial W_\beta}{\partial \nu} \nu_\alpha + \frac{\partial W_\beta}{\partial s} \tau_\alpha = \frac{\partial W_\beta}{\partial s} \tau_\alpha , \qquad (4.36)$$

where (τ_α) are components of the unit vector tangent to $\partial \omega$. Thus we have

$$|\nabla_y W|^2 = \left(\frac{\partial W}{\partial s}\right)^2 \qquad (4.37)$$

and

$$E(\Omega_\varepsilon) - E(\Omega) = \frac{1}{2} \varepsilon^2 \mu \int_{\partial \omega} \left(\frac{\partial W}{\partial s}\right)^2 y \cdot \nu \, ds + o(\varepsilon^2) . \qquad (4.38)$$

The constant functions do not satisfy the condition (3.26) in the problem $(\widetilde{P}_\omega)$. Therefore, $W \neq $ const and $\partial W/\partial s$ is not identically zero along $\partial \omega$. The condition (4.30) implies the inequality (4.34). ∎

By (4.36) and (4.37) the formula (4.27) can be rewritten in the form

$$G_{\alpha\beta} = -\frac{1}{2} \mu \int_{\partial \omega} \frac{\partial W_\alpha}{\partial s} \frac{\partial W_\beta}{\partial s} y \cdot \nu \, ds . \qquad (4.39)$$

Let us introduce the *force* fields

$$\sigma^\alpha(u) = \mu \frac{\partial u}{\partial y_\alpha} \qquad (4.40)$$

and denote $\sigma_\nu = \sigma^1 \nu_1 + \sigma^2 \nu_2$, $\sigma_\tau = \sigma^1 \tau_1 + \sigma^2 \tau_2$. The boundary condition (3.25) implies $\sigma_\nu(W_\beta) = 0$. Thus the formula (4.39) can be expressed in terms of the *force* fields as follows

$$G_{\alpha\beta} = -\frac{1}{2\mu} \int_{\partial \omega} \sigma_\tau(W_\alpha) \sigma_\tau(W_\beta) \, y \cdot \nu \, ds \qquad (4.41)$$

and the energy increment is given by

$$E(\Omega_\varepsilon) - E(\Omega) = \frac{1}{2\mu} \varepsilon^2 \int_{\partial \omega} [\sigma_\tau(W)]^2 \, y \cdot \nu \, ds + o(\varepsilon^2) . \qquad (4.42)$$

The result (4.32) has its counterpart in the literature. By applying the method of compound asymptotics Maz'ya and Nazarov (1987) found the expansion of the form

$$E(\Omega_\varepsilon) = E(\Omega) - \varepsilon^2 \sum_{\alpha,\beta=1}^{2} \epsilon_\alpha^0 m_{\alpha\beta} \epsilon_\beta^0 + o(\varepsilon^{2+0}) , \qquad (4.43)$$

where $m_{\alpha\beta}$ are the components of the mass matrix, see Eq. (3.16). The integral formulae for $(m_{\alpha\beta})$ are reported in Appendix A. The derivation of the formula (4.43) can be found in Appendix B.

Let us consider the relations between the matrices G and m. By Theorem 4.2 the star-shaped property of ω implies positive definiteness of the matrix $(-G)$. No obvious reason is seen why this matrix should always be positive definite. On the other hand, the matrix $(-m)$ is always positive definite, see Appendix A. Thus result (4.43) of Maz'ya and Nazarov (1987) means that drilling an opening always results in the increase of the energy, i.e. irrespective of its shape. On the other hand the Theorem 4.2 states that drilling a star-shaped opening increases the energy, and not more.

A natural question arises, whether the matrices G and m coincide for some shapes of ω. The partial answer is formulated below.

Theorem 4.3.
In the case when ω is an ellipse the matrices G and m coincide.

The proof is divided into small steps.

Step 1. Formulae for W_1, W_2. Let us consider problems ($\widetilde{P}_\omega^\beta$;3.24-3.26). We assume that the function of a complex variable $\zeta = \rho \cdot e^{i\vartheta}$

$$\wp(\zeta) = \ell \left(\zeta + \frac{c}{\zeta} + \frac{c_2}{\zeta^2} + \frac{c_3}{\zeta^3} + \dots \right) \tag{4.44}$$

depending on real parameters $\ell, c, c_i, i = 2, 3, \dots$, transforms the exterior of the unit circle in the complex plane onto the domain $\mathbb{R}^2 \setminus \overline{\omega}$.

The assumption (4.44) requires some explanations. With the choice of $\ell \cdot \exp(i\lambda)$ instead of ℓ it follows that the latter transformation works for the domain ω rotated by the angle λ. On the other hand, we can rotate the coordinate system (y_1, y_2) before the transformation since we deal with the only one domain ω. Therefore, without any loss of the generality, we can assume that $\lambda = 0$. The choice of $c \in \mathbb{R}$ is a standard assumption while selecting in a more general way c_i with complex values does not really change the formulae obtained in the sequel.

For $c_i = 0$, $i = 2, 3, \dots$, and $c \in (0, 1)$, ω is an ellipse with the major and minor axes coinciding with the coordinate system in the plane. In particular, we have

$$\ell = \frac{1}{2}(a+b), \quad c = \frac{a-b}{a+b}, \quad a > b,$$

where a, b are the half-lengths of the major and minor axes of the ellipse ω. For $c = 0$, $a = b$ and the ellipse ω becomes a circle. On the other hand, the ellipse tends to a cut (crack) with the limit passage $c \to 1$.

Schiffer and Szegö (1949) have shown that for $\ell \in \mathbb{R}$ the solutions W_1, W_2 of problem $(\widetilde{P}_\omega)$, for $\beta = 1, \beta = 2$ take the following form

$$W_1 = \ell \Re \left[\zeta + \frac{1}{\zeta} \right], \quad W_2 = -\ell \Re \left[i \left(\zeta - \frac{1}{\zeta} \right) \right], \quad \zeta = \wp^{-1}(z) \tag{4.45}$$

Step 2. $G_{\alpha\beta}$ **for an arbitrary opening** ω. The formulae for $G_{\alpha\beta}$ are given by (4.27), in which

$$y \cdot \nu = \Re(z \cdot \exp(-i\alpha)), \tag{4.46}$$

where $z = \wp(\zeta)$ and α is an angle between ν and the axis y_1. We make use of the following formula (see Muskhelishvili N.I. (1975), formula (49.3))

$$\exp(-i\alpha) = \frac{1}{\rho} \cdot \frac{1}{|\wp'(\zeta)|} \cdot \overline{\zeta} \cdot \overline{\wp'(\zeta)}, \tag{4.47}$$

where $\rho = |\zeta|$. Therefore,

$$y \cdot \nu = \frac{1}{\rho} \cdot \frac{1}{|\wp'(\zeta)|} \cdot \Re \left[\overline{\zeta} \cdot \overline{\wp'(\zeta)} \cdot \wp(\zeta) \right] \tag{4.48}$$

On the contour $\partial \omega$ we have $\rho = 1$, and ds is given by

$$ds = \sqrt{(dy_1)^2 + (dy_2)^2} = |dz| = |\wp'(\zeta) d\zeta| = |\wp'(\zeta)| \rho d\vartheta, \tag{4.49}$$

because $|d\zeta| = \rho d\vartheta$, and we evaluate

$$\boldsymbol{y} \cdot \boldsymbol{\nu} ds = \Re \left[\bar{\zeta} \cdot \overline{\wp'(\zeta)} \cdot \wp(\zeta) \right]_{\rho=1} d\vartheta . \qquad (4.50)$$

The next step is to evaluate the expressions $\|\nabla_y W_1\|^2$, $\nabla_y W_1 \cdot \nabla_y W_2$, $\|\nabla_y W_2\|^2$ on the contour $\rho = 1$.

For the function $f(z) = \ell \left(\zeta + \frac{1}{\zeta} \right) = W_1(y_1, y_2) + iV(y_1, y_2)$, $\zeta = \wp^{-1}(z)$, compare with (4.45), taking into account that $f(z)$ is a holomorphic function, it follows that

$$f'(z) = \frac{\partial W_1}{\partial y_1} - i \frac{\partial W_1}{\partial y_2}$$

hence

$$\|\nabla_y W_1\|^2 = |f'(z)|^2 = |f'(\zeta)|^2 \frac{1}{|\wp'(\zeta)|^2} , \qquad (4.51)$$

where $f'(\zeta) = \ell \left(1 - \frac{1}{\zeta^2} \right)$. Thus

$$\|\nabla_y W_1\|_{\rho=1}^2 = \left\{ \frac{4\ell^2 \sin^2 \vartheta}{|\wp'(\zeta)|^2} \right\}_{|\rho=1} \qquad (4.52)$$

In a similar way we evaluate

$$\{\nabla_y W_1 \cdot \nabla_y W_2\}_{\rho=1} = \left\{ -\frac{2\ell^2 \sin 2\vartheta}{|\wp'(\zeta)|^2} \right\}_{|\rho=1} \qquad (4.53)$$

and

$$\|\nabla_y W_2\|_{\rho=1}^2 = \left\{ \frac{4\ell^2 \cos^2 \vartheta}{|\wp'(\zeta)|^2} \right\}_{|\rho=1} \qquad (4.54)$$

Using (4.51)-(4.54) in (4.46) we find

$$G_{11} = -2\mu \ell^2 \int_0^{2\pi} \sin^2 \vartheta \, h(\vartheta) d\vartheta , \qquad (4.55)$$

$$G_{12} = 2\mu \ell^2 \int_0^{2\pi} \sin \vartheta \cos \vartheta \, h(\vartheta) d\vartheta , \qquad (4.56)$$

$$G_{22} = -2\mu \ell^2 \int_0^{2\pi} \cos^2 \vartheta \, h(\vartheta) d\vartheta , \qquad (4.57)$$

where

$$h(\vartheta) = \Re \left[\exp(-i\vartheta) \frac{\wp(\zeta)}{\wp'(\zeta)} \right]_{\rho=1}$$

Step 3. $G_{\alpha\beta}$ for an ellipse. For an ellipse

$$h(\vartheta) = \Re \left[\exp(-i\vartheta) \frac{\exp(i\vartheta) + c \exp(-i\vartheta)}{1 - c \exp(-2i\vartheta)} \right] = \frac{\beta}{\beta^2 \cos^2 \vartheta + \sin^2 \vartheta} ,$$

where we denote $\beta = \frac{1-c}{1+c}$. Therefore, simple calculations show that

$$G_{11} = -2\pi \mu \ell^2 (1-c), \quad G_{12} = 0, \quad G_{22} = -2\pi \mu \ell^2 (1+c) . \qquad (4.58)$$

The same formulae can be obtained for the mass matrix for an ellipse (Schiffer, Szegö 1949) which confirms the equality: $m_{\alpha\beta} = G_{\alpha\beta}$ for any ω in the form of an ellipse.

5. Final remarks

The method of compound asymptotics makes it possible to generalize the definition of the topological derivative to the general case of nucleation of non-circular openings in the Neumann problem. Similarly, one can introduce the topological derivative in the linear elasticity. The case of circular and ball openings was considered in Sokołowski and Żochowski (1999c). An extension to the case of openings of arbitrary shape in the plane elastic bodies is proposed in Lewiński and Sokołowski (1999). An alternative expression for the change of energy due to appearing of a void in an elastic body is reported in Movchan and Movchan (1995, Sec. 5.1.3, page 256). Both the expressions for the energy change involve fourth rank tensors, still denoted by G and m, respectively. In the case of a circular opening, $G = m$ and this tensor assumes an isotropic representation, see Lewiński and Sokołowski (1999, Eq. (6.39)). If an opening is elliptic, then the equality $G = m$ still holds, but the tensor ceases to be isotropic, see Lewiński and Sokołowski (2000).

The notion of the topological derivative can be generalized to the case of nucleation of an inclusion made from a material different than the virgin body. Also in such a case the compound asymptotic expansions can be helpful, see Movchan and Movchan (1995). Different techniques for modelling of inclusions in elasticity were used by Mura (1982).

APPENDIX A. The mass matrix

The components $m_{\alpha\beta}$ of the mass matrix in the representation (3.16) of the solutions of problems (P_ω) are given by the following integral formulae

i)
$$m_{\alpha\beta} = -\mu \int_{\partial\omega} \nu_\alpha W_\beta \, ds \tag{A.1}$$

ii)
$$m_{\alpha\beta} = -\mu \left[\delta_{\alpha\beta} |\omega| + \int_{\partial\omega} \nu_\alpha w_\beta \, ds \right] \tag{A.2}$$

iii)
$$m_{\alpha\beta} = -\mu \delta_{\alpha\beta} |\omega| - \mathfrak{M}_{\alpha\beta}, \tag{A.3}$$

where
$$\mathfrak{M}_{\alpha\beta} = \mu \int_{\mathbb{R}^2 \setminus \overline{\omega}} \nabla_y w_\alpha \cdot \nabla_y w_\beta \, dy. \tag{A.4}$$

The formulae (A.3), (A.4) imply the symmetry property: $m_{\alpha\beta} = m_{\beta\alpha}$. Symmetrization of (A.1) gives

$$m_{\alpha\beta} = -\frac{1}{2}\mu \int_{\partial\omega} (\nu_\alpha W_\beta + \nu_\beta W_\alpha) \, ds.$$

Symmetrization of (A.2) can be done in a similar manner.
Proof of the formula (A.1).
Let the domain ω be encompassed by a circle B_R with the boundary $\Gamma_R = \{y | \, \|y\| = R\}$; $R \gg \text{diam}(\omega)$. The functions w_β and W_β are harmonic, hence the following variational

equation hold

$$\mu \int_{B_R \setminus \overline{\omega}} \nabla_y w_\alpha \cdot \nabla_y \widetilde{\varphi}\, dy = \mu \int_{\Gamma_R} \frac{\partial w_\alpha}{\partial n} \widetilde{\varphi}\, ds - \mu \int_{\partial \omega} \widetilde{\varphi} \frac{\partial w_\alpha}{\partial \nu}\, ds, \qquad (A.5)$$

$$\mu \int_{B_R \setminus \overline{\omega}} \nabla_y W_\beta \cdot \nabla_y \widehat{\varphi}\, dy = \mu \int_{\Gamma_R} \widehat{\varphi} \frac{\partial W_\beta}{\partial n}\, ds - \mu \int_{\partial \omega} \widehat{\varphi} \frac{\partial W_\beta}{\partial \nu}\, ds, \qquad (A.6)$$

for $\widetilde{\varphi}$ and $\widehat{\varphi}$ sufficiently regular. Let us put $\widetilde{\varphi} = W_\beta$ and $\widehat{\varphi} = w_\alpha$. We find

$$J_{\beta\alpha} = \mu \int_{\partial \omega} \left(W_\beta \frac{\partial w_\alpha}{\partial \nu} - w_\alpha \frac{\partial W_\beta}{\partial \nu} \right) ds, \qquad (A.7)$$

where

$$J_{\beta\alpha} = \mu \int_{\Gamma_R} \left(W_\beta \frac{\partial w_\alpha}{\partial n} - w_\alpha \frac{\partial W_\beta}{\partial n} \right) ds. \qquad (A.8)$$

We shall prove that $J_{\beta\alpha} = m_{\alpha\beta}$. To this end let us recall the formula (3.16). It follows that

$$W_\beta = y_\beta + \sum_{\alpha=1}^{2} m_{\beta\alpha} \frac{\partial S}{\partial y_\alpha} + O\left(\frac{1}{\|y\|^2}\right), \qquad (A.9)$$

$$\frac{\partial w_\alpha}{\partial n} = \sum_{\sigma=1}^{2} m_{\alpha\sigma} \frac{\partial^2 S}{\partial y_\sigma \partial n} + O\left(\frac{1}{\|y\|^3}\right), \qquad (A.10)$$

$$w_\alpha = \sum_{\sigma=1}^{2} m_{\alpha\sigma} \frac{\partial S}{\partial y_\sigma} + O\left(\frac{1}{\|y\|^2}\right), \qquad (A.11)$$

$$\frac{\partial W_\beta}{\partial n} = n_\beta + \sum_{\sigma=1}^{2} m_{\beta\sigma} \frac{\partial^2 S}{\partial y_\sigma \partial n} + O\left(\frac{1}{\|y\|^3}\right). \qquad (A.12)$$

We substitute (A.9)–(A.12) into (A.8) and neglect the terms of order $O\left(\frac{1}{\|y\|^n}\right)$, $n \geq 2$. This leads to

$$J_{\beta\alpha} = \sum_{\sigma=1}^{2} m_{\alpha\sigma} \int_{\Gamma_R} \left[y_\beta \frac{\partial^2 \overline{S}}{\partial y_\sigma \partial n} - \frac{\partial \overline{S}}{\partial y_\sigma} \frac{\partial y_\beta}{\partial n} \right] ds + O\left(\frac{1}{R}\right), \qquad (A.13)$$

where $\overline{S} = \mu S$.

The weak formulation of the equation (3.13) reads

$$\int_{B_R} \nabla_y \overline{S} \cdot \nabla_y \varphi\, dy = \int_{\Gamma_R} \frac{\partial \overline{S}}{\partial n} \varphi\, ds + \langle \varphi, \delta \rangle_{B_R} \qquad (A.14)$$

for any sufficiently regular test function φ. The function $\dfrac{\partial \overline{S}}{\partial y_\sigma}$ satisfies the equation

$$\Delta_y \left(\frac{\partial \overline{S}}{\partial y_\sigma} \right) + \frac{\partial \delta(y)}{\partial y_\sigma} = 0. \qquad (A.15)$$

Let us introduce the mollifier function defined by

$$\phi_h(y) = h^{-2} \phi\left(\frac{y}{h}\right),$$

where h is a small positive number and

$$\phi(x) = \begin{cases} C \exp\left[-1/(1-\|x\|^2)\right] & \text{if } \|x\| < 1 \\ 0 & \text{if } \|x\| \geq 1 \end{cases}$$

and the constant C is chosen such that

$$\int_{\mathbb{R}^2} \phi(x)\,dx = 1\,.$$

Then

$$\int_{\mathbb{R}^2} \phi_h(y)\,dy = 1\,.$$

Let $\overline{S}_h(y)$ be a solution to the equation (A.14) with $\delta(y)$ replaced by $\phi_h(y)$. Since $\overline{S}(x-y)$ is the Green function, one can represent $\overline{S}_h(y)$ as follows

$$\overline{S}_h(y) = \int_{\mathbb{R}^2} \overline{S}(y-x)\phi_h(x)\,dx\,.$$

By the symmetry: $\overline{S}_h(y) = \overline{S}_h(-y)$, $\phi_h(x) = \phi_h(-x)$, whence we can rearrange the formula above to the form

$$\overline{S}_h(y) = \int_{\mathbb{R}^2} \phi_h(y-z)\overline{S}(z)\,dz\,.$$

According to Theorem 1.6.1 and Lemma 2.1.3 reported in Ziemer (1989) the functions $\overline{S}_h(y)$ tend to $\overline{S}(y)$ pointwise, as well as in Sobolev spaces, when $h \searrow 0$. We shall make use of this property in the sequel.

Note that the function $\dfrac{\partial \overline{S}_h}{\partial y_\sigma}$ satisfies the variational equation

$$\int_{B_R} \nabla_y\left(\frac{\partial \overline{S}_h}{\partial y_\sigma}\right) \cdot \nabla\varphi\,dy = \int_{\Gamma_R} \frac{\partial^2 \overline{S}_h}{\partial n \partial y_\sigma} \varphi\,ds + \int_{B_R} \varphi \frac{\partial \phi_h(y)}{\partial y_\sigma}\,dy \qquad (A.16)$$

for a sufficiently regular test function φ.

The function y_β is harmonic, hence

$$\int_{B_R} \nabla_y y_\beta \cdot \nabla\widetilde{\varphi}\,dy = \int_{\Gamma_R} \frac{\partial y_\beta}{\partial n} \widetilde{\varphi}\,ds \qquad (A.17)$$

for a sufficiently regular test function $\widetilde{\varphi}$.

Now, we set $\varphi = y_\beta$ in (A.16) and $\widetilde{\varphi} = \dfrac{\partial \overline{S}_h}{\partial y_\sigma}$ in (A.17) to make their left-hand sides identical. By equating the right-hand sides we find

$$K^h_{\sigma\beta} = L^h_{\sigma\beta}\,,$$

where

$$K^h_{\sigma\beta} = -\int_{B_R} y_\beta \frac{\partial \phi_h(y)}{\partial y_\sigma}\,dy \qquad (A.18a)$$

and
$$L^h_{\sigma\beta} = \int_{\Gamma_R} \left(\frac{\partial^2 \overline{S}_h}{\partial n \partial y_\sigma} y_\beta - \frac{\partial y_\beta}{\partial n} \frac{\partial \overline{S}_h}{\partial y_\sigma} \right) ds \ . \qquad (A.18b)$$

Note that $K^h_{\alpha\beta} = \delta_{\alpha\beta}$ for h sufficiently small and
$$\lim_{h \searrow 0} K^h_{\sigma\beta} = \int_{\Gamma_R} \left(\frac{\partial^2 \overline{S}}{\partial n \partial y_\sigma} y_\beta - \frac{\partial y_\beta}{\partial n} \frac{\partial \overline{S}}{\partial y_\sigma} \right) ds$$

since $\dfrac{\partial \overline{S}_h}{\partial y_\sigma}$ tend to $\dfrac{\partial \overline{S}}{\partial y_\sigma}$.

We conclude that
$$\int_{\Gamma_R} \left(\frac{\partial^2 \overline{S}}{\partial n \partial y_\sigma} y_\beta - \frac{\partial y_\beta}{\partial n} \frac{\partial \overline{S}}{\partial y_\sigma} \right) ds = \delta_{\sigma\beta} \ .$$

By (A.13) we find
$$J_{\beta\alpha} = \lim_{R \searrow 0} J_{\beta\alpha} = m_{\alpha\sigma} \delta_{\sigma\beta} = m_{\alpha\beta} \ , \qquad (A.19)$$

since $J_{\beta\alpha}$ does not depend on R, cf. (A.8). Thus we have
$$m_{\alpha\beta} = \mu \int_{\partial\omega} \left(W_\beta \frac{\partial w_\alpha}{\partial \nu} - w_\alpha \frac{\partial W_\beta}{\partial \nu} \right) ds \ . \qquad (A.20)$$

By using (3.11) and (3.25), (A.1) follows.

To prove (A.2) we substitute $W_\alpha = y_\alpha + w_\alpha$ and make use of the indentity (4.28).

Let us prove (A.3). First, we show that
$$\mathfrak{M}_{\alpha\beta} = \mu \int_{\partial\omega} w_\beta \nu_\alpha \, ds \ . \qquad (A.21)$$

Let us set $\widetilde{\varphi} = w_\beta$ in (A.5). Let us note that
$$\lim_{R \to \infty} \int_{\Gamma_R} w_\beta \frac{\partial w_\alpha}{\partial n} \, ds \to 0 \ , \qquad (A.22)$$

since $w_\beta = O\left(\dfrac{1}{\|y\|}\right)$, $\dfrac{\partial w_\alpha}{\partial n} = O\left(\dfrac{1}{\|y\|^2}\right)$. Thus, we arrive at
$$\mu \int_{\mathbb{R}^2 \setminus \overline{\omega}} \nabla_y w_\alpha \cdot \nabla_y w_\beta \, dy = -\mu \int_{\partial\omega} w_\beta \frac{\partial w_\alpha}{\partial \nu} \, ds \ . \qquad (A.23)$$

Substitution of $\partial w_\alpha / \partial \nu = -\nu_\alpha$ on $\partial \omega$ gives (A.21), hence (A.3) follows.

Let us note that the quadratic form
$$-\sum_{\alpha,\beta=1}^{2} q^\alpha m_{\alpha\beta} q^\beta = \mu |\omega| |q|^2 + \mu \int_{\mathbb{R}^2 \setminus \overline{\omega}} |\nabla_y w|^2 \, dy \ , \qquad (A.24)$$

where $w = q^1 w_1 + q^2 w_2$, is positive definite. We conclude that the matrix $(m_{\alpha\beta})$ is negative definite.

The present Appendix is based on the papers by: Schiffer and Szegö (1949), Mazja et al. (1991) and Movchan and Movchan (1995).

APPENDIX B. The Maz'ya-Nazarov formula for the energy of a body with a small opening

The aim of the present Appendix is to recall the derivation of the formula (4.43). The derivation given below repeats the arguments of the the paper by Maz'ya and Nazarov (1987).

The energy function $E(\Omega_\varepsilon)$ equals

$$E(\Omega_\varepsilon) = \int_{\partial\Omega} p(x)u_\varepsilon(x)\,ds \,. \tag{B.1}$$

The function u_ε is represented by (3.8) with $u^{(1)}$, $u^{(2)}$ defined by Eqs. (3.9)-(3.19). Let us note that the functions

$$T_\alpha(x) = z_\alpha(x) - \frac{1}{2\pi\mu}\frac{x_\alpha}{\|x\|^2} \tag{B.2}$$

satisfy the homogeneous boundary condition along $\partial\Omega$

$$\mu\frac{\partial T_\alpha}{\partial n} = 0 \quad \text{on} \quad \partial\Omega \tag{B.3}$$

Moreover, by (B.2) and (3.13)-(3.15) we have

$$\mu\Delta T_\alpha = \mu\Delta z_\alpha + \mu\Delta\left(\frac{\partial S}{\partial x_\alpha}\right)$$

Taking into account (3.18) and (3.13) one finds

$$\mu\Delta T_\alpha = -\frac{\partial\delta(x)}{\partial x_\alpha}$$

Thus T_α is a solution of the problem (B.5), (B.3).

Substitution of $u^{(1)}$ given by (3.9), with w_α given by (3.16), into Eq. (3.8) results in the following asymptotic expansion,

$$u_\varepsilon(x) = v(x) - \frac{\varepsilon^2}{2\pi\mu}\sum_{\alpha,\beta=1}^{2} m_{\alpha\beta}\epsilon_\beta^0\frac{x_\alpha}{\|x\|^2} + \varepsilon^2 u^{(2)}(x) + \varepsilon O\left(\frac{\varepsilon^2}{\|x\|^2}\right) + \varepsilon^2 O\left(\frac{\varepsilon}{\|x\|}\right) \,.$$

Taking into account (3.17) and (B.2) we reduce (B.6) to the form

$$u_\varepsilon(x) = v(x) + \varepsilon^2\sum_{\alpha,\beta=1}^{2} m_{\alpha\beta}\epsilon_\beta^0 T_\alpha(x) + \varepsilon O\left(\frac{\varepsilon^2}{\|x\|^2}\right) + \varepsilon^2 O\left(\frac{\varepsilon}{\|x\|}\right) \,.$$

Substitution of (B.7) into (B.1) gives

$$E(\Omega_\varepsilon) = E(\Omega) + \varepsilon^2\sum_{\alpha,\beta=1}^{2} m_{\beta\alpha}\epsilon_\beta^0\int_{\partial\Omega} p(x)T_\alpha(x)\,ds + o(\varepsilon^{2+0}) \,.$$

Let us write down the weak formulation of the problem (B.3), (B.5) and of the problem (P_0) from Sec. 3

$$\mu\int_\Omega \nabla T_\alpha \cdot \nabla\varphi\,dx = \left\langle\frac{\partial\delta}{\partial x_\alpha},\varphi\right\rangle_\Omega \,,$$

$$\mu\int_\Omega \nabla v \cdot \nabla\widetilde\varphi\,dx = \int_{\partial\Omega} p\widetilde\varphi(x)\,ds \,,$$

where $\langle\cdot,\cdot\rangle_\Omega$ denotes the duality pairing between $\mathcal{D}'(\Omega)$ and $\mathcal{D}(\Omega)$.

Substitution of $\varphi = v$ and $\widetilde{\varphi} = T_\alpha$ leads to the identity

$$\int_{\partial\Omega} pT_\alpha \, ds = \left\langle \frac{\partial \delta}{\partial x_\alpha}, v \right\rangle_\Omega$$

which gives

$$\int_{\partial\Omega} pT_\alpha \, ds = -\left\langle \delta, \frac{\partial v}{\partial x_\alpha} \right\rangle_\Omega = -\frac{\partial v}{\partial x_\alpha}(O) = -\epsilon_\alpha^0$$

The expansion (4.43) is now confirmed.

References

Lewiński T., Sokołowski J. (1999) Topological derivative for nucleation of non-circular voids. Rapport de Recherche. No 3798, INRIA – Lorraine. 37 pp.

Lewiński T., Sokołowski J. (2000) Energy change due to appearing of voids in an elastic medium. pp. 201-206. In: W. Szcześniak (Ed.) Theoretical Foundations of Civil Engineering – 8. Polish – Ukrainian Transactions. Oficyna Wydawnicza Politechniki Warszawskiej. Warsaw

Lewiński T., Sokołowski J., Żochowski A. (1999) Justification of the bubble method for the compliance minimization problems of plates and spherical shells. In: Proceedings of the WCSMO-3. Buffalo University, May 1999. CD ROM

Mazja W.G., Nasarow S.A., Plamenewski B.A. (1991) Asymptotische Theorie elliptischer Randwertaufgaben in singulär gestörten Gebieten I. Störungen isolierter Randsingularitäten. Akademie Verlag. Berlin

Maz'ya V.G., Nazarov S.A. (1987) The asymptotic behavior of energy integrals under small perturbations of the boundary near corner points and conical points. (in Russian) Trudy Moskovsk. Matem. Obshch. vol. **50**, See English translation. Trans. Moscow. Math. Soc. (1988), pp. 77-127

Movchan A.B., Movchan N.V. (1995) Mathematical Modelling of Solids with Nonregular Boundaries. CRC Press. Boca Raton

Mura T. (1982) Micromechanics of Defects in Solids. Martinus Nijhoff Publ. The Hague, Boston, London

Muskhelishvili N.I. (1975) Some basic problems of the mathematical theory of elasticity. Noordhoff, Leyden

Nazarov S. A. (1999) Asymptotic conditions at a point, selfadjoint extensions of operators, and the method of matched asymptotic expansions. Amer. Math. Soc. Transl. (2) vol **193**, pp. 77-125

Schiffer M., Szegö G. (1949) Virtual mass and polarization. Trans. Amer. Math. Soc. **67**, pp. 130-205

Schiffer M. M., (1956) Boundary value problems for elliptic partial differential equations. in: E. F. Beckenbach (Ed.). Modern Mathematics for the Engineer. Chapter 6. McGraw-Hill. New York, Toronto, London

Schumacher A. (1995) Topologieoptimierung von Bauteilstrukturen unter Verwendung von Lochpositionierungskriterien. Doctor's Thesis. University of Siegen. pp. 1-150

Sokołowski J., Zolesio J.-P. (1992) Introduction to Shape Optimization. Shape Sensitivity Analysis. Springer. Berlin

Sokołowski J., Żochowski A. (1999a) On Topological Derivative in Shape Optimization. SIAM Journal on Control and Optimization, Volume 37, Number 4, pp. 1251-1272

Sokołowski J., Żochowski A. (1999b) Topological derivative for optimal control problems. Control and Cybernetics 3(28), pp. 611-625.

Sokołowski J., Żochowski A. (1999c) Topological derivatives of shape functionals for elasticity systems. Les prépublications de l'Institut Élie Cartan. No 35. Université Henri Poincaré, Nancy 1. 16 pp.

Ziemer W.P. (1983) Weakly Differentiable Functions. Sobolev Spaces and Functions of Bounded Variation. Springer. New York

WARSAW UNIVERSITY OF TECHNOLOGY, CIVIL ENGINEERING FACULTY, INSTITUTE OF STRUCTURAL MECHANICS, AL. ARMII LUDOWEJ 16. 00-637 WARSZAWA, POLAND
E-mail address: T.Lewinski@il.pw.edu.pl

INSTITUT ELIE CARTAN, LABORATOIRE DE MATHÉMATIQUES, UNIVERSITÉ HENRI POINCARÉ NANCY I, B.P. 239, 54506 VANDOEUVRE LÈS NANCY CEDEX, FRANCE AND SYSTEMS RESEARCH INSTITUTE OF THE POLISH ACADEMY OF SCIENCES, UL. NEWELSKA 6, 01-447 WARSZAWA, POLAND
E-mail address: Jan.Sokolowski@loria.fr

Remarks on global uniqueness theorems for partial differential equations

Walter Littman

ABSTRACT. We discuss the transition from local to global versions of Holmgren type uniqueness theorems.

1. Introduction

A century ago E. Holmgren [H] proved his famous uniqueness theorem for systems of linear partial differential equations. It was remarkable in that it did not require a specific type of equation—elliptic, hyperbolic, etc., and although the coefficients were required to be real analytic, the solutions were not. According to this theorem, if the Cauchy data vanish on an analytic non characteristic hypersurface contained in an open set in R^n in which the solution exists, then in *some* neighborhood of that surface, the solution must vanish, provided the right hand side of the equation is zero. (In the second order case, for example, zero Cauchy data means that u and its normal derivative vanish.) This local version, and its proof, using the existence part of the Cauchy Kowalewski theorem applied to the adjoint equation, is essentially what is found in *most* standard texts, for example [G], [Tr]. On the other hand, some treatments, starting with the local version, obtain a global version by assuming the existence of an "analytic field" of noncharacteristic hypersurfaces. (See for example [CH] and [J2]).

Recent research in boundary control theory has exhibited a great search for uniqueness theorems. See for example [BLR], [LTY]. The type of theorem frequently needed is where Cauchy data zero is prescribed on the lateral boundary of a cylinder $Q = \Omega \times (0, T)$ in space time, and P is a hyperbolic operator defined in a somewhat larger set. If $Pu = 0$ in Q and u has Cauchy data zero on the lateral boundary of the cylinder Q, must u vanish in Q? One easy answer comes in [L] where P has real analytic coefficients and is strictly hyperbolic. We may assume that u vanishes outside of Ω for $0 < t < T$. Using analytic propagation of singularities (see [H2] sec. 7), it follows that u is analytic in a slightly fatter cylinder, and since it vanishes in an open set, then $u \equiv 0$ in that cylinder, provided T is large enough and there are no zero speeds of propagation. This is very useful in boundary control theory (see [L] for example). However, as a uniqueness theorem, it is not optimal for many domains, especially if they are thin in one direction, for example. Our aim is also to obtain more versatile tools than "analytic fields" in attacking problems of this kind

---

2000 *Mathematics Subject Classification*. Primary 35Bxx, 35Lxx, 51–XX.

and their generalizations. For other applications, see for example [I]. For related global uniqueness theorems, see also [I, p. 67], [KK], [LRS], [LTZ].

At a particular point where P is defined, we call a direction in space time ($\mathbf{R}^{n+1}$) *time like* with respect to P, if all normals to it point in noncharacteristic directions. See Fig. 5. Time like vectors fill out two solid convex open cones, one forward cone, containing the direction of the positive t axis, and a backward cone. We shall deal only with the former. A curve in space time will be called time like if its tangent vector is time like at all points. Time like curves can always be parametrized by the time variable. A hypersurface is called *spacelike* if its normals are timelike.

2. The "sweeping integral method" of F. John

An earlier paper of Fritz John [J1] gave a completely different proof of Holmgren's theorem. "He proved that integrals of the form

$$\int_{S_t} gu\,dS$$

are analytic functions of the parameter t, if g and Pu are analytic, S_t are analytic manifolds, depending analytically on t, ∂S_t is fixed or $u = 0$ at ∂S_t, and the normal bundle of S_t is noncharacteristic with respect to P." That was Hörmander's point of departure in [H2], where he significantly improved on John's approach. The important observation for our problem is that the S_t need not be hypersurfaces, but could be manifolds of *any* dimension from 1 to $n-1$. It turns out that if (in a linear hyperbolic equation) we are interested in uniqueness where Cauchy data is prescribed on a space like hypersurface, then taking S_t to be $n-1$ dimensional is appropriate - but in that case one need not rely on Holmgren's theorem. But if Cauchy data zero is prescribed on a time like hypersurface, such as the lateral boundary of Q (see previous section), *then one should let $\{S_t\}$ be a family of timelike curves*. This is one lesson from [J1].

3. Strings to spaghetti: $C_\lambda \to S_\lambda$. See Fig. 1

Definition: Let P be a linear partial differential operator defined in an open set V. Let Σ be a C^1 hypersurface dividing V into two disjoint open connected sets V_+ and V_-. We say that *there is unique continuation across* Σ if every solution (in a certain smoothness class, to be specified) to $Pu = 0$ defined in V, vanishing in V_+ or V_-, vanishes in some open subset of V containing Σ.

The classical example of this property is where P is a linear partial differential operator with analytic coefficients and Σ is non-characteristic. In that case, u need only be a distribution solution.

A more recent example is the following theorem due to Tataru [T1], [T2] and Hörmander [H3].

Theorem: *Suppose*

(2.1) $$P = \sum_{i,j=0}^{n} a^{ij} \partial_i \partial_j + b^i \partial_i + c$$

with real coefficients is hyperbolic with respect $x_0 \equiv t$. Assume that a^{ij} are independent of t and of class C^1 while the other coefficients may depend analytically on t and are in L^∞. Then there is unique continuation for H^1 solutions across noncharactieristic surfaces.

Our object is to describe a method that will transform local results into global results. Thus we obtain the following theorem:

Theorem 1: *Let*

$$Pu = u_{tt} - \sum_{i,j=1}^n a^{ij} u_{x_i x_j} + b^i u_{x_i} + cu$$

be defined in $Q^* = \Omega^* \times (0,T)$, where Ω^* is an open set in R^n containing the closure $\bar\Omega$ of the bounded C^1 domain Ω. Assume that the elliptic principal part is uniformly elliptic in Ω^* with C^1 coefficients independent of t, while the other coefficients are analytic in t with values in L^∞. Let

$$ds^2 = \Sigma a_{ij} dx_i dx_j$$

be the Riemannian metric associated with P i.e. the matrix $(a_{ij}) = (a^{ij})^{-1}$. Let Γ be an open nonempty subset of $\partial\Omega$. Assume $Pu = 0$ in Q, that $u \in H^1(Q)$, where $Q = \Omega \times (0,T)$, that u vanishes on $\partial\Omega \times (0,T)$ and that u can be continued by zero as a solution to $Pu = 0$ across $\Gamma \times (0,T)$. Then $u \equiv 0$ in Q provided

$$T > T_0 = 2\max\{\operatorname{dist}(x, \Gamma)\}$$

where the maximum is taken over $\bar\Omega$, and where "dist" is induced by the Riemannian metric ds.

Lemma: *Let $Pu = 0$ in an open set W in space time where P is given as above. Suppose there is a one parameter family of time like C^2 curves C_λ varying continuously with $\lambda \in [0, \lambda_1]$, such that the end points of each curve C_λ stay away from the support of u, and the initial curve C_0 does not intersect the support of u. Then the support of u does not intersect with an open set containing the set of points swept out by the C_λ, provided $u \in H^1$.*

Proof: The crux of the proof is to replace each curve with a tubular surface $S_\lambda = S_\lambda^\delta$ enclosing C_λ where each S_λ consists of points a distance δ from C_λ and has two boundary components, $\partial_1 S_\lambda$ and $\partial_2 S_\lambda$. For small δ the S_λ will be C^2 hypersurfaces immersed in $\mathbf{R}^{n+1}$, such that $\partial_1 S_\lambda$ and $\partial_2 S_\lambda$, as well as S_0, will be bounded away from the support of u. Furthermore, let us choose δ so small that the normals to all S_λ stay noncharacteristic. The result follows by contradiction. If it were false there must be a first value of λ where S_λ intersects the support of u. This would contradict the theorem of Tataru and Hörmander quoted earlier, according to which there is unique continuation across each S_λ. Note that to reach particular points in W it may be necessary to vary δ, so that the S_λ^δ stay in W.

Remark: The conclusion of the theorem is still valid if the C_λ's are piecewise C^1 and *uniformly time like*, (definition obvious), as a family. For then each C_λ can be replaced by a time like C^2 approximation $\hat C_\lambda$ resulting in a new family $\hat S_\lambda$ having the desired properties. Note that since small translations of these families will also have the required properties, the same points in space time can be reached.

4. Round trips. Fig. 2

Returning to the proof of Theorem 1 stated in section 3, we know that $Pu = 0$ in the cylinder Q, $u = 0$ on its lateral boundary, and that u has Cauchy data zero on $\Gamma \times (0, T)$ where Γ is an open subset of $\partial \Omega$. The last statement can be reformulated as follows: Let ω be an open subset of Ω^*, such that $\omega \supset \Gamma$, but $\omega \cap (\partial\Omega/\Gamma) = \emptyset$. Let $\Omega^+ = \Omega \cup \omega$. Thus Ω^+ is Ω (slightly) extended across Γ. Further, let $Q^+ = \Omega^+ \times (0,T)$. Thus we wish to show that for T large enough, if $Pu = 0$ in Q^+, $u = 0$ on the lateral boundary of Q, $u = 0$ in $Q^+ \backslash \bar{Q}$, then $u \equiv 0$ in Q. We will show this by showing that u must vanish in a neighborhood of the set $\Omega^+ \times \{\frac{T}{2}\}$. For then in the upper half of Q, u is the solution to a mixed problem with zero Cauchy data on the bottom, and zero Dirichlet data on the lateral boundary, implying that $u \equiv 0$ there. Similarly $u \equiv 0$ in the bottom half of Q, and we are through.

To show that u vanishes in $\Omega^+ \times \{\frac{T}{2}\}$, we introduce the notion of a *round trip*. Let $\Omega^e = \Omega^+ \backslash \bar{\Omega}$, and similarly $Q^e = Q^+ \backslash \bar{Q}$.

Recall that the cone of all characteristic directions, sometimes called the normal cone, $\sum_1^n a^{ij} \xi_i \xi_j = \tau^2$ has a dual, $\sum_1^n a_{ij} dx_i dx_j = dt^2$, called the ray cone or Monge cone. Here $(a_{ij}) = (a^{ij})^{-1}$. The directions of the ray cone determine the speed of propagation of a "light" signal in a given spatial direction at a point. For example in the direction of dx_1, $a_{11} dx_1^2 = dt^2$, thus the speed of propagation is $1/\sqrt{a_{11}}$, using Euclidean length. A timelike direction or displacement points into the interior of the Monge cone. We will only consider the forward Monge cone, the one pointing in the positive t direction.

We will regard a space time curve parametrized by t as the motion of a particle in space, and its space projection as the space path it travels on. The (forward) direction of this path (if non zero) determines a certain velocity, or speed of propagation of light (not necessarily of the particle) in that direction. We will call that speed the "speed limit" along that path. A time like curve thus represents a trip the particle made staying below the speed limit.

By a round trip we mean a time like curve in space time representing a particle starting at a point p in Ω^e, travelling to a point q in Ω, stopping at q for a while, then back tracking along the same spatial path at the same speeds in the reverse direction, and stopping at p. Notice that the "speed limit", in the reverse direction is the same as that in the forward direction. The whole trip may be parametrized by t, including the rest stop at q. The whole trip is represented by a time like curve in space time which may be smooth, except for the beginning and end of the "rest stop". However, the curve will be *uniformly time like*, since "staying put" is time like. By a *one way trip*, we simply mean the part of the trip from p to q.

Now from the round trip just constructed we can manufacture a one parameter family of time like curves á la the lemma of the previous section. Thus suppose the trip from p to q takes τ time units. Let us consider a second particle starting at p, keeping company with the first particle for $\lambda \tau$ units of time, then stopping and joining the first particle on its return trip. For every $\lambda \in [0,1]$ we get a time like curve satisfying the conditions of the lemma in section 3, taking account the remark following the lemma.

Theorem 2: *Let P be as in Theorem 1. Assume $Pu = 0$ in Q^+ and that u vanishes in Q^e, and $u = 0$ on the lateral boundary of Q, $u \in H^1(Q^+)$. Suppose every point*

q in Ω can be reached from some point p in Ω^e by a one way trip taking not more than time τ. Then if $T > 2\tau$, we conclude that $u = 0$ in Q.

Proof: Let us first construct the round trip path from p to q and back. Let $4\epsilon = T - 2\tau$. Then begin the trip from p to q at time ϵ, ending at time $\tau + \epsilon$. Wait 2ϵ time units then proceed back, arriving at p at time $T - \epsilon$. We consider the complete trip as occuring in the time interval $[0, T]$, with ϵ rest stops at start and finish. Note that at time $\frac{T}{2}$ (and neighboring times) the particle will be at the point q. Now construct the one parameter family of round trips - or time like curves C_λ and their smoothed out versions $\hat{C}_\lambda$ - as constructed above. From the lemma it follows that $u = 0$ in a neighborhood of the point $(q, \frac{T}{2})$ in Q for all q in Ω. It follows from the argument given in the first paragraph of this section that $u \equiv 0$ in Q.

Proof of Theorem 1. In Theorem 2 distances were measured in the Euclidean metric. If we replace that by the Riemannian metric ds then the speed of propagation of "light signals" becomes 1. Thus the "speed limit" of the proof of Theorem 2 is also 1. Therefore, given a point q in Ω, we can find a point p in Ω^e such that one can travel from p to q at unit speed (staying in Ω^+) in duration not exceeding

$$\max_{q \in \bar{\Omega}} \text{dist}(q, \Gamma) + \epsilon$$

which means we can accomplish the same travelling *below unit* speed in ϵ more time. The conclusion of Theorem 1 follows just as in the proof of Theorem 2, by taking the "rest stops" sufficiently small.

The method of proof of Theorems 1 and 2 can be extended to other situations as long as one has the local version of Holmgren's theorem. Even in the analytic case many results of this kind are new. Note that in all these theorems the null Dirichlet condition can be replaced by other homogeneous boundary conditions that result in a well posed mixed problem. Let us also note that the methods and results of this paper apply, with minor modifications to manifolds.

5. Time-dependent coefficients

Tataru's results for equation 2.1 hold even if the coefficients of the principal part depend analytically on $t = x_0$, with values in C^1 of the spatial variables. With that in mind, minor changes in the proof of Theorem 2 give a slightly less precise version for time dependent principal parts.

Given a point p in Ω^e and a point q in Ω^+ we construct a "round trip" as follows: Rest at p during time $[0, \epsilon]$, travel from p to q for τ time units, below the speed limit, rest at q for 2ϵ time units, return to p along the same space path without reaching the speed limit (τ time units), rest at p for ϵ time units. Notice that the speed limit may vary with time at a particular location and may be different on the return trip from what it was on the direct trip. With this modification of the notion of round trip we have

Theorem 2': *Let P be given by 2.1 with Lorentz signature in the principal part, where the coefficients depend analytically on time $t = x_0$ with values in C^1 of the space variables; the other coefficients as before. Assume $Pu = 0$ in Q^+, $u = 0$ in Q^e, and $u = 0$ on the lateral boundary of Q, while $u \in H^1(Q^+)$. Suppose for each* point q in Ω a round trip from some point p in Ω^e is possible, of total duration $< T$, provided ϵ is chosen sufficiently small. Then $u = 0$ in Q^+.

Note: A slightly more precise version of the above theorem possibly yielding a smaller value for T is obtained by replacing the hyperplane $t = \frac{T}{2}$ by a space-like hypersurface $t = \varphi(x)$ in Q^+ such that the roundtrip reaches q at time $\varphi(q) - \epsilon$ and leaves q at time $\varphi(q) + \epsilon$, returning to p at time $T - \epsilon$. During the direct trip $0 \leq t \leq \varphi(x(t))$, while $t \geq \varphi(x(t))$ during the return trip. Here the direct and return trips need not be of the same duration.

6. A direct application of the lemma

Let us illustrate an application of the lemma of Section 3 with a simple example. Let Ω be the unit disk $x^2 + y^2 < 1$ and $Q = \Omega \times (0, 1 + \epsilon)$. Then if $u_{tt} - \Delta u = 0$ in Q and u has Cauchy data zero on the lateral boundary of Q, then $u \equiv 0$ in Q by Theorem 1. In addition to assuming that $u = 0$ on the lateral boundary of Q, is it necessary that zero Neumann data be prescribed on the *whole* lateral boundary of Q? It turns out that one need only do this on an arbitrarily small part σ of Q if it is chosen with care. Cut Q by a time-like hyperplane π whose normal makes an angle of slightly more than 45° with the t axis, but is such that it intersects only the lateral boundary of Q and not the remaining boundary. See Fig. 3. Such a plane can be swept out by one parameter family of timelike straight lines satisfying the requirements of the lemma, provided u has Cauchy data zero on an arbitrarily small open set σ containing the ellipse of intersection of π with the lateral boundary of Q. This forces u to vanish near π, and hence, using earlier arguments in all of Q. See Fig. 4.

7. Higher order operators

The results can be extended to higher order operators, provided the coefficients are real analytic in all variables. Let us take P to be strictly hyperbolic, for simplicity and assume no zero speeds of propagation. At each point in space time there will be the "normal cone" of characteristic directions pointing "up", and having several nested sheets, non intersecting except at the origin. The dual cone, i.e. the ray cone, will also have several sheets. The open "core" K of the inner sheet will contain all "time like" directions.

Theorems similar to Theorems 2 and 2' can now be given for distribution solutions. One need however assume that $K \cap \{t = 1\}$ is star-shaped with respect to the origin, to allow the smoothing out of the curves $C_\lambda \to \hat{C}_\lambda$ without losing their time like character.

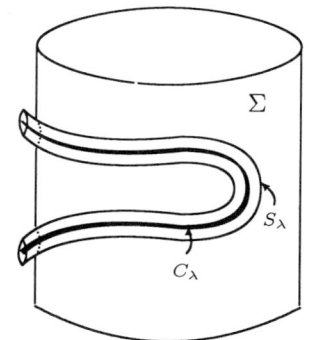

FIG. 1: STRINGS TO SPAGHETTI

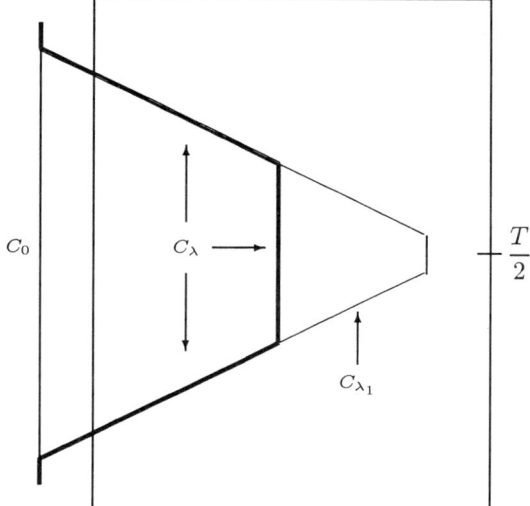

FIG. 2: ROUND TRIPS

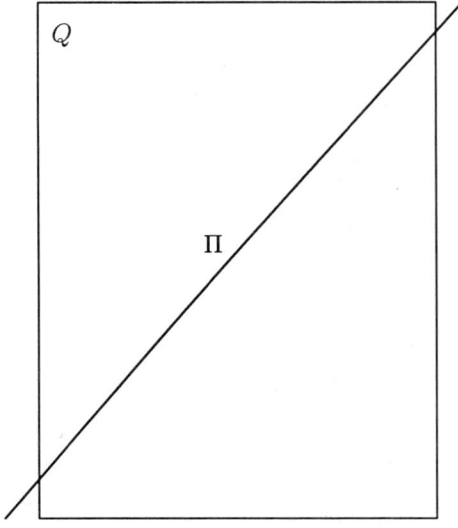

Fig. 3: (Schematic)

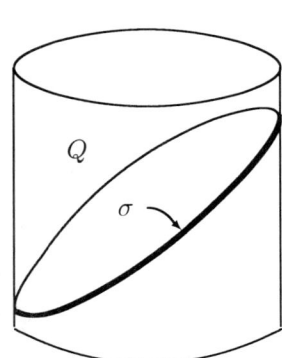

Fig. 4

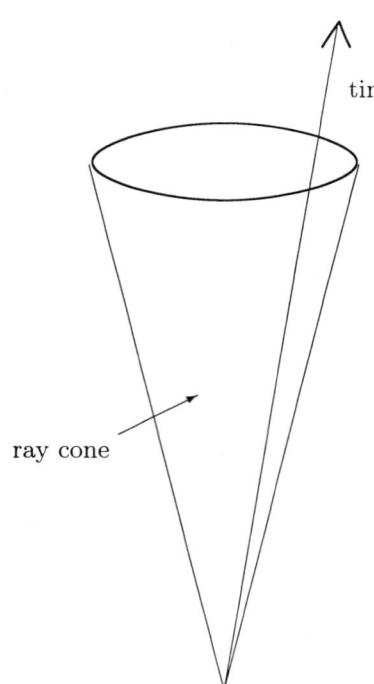

Fig. 5

References

[BLR] C. Bardos, G. Labeau and J. Rauch, Sharp sufficient condtions for observation, Control and Stabilization of the wave equation, SIAM J. Control Optim. 30 (5) (1992), 1024-1065.

[CH] R. Courant and D. Hilbert, Methods of Mathematical Physics, vol 2. Interscience Publishers, New York, 1962.

[G] P.R. Garabedian, Partial Differential Equations, John Wiley & Sons, Inc., New York, London, Sydney 1964.

[H] E. Holmgren, Über Systeme von Lineare Partielle Differentialgleichungen, Öfversigt af. Kongl. Vetenskaps-Akad. Förh. 58, (1901) 91-105.

[H1] L. Hörmander, *Linear Partial Differential Operators*, Springer-Verlag, New York, 1963.

[H2] L. Hörmander, Uniqueness theorems and wave front sets for solution to linear partial differential equations with analytic coefficients, CPAM 24, (1971), 671-704.

[H3] L. Hörmander, On the uniqueness of the Cauchy problem under partial analyticity assumptions, *Geometric Optics and Related Topics*, Ed. F. Columbin and N. Lerner, Birkhäuser, Boston 1997.

[I] V. Isakov, *Inverse Problems for Partial Differential Equations*, Springer Verlag, 1998.

[J1] F. John, On linear partial differential equations with analytic coefficients. Unique continuation of data, CPAM 2, (1949), 209-253.

[J2] F. John, *Partial Differential Equations*, Fourth edition, Springer Verlag, New York (1982).

[KK] M. Kazami and M. V. Klibanov, Stability estimates for ill-posed Cauchy problems involving hyperbolic euqations and inequalities, *Applicable Analysis* 50 (1993), 93–102.

[L] W. Littman, Near optimal time boundary controllability for a class of hyperbolic equations, LNCIS 178, Springer Verlag 1987, 272-284.

[LRS] M. M. Lavrentev, V. G. Romanov, S. P. Shishatskii, *Ill-posed Problems of Mathematical Physics and Analysis*, Volume 64, Translation of Math. Monographs, pp. 133–140.

[LTY] I. Lasiecka, R. Triggiani and P.F. Yao, Inverse/observability estimates for second-order hyperbolic equations with variable coefficients, J. Math. Anal. Appl. 235 (1999), 13-57.

[LTZ] I. Lasiecka, R. Triggiani, X. Zhang, Non-conservative wave equations with unobserved Neumann B.C.: Global uniqueness and observability in one shot.

[R] L. Robbiano, Théoréme d' unicité adapté an controlé de solutions des problémes hyperboliques, Comm. P.D.E. 16 (1991) 789-800.

[T1] D. Tataru, Unique continuation for solutions to PDE's; between Hörmander's theorem and Holmgren's theorem, Comm. P.D.E. 20 (1995) 855-884.

[T2] D. Tataru, Unique continuation for partially differential operators with partial analytic coefficients, J. Math. Pure Appl. 78 (1999), 505-521.

[Tr] F. Treves, *Basic Linear Partial Differential Equations*, Academic Press, New York, 1975.

SCHOOL OF MATHEMATICS, UNIVERSITY OF MINNESOTA, MINNEAPOLIS, MINNESOTA, 55455, USA

Evolution of a graph by Levi form

Zbigniew Slodkowski and Giuseppe Tomassini

1. Introduction

In [ST3] (cfr. also [ST2]) the notion of *evolution* of a compact subset of $\mathbb{C}^2$ by Levi form was introduced. One motivation behind this notion is to explore the possibility to obtain by an "analytic procedure" some information on various envelopes of K which are of interest in Complex Analysis. In this vein it would be pleasing to find an algorithm for this problem. Papers [ST3], [ST4], [ST5] quoted in the bibliography collect several results of a preliminary exploration in this direction but many natural questions remain unanswered.

Another motivation is to study what kind of singularities can be obtained by evolution, starting with regular sets.

In this paper we are dealing with the evolution of a compact subset K of $\mathbb{C}^2$ "with fixed part" K^*. Also in this case the evolving family $\{E_t(K, K^*)\}_{t \geq 0}$ is given as the zero sets of a weak solution of a parabolic problem for the Levi equation (cfr. Sec. 2).

An interesting particular case is when K is the graph Γ of a continuous function $u : \overline{D} \to \mathbb{R}$ where D is a bounded domain in $\mathbb{C} \times \mathbb{R}$ and $K^* = b\Gamma$ the boundary of Γ. In this situation we prove the following results: if D is strictly pseudoconvex (i.e. $D \times i\mathbb{R}$ is a strictly pseudoconvex domain in $\mathbb{C}^2$)

(a) $E_t(\Gamma, b\Gamma)$ is a graph for all $t \geq 0$ (Theorem 3.1);

(b) asymptotically $E_t(\Gamma, b\Gamma)$ approaches, in the C^0-topology, the Levi flat hypersurface with boundary $b\Gamma$ (Theorem 4.1).

One may compare (b) with the similar result stated by Huisken for minimal surfaces ([H]).

2. Generalities

Let K be a compact of $\mathbb{C}^2$, $K = g^{-1}(0)$ where g is continuous on whole $\mathbb{C}^2$ and constant $(=C)$ for $|z| \gg 0$, $z = (z_1, z_2) \in \mathbb{C}^2$. Following [ST3], (cfr. also [ST2]) the evolution of K by Levi form is by definition the family of the level sets

---

1991 *Mathematics Subject Classification.* Primary 32W50, 32T05; Secondary 53C44.
Partially supported by an NSF Grant.
Support by the MURST project "Geometric Properties of Real and Complex Manifolds.

$K_t = \{z \in \mathbb{C}^2 : u(z,t) = 0\}$ where $u \in C^0(\mathbb{C}^2 \times [0, +\infty))$ is the weak solution of the parabolic problem

(2.1) $$\begin{cases} u_t = \mathcal{L}(u) & \text{in } \mathbb{C}^2 \times (0, +\infty) \\ u = g & \text{on } \mathbb{C}^2 \times \{0\} \\ u = C & \text{for } |z| + t \gg 0 \end{cases}$$

for the Levi operator

$$\mathcal{L}(u) = (\delta_{\alpha\beta} - |\partial u|^{-2} u_{\bar{\alpha}} u_\beta) u_{\alpha\bar{\beta}}$$

($u_\alpha = \partial u/\partial z_\alpha$, $u_{\bar{\alpha}} = \partial u/\partial \bar{z}_\alpha$, $u_{\alpha\bar{\beta}} = \partial^2 u/\partial z_\alpha \partial \bar{z}_\beta$, $\alpha, \beta = 1, 2$, $|\partial u|^2 = |u_1|^2 + |u_2|^2$).

REMARK 2.1. In (2.1) the condition $u = C$ for $|z| + t \gg 0$ can be replaced by the following one: u is bounded. This follows from the comparison principle and the existence theorem in [ST3].

The evolution $\{K_t\}_{t \geq 0}$ is also governed by the parabolic problem

(2.2) $$\begin{cases} v_t = \mathcal{L}(v) & \text{in } \Omega \times (0, +\infty) \\ v = g & \text{on } \Omega \times \{0\} \\ v = C & \text{on } b\Omega \times [0, +\infty) \end{cases}$$

where Ω is any bounded strictly pseudoconvex domain containing K and such that $g = C$ in a neighbourhood of $b\Omega$.

We sketch the proof of this fact when Ω is the ball $B = B(0, R)$ of $\mathbb{C}^2$.

First of all we observe that for the problem (2.2) we have unicity and that the level sets $K_t = \{z \in \mathbb{C}^2 : v(z,t) = 0\}$ of a weak solution v do not depend on g, provided $g^{-1}(0) = K$.

Let $v \in C^0(\overline{B} \times [0, +\infty))$ be a weak solution of (2.2) ([ST3]); we may assume that $C = \max_{\overline{B}} g$ and $C + |z|^2 - R^2 \leq g$ on $\overline{B}$. Let

$$\psi(z, t) = C + \min(0, C + |z|^2 - R^2 + t).$$

ψ is a continuous weak subsolution of $\psi_t = \mathcal{L}(\psi)$ which is constant ($=C$) for $t \gg 0$. In view of the comparison principle we have $\psi \leq v \leq C$ and consequently $v = C$ for $t \gg 0$. Now we extend v on all of $\mathbb{C}^2 \times [0, +\infty)$ by setting $v = C$ outside of $\overline{B} \times [0, +\infty)$. v is a weak solution of (2.1) (v is uniform limit of weak solutions $v_\epsilon = \lambda_\epsilon \circ v$, $\epsilon \searrow 0$, where $\lambda_\epsilon = \lambda_\epsilon(s)$ is smooth, $\lambda_\epsilon(s) = s$ for $0 \leq s \leq C - 2\epsilon/3$ and $\lambda_\epsilon = C$ for $s \geq C - \epsilon/3$, [ST3]). Since the family $\{K_t\}_{t \geq 0}$ is actually independent on g ([ST3]) this proves our claim.

PROPOSITION 2.2. *Let (K^*, K) be a pair of compact sets in $\mathbb{C}^2$ such that $K^* \subset K$, $K \neq \emptyset$. Suppose that there is a bounded strictly pseudoconvex domain $\Omega \subset \mathbb{C}^2$ such that $K \setminus K^* \subseteq \Omega$, $K^* \subseteq b\Omega$. Let $g : \overline{\Omega} \to \mathbb{R}$ be a continuous function such that $g^{-1}(0) = K$. Then there is a unique continuous weak solution $u : \overline{\Omega} \times [0, +\infty)$ of the following parabolic problem*

(2.3) $$\begin{cases} u_t = \mathcal{L}(u) & \text{in } \Omega \times (0, +\infty) \\ u = g & \text{on } \overline{\Omega} \times \{0\} \\ u(z, t) = g(z) & \text{for } z \in b\Omega \times (0, +\infty). \end{cases}$$

Moreover u is bounded and uniformly continuous in $\overline{\Omega} \times [0, +\infty)$.
The set

$$X = \{(z, t) \in \overline{\Omega} \times [0, +\infty) : u(z, t) = 0\}$$

satisfies
$$X \cap (\overline{\Omega} \times \{0\}) = K \times \{0\}, \quad X \cap (b\Omega \times [0, +\infty)) = K^* \times [0, +\infty)$$
and is independent on the choice of g and Ω.

In the context of the proposition we define
$$E_t(K, K^*) = \{z \in \mathbb{C}^2 : (z, t) \in X \text{ i.e. } u(z, t) = 0\}.$$
The family $\{E_t(K, K^*)\}_{t \geq 0}$ is said to be the *evolution* of K mod K^* (*by Levi form*).

REMARK 2.3. We note that the strict pseudoconvexity condition for Ω can be much relaxed. In particular the following condition (C) suffices: for all $\zeta \in b\Omega$ there is a ball $B(\zeta, r)$ and a strictly plurisubharmonic function ψ on $B(\zeta, r)$ such that $\psi(\zeta) = 0$ and $\psi(z) < 0$ for $z \in B(\zeta, r) \cap (\overline{\Omega} \setminus \{\zeta\})$.

For the proof of independence of the evolution of the pair (K, K^*) on Ω we have to discuss a local maximum property of the level sets u=const.
For an open set V in $\mathbb{C}^2 \times (0, +\infty)$ denote
$$\mathcal{P}_\mathcal{L} = \mathcal{P}_\mathcal{L}(V) = \{\psi \in C^2(V) : \psi_t \leq \mathcal{L}(\psi)\}.$$
Let $Z \subseteq \mathbb{C}^2 \times (0, +\infty)$ be locally closed. We say that Z has *local maximum property (relative to $\mathcal{P}_\mathcal{L}$)* if for every open set V in $\mathbb{C}^2 \times (0, +\infty)$ such that $\overline{V} \cap Z$ is closed and $\overline{V}$ is compact, and for every $\psi \in \mathcal{P}_\mathcal{L}(V')$, V' neighbourhood of $\overline{V}$ it holds:
$$\max_{\overline{V} \cap Z} \psi = \max_{b V \cap Z} \psi.$$

LEMMA 2.4. *Let $W \subseteq \mathbb{C}^2 \times (0, +\infty)$ be open and $u : W \to \mathbb{R}$ a weak solution of the parabolic equation $u_t = \mathcal{L}(u)$. Let $Z = \{u = 0\}$. Then*

(a) Z has local maximum property.

(b) For every $c > 0$, $Z^c = \{(z, t) \in Z : t \leq c\}$, has local maximum property.

Proof. Suppose the claim (a) is false, i.e.
$$\max_{\overline{V} \cap Z} \psi > \max_{bV \cap Z} \psi,$$
for some $\psi \in \mathcal{P}_\mathcal{L}$. Then there is $\epsilon > 0$ small enough so that the function $\psi^\epsilon = \psi - \epsilon t$ still satsfies
$$\max_{\overline{V} \cap Z} \psi^\epsilon > \max_{bV \cap Z} \psi^\epsilon,$$
and, in addition $\psi_t^\epsilon < \mathcal{L}(\psi^\epsilon)$ in $\overline{V}$.

Let (z°, t°) denote the point where ψ^ϵ takes maximum value, say M, relative to $\overline{V} \cap Z$. Clearly $(z^\circ, t^\circ) \in V \cap Z$, and
$$\{(z,t) \in V : \psi^\epsilon(z,t) > M\} \subset V \setminus Z \;=\; \{(z,t) \in V : u(z,t) \neq 0\}$$
$$= \{(z,t) \in V : u(z,t)^2 > 0\}.$$
We apply now [ST4]. If we set $\phi = \psi^\epsilon - M$ and $U = u^2$, then U is still a weak solution of the parabolic problem, $\phi(z^\circ, t^\circ) = U(z^\circ, t^\circ)$ and
$$\{(z,t) \in V : \phi(z,t) > 0\} \subset \{(z,t) \in V : U(z,t) > 0\}.$$
By Lemma 1.1 in [ST4] we obtain
$$\psi_t^\epsilon(z^\circ, t^\circ) = \phi_t(z^\circ, t^\circ) \geq \mathcal{L}(\phi)(z^\circ, t^\circ) = \mathcal{L}(\psi^\epsilon)(z^\circ, t^\circ)$$

which is a contradiction.

In order to prove (b) fix $c > 0$ and consider ϕ as in definition of local maximum property. Let $\rho(t) = 0$ for $t \leq c$ and $\rho(t) = -(c-t)^3$ for $t > c$ and, for $N > 0$, $(z,t) \in \overline{V}$, $\psi^N(z,t) = \psi(z,t) + N\rho(t)$. Clearly $\psi^N \in \mathcal{P}_\mathcal{L}$ and so, by part (a),

$$\max_{\overline{V} \cap X} \psi^N = \max_{bV \cap X} \psi^N.$$

Observe, however, that

$$\lim_{N \to +\infty} \psi^N(z,t) = -\infty$$

if $t > c$ and

$$\psi(z,t)^N = \psi(z,t)$$

for $(z,t) \in X^c$, thus

$$\lim_{N \to +\infty} \max_{\overline{V} \cap X} \psi^N = \max_{\overline{V} \cap X^c} \psi.$$

The same being true for $bV \cap X^c$, we conclude that

$$\max_{\overline{V} \cap X^c} \psi^N = \max_{bV \cap X^c} \psi^N. \qquad \square$$

Proof of Proposition 2.1 (Sketch). The existence, uniqueness and uniform continuity of u for domains $\Omega \times [0, +\infty)$ where Ω satisfies condition (C) of the last remark, is proven in much the same way as corresponding results in [ST5] (using the Perron method, comparison principle and a uniform continuous version of Walsh Lemma for unbounded domains [ST5]). The independence of the zero set $\{u = 0\}$ on the choice of g satisfying $g^{-1}(0) = K$ is essentially the argument of Evans and Spruck [ES 1] (cfr. also [ST5]).

It remains to show independence of $X = u^{-1}(0)$ on the choice of domain Ω satisfying the condition of the Lemma 2.2.

Suppose Ω_1, Ω_2 are such domains. Let $\Omega_0 := \Omega_1 \cap \Omega_2$. Then Ω_0 satisfies condition (C) of the remark and also $K \setminus K^* \subseteq \Omega_0$, $K^* \subseteq b\Omega_0$. In view of the above sketchy remarks, for each of these sets we have unique (independent on respective u) "evolution hypersurface" i.e X_j, where $j = 0, 1, 2$, $X_j \subseteq \overline{\Omega}_j \times [0, +\infty)$ and

$$X_j \cap (\Omega_j \times \{0\}) = K \times \{0\}, \quad X_j \cap (b\Omega_j \times [0, +\infty)) = K^* \times [0, +\infty).$$

We will show that $X_1 = X_0$ and this will imply that $X_1 = X_2$, as required.

Let g, u be as in the lemma, for the domain Ω_1, so that $X_1 = u^{-1}(0)$. Let now $g_0 = g_{|\overline{\Omega}_0}$ and let $u_0 \in C^0\left(\overline{\Omega}_0 \times [0, +\infty)\right)$ be the corresponding solution of the parabolic problem so that $X_0 = u_0^{-1}(0)$.

Assertion 1. $X_1 \subseteq \overline{\Omega}_0 \times [0, +\infty)$.

Since Ω_0 is the intersection of two strictly pseudoconvex domains Ω_1, Ω_2, there is a neighbourhood N of $\overline{\Omega}_0$ and a continuous plurisubharmonic function $\phi : N \to \mathbb{R}$ such that $\overline{\Omega}_0 = \{\phi \leq 0\}$. Suppose $X_1 \not\subseteq \overline{\Omega}_0$, then there exists $c > 0$ such that $X^c \not\subseteq \overline{\Omega}_0$ but $X_1^c \subset N$. (Note that $c \mapsto X^c$ is an upper semicontinuous correspondence and $X_1^\circ = K \subset N$.)

Let $\tilde{\phi}(z,t) = \phi(z)$. Clearly $u \in \mathcal{P}_\mathcal{L}$. Denote

$$M = \max_{X^c} \tilde{\phi}, \quad F = \left\{(z,t) \in X^c : \tilde{\phi}(z,t) = M\right\}.$$

Then $M > 0$, F is compact, $F \cap (K^* \times \{0\}) = \emptyset$. Choose V, a neighbourhood of F such that $\overline{V}$ is compact, $\overline{V} \subset N \setminus \overline{\Omega}_0 \times (0, +\infty)$. Then

$$M = \max_{X_1^c \cap \overline{V}} \tilde{\phi} > \max_{X_1^c \cap bV} \tilde{\phi}$$

which contradicts the local maximum property (b) of Lemma 2.2.

Assertion 2. $X_1 \subseteq (\Omega_0 \cup K^*) \times [0, +\infty)$.

Suppose $(z^\circ, t^\circ) \in X_1 \cap (b\Omega_0 \setminus K^*) \times [0, +\infty)$. Then $z^* \in b\Omega_1$ or $t^\circ \in b\Omega_2$. In either case there is a C^2 strictly plurisubharmonic function $v = v(z)$ in a neighbourhood of z^* such that $v(z^*) = 0$, $v(z) < 0$ for $z \in B(z^*, r) \cap (\overline{\Omega}_0 \setminus \{z^*\})$. Since v is strictly plurisubharmonic, there is an $\epsilon > 0$, small enough so that the function $\psi^\epsilon(z, t) = v(z) - \epsilon(t - t^*)^2$ is of the class $\mathcal{P}_\mathcal{L}$ in $V = B \times (t^\circ - r, t^* + r)$. Observe now that $\psi^*(z^\circ, t^\circ) = 0$ while $\psi^\epsilon(z, t) < 0$ for $(z, t) \in X_1 \cap V \setminus \{(z^\circ, t^\circ)\}$. This contradicts again the local maximum property (a) of Lemma 2.2. Thus $X_1 \cap (b\Omega_0 \setminus K^*) \times [0, +\infty) = \emptyset$.

We can show now that $X_0 = X_1$.

Fix $c > 0$ and let
$$W^c = \Omega_0 \times (0, c), \quad \Sigma^c = (\overline{\Omega}_0 \times \{0\}) \cup (b\Omega_0 \times [0, c]).$$

Denote $U^c := u_{|\overline{W}^c}$. Then u_0, U^c are continuous weak solutions in $\overline{W}$ of the parabolic equation. By Assertions 1, 2
$$u^{-1}(0) \cap \Sigma^c = (U^c)^{-1}(0) \cap \Sigma^c.$$

Hence similarly as in [ES 1] there are continuous increasing functions $\chi_1, \chi_2 : \mathbb{R} \to \mathbb{R}$, with $\chi_j(0) = 0$, $j = 1, 2$, such that
$$\chi_1 \circ u_0 \leq U^c \leq \chi_2 \circ u_0$$
on Σ^c.

Since $\chi_j \circ u_0$, $j = 1, 2$ are weak solutions of the parabolic equation, the comparison principle implies that
$$\chi_1 \circ u_0 \leq U^c \leq \chi_2 \circ u_0$$
in W and so
$$X_0^c = (u_0)^{-1}(0) = (U^c)^{-1}(0) = X_1^c,$$
for every $c > 0$. Thus $X_0 = X_1$. □

A particularly important case will be considered in the sequel, when K is a compact hypersurface (in particular a graph) with boundary bK and $K^* = bK$.

THEOREM 2.5. *Let Ω be a bounded strictly pseudoconvex domain of $\mathbb{C}^2$, $K \subset \overline{\Omega}$, $K^* \subset b\Omega$ compact sets such that: $K^* \subset K$, $K \setminus K^* \subset \Omega$ and separates Ω. Let $\{E_t(K, K^*)\}_{t \geq 0}$ be the evolution of K mod K^*. Then for every t, $E_t(K, K^*) \setminus K^*$ separates Ω.*

Proof. Choose $g \in C^0(\overline{\Omega})$ such that $g^{-1}(0) = K$; $\Omega \setminus K = \{g > 0\} \cup \{g < 0\}$ and we choose ζ_1, ζ_2 such that $g(\zeta_1) > 0$, $g(\zeta_2) < 0$. Let u be the weak solution of (2.3). Then $E_t(K, K^*) = \{u(\cdot, t) = 0\}$ and $\Omega \setminus \{E_t(K, K^*)\}_{t \geq 0}$ is a union $\{u(\cdot, t) > 0\} \cup \{u(\cdot, t) < 0\}$ of nonempty subsets. □

PROPOSITION 2.6. *In the context of the theorem*
$$\limsup_{t \to +\infty} E_t(K, K^*) = K^\infty$$
where $K^\infty \setminus K^*$ is pseudoconcave i.e. has local maximum property with respect to the functions $|P|$, $P \in \mathbb{C}[z_1, z_2]$. Furthermore $K^\infty \setminus K^*$ separates Ω.

For the proof we need the following elementary

LEMMA 2.7. *Let $\{X_t\}_{t \in T}$, where T is a (direct) partially ordered set, be a family of relatively closed subsets of an open subset W in $\mathbb{C}^2 \times (0, +\infty)$. Assume all X_t have local maximum property relative to $\mathcal{P}_\mathcal{L}$. Then*

$$\limsup_{t \to +\infty} X_t = \bigcap_{t^\circ} \overline{\bigcup_{t \geq t^\circ} X_t}$$

has local maximum property relative to $\mathcal{P}_\mathcal{L}$ provided it is nonempty.

The proof of this fact is a straightforward consequence of the definition.

Proof of Proposition 2.4 Let us prove that $K^\infty \setminus K^*$ is pseudoconcave.

Let $W = \Omega \times (0, +\infty)$ and u be the solution of the parabolic problem (2.3). We know that u is uniformly continuous in $\overline{W}$.

Denote

$$X = \{(z, t) \in \Omega \times (0, +\infty) : u(z, t) = 0\}$$

and

$$X^h = \{(z, t) \in \Omega \times (0, +\infty) : u^h(z, t) = 0\}$$

where $u^h(z, t) = u(z, t + h)$, $h > 0$.

Since the parabolic equation $u_t = \mathcal{L}(u)$ is invariant with respect to time shift $t \mapsto t + h$, $h \geq 0$, we obtain that $\{X^h \cap W\}_{h > 0}$ is a family of sets with local maximum property relative to $\mathcal{P}_\mathcal{L}$ defined above. Let

$$X^\infty = \limsup_{h \to +\infty} X^h.$$

By Lemma 2.5, $X^\infty \cap W$ has local maximum property relative to $\mathcal{P}_\mathcal{L}$ provide $X^\infty \cap W \neq \emptyset$. On the other hand

$$X^h \cap (\mathbb{C}^2 \times \{t\}) = E_{t+h}(K, K^*) \times \{t\},$$

and

$$\limsup_{h \to +\infty} E_{t+h}(K, K^*) = K^\infty,$$

for each $t > 0$. It is easy to conclude that $X^\infty = K^\infty \times (0, +\infty)$ and so the set $(K^\infty \setminus K^*) \times (0, +\infty)$ has local maximum property relative to the class of subsolutions $\mathcal{P}_\mathcal{L}$.

Suppose now that $K^\infty \setminus K^*$ is not a local maximum set relative to polynomials. By [S] there is a point $z^0 \in K^\infty \setminus K^* \subset \Omega$ and a strictly plurisubharmonic function $\rho \in C^2(B(z^0, r))$, $r > 0$, such that $\rho(z^0) = 0$ and $\rho(z) < 0$ for $z \in K^\infty \cap (B(z^0, r) \setminus \{z^0\})$. Choose a small $\epsilon > 0$ such that the function $\psi(z, t) = \rho(z) - \epsilon(t - t^0)^2$ satisfies $\mathcal{L}(\psi) - \psi_t > 0$ in $B(z^0, r) \times (t^0 - r, t^0 + r)$, i.e. $\psi \in \mathcal{P}_\mathcal{L}$ in a neighbourhood of (z^0, t^0). Owing to the properties of ρ,

$$\psi_{|(K^\infty \setminus K^*) \times (0, +\infty)} = \psi_{|X^\infty \cap W}$$

has strict local maximum at (z^0, t^0): contradiction. □

3. Evolution of graphs

Now we suppose that K is the graph Γ of a continuous function $v : \overline{D} \to \mathbb{R}$ where D is a bounded domain of $\mathbb{C} \times \mathbb{R}$. Let $b\Gamma$ be the boundary of Γ. We have the following theorem:

THEOREM 3.1. *If $D \times i\mathbb{R}$ is strictly pseudoconvex then $E_t(\Gamma, b\Gamma)$ is a graph for every $t \geq 0$.*

We need the following general fact.

PROPOSITION 3.2. *Let $\Omega \subset \mathbb{C}^2$ be a bounded strictly pseudoconvex domain and K, K' disjoint compact subsets of Ω. Let $K \cap b\Omega = K^*$, $K' \cap b\Omega = K'^*$. Then*
$$E_t(K, K^*) \cap E_t(K', K'^*) = \emptyset$$
for every $t > 0$.

Proof. To prove this take a continuous function $g : \overline{\Omega} \to \mathbb{R}$ such that $g^{-1}(0) = K$, $g^{-1}(1) = K'$ and solve the problem
$$\begin{cases} u_t & = \mathcal{L}(u) \text{ in } \Omega \times (0, +\infty) \\ u & = g \text{ on } \overline{\Omega} \times \{0\} \\ u(z, t) & = g(z) \text{ for } z \in b\Omega \times (0, +\infty). \end{cases}$$

We have
$$E_t(K, K^*) = \{u(\cdot, t) = 0\}, \quad E_t(K', K'^*) = \{u(\cdot, t) = 1\}$$
and consequently $E_t(K, K^*), E_t(K', K'^*)$ are disjoint for every $t > 0$. □

REMARK 3.3. We do not know if the same is true if we have two different strictly pseudoconvex domains Ω, Ω' with $K \subseteq \Omega$, $K' \subseteq \Omega'$.

Proof of Theorem 3.1 In our situation $K = \Gamma$ and $D \times i\mathbb{R}$ is a strictly pseudoconvex domain in $\mathbb{C}^2$. We consider translations $T_h : \mathbb{C}^2 \to \mathbb{C}^2$ of the form $(z_1, z_2) \mapsto (z_1, z_2 + ih)$, $h \in \mathbb{R}$.
For fixed $h > 0$, consider a bounded strictly pseudoconvex domain Ω and a large enough number M such that
$$\overline{D} \times i\mathbb{R} \supset \overline{\Omega} \supset \overline{D} \times [-iM, iM]$$
$$\supset \Gamma \cup T_h(\Gamma).$$

Then we can consider the evolutions of Γ and $T_h(\Gamma)$ (mod $b\Gamma$) with such $\overline{\Omega}$ and they must be disjoint. (The evolution is independent on the specific choice of such Ω.) □

Given a continuous graph $\Gamma_\circ$ over D the "evolution of $\Gamma_\circ$ with fixed boundary" i.e. the evolution $\Gamma_\circ$ mod $b\Gamma_\circ$ reduces to a parabolic problem for the (graph) Levi operator $\mathcal{L}_\circ$.
Let us fix some notation: $z_1 = x_1 + ix_2$, $z_2 = x_3 + ix_4$, $z = (z_1, z_2)$, $z = x = (x_1, x_2, x_3, x_4)$, $x' = (x_1, x_2, x_3)$. If D is a bounded domain in $\mathbb{C} \times \mathbb{R} = \{x_4 = 0\} \subset \mathbb{C}^2$, $D \times i\mathbb{R}$ the domain $\{z \in \mathbb{C}^2 : (z_1, x_3) \in D\}$, the Levi operator for graphs over D is
$$\begin{aligned}\mathcal{L}_\circ(u) &= \frac{1}{4}(1 + |Du|^2)^{-1}\{(1 + u_3^2)(u_{11} + u_{22}) + (u_1^2 + u_2^2)u_{33} \\ &+ 2(u_2 - u_1 u_3)u_{13} - 2(u_1 + u_2 u_3)u_{23}\}\end{aligned}$$
where $u_j = \partial u / \partial x_j$, $u_{ij} = \partial^2 u / \partial x_i \partial x_j$, $i, j = 1, 2, 3$ ([ST1]) .
For smooth functions $u = u(x_1, x_2, x_3)$ we have $\mathcal{L}(x_4 - u) = -\mathcal{L}_\circ(u)$.

LEMMA 3.4. *Let u be continuous in a domain $D \subseteq \mathbb{C} \times \mathbb{R}$. Then $x_4 - u$ is a weak solution of $v_t = \mathcal{L}(v)$ in $D \times i\mathbb{R} \times \mathbb{R}$ if and only if u is a weak solution of $u_t = \mathcal{L}_\circ(u)$ in $D \times \mathbb{R}$.*

Proof. If $v = x_4 - u$ is a weak solution of $v_t = \mathcal{L}(v)$ in $D \times i\mathbb{R} \times \mathbb{R}$ then is immediate to check that u is a weak solution of $u_t = \mathcal{L}_\circ(u)$ in $D \times \mathbb{R}$.
Conversely, let us suppose that u is a weak solution of $u_t = \mathcal{L}_\circ(u)$ and let $\phi = \phi(x,t)$ be smooth and such that $x_4 - u - \phi$ has a local maximum at (x^0, t^0). We may assume that $(x^0, t^0) = (0, 0)$ and that $u(0,0) = \phi(0,0) = 0$. Since, locally at $(0,0)$, $x_4 - u \le \phi$ we have $\phi_{x_4}(0,0) = 1$. In particular $\phi = 0$ is a (local) graph $x_4 = g$ and $\phi = \lambda(x_4 - g)$ where λ is smooth and $\lambda(0,0) = 1$. Moreover, since the operator $\mathcal{L}$ is invariant with respect to unitary transformations of $\mathbb{C}^2$, we may assume that $D_{x'}g(0,0) = 0$. In this situation we have

$$-u(x',t) \le -\lambda(x',0,t)g(x',t)$$

and
$$\mathcal{L}_\circ(-\lambda g)(0,0) = -\frac{1}{4}\{g_{x_1 x_1}(0,0) + g_{x_2 x_2}(0,0)\} = \mathcal{L}_\circ(-g)(0,0).$$

Furthermore
$$\phi_t(0,0) = g_t(0,0), \quad \mathcal{L}(\phi)(0,0) = \mathcal{L}_\circ(-g)(0,0).$$

Since $-u$ is a weak solution of $w_t = \mathcal{L}_\circ(w)$
$$-g_t(0,0) \le \mathcal{L}_\circ(-g)(0,0).$$

From this, in view of the above identities, we obtain
$$\phi_t(0,0) = -g_t(0,0) =\le -\mathcal{L}_\circ(-g)(0,0) = \mathcal{L}_\circ(-g)(0,0) = \mathcal{L}(\phi)(0,0).$$

This proves that $x_4 - u$ is a weak subsolution.
 Similarly we prove that $x_4 - u$ is a weak supersolution.
 Therefore $x_4 - u$ is a weak solution of $v_t = \mathcal{L}(v)$. □

Using the same method as in [ES 2] it is possible to prove the following

LEMMA 3.5. *Let $u = u(z,t)$ be a local weak solution of $u_t = \mathcal{L}(u)$ and assume that, locally at (z°, t°), $u = 0$ is a graph $x_4 = v(x',t)$. Then v is a weak solution of $v_t = \mathcal{L}_\circ(v)$.*

Now we are in position to prove the following

THEOREM 3.6. *Let $\Gamma_\circ$ be the graph of a continuous function $u_\circ : \overline{D} \to \mathbb{R}$. Suppose that D is strictly pseudoconvex. Then the evolution of $\Gamma_\circ$ with fixed boundary is governed by the following parabolic problem*

(3.1) $$\begin{cases} u_t = \mathcal{L}_\circ(u) & \text{in } D \times (0, +\infty) \\ u = u_\circ & \text{on } \overline{D} \times \{0\} \\ u = u_\circ & \text{on } bD \times [0, +\infty). \end{cases}$$

Proof. Let the evolution be defined by the zero set $\{v = 0\}$ where v is the weak solution of the parabolic problem (2.3).
 In view of Theorem 3.1 every $E_t(\Gamma, b\Gamma)$, $t \ge 0$, is a graph, a priori over a subset of $\overline{D}$, but in view of Theorem 2.3 it separates $D \times \mathbb{R}$ so is the graph over $\overline{D}$ say of a continuous function $u^t = u^t(x')$. Define $u : \overline{D} \times (0,+\infty) \to \mathbb{R}$ by $u(x',t) = u^t(x')$. u is continuous: if $(x'^n, t^n) \to (y', \theta)$ then the sequence $(x'^n, u^{t^n}(x'^n), t^n)$ tends to a point (y', y_4, θ) which lies on the graph of u^t. In particular $y_4 = u(y', \theta)$. Thus
$$E_t(\Gamma, b\Gamma) = \{x_4 = u(x',t)\}.$$

Owing to Lemma 3.3 u is a weak solution of $u_t = \mathcal{L}_\circ(u)$ which satisfies all conditions in (3.1) and this concludes the proof. □

4. Limit for solutions

Now we study the asymptotic behaviour of the weak solution u of (3.1).
Since D is strictly pseudoconvex by hypothesis there exists a smooth function $\rho = \rho(x')$ on a neighbourhood U of $\overline{D}$ with the following properties: ρ is strictly plurisubharmonic in $D \times i\mathbb{R}$ and $D = \{\rho < 0\}$, $d\rho \neq 0$ on bD.

Let $g_\circ := u_{\circ|bD}$ and w be the weak solution of the problem: $\mathcal{L}_\circ(w) = 0$ in D, $w = g_\circ$ on bD. The graph Γ of w is the Levi flat hypersurface over D with boundary γ, the graph of $g_\circ$ ([BG], [A], [ST1], [CS], [E]).

THEOREM 4.1. *Let $u \in C^0(\overline{D} \times [0,+\infty))$ be the weak solution of (3.1). Then*
$$\lim_{t \to +\infty} u(\cdot, t) = w$$
in $C^0(\overline{D})$. In particular, if $\Gamma_\circ$ denotes the graph of $u_\circ$, $E_t(\Gamma_\circ, b\Gamma_\circ) \to \Gamma$ as $t \to +\infty$ in the C^0-topology.

Proof. We divide the proof in several steps.

First of all we construct two continuous barriers $\delta^\pm$: $\delta^- \leq u_\circ \leq \delta^+$ in D, $\delta^- = \delta^+ = g_\circ$ on bD and $\mathcal{L}_\circ(\delta^-) \geq 0$, $\mathcal{L}_\circ(\delta^+) \leq 0$ in D. This is easily done using the functions $\delta^\pm = u_\circ \pm \lambda\rho$ where λ is a suitable positive constant.

Next we consider the weak solutions $u^\pm$ of (3.1) corresponding respectively to the boundary values $\delta^\pm$ on D and $g_\circ$ on $bD \times [0,+\infty)$. $u^\pm$ are bounded by virtue of the maximum principle and uniformly continuous because of Walsh's Lemma. Moreover u^+ (u^-) is non-increasing (non-decreasing) in t since $\mathcal{L}_\circ(\delta^+) \leq 0$ ($\mathcal{L}_\circ(\delta^-) \geq 0$) ([ST5]). It follows that $\lim_{t \to +\infty} u^\pm(x,t) := \tilde{u}^\pm(x)$ exists pointwise.

Now define functions $u_h^\pm(\cdot, t) = u^\pm(\cdot, t+h)$ for each positive h. These functions are still weak solutions (with different boundary values). Moreover, since $u^\pm$ are bounded, the sets $\{u_h^\pm\}_{h \geq 0}$ are equicontinuous and
$$\tilde{u}^\pm(x) = \lim_{t \to +\infty} u^\pm(x,t) = \lim_{t \to +\infty} u_h^\pm(x,t)$$
for every $x \in D$. It follows that $\tilde{u}^\pm$ are continuous in D, $\tilde{u}^\pm = \phi_\circ$ on bD and $\mathcal{L}_\circ(u^\pm) = 0$ in D and consequently (by uniqueness) $\tilde{u}^+ = \tilde{u}^- = w$ in D. Consider now the weak solution of the parabolic problem (3.1). By virtue of the comparison principle we have
$$u^-(\cdot, t) \leq u(\cdot, t) \leq u^+(\cdot, t)$$
and from this, letting $t \to +\infty$ we obtain
$$w(x) = \lim_{t \to +\infty} u^-(x,t) \leq \liminf_{t \to +\infty} u(x,t) \leq \limsup_{t \to +\infty} u(x,t) = \lim_{t \to +\infty} u^+(x,t) = w(x)$$
for every $x \in D$, so
$$\liminf_{t \to +\infty} u(\cdot, t) = \limsup_{t \to +\infty} u(\cdot, t) = \lim_{t \to +\infty} u(\cdot, t) = w$$
in $C^0(D)$. □

References

[A]   H. ALEXANDER, *Polynomial hulls of graphs*, Pacific J. Math. **147** (1991), 201-212.
[BG] E. BEDFORD and B. GAVEAU, *Envelopes of holomorphy of certain 2-spheres in $\mathbb{C}^2$*, Amer. J. Math. **105** (1983), 975-1009.

[CS]   E. M. CHIRKA and N. V. SHCHERBINA, *Pseudoconvexity of rigid domains and foliations of hulls of graphs*, Ann. Scuola Norm. Sup. Pisa Cl. Sci. (4), **XXI** (1995), 707-735.

[E]   Y. ELIASHBERG, *Filling by holomorphic discs and its applications*, London Math. Soc. Lecture Note Ser., **151** (1991), 45-67.

[ES 1]   L. C. EVANS and J. SPRUCK, *Motion of level sets by mean curvature. I*, J. Differential Geometry **33**, n. 4 (1991), 635-681.

[ES 2]   L. C. EVANS and J. SPRUCK, *Motion of Level Sets by Mean Curvature. III*, J. Diff. Eq., **77** (1989), 369-376.

[H]   G. HUISKEN, *Non-parametric Mean Curvature Evolution with Boundary Conditions*, J. Diff. Eq., **77** (1989), 369-376.

[S]   Z. SLODKOWSKI, *Local maximum property and q-plurisubharmonic functions in uniform algebras*, J. Math. Anal. Appl., **115** (1986), 105-130.

[ST1]   Z. SLODKOWSKI and G. TOMASSINI, *Weak Solutions for the Levi Equation and Envelope of Holomorphy*, J. Funct. Anal., **101**, No. 2 (1991), 392-407.

[ST2]   Z. SLODKOWSKI and G. TOMASSINI, *Levi equation and evolution of subsets of $\mathbb{C}^2$*, Rend. Mat. Acc. Lincei s. 9, v. **7**(1996), 235-239.

[ST3]   Z. SLODKOWSKI and G. TOMASSINI, *Evolution of subsets of $\mathbb{C}^2$ and Parabolic Problem for the Levi Equation*, Ann. Scuola Norm. Sup. Pisa Cl Sci. (**4**) Vol XXV (1997), 757-784.

[ST4]   Z. SLODKOWSKI and G. TOMASSINI, *Evolution of special subsets of $\mathbb{C}^2$*, (to appear in Adv. in Math.).

[ST5]   Z. SLODKOWSKI and G. TOMASSINI, *Stein hull and evolution*, (preprint).

DEPARTMENT OF MATHEMATICS, UNIVERSITY OF ILLINOIS AT CHICAGO, 851 SOUTH MORGAN STREET, CHICAGO, ILLINOIS 60607, USA
E-mail address: zbigniew@uic.edu

SCUOLA NORMALE SUPERIORE DI PISA, PIAZZA DEI CAVALIERI 7, 56126 PISA, ITALY
E-mail address: tomassini@sns.it

OBSERVABILITY INEQUALITIES FOR THE EULER-BERNOULLI PLATE WITH VARIABLE COEFFICIENTS*

Peng-Fei Yao

Abstract We consider some observability inequalities from the boundary for the Euler-Bernoulli plate with variable coefficients by geometry methods. At first, estimates of two kinds are established by a geometric multiplier method, subject to some checkable geometric conditions, in the case that no boundary conditions are imposed. Then our estimates yield continuous observability inequalities for two kinds of boundary conditions which have a physical meaning.

Keywords Euler-Bernoulli plate, observability inequality, Riemannian manifold, Hodge-Laplace operator

1 Introduction; Statement of Main Results

The purpose of this paper is to present some observability inequalities from the boundary for the Euler-Bernoulli equation with variable coefficients. This problem has been well understood in the case of constant coefficients, Lagnese, Lions [4], and Lasiecka, Triggiani [5]. It is hard to handle the variably coefficient case in which some special tools are often needed. We note the pseudo-differential method, Tataru [8], which is one of the useful tools in handling variable coefficient problems. But we here use the geometric approach to handle our problems. This approach was introduced into the boundary control problems by Yao [10] in the case of the wave equation with variable coefficients and is extended to cope with the first-order term in Lasiecka, Triggiani and Yao [6]. For a brief comparison of different methods, we also refer to [6].

---

*This work is supported by National Natural Science Foundation and the Key Project of China.

AMS(MOS) subject classifications 35L35, 35Q72, 93B05, 93B07

We just consider a 4th-order operator with variable coefficients principal part for the Euler-Bernoulli equation in the two cases of clamped and hinged boundary conditions. The following multipliers are used

$$H(w) \quad \text{and} \quad H(\mathcal{A}w)$$

where H is a vector field and $\mathcal{A}$ is a 2nd-order operator defined by (1.2) below. They are counterparts of multipliers $m(w)$ and $m(\Delta w)$, respectively, where $m = x - x_0$ and Δ is the classical Laplace in the Euclidean metric in the constant coefficient case. For the classical case, we refer to Lasiecka and Triggiani [5].

Let Ω be an open bounded domain in $I\!R^n$, with smooth boundary $\Gamma = \partial \Omega$. We shall consider regular solutions of

$$w_{tt} + \mathcal{A}^2 w = 0 \quad \text{in} \quad (0, \infty) \times \Omega \tag{1.1}$$

and establish some priori estimates. In (1.1) we have set

$$\mathcal{A}w = \sum_{ij=1}^{n} \frac{\partial}{\partial x_i}\left(a_{ij}(x)\frac{\partial w}{\partial x_j}\right), \quad w \in C_0^\infty(I\!R^n), \; x \in I\!R^n, \tag{1.2}$$

where $a_{ij} = a_{ji}$ are C^∞ functions in $I\!R^n$, and

$$\sum_{ij=1}^{n} a_{ij}(x)\xi_i\xi_j > 0, \quad x \in I\!R^n, \; \xi = (\xi_1, \xi_2, \cdots, \xi_n)^\tau \in I\!R^n. \tag{1.3}$$

We now introduce a Riemannian manifold as in Yao [10] which we shall work on. Let $I\!R^n$ have the usual topology and $x = (x_1, x_2, ..., x_n)$ be the natural coordinate system. For each $x \in I\!R^n$, define the inner product and norm over the tangent space $I\!R^n_x = I\!R^n$ by $g(X,Y) = \langle X, Y \rangle_g = \sum_{ij=1}^n g_{ij}(x)\alpha_i\beta_j$ and $|X|_g = \langle X, X \rangle_g^{1/2}$, respectively, where $(g_{ij}) = (a_{ij})^{-1}$, $X = \sum_{i=1}^n \alpha_i \frac{\partial}{\partial x_i}$ and $Y = \sum_{i=1}^n \beta_i \frac{\partial}{\partial x_i}$. It is easy to check that $(I\!R^n, g)$ is a Riemannian manifold with metric g. Denote the Levi-Civita connection in the metric g by D. Let H be a vector field on $(I\!R^n, g)$. The covariant differential DH of H determines a bilinear form on $I\!R^n_x \times I\!R^n_x$, for each $x \in I\!R^n$, by $DH(X, Y) = \langle D_X H, Y \rangle_g$, for all $X, Y \in I\!R^n_x$, where $D_X H$ is the covariant derivative of vector field H with respect to X.

We may now state our main assumptions, imposed throughout this paper upon the differential operator A in (1.2).

Assumption (H1). There exists a vector field H on the Riemannian manifold $(I\!R^n, g)$ such that

$$DH(X, X) = \langle D_X H, X \rangle_g \geq a|X|_g^2 \quad \forall X \in I\!R^n \; x \in \Omega, \tag{1.4}$$

where $a > 0$ is a constant.

Or we make the following

Assumption (H2). There is a vector field H on the Riemannian manifold $(\mathbb{R}^n, g)$ such that

$$DH(X, X) = b(x)|X|_g^2 \quad \forall X \in R_x^n, \; x \in \Omega, \tag{1.5}$$

where $b(x)$ is a function on Ω such that

$$b_0 = \min_{x \in \Omega} b(x) > 0. \tag{1.6}$$

It is clear that if the vector field H meets conditions (1.5) and (1.6), it will meet condition (1.4) with $a = b_0$ as well, so assumption (H2) is stronger than assumption (H1). In the case of constant coefficients where $a_{ij}(x) = \delta_{ij}$, for any x_0 to be fixed, the radial field $H = x_0 - x$ meets assumption (H2) with $b(x) = 1$.

In general, it is not easy to find a vector field verifying even assumption (H1) in the case of variable coefficients. Some corollaries on how to check assumption (H1) are presented in Yao [10]. A number of nontrivial examples, where this assumption is verified to hold by using the Riemann geometry, are also given in the same paper.

For $n = 2$, a real plate, it has been proved by Yao [11] that there is always a vector field such that condition (1.5) holds for any $\Omega \subset \mathbb{R}^n$.

In the end of subsection 2.1 we shall show how to verify assumption (H2) in some cases and some examples will also be given.

Main results. Set

$$Q = \Omega \times (0, T) \quad \text{and} \quad \Sigma = \Gamma \times (0, T). \tag{1.7}$$

We introduce the energy of system (1.1) by

$$E(t) = \frac{1}{2} \int_\Omega [w_t^2 + (\mathcal{A}w)^2] \, dx \tag{1.8}$$

and set

$$L(t) = \int_\Omega (w^2 + |Dw|_g^2) \, dx \tag{1.9}$$

to be the lower order term with respect to the energy $E(t)$, where w is a solution to equation (1.1).

Theorem 1.1 *Let a vector field H be such that assumption (H2) holds with b_0 defined by (1.6). Let w be a solution to equation (1.1) such that all the terms on the left hand side of inequality (1.10) below are well defined.*

Then, there is a constant $C > 0$, which is independent of w, of time $T > 0$ and of $\epsilon > 0$, such that

$$BT1|_\Sigma + \frac{C}{\epsilon}[L(0) + L(T) + \int_0^T L(t)dt]$$
$$+\epsilon[E(0) + E(T)] \geq (2b_0 - \epsilon) \int_0^T E(t)\,dt \qquad (1.10)$$

where

$$BT1|_\Sigma = \int_\Sigma \left[\mathcal{A}w\left(\frac{\partial H(w)}{\partial v_\mathcal{A}} + \frac{\partial(hw)}{\partial v_\mathcal{A}}\right) - (hw + H(w))\frac{\partial(\mathcal{A}w)}{\partial v_\mathcal{A}}\right]d\Sigma$$
$$+\frac{1}{2}\int_\Sigma [w_t^2 - (\mathcal{A}w)^2]H \cdot v\, d\Sigma, \qquad (1.11)$$

$$h = \frac{1}{2}(div H - 2b_0), \qquad (1.12)$$

and $\frac{\partial w}{\partial v_\mathcal{A}} = \sum_{ij=1}^n a_{ij}\frac{\partial w}{\partial x_j}v_i$ is the so called co-normal derivative and $v = (v_1, v_2, \cdots, v_n)$ is the unit normal of Γ pointing towards the exterior of Ω.

If instead of (1.8) we define an energy by

$$E_{1/4}(t) = \frac{1}{2}\int_\Omega (|Dw_t|_g^2 + |D(\mathcal{A}w)|_g^2)\,dx, \qquad (1.13)$$

we shall have the estimates below, where the lower order term becomes

$$L_{1/4}(t) = \int_\Omega (w^2 + w_t^2 + |Dw|_g^2 + |D^2w|_g^2)\,dx \qquad (1.14)$$

and w is a solution to equation (1.1). In expression (1.14), $|D^2w|_g$ for each $x \in \Omega$ has been defined to be the norm of D^2w as a 2-order tensor over R_x^n to be given by formula (2.2.6) in subsection 2.2 below.

Theorem 1.2 Let there be a vector field H such that inequality (1.4) in (H.1) holds with a defined by (1.4). Let w be a solution to equation (1.1). Then, there is a constant $C > 0$, which is independent of w, of time $T > 0$, and of $\epsilon > 0$, such that

$$BT2|_\Sigma + \frac{C}{\epsilon}[L_{1/4}(0) + L_{1/4}(T) + \int_0^T L_{1/4}(t)dt]$$
$$+\epsilon[E_{1/4}(0) + E_{1/4}(T)] \geq (2a - \epsilon)\int_0^T E_{1/4}(t)\,dt \qquad (1.15)$$

where

$$BT2|_\Sigma = \int_\Sigma [qw_t + H(w_t)]\frac{\partial w_t}{\partial v_\mathcal{A}} d\Sigma + \int_\Sigma [H(\mathcal{A}w) + q\mathcal{A}w]\frac{\partial(\mathcal{A}w)}{\partial v_\mathcal{A}} d\Sigma$$
$$- \int_\Sigma [w_t \mathcal{A}w_t + \frac{1}{2}(|Dw_t|_g^2 + |D(\mathcal{A}w)|_g^2)] H \cdot v\, d\Sigma$$
$$- \frac{1}{2}\int_\Sigma [w_t^2 + (\mathcal{A}w)^2] Dq \cdot v\, d\Sigma, \tag{1.16}$$

$$q = \frac{1}{2} div H. \tag{1.17}$$

Control problems. Let Γ_0, the controlled portion of Γ, be relatively open, and $\Gamma_1 = \Gamma/\Gamma_0$. Set

$$\Sigma_0 = \Gamma_0 \times (0,T) \quad \text{and} \quad \Sigma_1 = \Gamma_1 \times (0,T). \tag{1.18}$$

Case I. First, we consider the exact controllability problem in the unknown ϕ

$$\begin{cases} \phi_{tt} + \mathcal{A}^2\phi = 0 & \text{in } Q, \\ \phi(0) = \phi^0 \quad \phi_t(0) = \phi^1 & \text{in } \Omega, \\ \phi = \dfrac{\partial \phi}{\partial v_\mathcal{A}} = 0 & \text{on } \Gamma_1 \times (0,T), \\ \phi = u \quad \dfrac{\partial \phi}{\partial v_\mathcal{A}} = v & \text{on } \Gamma_0 \times (0,T), \end{cases} \tag{1.19}$$

with controls u and v. The dual version for the above problem is in w

$$\begin{cases} w_{tt} + \mathcal{A}^2 w = 0 & \text{in } Q, \\ w(0) = w^0 \quad w_t(0) = w^1 & \text{in } \Omega, \\ w = \dfrac{\partial w}{\partial v_\mathcal{A}} = 0 & \text{on } \Gamma \times (0,T). \end{cases} \tag{1.20}$$

In the case of constant coefficients where $a_{ij} = \delta_{ij}$, one control function, $\frac{\partial \phi}{\partial v_\mathcal{A}} = v$, is enough for problem (1.19), Lagnese and Lions [4]. We here add another control function, $\phi = u$, in order to avoid the following uniqueness assumption: problem

$$\begin{cases} \mathcal{A}^2 w = \lambda w & \text{on } \Omega, \\ w = \dfrac{\partial w}{\partial v_\mathcal{A}} = \mathcal{A}w = 0 & \text{on } \Gamma_0 \end{cases} \tag{1.21}$$

has the unique zero solution where λ is a complex number. The above uniqueness result holds true if $a_{ij} = \delta_{ij}$, Lagnese and Lions [4]. Another kind of uniqueness result is given by Lasiecka and Triggiani [5], that is,

$$w = \frac{\partial w}{\partial v} = \frac{\partial \Delta w}{\partial v} = 0 \quad \text{on} \quad \Gamma_0$$

imply $w = 0$ for the classical case where $A = \Delta$. In general, we do not know if the uniqueness in (1.21) is true because it is not a Cauchy problem.

Continuous observability inequality in Case I. It is easily checked that exact controllability problem (1.19) then leads to the following observability inequality: for any $T > 0$, to seek constant $C_T > 0$ such that

$$\int_{\Sigma_0} \left[\left(\frac{\partial \mathcal{A} w}{\partial \nu_\mathcal{A}}\right)^2 + (\mathcal{A} w)^2\right] d\Sigma \geq C_T E(0) \tag{1.22}$$

where w is a solution to problem (1.20).

By a standard compactness/uniqueness argument, the lower order term in inequality (1.10) can be absorbed so that we have the following

Theorem 1.3 *(Case I) Let assumption (H2) hold. Then for any $T > 0$, there is $C_T > 0$ such that observability inequality (1.22) holds, where*

$$\Gamma_0 = \{\, x \,|\, x \in \Gamma,\ H(x) \cdot v > 0 \,\}. \tag{1.23}$$

Case II. Finally, we consider exact controllability problem

$$\begin{cases} \phi_{tt} + \mathcal{A}^2 \phi = 0 & \text{in } Q, \\ \phi(0) = \phi^0 \quad \phi_t(0) = \phi^1 & \text{in } \Omega, \\ \phi = \mathcal{A}\phi = 0 & \text{on } \Gamma_1 \times (0, T), \\ \phi = u \quad \mathcal{A}\phi = v & \text{on } \Gamma_0 \times (0, T), \end{cases} \tag{1.24}$$

with controls u and v. The dual version for problem (1.28) is in w

$$\begin{cases} w_{tt} + \mathcal{A}^2 w = 0 & \text{in } Q, \\ w(0) = w^0 \quad w_t(0) = w^1 & \text{in } \Omega, \\ w = \mathcal{A} w = 0 & \text{on } \Gamma \times (0, T). \end{cases} \tag{1.25}$$

Continuous observability inequality in Case II. Now the observability inequality is: to obtain $C_T > 0$ for any $T > 0$ given such that

$$\int_{\Sigma_0} \left[\left(\frac{\partial w}{\partial \nu_\mathcal{A}}\right)^2 + \left(\frac{\partial (\mathcal{A} w)}{\partial \nu_\mathcal{A}}\right)^2\right] d\Sigma \geq C_T [\|w^0\|^2_{H^1_0(\Omega)} + \|w^1\|^2_{H^{-1}(\Omega)}] \tag{1.26}$$

where w is a solution to problem (1.29).

Theorem 1.4 *(Case II) Let assumption (H1) hold. Then for any $T > 0$, there is $C_T > 0$ such that observability inequality (1.26) holds, where*

$$\Gamma_0 = \{\, x \,|\, x \in \Gamma,\ H(x) \cdot v > 0 \,\}. \tag{1.27}$$

2 Geometric Conditions; Proofs of Main Results

2.1. Geometric Conditions. We shall give two propositions on the exisitence of a vector field to meet assumption (H2) and some examples.

Proposition 2.1 *Let $(\mathbb{R}^n, g)$ be of everywhere constant sectional curvature K. Given $x_0 \in \mathbb{R}^n$. Denote by $\rho = \rho(x)$ the distance function from x to x_0 in metric g (in general $\rho(x) \neq |x - x_0|^2$). Set*

$$H = h'(\rho) D\rho \tag{2.1.1}$$

where function $h(t)$ is defined by

$$h(t) = \begin{cases} -\cos\sqrt{K}t & K > 0, \\ \dfrac{1}{2}t^2 & K = 0, \\ \cosh\sqrt{-K}t & K < 0. \end{cases} \tag{2.1.2}$$

Then

$$DH(X, X) = \begin{cases} K|X|_g^2 \cos\sqrt{K}\rho(x) & K > 0, \\ |X|_g^2 & K = 0, \\ -K|X|_g^2 \cosh\sqrt{-K}\rho(x) & K < 0, \end{cases} \tag{2.1.3}$$

for $X \in \mathbb{R}_x^n$.

Proof. By Wu, Shen and Yu [9], we have

$$D^2\rho(Y, Y) = \begin{cases} \sqrt{K}\cot(\sqrt{K}\rho(x))|Y|_g^2 & K > 0, \\ \dfrac{1}{\rho(x)}|Y|_g^2 & K = 0, \\ \sqrt{-K}\coth(\sqrt{K}\rho(x))|Y|_g^2 & K < 0, \end{cases} \tag{2.1.4}$$

for all $Y \in \mathbb{R}_x^n$ and $Y \perp (D\rho)(x)$.

Let X be a vector field on $(\mathbb{R}^n, g)$. We have

$$\begin{aligned} D^2 h(\rho)(X, X) &= XXh(\rho) - (D_X X)h(\rho) \\ &= h'(\rho)[XX\rho - (D_X X)\rho] + h''(\rho)(X\rho)^2 \\ &= h'(\rho)D^2\rho(X, X) + h''(\rho)\langle X, D\rho\rangle_g^2. \end{aligned} \tag{2.1.5}$$

Relations (2.1.4) and (2.1.5) together with (2.1.2) give (2.1.3) since $D^2\rho(X, X) = D^2\rho(Y, Y)$, where $X = \langle X, D\rho\rangle_g D\rho + Y$, $Y \perp D\rho$. □

Proposition 2.2 Let there be a metric $\hat{g}$ such that $(\mathbb{R}^n, \hat{g})$ has zero curvature and such that there is a function u on $\mathbb{R}^n$ to meet relation

$$g = e^{2u}\hat{g} \quad \forall\, x \in \mathbb{R}^n. \tag{2.1.6}$$

Given $x_0 \in \mathbb{R}^n$. Denote by $\hat{\rho}(x)$ the distance function from x_0 to x in metric $\hat{g}$. Set

$$H = \hat{\rho}\hat{D}\hat{\rho} \tag{2.1.7}$$

where $\hat{D}$ is the Levi-Civita connection on $(\mathbb{R}^n, \hat{g})$. Then

$$DH(X, X) = [1 + H(u)]|X|_g^2 \quad \forall\, X \in \mathbb{R}_x^n,\ x \in \mathbb{R}^n. \tag{2.1.8}$$

Proof. Denote by Γ_{ij}^k and $\hat{\Gamma}_{ij}^k$ the coefficients of the connections D and $\hat{D}$, respectively. Set

$$\tilde{g}_{ij} = \tilde{g}(\frac{\partial}{\partial x_i}, \frac{\partial}{\partial x_j}),$$

and we have

$$g_{ij} = e^{2u}\tilde{g}_{ij}. \tag{2.1.9}$$

By relation (2.1.9), we have

$$\Gamma_{ij}^k = \frac{1}{2}\sum_{l=1}^n g^{kl}\left(\frac{\partial g_{il}}{\partial x_j} + \frac{\partial g_{jl}}{\partial x_i} - \frac{\partial g_{ij}}{\partial x_l}\right)$$

$$= \tilde{\Gamma}_{ij}^k + \delta_{ki}\frac{\partial u}{\partial x_j} + \delta_{kj}\frac{\partial u}{\partial x_i} - \tilde{g}_{ij}\sum_{l=1}^n \tilde{g}^{kl}\frac{\partial u}{\partial x_l}. \tag{2.1.10}$$

Suppose that $H = \sum_{i=1}^n h_i \frac{\partial}{\partial x_i}$. Let $X = \sum_{i=1}^n X_i \frac{\partial}{\partial x_i}$ be a vector field on $\mathbb{R}^n$. It follows from (2.1.10) that

$$D_X H = \sum_{j=1}^n X(h_j)\frac{\partial}{\partial x_j} + \sum_{ij=1}^n X_i h_j D_{\frac{\partial}{\partial x_i}}\frac{\partial}{\partial x_j}$$

$$= \sum_{j=1}^n X(h_j)\frac{\partial}{\partial x_j} + \sum_{k=1}^n \sum_{ij=1}^n X_i h_j \Gamma_{ij}^k \frac{\partial}{\partial x_k}$$

$$= \tilde{D}_X H + H(u)X + X(u)H - \langle X, H\rangle_g Du. \tag{2.1.11}$$

Formula (2.1.11) yields

$$DH(X, X) = \langle \tilde{D}_X H, X\rangle_g + H(u)|X|_g^2$$
$$= e^{2u}\tilde{D}H(X, X) + H(u)|X|_g^2, \quad \forall\, X \in \mathbb{R}_x^n. \tag{2.1.12}$$

On the other hand, we have

$$\tilde{D}H(X,X) = |X|_{\tilde{g}}^2, \quad \forall X \in \mathbb{R}_x^n, \tag{2.1.13}$$

since $(\mathbb{R}^n, \tilde{g})$ is of zero curvature and $H = \tilde{\rho}\nabla_{\tilde{g}}\tilde{\rho}$. Finally, it follows from (2.1.12) and (2.1.13) that

$$DH(X,X) = [1 + H(u)]|X|_g^2, \quad \forall X \in \mathbb{R}_x^n. \quad \square \tag{2.1.14}$$

Example 2.1. Let $\mathcal{A}$ be defined by

$$\mathcal{A}p = \frac{\partial}{\partial x}(e^{-x^2 - y^2}\frac{\partial p}{\partial x}) + \frac{\partial}{\partial y}(e^{-x^2 - y^2}\frac{\partial p}{\partial y}).$$

Then the Riemann manifold $(\mathbb{R}^2, g)$ has a metric

$$g = e^{x^2 + y^2}(dx^2 + dy^2).$$

Thus we have

$$g = e^{2u}\tilde{g},$$

where $u = \frac{1}{2}(x^2 + y^2)$, $\tilde{g} = dx^2 + dy^2$, and $(\mathbb{R}^n, \tilde{g})$ is of zero curvature. Take $\tilde{\rho}(x) = x^2 + y^2$. Then $H = x\frac{\partial}{\partial x} + y\frac{\partial}{\partial y}$, and

$$1 + H(u) = 1 + x^2 + y^2 \geq 1, \quad \forall x \in \mathbb{R}^2.$$

It follows from Proposition 2.2 that, for any bounded domain $\Omega \subset \mathbb{R}^2$, assumption (H2) holds. By Wu, Shen and Yu [9], however, the curvature of $(\mathbb{R}^2, g)$ is

$$k(x,y) = -2e^{-x^2 - y^2}, \quad (x,y) \in \mathbb{R}^2.$$

Example 2.2. Let $a > 0$ and $b > 0$ be constants. Consider the operator on $\mathbb{R}^2$

$$\mathcal{A}p = \frac{\partial}{\partial x}\left((1 + ax^2 + by^2)\frac{\partial p}{\partial x}\right) + \frac{\partial}{\partial y}\left((1 + ax^2 + by^2)\frac{\partial p}{\partial y}\right).$$

Then the Riemann manifold $(\mathbb{R}^2, g)$ has a metric

$$g = \frac{1}{1 + ax^2 + by^2}(dx^2 + dy^2). \tag{2.1.15}$$

For any $(x,y) \in \mathbb{R}^2$, the curvature is
$$k(x,y) = \frac{a+b+a(b-a)x^2+b(a-b)y^2}{1+ax^2+by^2}.$$

By (2.1.15), we have
$$g = e^{2u}\tilde{g},$$
where $u = -\frac{1}{2}\log(1+x^2+y^2)$, $\tilde{g} = dx^2+dy^2$, and $(\mathbb{R}^n, \tilde{g})$ is of zero curvature. Take $\tilde{\rho}(x) = x^2+y^2$. Then $H = x\frac{\partial}{\partial x} + y\frac{\partial}{\partial y}$, and
$$1 + H(u) = \frac{1}{1+ax^2+by^2} > 0, \quad \forall (x,y) \in \mathbb{R}^2.$$

It follows from Proposition 2.2 that, for any bounded domain $\Omega \subset \mathbb{R}^2$, assumption (H2) holds.

2.2. Some Multiplier Identities. As usual in the control theory, by multiplying equation (1.1) by different multipliers, we can obtain different identities which play an important role in deriving the estimates. Here we shall consider the following multipliers respectively

$$H(w) \quad \text{and} \quad H(\mathcal{A}w). \tag{2.2.1}$$

With two metrics on $\mathbb{R}^n$ in mind, one the Euclidean metric and the other the Riemann metric g, we have to deal with various notations carefully.

We recall that $E_1, E_2, \cdots, E_n$ is a frame field normal at x (Wu, Shen and Yu [9]) on the Riemann manifold $(\mathbb{R}^n, g)$ if and only if it is a local basis for vector fields with $\langle E_i, E_j \rangle_g = \delta_{ij}$ in some neighborhood of x and with $(D_{E_i}E_j)(x) = 0$, for $1 \le i, j \le n$.

Let H be a vector field on $\mathbb{R}^n$ and $h \in C^2(\overline{\Omega})$. We have the following formulae for divergence in the Euclidean metric

$$div(hH) = hdiv(H) + H(h) \tag{2.2.2}$$

and

$$\int_\Omega div(H)dx = \int_\Gamma H \cdot v \, d\sigma \tag{2.2.3}$$

where $d\sigma$ is the Euclidean surface element on Γ.

By Yao [10, Lem. 2.1], we have

$$\langle Dh, D(H(h))\rangle_g = DH(Dh, Dh) + \frac{1}{2}div(|Dh|_g^2 H) -$$
$$- \frac{1}{2}|Dh|_g^2 div H, \quad h \in C^2(\mathbb{R}^n), x \in \mathbb{R}^n. \tag{2.2.4}$$

By Yao [10, Lem. 2.1] again we also have

$$\mathcal{A}(ph) = divD(ph) = p\mathcal{A}h + 2\langle Dp, Dh\rangle_g + h\mathcal{A}p \quad \forall p, h \in C^2(\mathbb{R}^n). \quad (2.2.5)$$

Give $x \in \mathbb{R}^n$. Denote the set of all covariant tensors of order 2 on $\mathbb{R}^n_x$ by $T^2(\mathbb{R}^n_x)$. Then $T^2(\mathbb{R}^n_x)$ is an inner product space of dimension n^2 with inner product

$$\langle F, G\rangle_{T^2(\mathbb{R}^n_x)} = \sum_{ij=1}^{n} F(e_i, e_j)G(e_i, e_j), \quad \forall F, G \in T^2(\mathbb{R}^n_x), \quad (2.2.6)$$

where $e_1, e_2, \cdots, e_n$ is an orthonormal basis of $(\mathbb{R}^n_x, g)$. It is easily checked that $\langle \cdot, \cdot\rangle_{T^2(\mathbb{R}^n_x)}$ is independent of the choice of the orthonormal basis of $(\mathbb{R}^n_x, g)$.

We denote by $\mathcal{X}(\mathbb{R}^n)$ the set of all vector fields on $\mathbb{R}^n$. Denote by $\triangle$: $\mathcal{X}(\mathbb{R}^n) \to \mathcal{X}(\mathbb{R}^n)$ the Hodge-Laplace operator. Let H be a vector field on $(\mathbb{R}^n, g)$. We have the following Weitzenboc formula (Wu, Shen and Yu [9])

$$\Delta(H(h)) = (\triangle H)(h) + 2\langle DH, D^2h\rangle_{T^2(\mathbb{R}^n_x)} + H(\Delta h) + Ric(H, Dh)$$

$$\forall h \in C^2(\mathbb{R}^n), \quad (2.2.7)$$

where Δ: $C^2(\mathbb{R}^n) \to C^2(\mathbb{R}^n)$ is the Laplace operator in the Riemann metric g and $Ric(\cdot, \cdot) \in T^2(\mathbb{R}^n_x)$ is the Ricci curvature tensor of the Riemann metric g. It is well known that in the natural coordinate system $x = (x_1, x_2, \cdots, x_n)$

$$\Delta h = \frac{1}{\sqrt{\mathcal{G}(x)}} \sum_{ij=1}^{n} \frac{\partial}{\partial x_i}\left(\sqrt{\mathcal{G}(x)} a_{ij}(x) \frac{\partial h}{\partial x_j}\right), \quad \forall h \in C^2(\mathbb{R}^n), \quad (2.2.8)$$

where $\mathcal{G}(x) = det(g_{ij})$. We thus have from Yao [10, Lem. 2.1]

$$\Delta h = \sum_{ij=1}^{n} \frac{\partial}{\partial x_i}\left(a_{ij}(x)\frac{\partial h}{\partial x_j}\right) + \frac{1}{\sqrt{\mathcal{G}(x)}}\sum_{ij=1}^{n} a_{ij}(x)\frac{\partial \sqrt{\mathcal{G}(x)}}{\partial x_i}\frac{\partial h}{\partial x_j}$$

$$= \mathcal{A}h - \langle Dh, Df\rangle_g, \quad \forall h \in C^2(\mathbb{R}^n), \quad (2.2.9)$$

where

$$f(x) = \frac{1}{2}\log \det(a_{ij}(x)). \quad (2.2.10)$$

It follows from (2.2.9) that

$$\mathcal{A} = \Delta + Df. \quad (2.2.11)$$

We shall write (w, u) for $\int_\Omega wu\,dx$ and $\|w\|$ for $(w, w)^{1/2}$.

Proposition 2.3 Let H is a vector field on $(\mathbb{R}^n, g)$. Assume that w solves problem (1.1). Let f be defined by (2.2.10). Then

$$\int_\Sigma \left[\mathcal{A}w \frac{\partial(H(w))}{\partial \nu_\mathcal{A}} - H(w) \frac{\partial(\mathcal{A}w)}{\partial \nu_\mathcal{A}} \right] d\Sigma + \frac{1}{2} \int_\Sigma [w_t^2 - (\mathcal{A}w)^2] H \cdot \nu \, d\Sigma$$
$$= \frac{1}{2} \int_Q [w_t^2 - (\mathcal{A}w)^2] \mathrm{div}\, H \, dQ + 2 \int_Q \mathcal{A}w \langle DH, D^2 w \rangle_{T^2(\mathbb{R}^n_x)} \, dQ$$
$$+ \int_Q \mathcal{A}w [(\mathcal{A}H)(w) + Ric(H, Dw) - D^2 f(H, Dw)] \, dQ$$
$$+ (w_t, H(w)) \Big|_0^T \tag{2.2.12}$$

where $\mathcal{A}H \in \mathcal{X}(\mathbb{R}^n)$ is defined by $\mathcal{A}H = \triangle H + D_{Df} H$ and $D^2 f$ is the Hessian of f in the Riemann metric g.

Proof. We multiply (1.1) by $H(w)$ and integrate by parts.

First, by the formula for divergence, we obtain

$$\int_Q w_{tt} H(w) \, dQ$$
$$= (w_t, H(w)) \Big|_0^T - \int_Q w_t H(w_t) \, dQ$$
$$= (w_t, H(w)) \Big|_0^T - \frac{1}{2} \int_\Sigma w_t^2 H \cdot \nu \, d\Sigma + \frac{1}{2} \int_Q w_t^2 \mathrm{div}\, H \, dQ. \tag{2.2.13}$$

Next, we compute term $\int_Q \mathcal{A}^2 w H(w) dQ$. To this end we need some formulae further.

Given $x \in \mathbb{R}^n$. Let $E_1, E_2, \cdots, E_n$ be a frame field normal at x on the Riemann manifold $(\mathbb{R}^n, g)$. Let $H = \sum_{i=1}^n h_i E_i$. Then $H(w) = \sum_{i=1}^n h_i E_i(w)$, where $E_i(w)$ is the covariant differential of w with respect to E_i in the Riemann metric g. By (2.2.11) we obtain

$$H(\mathcal{A}w) = H(\Delta w) + H(\langle Df, Dw \rangle_g)$$
$$= H(\Delta w) + \sum_{ij=1}^n E_j E_i(f) E_i(w) h_j + \sum_{ij=1}^n E_j E_i(w) h_j E_i(f)$$
$$= H(\Delta w) + D^2 f(H, Dw) + D^2 w(H, Df), \tag{2.2.14}$$

where $E_j E_i(w)$ and $E_j E_i(f)$ are the second covariant differentials of w and f, respectively, at x. In addition, we have

$$\langle Df, D(H(w)) \rangle_g$$
$$= \sum_{j=1}^n E_j(f) E_j(H(w)) = \sum_{ij=1}^n E_j(f) [E_j(h_i) E_i(w) + h_i E_j E_i(w)]$$
$$= DH(Dw, Df) + D^2 w(H, Df). \tag{2.2.15}$$

With the above formulae (2.2.14) and (2.2.15), we obtain from (2.2.11) and (2.2.7)

$$\begin{aligned}
\mathcal{A}(H(w)) &= \Delta(H(w)) + \langle Df, D(H(w))\rangle_g \\
&= (\triangle H)(w) + 2\langle DH, D^2w\rangle_{T^2(\mathbb{R}^n_x)} \\
&\quad + H(\Delta w) + Ric(H, Dw) + DH(Dw, Dw) + D^2w(H, Df) \\
&= H(\mathcal{A}w) + 2\langle DH, D^2w\rangle_{T^2(\mathbb{R}^n_x)} + (\mathcal{A}H)(w) \\
&\quad + Ric(H, Dw) - D^2f(H, Dw)
\end{aligned} \quad (2.2.16)$$

where we have denoted $\mathcal{A}H = \triangle H + D_{Df}H$. Furthermore we also need the following

$$\int_\Omega H((\mathcal{A}w)^2)\,dx = \int_\Gamma (\mathcal{A}w)^2 H \cdot v\,d\Gamma - \int_\Omega (\mathcal{A}w)^2 div H\,dx \quad (2.2.17)$$

which comes from the formula of divergence.

We are now ready for our computing. In fact, it follows from (2.2.16) and (2.2.17) and the Green formula that

$$\begin{aligned}
\int_Q & \mathcal{A}^2 w H(w)\,dQ \\
&= -\int_Q \langle D(\mathcal{A}w), D(H(w))\rangle_g\,dQ + \int_\Sigma H(w)\frac{\partial(\mathcal{A}w)}{\partial v_\mathcal{A}}\,d\Sigma \\
&= \int_Q \mathcal{A}w\mathcal{A}(H(w))\,dQ + \int_\Sigma \left[H(w)\frac{\partial(\mathcal{A}w)}{\partial v_\mathcal{A}} - \mathcal{A}w\frac{\partial(H(w))}{\partial v_\mathcal{A}}\right]d\Sigma \\
&= \frac{1}{2}\int_\Sigma \left[(\mathcal{A}w)^2 H \cdot v + 2H(w)\frac{\partial(\mathcal{A}w)}{\partial v_\mathcal{A}} - 2\mathcal{A}w\frac{\partial(H(w))}{\partial v_\mathcal{A}}\right]d\Sigma \\
&\quad -\frac{1}{2}\int_Q (\mathcal{A}w)^2 dviH\,dQ + 2\int_Q \mathcal{A}w\langle DH, D^2w\rangle_{T^2(\mathbb{R}^n_x)}\,dQ \\
&\quad + \int_Q \mathcal{A}w[(\mathcal{A}H)(w) + Ric(H, Dw) - D^2f(H, Dw)]\,dQ. \quad (2.2.18)
\end{aligned}$$

Finally, by combining (2.2.13) with (2.2.18), we obtain identity (2.2.12). □

Proposition 2.4 *Let H be a vector field and w be a solution to problem (1.1). Set $p = divH$. Then*

$$\begin{aligned}
\int_\Sigma & [pw_t + H(w_t)]\frac{\partial w_t}{\partial v_\mathcal{A}}\,d\Sigma + \int_\Sigma H(\mathcal{A}w)\frac{\partial(\mathcal{A}w)}{\partial v_\mathcal{A}}\,d\Sigma - \\
&- \int_\Sigma [w_t\mathcal{A}w_t + \frac{1}{2}(|Dw_t|^2_g + |D(\mathcal{A}w)|^2_g)]H \cdot v\,d\Sigma - \frac{1}{2}\int_\Sigma w_t^2 Dp \cdot v\,d\Sigma \\
&= -(w_t, H(\mathcal{A}w))\Big|_0^T + \int_Q [DH(Dw_t, Dw_t) + DH(D(\mathcal{A}w), D(\mathcal{A}w))]\,dQ \\
&\quad + \frac{1}{2}\int_Q [|Dw_t|^2_g - |D(\mathcal{A}w)|^2_g]p\,dQ - \frac{1}{2}\int_Q w_t^2 \mathcal{A}p\,dQ. \quad (2.2.19)
\end{aligned}$$

Proof. We multiply (1.1) from both sides by $H(\mathcal{A}w)$ and integrate by parts.

First, we give two formulae we will need later. By identity (2.2.4), we have

$$\int_\Omega \mathcal{A}w_t H(w_t)\,dx = -\int_\Omega \langle Dw_t, D(H(w_t))\rangle_g\,dx + \int_\Gamma H(w_t)\frac{\partial w_t}{\partial v_\mathcal{A}}\,d\Gamma$$

$$= -\int_\Omega [DH(Dw_t, Dw_t) + \frac{1}{2}\mathrm{div}(|Dw_t|_g^2 H) - \frac{1}{2}|Dw_t|_g^2 p]\,dx$$

$$+ \int_\Gamma H(w_t)\frac{\partial w_t}{\partial v_\mathcal{A}}\,d\Gamma$$

$$= -\int_\Omega DH(Dw_t, Dw_t)\,dx - \frac{1}{2}\int_\Gamma |Dw_t|_g^2 H\cdot v\,d\Gamma$$

$$+ \frac{1}{2}\int_\Omega |Dw_t|_g^2 p\,dx + \int_\Gamma H(w_t)\frac{\partial w_t}{\partial v_\mathcal{A}}\,d\Gamma \quad (2.2.20)$$

In addition, by the Green formula, we obtain

$$\int_\Omega pw_t \mathcal{A}w_t\,dx = -\frac{1}{2}\int_\Omega Dp(w_t^2)\,dx - \int_\Omega |Dw_t|_g^2 p\,dx + \int_\Gamma pw_t\frac{\partial w_t}{\partial v_\mathcal{A}}\,d\Gamma$$

$$= -\frac{1}{2}\int_\Gamma w_t^2 Dp\cdot v\,d\Gamma + \frac{1}{2}\int_\Omega w_t^2 \mathcal{A}p\,dx - \int_\Omega |Dw_t|_g^2 p\,dx$$

$$+ \int_\Gamma pw_t\frac{\partial w_t}{\partial v_\mathcal{A}}\,d\Gamma. \quad (2.2.21)$$

With formulae (2.2.20) and (2.2.21), we obtain in one hand

$$\int_Q w_{tt} H(\mathcal{A}w)\,dQ = (w_t, H(\mathcal{A}w))\Big|_0^T - \int_Q [H(w_t \mathcal{A}w_t) - \mathcal{A}w_t H(w_t)]\,dQ$$

$$= (w_t, H(\mathcal{A}w))\Big|_0^T - \int_\Sigma w_t \mathcal{A}w_t H\cdot v\,d\Sigma$$

$$+ \int_Q [w_t \mathcal{A}w_t p + \mathcal{A}w_t H(w_t)]\,dQ$$

$$= (w_t, H(\mathcal{A}w))\Big|_0^T - \int_\Sigma [w_t \mathcal{A}w_t + \frac{1}{2}|Dw_t|_g^2]H\cdot v\,d\Sigma$$

$$+ \int_\Sigma [H(w_t) + pw_t]\frac{\partial w_t}{\partial v_\mathcal{A}}\,d\Sigma - \frac{1}{2}\int_\Sigma w_t^2 Dp\cdot v\,d\Sigma -$$

$$- \int_Q DH(Dw_t, Dw_t)\,dQ$$

$$+ \frac{1}{2}\int_Q [w_t^2 \mathcal{A}p - |Dw_t|_g^2 p]\,dQ. \quad (2.2.22)$$

By identity (2.2.4) again, we have on the other hand

$$\int_Q \mathcal{A}^2 w H(\mathcal{A}w)\,dQ$$

$$= -\int_Q \langle D(\mathcal{A}w), D(H(\mathcal{A}w))\rangle_g \, dQ + \int_\Sigma H(\mathcal{A}w) \frac{\partial(\mathcal{A}w)}{\partial v_\mathcal{A}} \, d\Sigma$$

$$= -\int_Q [DH(D(\mathcal{A}w), D(\mathcal{A}w)) - \frac{1}{2}|D(\mathcal{A}w)|_g^2 p] \, dQ$$

$$+ \int_\Sigma \left[H(\mathcal{A}w)\frac{\partial(\mathcal{A}w)}{\partial v_\mathcal{A}} - \frac{1}{2}|D(\mathcal{A}w)|_g^2 H \cdot v\right] d\Sigma. \tag{2.2.23}$$

Finally, identity (2.2.19) follows by combining (2.2.22) and (2.2.23) with (1.1). □

Proposition 2.5 *Let H be a vector field and w be a solution to problem (1.1). Assume that $h \in C^2(\mathbb{R}^n)$. Then*

(i) $\int_Q [w_t^2 - (\mathcal{A}w)^2] h \, dQ = \int_Q \mathcal{A}w [2Dh(w) + w\mathcal{A}h] \, dQ$

$$+ (w_t, wh)\Big|_0^T + \int_\Sigma \left[hw\frac{\partial(\mathcal{A}w)}{\partial v_\mathcal{A}} - \mathcal{A}w \frac{\partial(hw)}{\partial v_\mathcal{A}}\right] d\Sigma. \tag{2.2.24}$$

(ii) $\int_Q h[|Dw_t|_g^2 - |D(\mathcal{A}w)|_g^2] \, dQ$

$$= \int_\Sigma h \left[w_t \frac{\partial w_t}{\partial v_\mathcal{A}} - \mathcal{A}w \frac{\partial(\mathcal{A}w)}{\partial v_\mathcal{A}}\right] d\Sigma + \frac{1}{2}\int_Q [w_t^2 - (\mathcal{A}w)^2]\mathcal{A}h \, dQ$$

$$- (hw_t, \mathcal{A}w)\Big|_0^T + + \frac{1}{2}\int_\Sigma [(\mathcal{A}w)^2 - w_t^2] Dh \cdot v \, d\Sigma. \tag{2.2.25}$$

Proof. (i). By the Green formula, we obtain by formula (2.2.5)

$$(w_t, wh)\Big|_0^T = \int_0^T [(w_{tt}, wh) + (w_t, w_t p)] \, dt$$

$$= \int_Q [-wh\mathcal{A}^2 w + w_t^2 h] \, dQ$$

$$= -\int_Q \mathcal{A}(wh)\mathcal{A}w \, dQ + \int_Q w_t^2 h \, dQ$$

$$+ \int_\Sigma \left[\mathcal{A}w \frac{\partial(wh)}{\partial v_\mathcal{A}} - wh\frac{\partial(\mathcal{A}w)}{\partial v_\mathcal{A}}\right] d\Sigma$$

$$= \int_Q [w_t^2 - (\mathcal{A}w)^2] h \, dQ - \int_Q \mathcal{A}w [w\mathcal{A}h + 2Dh(w)] \, dQ$$

$$+ \int_\Sigma \left[\mathcal{A}w \frac{\partial(wh)}{\partial v_\mathcal{A}} - wh\frac{\partial(\mathcal{A}w)}{\partial v_\mathcal{A}}\right] d\Sigma. \tag{2.2.26}$$

This gives identity (2.2.24).

(ii). Similar to (2.2.26), we have

$$(w_t, h\mathcal{A}w)\Big|_0^T = \int_Q [-h\mathcal{A}^2 w \mathcal{A}w + hw_t \mathcal{A}w_t] \, dQ$$

$$= \int_Q [\langle D(\mathcal{A}w), D(h\mathcal{A}w)\rangle_g - \langle Dw_t, D(hw_t)\rangle_g] \, dQ$$
$$+ \int_\Sigma h \left[w_t \frac{\partial w_t}{\partial \nu_\mathcal{A}} - \mathcal{A}w \frac{\partial(\mathcal{A}w)}{\partial \nu_\mathcal{A}} \right] d\Sigma$$
$$= \int_Q [|D(\mathcal{A}w)|_g^2 - |Dw_t|_g^2] h \, dQ + \frac{1}{2} \int_Q [w_t^2 - (\mathcal{A}w)^2] \mathcal{A}h \, dQ$$
$$+ \frac{1}{2} \int_\Sigma \left[(\mathcal{A}w)^2 - w_t^2 \right] Dh \cdot \nu \, d\Sigma$$
$$+ \int_\Sigma h \left[w_t \frac{\partial w_t}{\partial \nu_\mathcal{A}} - \mathcal{A}w \frac{\partial(\mathcal{A}w)}{\partial \nu_\mathcal{A}} \right] d\Sigma. \tag{2.2.27}$$

That completes our proof. □

2.3. Proofs of Main Results.

Proof of Theorem 1.1. First, we need several relations to prepare for getting estimate (1.10).

Let us define $T \in T^2(\mathbb{R}_x^n)$ for $x \in \Omega$ by

$$T(X, Y) = DH(X, Y) + DH(Y, X) \quad \forall \, X, Y \in \mathbb{R}_x^n. \tag{2.3.1}$$

It is clear that $T(\cdot, \cdot)$ is symmetric. Thus it is easily checked that condition (1.5) yields

$$DH(X, Y) + DH(Y, X) = 2b(x)\langle X, Y\rangle_g \quad \forall \, X, Y \in \mathbb{R}_x^n, \, x \in \Omega. \tag{2.3.2}$$

For $x \in \Omega$ given, let $e_1, e_2, \cdots, e_n$ be an orthonormal basis of $(\mathbb{R}_x^n, g(x))$. By using the symmetry of D^2w as a 2-order tensor on $\mathbb{R}_x^n$ and from identity (2.3.2), we then obtain

$$\langle DH, D^2w\rangle_{T^2(\mathbb{R}_x^n)} = \sum_{ij=1}^n DH(e_i, e_j) D^2w(e_i, e_j)$$
$$= \frac{1}{2} \sum_{ij=1}^n [DH(e_i, e_j) + DH(e_j, e_i)] D^2w(e_i, e_j)$$
$$= b(x)\Delta w = b(x)\mathcal{A}w - b(x)Df(w). \tag{2.3.3}$$

For any $\epsilon > 0$ given, we have

$$|(w_t, H(w))| \leq \epsilon \int_\Omega w_t^2 \, dx + \frac{1}{4\epsilon} \int_\Omega |H(w)|^2 \, dx$$
$$\leq \epsilon E(t) + \frac{1}{4\epsilon} \sup_{x \in \Omega} |H|_g^2 L(t). \tag{2.3.4}$$

The same computing yields

$$|(w_t, hw)| \leq \epsilon E(t) + \frac{1}{4\epsilon} \sup_{x \in \Omega} h^2 L(t). \tag{2.3.5}$$

Furthermore, we have

$$|(\mathcal{A}H)(w) + Ric(H, Dw) - D^2 f(H, Dw)|$$
$$\leq (|\mathcal{A}H|_g + |Ric|_g |H|_g + |D^2 f|_g |H|_g)|Dw|_g. \tag{2.3.6}$$

This gives

$$\left| \int_Q \mathcal{A}w[(\mathcal{A}H)(w) + Ric(H, Dw) - D^2 f(H, Dw)] \, dQ \right|$$
$$\leq \frac{1}{4\epsilon} \sup_{x \in \Omega}(|\mathcal{A}H|_g + |Ric|_g |H|_g + |D^2 f|_g |H|_g)^2 \int_0^T L(t)$$
$$+ \epsilon \int_0^T E(t) \, dt. \tag{2.3.7}$$

Similarly, we obtain

$$\left| \int_Q \mathcal{A}w[2Dh(w) + w\mathcal{A}h] \, dQ \right| \leq \epsilon \int_0^T E(t) \, dt + \frac{1}{4\epsilon} \sup_{x \in \Omega}(2|Dh|_g + |\mathcal{A}h|)^2 \int_0^T L(t). \tag{2.3.8}$$

Now we are ready to obtain estimate (1.10).

Let $BT1|_\Sigma$ be defined by (1.11). By identities (2.2.12) and (2.2.24) and relations (2.3.3)-(2.3.8), we then obtain

$$BT1|_\Sigma = (w_t, H(w)) \Big|_0^T + \frac{1}{2} \int_Q [w_t^2 - (\mathcal{A}w)^2] \operatorname{div} H \, dQ$$
$$+ 2 \int_Q \mathcal{A}w \langle DH, D^2 w \rangle_{T^2(\mathbb{R}_x^n)} \, dQ + \int_\Sigma \left[\mathcal{A}w \frac{\partial(hw)}{\partial v_\mathcal{A}} - hw \frac{\partial(\mathcal{A}w)}{\partial v_\mathcal{A}} \right] d\Sigma$$
$$+ \int_Q \mathcal{A}w[(\mathcal{A}H)(w) + Ric(H, Dw) - D^2 f(H, Dw)] \, dQ$$
$$\geq 2b_0 \int_Q (\mathcal{A}w)^2 \, dQ + 2 \int_Q b(x) \mathcal{A}w Df(w) \, dQ$$
$$+ \frac{1}{2} \int_Q [w_t^2 - (\mathcal{A}w)^2] \operatorname{div} H \, dQ + \int_\Sigma \left[\mathcal{A}w \frac{\partial(\mathcal{A}w)}{\partial v_\mathcal{A}} - hw \frac{\partial(\mathcal{A}w)}{\partial v_\mathcal{A}} \right] d\Sigma$$
$$- \epsilon[E(0) + E(T)] - \frac{C}{\epsilon}[L(0) + L(T)] - \epsilon \int_0^T E(t) \, dt - \frac{C}{\epsilon} \int_0^T L(t)$$
$$\geq b_0 \int_Q [w_t^2 + (\mathcal{A}w)^2] \, dQ + \int_Q [w_t^2 - (\mathcal{A}w)^2] h \, dQ$$
$$+ \int_\Sigma \left[\mathcal{A}w \frac{\partial(\mathcal{A}w)}{\partial v_\mathcal{A}} - hw \frac{\partial(\mathcal{A}w)}{\partial v_\mathcal{A}} \right] d\Sigma - \epsilon[E(0) + E(T)]$$

$$-\frac{C}{\epsilon}[L(0)+L(T)] - 2\epsilon \int_0^T E(t)\,dt - \frac{2C}{\epsilon}\int_0^T L(t)$$

$$\geq (2b_0 - 3\epsilon)\int_0^T E(t)\,dt - 3\epsilon[E(0)+E(T)]$$

$$-\frac{3C}{\epsilon}[L(0)+L(T)+\int_0^T L(t)\,dt], \tag{2.3.9}$$

where $h = (\text{div}H)/2 - b_0$ and $C > 0$ is a constant, independent of w and time $t > 0$. □

Proof of Theorem 1.2. For convenience, this time we denote by $lot(w)$ all the lower order terms with respect to energy $E_{1/4}(t)$, that is, all the terms such that there is $C > 0$, independent of w and time t and $\epsilon > 0$, satisfying

$$|lot(w)| \leq \epsilon \int_0^T E_{1/4}(t)\,dt + \frac{C}{\epsilon}\int_0^T L_{1/4}(t)\,dt. \tag{2.3.10}$$

In general $lot(w)$ may change from line to line.

Let $BT2|_\Sigma$ be defined by (1.16). By identities (2.2.19) and (2.2.25) and condition (1.4), we have

$$BT2|_\Sigma$$
$$= \int_\Sigma [pw_t + H(w_t)]\frac{\partial w_t}{\partial v_\mathcal{A}}\,d\Sigma + \int_\Sigma H(\mathcal{A}w)\frac{\partial(\mathcal{A}w)}{\partial v_\mathcal{A}}\,d\Sigma -$$
$$-\int_\Sigma [w_t \mathcal{A}w_t + \frac{1}{2}(|Dw_t|_g^2 + |D(\mathcal{A}w)|_g^2)]H\cdot v\,d\Sigma - \frac{1}{2}\int_\Sigma w_t^2 Dp\cdot v\,d\Sigma$$
$$-\frac{1}{2}\int_\Sigma p\frac{\partial w_t}{\partial v_\mathcal{A}}\,d\Sigma + \frac{1}{2}\int_\Sigma p\mathcal{A}w\frac{\partial \mathcal{A}w}{\partial v_\mathcal{A}}\,d\Sigma + \frac{1}{4}\int_\Sigma [w_t^2 - (\mathcal{A}w)^2]Dp\cdot v\,d\Sigma$$
$$= -(w_t, H(\mathcal{A}w))\Big|_0^T + \int_Q [DH(Dw_t, Dw_t) + DH(D(\mathcal{A}w), D(\mathcal{A}w))]\,dQ$$
$$+\frac{1}{2}\int_Q [|Dw_t|_g^2 - |D(\mathcal{A}w)|_g^2]p\,dQ + lot(w)$$
$$-\frac{1}{2}\int_\Sigma p\frac{\partial w_t}{\partial v_\mathcal{A}}\,d\Sigma + \frac{1}{2}\int_\Sigma p\mathcal{A}w\frac{\partial \mathcal{A}w}{\partial v_\mathcal{A}}\,d\Sigma + \frac{1}{4}\int_\Sigma [w_t^2 - (\mathcal{A}w)^2]Dp\cdot v\,d\Sigma$$
$$\geq a\int_Q (|DW_t|_g^2 + |D(\mathcal{A}w)|_g^2)\,dQ - (w_t, H(\mathcal{A}w))\Big|_0^T$$
$$-(hw_t, \mathcal{A}w)\Big|_0^T + lot(w), \tag{2.3.11}$$

where $h = p/2$.

Furthermore, for any $\epsilon > 0$ we may obtain

$$|(w_t, H(\mathcal{A}w))| \leq \int_\Omega |w_t||H|_g|D(\mathcal{A}w)|\,dx$$

$$\le \epsilon \int_\Omega |D(\mathcal{A}w)|_g^2 \, dx + \frac{1}{4\epsilon} \sup_{x \in \Omega} |H|_g^2 \int_\Omega w_t^2 \, dx$$
$$\le \epsilon E_{1/4}(t) + \frac{C}{\epsilon} L_{1/4}(t). \tag{2.3.12}$$

Similarly, it is easy to check that there is $C > 0$, independent of w and time t, such that
$$(hw_t, \mathcal{A}w)| \le CL_{1/4}(t). \tag{2.3.13}$$

Finally, after using inequalities (2.3.13) and (2.3.12) in (2.3.11), we obtain the desired estimate. □

In order to complete the proofs below we need the following lemma.

Lemma 2.1 Let H be a vector field on $\overline{\Omega}$. Given $u \in H_0^2(\Omega)$. Then

(1) $Du = 0$, in particular $H(u) = 0$ on Γ; \hfill (2.3.14)

(2) $\dfrac{\partial(H(u))}{\partial v_\mathcal{A}} = (H \cdot v)\mathcal{A}u$ on Γ. \hfill (2.3.15)

Proof. Set $X = v_\mathcal{A}/|v_\mathcal{A}|_g$, the unit normal along Γ in metric g. Let z_i be the tangent vector fields along Γ such that $X, z_1, z_2, \cdots, z_{n-1}$ to an orthonormal basis along Γ in $(\mathbb{R}^n, g)$. Then $u|_\Gamma = 0$ and $\frac{\partial u}{\partial v_\mathcal{A}}u|_\Gamma = 0$ imply that $\langle z_i, Du \rangle_g = 0$ and $\langle Du, X \rangle_g = 0$, respectively. We then have

$$Du = \langle Du, X \rangle_g \frac{v_\mathcal{A}}{|v_\mathcal{A}|} + \sum_{i=1}^{n-1} \langle Du, z_i \rangle_g z_i = 0 \quad \text{on} \quad \Gamma. \tag{2.3.16}$$

It follows from (2.3.16) that
$$D_{z_i} Du = 0 \quad \text{on} \quad \Gamma \tag{2.3.17}$$

for $1 \le i \le n-1$. By relations (2.3.16) and (2.3.17), we obtain

$$X(H(u)) = X\langle H, Du \rangle_g = \langle D_X H, Du \rangle_g + \langle H, D_X Du \rangle_g = D^2 u(X, H)$$
$$= \langle H, X \rangle_g D^2 u(X, X) + \sum_{i=1}^{n-1} \langle H, z_i \rangle_g D^2 u(X, z_i)$$
$$= \langle H, X \rangle_g [D^2 u(X, X) + \sum_{i=1}^{n-1} D^2 u(X, z_i)] = \langle H, X \rangle_g \Delta u$$
$$= \langle H, X \rangle_g (\Delta u + \langle Df, Du \rangle_g) = \langle H, X \rangle_g \mathcal{A}u,$$

that is,
$$\frac{\partial(H(u))}{\partial v_\mathcal{A}} = (H \cdot v)\mathcal{A}u \quad \text{on} \quad \Gamma. \quad \square$$

Proof of Theorem 1.3. Let w solve problem (1.20). We then have

$$E(t) = E(0) \qquad (2.3.18)$$

for $t > 0$. It then follows from the boundary condition in (1.20) and Lemma 2.1 that

$$Dw = 0 \quad \text{on} \quad \Gamma \quad \text{so that} \quad H(w) = 0 \quad \text{on} \quad \Gamma \qquad (2.3.19)$$

and

$$\frac{\partial (H(u))}{\partial v_{\mathcal{A}}} = (H \cdot v)\mathcal{A}u \quad \text{on} \quad \Gamma. \qquad (2.3.20)$$

By relations (2.3.19), (2.3.20), and (1.11), we obtain

$$BT1|_\Sigma = \frac{1}{2}\int_\Sigma (\mathcal{A}w)^2 H \cdot v \, d\Sigma. \qquad (2.3.21)$$

Take $\epsilon > 0$ such that $2b_0 T > \epsilon(T+2)$ and set $c_0 = \sup_{x \in \Gamma_0}(H \cdot v)$. After using relations (2.3.18)–(2.3.21) in estimate (1.10), we obtain

$$c_0 \int_{\Sigma_0} \left[\left(\frac{\partial \mathcal{A}w}{\partial v_{\mathcal{A}}}\right)^2 + (\mathcal{A}w)^2\right] d\Sigma + \frac{2C}{\epsilon}[L(0) + L(T) + \int_0^T L(t) dt]$$
$$\geq 2BT1|_\Sigma + \frac{2C}{\epsilon}[L(0) + L(T) + \int_0^T L(t) dt]$$
$$\geq (2b_0 T - \epsilon T - 2\epsilon) E(0). \qquad (2.3.22)$$

By a compactness/uniqueness argument, the lower order term in inequality (2.3.22) can be absorbed if the following uniqueness result holds: problem

$$\begin{cases} \mathcal{A}^2 w = \lambda w \quad \text{on} \quad \Omega, \\ w = \dfrac{\partial w}{\partial v_{\mathcal{A}}} = \mathcal{A}w = \dfrac{\partial \mathcal{A}w}{\partial v_{\mathcal{A}}} = 0 \quad \text{in} \quad \Gamma_0 \end{cases} \qquad (2.3.23)$$

has the unique zero solution. Fortunately, the above uniqueness results follows from Shirota [7] or Hörmander [1]. □

Proof of Theorem 1.4. We shall finish the proof by several steps.
Consider the operator on $L^2(\Omega)$ defined by

$$\mathcal{D}(B) = \left\{ u \mid u \in H^4(\Omega), u|_\Gamma = \mathcal{A}u|_\Gamma = 0 \right\},$$

$$Bu = \mathcal{A}^2 u, \quad u \in \mathcal{D}(B). \qquad (2.3.24)$$

It is well known that B is a positive selfadjoint operator on $L^2(\Omega)$ and its spectrum consists of eigenvalues:

$$0 < \lambda_1 < \lambda_2 < \cdots < \lambda_k < \cdots,$$

with $\lim_{k\to\infty} \lambda_k = \infty$. Denote by Z_k the eigenspaces of B corresponding to λ_k. Then
$$\dim Z_k < \infty, \quad \forall k \geq 1.$$
It is easily checked that
$$\mathcal{D}(B^{3/4}) = \left\{ u \mid u \in H^3(\Omega),\ u|_\Gamma = \mathcal{A}u|_\Gamma = 0 \right\}, \tag{2.3.25}$$

$$\mathcal{D}(B^{1/4}) = H_0^1(\Omega). \tag{2.3.26}$$

Let w be the solution to problem (1.25) with its initial data $(w^0, w^1) \in \mathcal{D}(B^{3/4}) \times H_0^1(\Omega)$. We then have
$$E_{1/4}(t) = E_{1/4}(0) \tag{2.3.27}$$
for $t > 0$.

Step I. First, we shall prove that there is $C_T > 0$ for any $T > 0$ given such that
$$\int_{\Sigma_0} \left[\left(\frac{\partial w_t}{\partial v_\mathcal{A}}\right)^2 + \left(\frac{\partial(\mathcal{A}w)}{\partial v_\mathcal{A}}\right)^2 \right] d\Sigma \geq C_T E_{1/4}(0) \ \forall\ (w^0, w^1) \in \mathcal{D}(B^{3/4}) \times H_0^1(\Omega). \tag{2.3.28}$$

By boundary conditions $w|_\Gamma = \mathcal{A}w|_\Gamma = 0$, we obtain
$$H(w_t) = \langle H, Dw_t \rangle_g = \frac{H \cdot v}{|v_\mathcal{A}|_g^2} \frac{\partial w_t}{\partial v_\mathcal{A}} \tag{2.3.29}$$

and
$$Dw_t = \langle Dw_t, \frac{v_\mathcal{A}}{|v_\mathcal{A}|_g} \rangle_g \frac{v_\mathcal{A}}{|v_\mathcal{A}|_g} = \frac{1}{|v_\mathcal{A}|_g^2} \frac{\partial w_t}{\partial v_\mathcal{A}} v_\mathcal{A}. \tag{2.3.30}$$

The same reason yields
$$H(\mathcal{A}w) = \langle H, D(\mathcal{A}w) \rangle_g = \frac{H \cdot v}{|v_\mathcal{A}|_g^2} \frac{\partial(\mathcal{A}w)}{\partial v_\mathcal{A}} \tag{2.3.31}$$

and
$$D(\mathcal{A}w) = \frac{1}{|v_\mathcal{A}|_g^2} \frac{\partial w_t}{\partial v_\mathcal{A}} v_\mathcal{A}. \tag{2.3.32}$$

Let $BT2$ be defined by (1.16). After using relations (2.3.29)-(2.3.32) in expression (1.16), we have
$$BT2|_\Sigma = \frac{1}{2} \int_\Sigma \frac{1}{|v_\mathcal{A}|_g^2} \left[\left(\frac{\partial w_t}{\partial v_\mathcal{A}}\right)^2 + \left(\frac{\partial(\mathcal{A}w)}{\partial v_\mathcal{A}}\right)^2 \right] H \cdot v\, d\Sigma. \tag{2.3.33}$$

Set $c_0 = \sup_{x\in\Gamma_0}(H\cdot v)/(2|v_\mathcal{A}|_g^2)$ and take $\epsilon > 0$ to be such that $2aT > \epsilon T + 2\epsilon$. By estimate (1.15) and expression (2.3.33), we obtain

$$c_0 \int_\Sigma \left[\left(\frac{\partial w_t}{\partial v_\mathcal{A}}\right)^2 + \left(\frac{\partial(\mathcal{A}w)}{\partial v_\mathcal{A}}\right)^2\right] d\Sigma + \frac{C}{\epsilon}[L_{1/4}(0) + L_{1/4}(T) + \int_0^T L_{1/4}(t)dt]$$

$$\geq BT2|\Sigma + \frac{C}{\epsilon}[L_{1/4}(0) + L_{1/4}(T) + \int_0^T L_{1/4}(t)dt]$$

$$\geq (2aT - \epsilon T - 2\epsilon)E_{1/4}(0). \qquad (2.3.34)$$

We now get to inequality (2.3.28) by absorbing the lower order terms in (2.3.34) through the compactness/uniqueness argument.

Step II. We are now to prove that there is $C_T > 0$ such that

$$\int_{\Sigma_0} \left(\frac{\partial(\mathcal{A}w)}{\partial v_\mathcal{A}}\right)^2 d\Sigma \geq C_T E_{1/4}(0) \qquad (2.3.35)$$

where w is the solution to problem (1.25) with the initial data $(w^0, w^1) \in \mathcal{D}(B^{3/4}) \times H_0^1(\Omega)$.

We consider the complex Hilbert space this time and use some ideas as in Komornik [3]. Denote by $N(\lambda)$ the number of eigenvalues $\leq \lambda$ of the operator

$$B^{1/2}u = -\mathcal{A}u \quad \text{and} \quad \mathcal{D}(B^{1/2}) = H_0^1(\Omega) \cap H^2(\Omega). \qquad (2.3.36)$$

By Hömander [2, Cor. 17.5.8], we know that there $c_3, c_4 > 0$ such that

$$|N(\lambda) - c_3\lambda^{n/2}| \leq c_4\lambda^{(n-1)/2}\log\lambda \qquad (2.3.37)$$

for any $\lambda > 0$. Set $\lambda = \lambda_k^{1/2}$ in (2.3.37), and we have

$$k \leq \lambda_k^{n/4}\left(c_3 + c_4\frac{\log\lambda_k}{2\lambda_k^{1/4}}\right), \quad k = 1, 2, \cdots. \qquad (2.3.38)$$

It follows from (2.3.38) that, for k large enough,

$$\lambda_k \geq c_5 k^{4/n}$$

for some constant $c_5 > 0$, since $\lim_{k\to\infty}(\log\lambda_k)/\lambda_k^{1/4} = 0$. We thus have

$$\sum_{k=1}^\infty (\sqrt{\lambda_k})^{-d} < \infty, \qquad (2.3.39)$$

for every $d > n/2$.

Denote
$$\tilde{Z}_k^+ = \left\{ (u, \sqrt{\lambda_k} i u) \mid u \in Z_k \right\},$$
$$\tilde{Z}_k^- = \left\{ (u, -\sqrt{\lambda_k} i u) \mid u \in Z_k \right\},$$

and then set
$$\tilde{Z}^\pm = \text{the linear hull of } \tilde{Z}_k^\pm.$$

It is easily checked that, for any $(w^0, w^1) \in \tilde{Z}^\pm$, w is the solution of problem to (1.25) if and only if it is also the solution of problem

$$\begin{cases} w_t \pm i\mathcal{A}w = 0 & \text{in } Q, \\ w = 0 & \Gamma \times (0, T), \\ w(0) = w^0. \end{cases} \quad (2.3.40)$$

By inequality (2.3.28) and equation (2.3.40), we have

$$\int_{\Sigma_0} \left| \frac{\partial (\mathcal{A}w)}{\partial v_\mathcal{A}} \right|^2 d\Sigma = \int_{\Sigma_0} \left| \frac{\partial w_t}{\partial v_\mathcal{A}} \right|^2 d\Sigma \geq \frac{C_T}{2} E_{1/4}(0) \quad (2.3.41)$$

for any $(w^0, w^1) \in \tilde{Z}^\pm$. By relations (2.3.39) and (2.3.41), we may apply Komornik [3, Thm. 5.5]. and inequality (2.3.35) follows.

Step III. Finally, we are ready to prove inequality (1.26).

It is easily checked from (2.3.36) that $(w^0, w^1) \in H_0^1(\Omega) \times H^{-1}(\Omega)$ if and only if
$$(B^{-1/2} w^0, B^{-1/2} w^1) \in \mathcal{D}(B^{3/4}) \times H_0^1(\Omega)$$
and there are constants $c_6, c_7 > 0$ such that

$$c_6(\|w^0\|_{H_0^1(\Omega)}^2 + \|w^1\|_{H^{-1}(\Omega)}^2)$$
$$\leq \| |D(B^{-1/2} w^1)|_g \|^2 + \| |D(\mathcal{A} B^{-1/2} w^0)|_g \|^2$$
$$\leq c_7(\|w^0\|_{H_0^1(\Omega)}^2 + \|w^1\|_{H^{-1}(\Omega)}^2), \quad \forall w^0 \in H_0^1(\Omega), w^1 \in H^{-1}(\Omega) \quad (2.3.42)$$

Inequality (1.26) follows from (2.3.42) and (2.3.35) since $B^{-1/2} w$ are the solution of problem (1.25) for the initial data $(B^{-1/2} w^0, B^{-1/2} w^1)$, and $\mathcal{A}(B^{-1/2} w) = -w$, $\forall (w^0, w^1) \in H_0^1(\Omega) \times H^{-1}(\Omega)$.

References

[1] L. Hörmander, On the uniqueness of the Cauchy problem II, Math. Scand., 7, pp. 177-190 (1959).

[2] L. Hömander, The Analysis of Linear Partial Differential Operators III, Pseudo-Differential Operators, Springer–Verlag, Berlin Heidelberg 1985.

[3] V. Komornik, Exact Controllability and Stabilization, Research in Applied Mathematics (Series Editors: P.G. Ciarlet and J. Lions), 1994, Masson/John Wiley co-publication.

[4] Lagnese and Lions, Modelling, Analysis and Control of Thin Plates, Collection RMA, Masson, Paris, 1988.

[5] I. Lasiecka, and R.Triggiani, Exact controllability of the Euler-Bernoulli equation with controls in the Dirichlet and Neumann boundary conditions: a nonconservative case, SIAM J. Control and Optimization, Vol. 27, No. 2, pp. 330-373, March 1989.

[6] I. Lasiecka and R.Triggiani and P. F. Yao, Exact controllability for second-order hyperbolic equations with variable coefficient-principal part and first-order terms, Nonlinear Analysis, Theory, Methods & Applications, Vol. 30 (1997), No. 1, 111-122.

[7] T. Shirota, A remark on the unique continuation theorem for certain fourth order elliptic equations, Proc. Japan Acad., 36, pp. 571-573 (1960).

[8] D. Tataru, Boundary controllability for Conservative PDEs, Appl. Math. Optim., 31: 257–295 (1995).

[9] H. Wu, C. L. Shen, and Y. L. Yu, A Introduction to Riemannian Geometry (in chinese), Univ. of Beijing, 1989.

[10] P. F. Yao, On the observability inequality for exact controllability of wave equations with variable coefficients, SIAM J. Contr. and Optim, Vol.37, No. 5, pp 1568-1599, 1999.

[11] —, Observability inequalities for shallow shells, SIAM J. Contr. and Optim, 2000.

Institute of Systems Science
Academy of Mathematics and Systems Science
Chinese Academy of Sciences
Beijing 100080, P. R. China
Email: pfyao@iss03.iss.ac.cn